WINSTON S. CHURCHILL

THE SECOND WORLD WAR

VOLUME I

THE GATHERING STORM

HOUGHTON MIFFLIN COMPANY BOSTON

Volume I: The Gathering Storm

Winston Churchill (1874–1965) was the elder son of Lord Randolph Churchill and his American wife, Jennie Jerome. In 1908, he married Clementine Ogilvy, who gave him life-long support, and they had four daughters and one son.

Churchill entered the army in 1895, served in Cuba, India, Egypt, and the Sudan and his first publications were *The Story of the Malakand Field Force* (1898), *The River War* (1899) and *Savrola* (1900), his only novel. On a special commission for the *Morning Post*, he became involved in the Boer War, was taken prisoner and escaped. His experiences led to the writing of *London to Ladysmith, via Pretoria* and *Ian Hamilton's March*, both published in 1900.

He began his erratic political career in October 1900, when he was elected Conservative M.P. for Oldham. Four years later, however, he joined the Liberal party. In 1906, he became Under-Secretary of State for the colonies and showed his desire for reform in such writings as *My African Journey* (1908). He became President of the Board of Trade in 1908 and Home Secretary in 1910 and, together with Lloyd George, introduced social legislation which helped form much of the basis for modern Britain. Because he foresaw the possibilities of war with Germany after the Agadir crisis, he was made First Lord of the Admiralty in October 1911. He achieved major changes, including that of modernising and preparing the Royal Navy for war, despite unpopularity in the Cabinet because of the cost involved. In May 1915, however, pressurised by the Opposition, he left the Admiralty and served for a time as Lieutenant-Colonel in France. Lloyd George appointed him Minister of Munitions in July 1917 and Secretary of State for War and Air the following year. In 1924, he rejoined the Conservative party and was made Chancellor of the Exchequer by Baldwin. He resigned in January 1931 and, during the 1930s, wrote numerous books, amongst which were *My Early Life* (1930), *Thoughts and Adventures* (1932) and *Great Contemporaries* (1937). Churchill was again asked to take office in September 1939, after the German invasion of Poland and, when Chamberlain was forced to retire because of the Labour party's refusal to serve under him, Churchill became Prime Minister (May 1940–May 1945). From 1945, he spent most of his time writing *The Second World War* and returned to office in 1951. In 1953, he accepted the garter and also won the Nobel prize for literature. In April 1955, however, owing to increasing illness, he resigned as Prime Minister, although he continued to write. *A History of the English-speaking Peoples* (1956–8) is his major work of this time. He died at the age of ninety.

Moral of the Work

IN *WAR:* RESOLUTION

IN *DEFEAT:* DEFIANCE

IN *VICTORY:* MAGNANIMITY

IN *PEACE:* GOODWILL

ACKNOWLEDGMENTS

I HAVE been greatly assisted in the establishment of the story in its military aspect by Lieutenant-General Sir Henry Pownall; in naval matters by Commodore G. R. G. Allen; and on European and general questions by Colonel F. W. Deakin, of Wadham College, Oxford, who also helped me in my work *Marlborough: His Life and Times*. I have had much assistance from Sir Edward Marsh in matters of diction. I must in addition make my acknowledgments to the very large numbers of others who have kindly read these pages and commented upon them.

Lord Ismay has also given me his invaluable aid, and with my other friends will continue to do so in the future.

I record my obligations to His Majesty's Government for permission to reproduce the text of certain official documents of which the Crown Copyright is legally vested in the Controller of His Majesty's Stationery Office.

INTRODUCTION

WINSTON CHURCHILL began to write the first of what were to be the six volumes of *The Second World War* in 1946. It was a work he had expected to postpone to a later stage of his life, since he had looked forward in 1945 to extending his wartime leadership into the peace. The rejection of his party by the electorate was a heavy blow, which might have dulled his urge to write. But resilience was perhaps the most pronounced of his traits of character, and he had already written the history of another great war in which he had been a principal actor. Once committed to the task, he attacked it with an energy, enthusiasm and power of organisation which would have been remarkable in a professional historian of half his age.

His five-volume history of the First World War, *The World Crisis*, had drawn heavily on the evidence he had submitted to the Dardanelles Committee and on episodic accounts written for newspapers. Its origins were therefore in political debate and in journalism. He set about composing *The Second World War* in an entirely different manner—different, too, from the way in which he had written his great life of Marlborough. Then his technique had been to dictate long passages of narrative, later correcting points of detail in consultation with experts. Now he began by assembling a team of advisers and collecting the documents on which the writing was to be based. The documents were set up in print by the publishers, Cassell, while the advisers worked on the chronologies into which they would fit. Churchill meanwhile prepared himself by dictating recollections of what he had identified as key eposides. They consisted partly of firm impressions and partly of queries to his team about dates, times, places, and personalities. He also wrote copiously to fellow-actors, begging of them their own papers and recollections, and inviting their comments on what he proposed to say. When documents, chronologies, corrections, and comments were collated, he began to write. The bulk of the writing, which was completed in 1953, was done by his normal method of dictation; however, long passages of the first volume, which is very much an *apologia pro vita sua*, were composed in his own hand.

Churchill was not, did not aspire to be, and would very probably have despised the label of, a scientific historian. Like Clarendon and Macaulay, he saw history as a branch of moral philosophy. Indeed, he gave his history a Moral. Its phrases have become some of the most famous words he pronounced—"In War: Resolution; In Defeat: Defiance; In Victory: Magnanimity; In Peace: Goodwill." Each of the component volumes was also given a Theme—"How the English-speaking peoples through their unwisdom, carelessness and good nature allowed the wicked to rearm" is that of the first—which the author believed encapsulated the period with which the volume dealt, but which he also organised his material to illustrate. He justified this method by comparing it to that of Defoe's *Memoirs of a Cavalier*, "in which the author hangs the chronicle and discussion of great military and political events upon the thread of the personal experience of an individual."

The history is, indeed, intensely personal. Explicitly so, because Churchill asks the reader to regard it as a continuation of *The World Crisis*, the two together forming both "an account of another Thirty Years War" and an expression of his "life-effort" on which he was "content to be judged." Implicitly so, because he related many of the major episodes of the war autobiographically. An excellent example is his account of the air fighting on September 15, 1940, which is regarded as the crisis of the Battle of Britain. He was lunching at Chequers and decided, since the weather seemed to favour a German attack, to spend the afternoon at the Headquarters of the R.A.F. No. 11 Group. He and his wife at once drove there, were given seats in the command room from which the British fighters were controlled, and watched the development of the action:

Presently the red bulbs showed that the majority of our squadrons were engaged. A subdued hum arose from the floor, where the busy plotters pushed their discs to and fro in accordance with the swiftly-changing situation. . . . In a little while all our squadrons were fighting, and some had already begun to return for fuel. . . . I became conscious of the anxiety of the Commander. Hitherto I had watched in silence. I now asked, "What other reserves have we?" "There are none," said Air Marshal Park. In an account which he wrote afterwards he said that at this I "looked grave". Well I might. The odds were great; our margins small; the stakes infinite. . . . Then it appeared that the enemy were going home. No new attack appeared. In another ten minutes the action was ended. We climbed again the stairways that led to the surface, and almost as we emerged the "All Clear" sounded. . . . It was 4.30 p.m. before I got back to Chequers, and I immediately went to bed [an unvarying wartime habit]. I did not wake till eight. When I rang my Principal Private Secretary came in with the evening budget of news from all over the world. It was repellent. . . . "However," he said, as he finished his account, "all is redeemed by the air. We have shot down one hundred and eighty-three for the loss of under forty."

This account is both unique—neither Roosevelt, Stalin, nor Hitler left any first-hand narrative of their involvement in the direction of the war—and acutely revealing. Churchill was fascinated by military operations and followed their progress very closely. But he forbore absolutely to intervene in their control at the hour-by-hour and unit-by-unit level adopted by Hitler. He warned and advised, encouraged and occasionally excoriated. He appointed and removed commanders. But he did not presume to do their job. Another chapter conveys the extent of his forbearance. It comes in Volume IV and concerns the fall of Singapore in February 1942. Very properly, Churchill was not merely disheartened but outraged by the failure of the Malaya garrison, under its commander, General Percival, to halt a Japanese invading force which it outnumbered. When it became clear that Percival was about to be defeated, outrage mingled with desperation and disbelief. Breaking a rule, he signalled Wavell, the Supreme Commander, to urge that the newly arrived 18th Division fight "to the bitter end" and that "commanders and senior officers should die with their troops." In the event, the 18th Division was captured by the Japanese almost intact and General Percival marched into enemy lines under a white flag. By not one immoderate word does the author convey in his narrative how deeply he—and, he felt, his country—were wounded by this humiliating and disastrous episode.

The restraint shown in the Singapore chapter was determined by another principle which he had adopted: that of "never criticising any measure of war or policy after the event unless I had before expressed publicly or formally my opinion or warning about it." The effect is to invest the whole history with those qualities of magnanimity and good will by which he set such store, and the more so as it deals with personalities. The volumes are not only a chronicle of events. They are a record of meetings, debate, and disagreements with a world of people. Some were friends with whom he was forced to differ. Some were with opponents or future enemies with whom he nevertheless succeeded in making common cause, Stalin foremost among them. The descriptions of his personal relationships with these men would alone assure the permanent value of this history to our understanding of the Second World War.

But the value of these volumes is assured in a host of other ways. They have their defects: they take no account, because they could not, of the then still secret Ultra intelligence available throughout the war to the Prime Minister (though a no longer cryptic reference on page 295 of Volume I alludes to its significance); and the first volume, in particular, may be judged excessively personal in its interpretation of the policies of the author's opponents. But these deficiencies do not detract from the history's monumental quality. It is an extraordinary achievement, extraordinary in its sweep and comprehensiveness, balance and literary effect; extraordinary in the singularity of its point of view; extraordinary as the labour of a man, already old, who still had ahead of him a career large enough to crown most other statesmen's lives; extraordinary as a contribution to the memorabilia of the English-speaking peoples. It is a great history and will continue to be read as long as Churchill and the Second World War are remembered.

JOHN KEEGAN

PREFACE

I MUST regard these volumes as a continuation of the story of the First World War which I set out in *The World Crisis*, *The Eastern Front*, and *The Aftermath*. Together, if the present work is completed, they will cover an account of another Thirty Years War.

I have followed, as in previous volumes, the method of Defoe's *Memoirs of a Cavalier*, as far as I am able, in which the author hangs the chronicle and discussion of great military and political events upon the thread of the personal experiences of an individual. I am perhaps the only man who has passed through both the two supreme cataclysms of recorded history in high executive office. Whereas however in the First World War I filled responsible but subordinate posts, I was in this second struggle with Germany for more than five years the head of His Majesty's Government. I write therefore from a different standpoint and with more authority than was possible in my earlier books.

Nearly all my official work was transacted by dictation to secretaries. During the time I was Prime Minister I issued Memoranda, Directives, Personal Telegrams, and Minutes which amount to nearly a million words. These documents, composed from day to day under the stress of events and with the knowledge available at the moment, will no doubt show many shortcomings. Taken together, they nevertheless give a current account of these tremendous events as they were viewed at the time by one who bore the chief responsibility for the war and policy of the British Empire and Commonwealth. I doubt whether any similar record exists or has ever existed of the day-to-day conduct of war and administration. I do not describe it as history, for that belongs to another generation. But I claim with confidence that it is a contribution to history which will be of service to the future.

These thirty years of action and advocacy comprise and express my life-effort, and I am content to be judged upon them. I have adhered to my rule of never criticising any measure of war or policy after the event unless I had before expressed publicly or formally my opinion or warning about it. Indeed in the after-light I have softened many of the severities of contemporary controversy. It has given me pain to record these disagreements with so many men whom I liked or respected; but it would be wrong not to lay the lessons of the past before the future. Let no one look down on those honourable, well-meaning men whose actions are chronicled in these pages without searching his own heart, reviewing his own discharge of public duty, and applying the lessons of the past to his future conduct.

It must not be supposed that I expect everybody to agree with what I say, still less that I only write what will be popular. I give my testimony according to the lights I follow. Every possible care has been taken to verify the facts; but much is constantly coming to light from the disclosure of captured documents or other revelations which may present a new aspect to the conclusions which I have drawn. This is why it is important to rely upon authentic contemporary records and the expressions of opinion set down when all was obscure.

One day President Roosevelt told me that he was asking publicly for suggestions about what the war should be called. I said at once "the Unnecessary War." There never was a war more easy to stop than that which has just wrecked what was left of the world from the previous struggle. The human tragedy reaches its climax in the fact that after all the exertions and sacrifices of hundreds of millions of people and the victories of the Righteous Cause we have still not found Peace or Security, and that we lie in the grip of even worse perils than those we have surmounted. It is my earnest hope that pondering upon the past may give guidance in days to come, enable a new generation to repair some of the errors of former years, and thus govern, in accordance with the needs and glory of man, the awful unfolding scene of the future.

WINSTON SPENCER CHURCHILL

Chartwell,
    Westerham,
        Kent
March 1948

NOTE TO THE NEW EDITION

The opportunity of a reprint enables various errors in detail to be corrected. I am grateful to those who have drawn attention to these and offered suggestions for improvement. I must express my appreciation also of the generous reception given to the work, and extend my cordial thanks to the many persons who have written to me concerning it.

For this new issue the publishers have found it possible to reset the entire book in a larger type, a change which it is hoped will be found agreeable to all readers.

WINSTON S. CHURCHILL

Chartwell,
  June 14, 1949

Theme of the Volume.

HOW THE ENGLISH-SPEAKING PEOPLES

THROUGH THEIR UNWISDOM

CARELESSNESS AND GOOD NATURE

ALLOWED THE WICKED

TO REARM

THE GATHERING STORM

———— ·{◊}· ————

BOOK I

From War to War

1919–1939

BOOK II

The Twilight War

September 3, 1939—
May 10, 1940

TABLE OF CONTENTS

BOOK I

FROM WAR TO WAR
1919–1939

BOOK II

THE TWILIGHT WAR
September 3, 1939–May 10, 1940

MAPS AND DIAGRAMS

BOOK I

FROM WAR TO WAR
1919–1939

CHAPTER I

THE FOLLIES OF THE VICTORS

1919–1929

*The War to End War – A Blood-drained France – The Rhine Fron-
tier – The Economic Clauses of the Versailles Treaty – Ignorance about
Reparations – Destruction of the Austro-Hungarian Empire by the
Treaties of St. Germain and of Trianon – The Weimar Republic –
The Anglo-American Guarantee to France Repudiated by the United
States – The Fall of Clemenceau – Poincaré Invades the Ruhr – The
Collapse of the Mark – American Isolation – End of the Anglo-
Japanese Alliance – Anglo-American Naval Disarmament – Fascism
the Child of Communism – How Easy to Prevent a Second Armaged-
don – The One Solid Security for Peace – The Victors Forget – The
Vanquished Remember – Moral Havoc of the Second World War –
Failure to Keep Germany Disarmed the Cause.*

AFTER the end of the World War of 1914 there was a deep
conviction and almost universal hope that peace would
reign in the world. This heart's desire of all the peoples
could easily have been gained by steadfastness in righteous convictions, and by reasonable common sense and prudence. The
phrase "the war to end war" was on every lip, and measures had
been taken to turn it into reality. President Wilson, wielding, as
was thought, the authority of the United States, had made the
conception of a League of Nations dominant in all minds. The
British Delegation at Versailles moulded and shaped his idea into
an Instrument which will for ever constitute a milestone in the
hard march of man. The victorious Allies were at that time all-
powerful, so far as their outside enemies were concerned. They
had to face grave internal difficulties and many riddles to which
they did not know the answer, but the Teutonic Powers in the

3

great mass of Central Europe which had made the upheaval were prostrate before them, and Russia, already shattered by the German flail, was convulsed by civil war and falling into the grip of the Bolshevik or Communist Party.

*　　*　　*　　*　　*

In the summer of 1919 the Allied Armies stood along the Rhine, and their bridgeheads bulged deeply into defeated, disarmed, and hungry Germany. The chiefs of the victor Powers debated and disputed the future in Paris. Before them lay the map of Europe to be redrawn almost as they might resolve. After fifty-two months of agony and hazards the Teutonic coalition lay at their mercy, and not one of its four members could offer the slightest resistance to their will. Germany, the head and front of the offence, regarded by all as the prime cause of the catastrophe which had fallen upon the world, was at the mercy or discretion of conquerors, themselves reeling from the torment they had endured. Moreover, this had been a war not of Governments, but of peoples. The whole life-energy of the greatest nations had been poured out in wrath and slaughter. The war leaders assembled in Paris had been borne thither upon the strongest and most furious tides that have ever flowed in human history. Gone were the days of the treaties of Utrecht and Vienna, when aristocratic statesmen and diplomats, victor and vanquished alike, met in polite and courtly disputation, and, free from the clatter and babel of democracy, could reshape systems upon the fundamentals of which they were all agreed. The peoples, transported by their sufferings and by the mass teachings with which they had been inspired, stood around in scores of millions to demand that retribution should be exacted to the full. Woe betide the leaders now perched on their dizzy pinnacles of triumph if they cast away at the conference table what the soldiers had won on a hundred blood-soaked battlefields.

France, by right alike of her efforts and her losses, held the leading place. Nearly a million and a half Frenchmen had perished defending the soil of France on which they stood against the invader. Five times in a hundred years, in 1814, 1815, 1870, 1914, and 1918, had the towers of Nôtre Dame seen the flash of Prussian guns and heard the thunder of their cannonade. Now for four horrible years thirteen provinces of France had lain in the rigorous

grip of Prussian military rule. Wide regions had been systematically devastated by the enemy or pulverised in the encounter of
the armies. There was hardly a cottage or a family from Verdun
to Toulon that did not mourn its dead or shelter its cripples. To
those Frenchmen—and there were many in high authority—who
had fought and suffered in 1870 it seemed almost a miracle that
France should have emerged victorious from the incomparably
more terrible struggle which had just ended. All their lives they
had dwelt in fear of the German Empire. They remembered the
preventive war which Bismarck had sought to wage in 1875; they
remembered the brutal threat which had driven Delcassé from
office in 1905; they had quaked at the Moroccan menace in 1906,
at the Bosnian dispute of 1908, and at the Agadir crisis of 1911.
The Kaiser's "mailed fist" and "shining armour" speeches might
be received with ridicule in England and America: they sounded a
knell of horrible reality in the hearts of the French. For fifty years
almost they had lived under the terror of the German arms. Now,
at the price of their life-blood, the long oppression had been rolled
away. Surely here at last was peace and safety. With one passionate spasm the French people cried "Never again!"

But the future was heavy with foreboding. The population of
France was less than two-thirds that of Germany. The French
population was stationary, while the German grew. In a decade
or less the annual flood of German youth reaching the military
age must be double that of France. Germany had fought nearly
the whole world, almost single-handed, and she had almost conquered. Those who knew the most knew best the several occasions when the result of the Great War had trembled in the
balance, and the accidents and chances which had turned the fateful scale. What prospect was there in the future that the Great
Allies would once again appear in their millions upon the battlefields of France or in the East? Russia was in ruin and convulsion,
transformed beyond all semblance of the past. Italy might be
upon the opposite side. Great Britain and the United States were
separated by the seas or oceans from Europe. The British Empire
itself seemed knit together by ties which none but its citizens could
understand. What combination of events could ever bring back
again to France and Flanders the formidable Canadians of the
Vimy Ridge; the glorious Australians of Villers-Bretonneux; the
dauntless New Zealanders of the crater-fields of Passchendaele;

the steadfast Indian Corps which in the cruel winter of 1914 had held the line by Armentières? When again would peaceful, careless, anti-militarist Britain tramp the plains of Artois and Picardy with armies of two or three million men? When again would the ocean bear two millions of the splendid manhood of America to Champagne and the Argonne? Worn down, doubly decimated, but undisputed masters of the hour, the French nation peered into the future in thankful wonder and haunting dread. Where then was that SECURITY without which all that had been gained seemed valueless, and life itself, even amid the rejoicings of victory, was almost unendurable? The mortal need was Security at all costs and by all methods, however stern or even harsh.

*　*　*　*　*

On Armistice Day the German armies had marched homeward in good order. "They fought well." said Marshal Foch, Generalissimo of the Allies, with the laurels bright upon his brow, speaking in soldierly mood: "let them keep their weapons." But he demanded that the French frontier should henceforth be the Rhine. Germany might be disarmed; her military system shivered in fragments; her fortresses dismantled: Germany might be impoverished; she might be loaded with measureless indemnities; she might become a prey to internal feuds: but all this would pass in ten years or in twenty. The indestructible might "of all the German tribes" would rise once more and the unquenched fires of warrior Prussia glow and burn again. But the Rhine, the broad, deep, swift-flowing Rhine, once held and fortified by the French Army, would be a barrier and a shield behind which France could dwell and breathe for generations. Very different were the sentiments and views of the English-speaking world, without whose aid France must have succumbed. The territorial provisions of the Treaty of Versailles left Germany practically intact. She still remained the largest homogeneous racial block in Europe. When Marshal Foch heard of the signing of the Peace Treaty of Versailles he observed with singular accuracy: "This is not Peace. It is an Armistice for twenty years."

*　*　*　*　*

The economic clauses of the Treaty were malignant and silly to an extent that made them obviously futile. Germany was condemned to pay reparations on a fabulous scale. These dictates

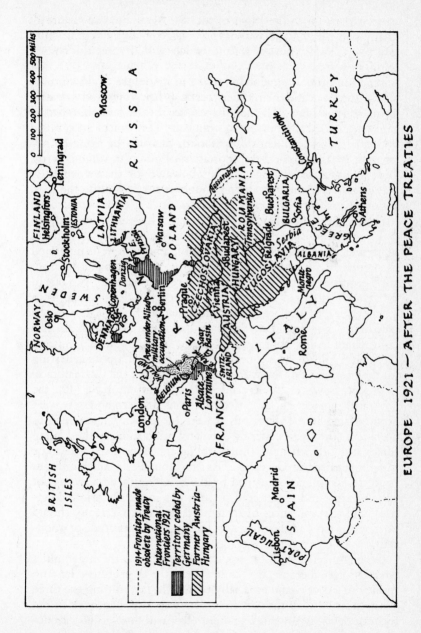

EUROPE 1921 — AFTER THE PEACE TREATIES

Legend:
- ─·─·─ 1914-Frontiers made obsolete by Treaty
- ─── International Frontiers 1921
- Territory ceded by Germany
- Former Austria-Hungary

7

gave expression to the anger of the victors, and to the failure of their peoples to understand that no defeated nation or community can ever pay tribute on a scale which would meet the cost of modern war.

The multitudes remained plunged in ignorance of the simplest economic facts, and their leaders, seeking their votes, did not dare to undeceive them. The newspapers, after their fashion, reflected and emphasised the prevailing opinions. Few voices were raised to explain that payment of reparations can only be made by services or by the physical transportation of goods in wagons across land frontiers or in ships across salt water; or that when these goods arrive in the demanding countries they dislocate the local industry except in very primitive or rigorously-controlled societies. In practice, as even the Russians have now learned, the only way of pillaging a defeated nation is to cart away any movables which are wanted, and to drive off a portion of its manhood as permanent or temporary slaves. But the profit gained from such processes bears no relation to the cost of the war. No one in great authority had the wit, ascendancy, or detachment from public folly to declare these fundamental, brutal facts to the electorates; nor would anyone have been believed if he had. The triumphant Allies continued to assert that they would squeeze Germany "till the pips squeaked". All this had a potent bearing on the prosperity of the world and the mood of the German race.

In fact, however, these clauses were never enforced. On the contrary, whereas about £1,000 millions of German assets were appropriated by the victorious Powers, more than £1,500 millions were lent a few years later to Germany principally by the United States and Great Britain, thus enabling the ruin of the war to be rapidly repaired in Germany. As this apparently magnanimous process was still accompanied by the machine-made howlings of the unhappy and embittered populations in the victorious countries, and the assurances of their statesmen that Germany should be made to pay "to the uttermost farthing", no gratitude or goodwill was to be expected or reaped.

Germany only paid, or was only able to pay, the indemnities later extorted because the United States was profusely lending money to Europe, and especially to her. In fact, during the three years 1926 to 1929 the United States was receiving back in the form of debt-instalment indemnities from all quarters about one-fifth of the money which she was lending to Germany with no chance of repayment. However, everybody seemed pleased and appeared to think this might go on for ever.

History will characterise all these transactions as insane. They helped to breed both the martial curse and the "economic blizzard", of which more later. Germany now borrowed in all directions, swallowing greedily every credit which was lavishly offered her. Misguided sentiment about aiding the vanquished nation, coupled with a profitable rate of interest on these loans, led British investors to participate, though on a much smaller scale than those of the United States. Thus Germany gained about fifteen hundred million pounds sterling in loans as against the one thousand millions of indemnities which she paid in one form or another by surrender of capital assets and *valuta* in foreign countries, or by juggling with the enormous American loans. All this is a sad story of complicated idiocy in the making of which much toil and virtue was consumed.

*   *   *   *   *

The second cardinal tragedy was the complete break-up of the Austro-Hungarian Empire by the Treaties of St. Germain and Trianon. For centuries this surviving embodiment of the Holy Roman Empire had afforded a common life, with advantages in trade and security, to a large number of peoples, none of whom in our own time had the strength or vitality to stand by themselves in the face of pressure from a revivified Germany or Russia. All these races wished to break away from the Federal or Imperial structure, and to encourage their desires was deemed a liberal policy. The Balkanisation of South-eastern Europe proceeded apace, with the consequent relative aggrandisement of Prussia and the German Reich, which, though tired and war-scarred, was intact and locally overwhelming. There is not one of the peoples or provinces that constituted the Empire of the Habsburgs to whom gaining their independence has not brought the tortures which ancient poets and theologians had reserved for the damned. The noble capital of Vienna, the home of so much long-defended culture and tradition, the centre of so many roads, rivers, and railways, was left stark and starving, like a great emporium in an impoverished district whose inhabitants have mostly departed.

The victors imposed upon the Germans all the long-sought

ideals of the liberal nations of the West. They were relieved from
the burden of compulsory military service and from the need of
keeping up heavy armaments. The enormous American loans
were presently pressed upon them, though they had no credit. A
democratic constitution, in accordance with all the latest improvements, was established at Weimar. Emperors having been driven
out, nonentities were elected. Beneath this flimsy fabric raged the
passions of the mighty, defeated, but substantially uninjured
German nation. The prejudice of the Americans against
monarchy, which Mr. Lloyd George made no attempt to counter-
act, had made it clear to the beaten Empire that it would have
better treatment from the Allies as a republic than as a monarchy.
Wise policy would have crowned and fortified the Weimar
Republic with a constitutional sovereign in the person of an infant
grandson of the Kaiser, under a Council of Regency. Instead, a
gaping void was opened in the national life of the German people.
All the strong elements, military and feudal, which might have
rallied to a constitutional monarchy and for its sake respected
and sustained the new democratic and Parliamentary processes
were for the time being unhinged. The Weimar Republic, with
all its liberal trappings and blessings, was regarded as an imposition of the enemy. It could not hold the loyalties or the imagination of the German people. For a spell they sought to cling as in
desperation to the aged Marshal Hindenburg. Thereafter mighty
forces were adrift, the void was open, and into that void after a
pause there strode a maniac of ferocious genius, the repository
and expression of the most virulent hatreds that have ever cor-
roded the human breast—Corporal Hitler.

★　　★　　★　　★　　★

France had been bled white by the war. The generation that
had dreamed since 1870 of a war of revenge had triumphed, but
at a deadly cost in national life-strength. It was a haggard France
that greeted the dawn of victory. Deep fear of Germany pervaded the French nation on the morrow of their dazzling success.
It was this fear that had prompted Marshal Foch to demand the
Rhine frontier for the safety of France against her far larger
neighbour. But the British and American statesmen held that the
absorption of German-populated districts in French territory was
contrary to the Fourteen Points and to the principles of

nationalism and self-determination upon which the Peace Treaty was to be based. They therefore withstood Foch and France. They gained Clemenceau by promising, first, a joint Anglo-American guarantee for the defence of France; secondly, a demilitarised zone; and, thirdly, the total, lasting disarmament of Germany. Clemenceau accepted this in spite of Foch's protests and his own instincts. The Treaty of Guarantee was signed accordingly by Wilson and Lloyd George and Clemenceau. The United States Senate refused to ratify the treaty. They repudiated President Wilson's signature. And we, who had deferred so much to his opinions and wishes in all this business of peace-making, were told without much ceremony that we ought to be better informed about the American Constitution.

In the fear, anger, and disarray of the French people the rugged, dominating figure of Clemenceau, with his world-famed authority, and his special British and American contacts, was incontinently discarded. "Ingratitude towards their great men," says Plutarch, "is the mark of strong peoples." It was imprudent for France to indulge this trait when she was so grievously weakened. There was little compensating strength to be found in the revival of the group intrigues and ceaseless changes of Governments and Ministers which were the characteristic of the Third Republic, however profitable or diverting they were to those engaged in them.

Poincaré, the strongest figure who succeeded Clemenceau, attempted to make an independent Rhineland under the patronage and control of France. This had no chance of success. He did not hesitate to try to enforce reparations on Germany by the invasion of the Ruhr. This certainly imposed compliance with the Treaties on Germany; but it was severely condemned by British and American opinion. As a result of the general financial and political disorganisation of Germany, together with reparation payments during the years 1919 to 1923, the mark rapidly collapsed. The rage aroused in Germany by the French occupation of the Ruhr led to a vast, reckless printing of paper notes with the deliberate object of destroying the whole basis of the currency. In the final stages of the inflation the mark stood at forty-three million millions to the pound sterling. The social and economic consequences of this inflation were deadly and far-reaching. The savings of the middle classes were wiped out, and a natural following was thus provided for the banners of National Socialism. The whole structure of German industry was distorted by the growth of mushroom trusts. The entire working capital of the country disappeared. The internal national debt and the debt of industry in the form of fixed capital charges and mortgages were of course simultaneously liquidated or repudiated. But this was no compensation for the loss of working capital. All led directly to the large-scale borrowings of a bankrupt nation abroad which were the feature of ensuing years. German sufferings and bitterness marched forward together—as they do to-day.

The British temper towards Germany, which at first had been so fierce, very soon went as far astray in the opposite direction. A rift opened between Lloyd George and Poincaré, whose bristling personality hampered his firm and far-sighted policies. The two nations fell apart in thought and action, and British sympathy or even admiration for Germany found powerful expression.

★   ★   ★   ★   ★

The League of Nations had no sooner been created than it received an almost mortal blow. The United States abandoned President Wilson's offspring. The President himself, ready to do battle for his ideals, suffered a paralytic stroke just as he was setting forth on his campaign, and lingered henceforward a futile wreck for a great part of two long and vital years, at the end of which his party and his policy were swept away by the Republican Presidential victory of 1920. Across the Atlantic on the morrow of the Republican success isolationist conceptions prevailed. Europe must be left to stew in its own juice, and must pay its lawful debts. At the same time tariffs were raised to prevent the entry of the goods by which alone these debts could be discharged. At the Washington Conference of 1921 far-reaching proposals for naval disarmament were made by the United States, and the British and American Governments proceeded to sink their battleships and break up their military establishments with gusto. It was argued in odd logic that it would be immoral to disarm the vanquished unless the victors also stripped themselves of their weapons. The finger of Anglo-American reprobation was presently to be pointed at France, deprived alike of the Rhine frontier and of her Treaty guarantee, for maintaining, even on a greatly reduced scale, a French Army based upon universal service.

The United States made it clear to Britain that the continuance of her alliance with Japan, to which the Japanese had punctiliously conformed, would constitute a barrier in Anglo-American relations. Accordingly this alliance was brought to an end. The annulment caused a profound impression in Japan, and was viewed as the spurning of an Asiatic Power by the Western world. Many links were sundered which might afterwards have proved of decisive value to peace. At the same time, Japan could console herself with the fact that the downfall of Germany and Russia had, for a time, raised her to the third place among the world's naval Powers, and certainly to the highest rank. Although the Washington Naval Agreement prescribed a lower ratio of strength in capital ships for Japan than for Britain and the United States (five: five: three), the quota assigned to her was well up to her building and financial capacity for a good many years, and she watched with an attentive eye the two leading naval Powers cutting each other down far below what their resources would have permitted and what their responsibilities enjoined. Thus, both in Europe and in Asia, conditions were swiftly created by the victorious Allies which, in the name of peace, cleared the way for the renewal of war.

While all these untoward events were taking place, amid a ceaseless chatter of well-meant platitudes on both sides of the Atlantic, a new and more terrible cause of quarrel than the Imperialism of Czars and Kaisers became apparent in Europe. The Civil War in Russia ended in the absolute victory of the Bolshevik Revolution. The Soviet armies which advanced to subjugate Poland were indeed repulsed in the Battle of Warsaw, but Germany and Italy nearly succumbed to Communist propaganda and designs, and Hungary actually fell for a while under the control of the Communist dictator Bela Kun. Although Marshal Foch wisely observed that "Bolshevism had never crossed the frontiers of victory", the foundations of European civilisation trembled in the early post-war years. Fascism was the shadow or ugly child of Communism. While Corporal Hitler was making himself useful to the German officer-class in Munich by arousing soldiers and workers to fierce hatred of Jews and Communists, on whom he laid the blame for Germany's defeat, another adventurer, Benito Mussolini, provided Italy with a new theme of government which, while it claimed to save the Italian people from Communism, raised himself to dictatorial power. As Fascism sprang from Communism, so Nazism developed from Fascism. Thus were set on foot those kindred movements which were destined soon to plunge the world into even more hideous strife, which none can say has ended with their destruction.

★   ★   ★   ★   ★

Nevertheless one solid security for peace remained. Germany was disarmed. All her artillery and weapons were destroyed. Her fleet had already sunk itself in Scapa Flow. Her vast army was disbanded. By the Treaty of Versailles only a professional long-service army, not exceeding one hundred thousand men, and unable on this basis to accumulate reserves, was permitted to Germany for purposes of internal order. The annual quotas of recruits no longer received their training; the cadres were dissolved. Every effort was made to reduce to a tithe the Officer Corps. No military air force of any kind was allowed. Submarines were forbidden, and the German Navy was limited to a handful of vessels under 10,000 tons. Soviet Russia was barred off from Western Europe by a cordon of violently anti-Bolshevik States, who had broken away from the former Empire of the Czars in its new and more terrible form. Poland and Czechoslovakia raised independent heads, and seemed to stand erect in Central Europe. Hungary had recovered from her dose of Bela Kun. The French Army, resting upon its laurels, was incomparably the strongest military force in Europe, and it was for some years believed that the French Air Force was also of a high order.

Up till the year 1934 the power of the conquerors remained unchallenged in Europe, and indeed throughout the world. There was no moment in these sixteen years when the three former Allies, or even Britain and France with their associates in Europe, could not in the name of the League of Nations and under its moral and international shield have controlled by a mere effort of the will the armed strength of Germany. Instead, until 1931 the victors, and particularly the United States, concentrated their efforts upon extorting by vexatious foreign controls their annual reparations from Germany. The fact that these payments were made only from far larger American loans reduced the whole process to the absurd Nothing was reaped except ill-will. On the other hand, the strict enforcement at any time till 1934 of the Disarmament Clauses of the Peace Treaty would have guarded indefinitely, without violence or bloodshed, the peace and safety of mankind. But this was neglected while the infringements remained petty, and shunned as they assumed serious proportions. Thus the final safeguard of a long peace was cast away. The crimes of the vanquished find their background and their explanation, though not, of course, their pardon, in the follies of the victors. Without these follies crime would have found neither temptation nor opportunity.

*   *   *   *   *

In these pages I attempt to recount some of the incidents and impressions which form in my mind the story of the coming upon mankind of the worst tragedy in its tumultuous history. This presented itself not only in the destruction of life and property inseparable from war. There had been fearful slaughters of soldiers in the First World War, and much of the accumulated treasure of the nations was consumed. Still, apart from the excesses of the Russian Revolution, the main fabric of European civilisation remained erect at the close of the struggle. When the storm and dust of the cannonade passed suddenly away, the nations, despite their enmities, could still recognise each other as historic racial personalities. The laws of war had on the whole been respected. There was a common professional meeting-ground between military men who had fought one another. Vanquished and victors alike still preserved the semblance of civilised States. A solemn peace was made which, apart from unenforceable financial aspects, conformed to the principles which in the nineteenth century had increasingly regulated the relations of enlightened peoples. The reign of law was proclaimed, and a World Instrument was formed to guard us all, and especially Europe, against a renewed convulsion.

In the Second World War every bond between man and man was to perish. Crimes were committed by the Germans under the Hitlerite domination to which they allowed themselves to be subjected which find no equal in scale and wickedness with any that have darkened the human record. The wholesale massacre by systematised processes of six or seven millions of men, women, and children in the German execution camps exceeds in horror the rough-and-ready butcheries of Genghis Khan, and in scale reduces them to pigmy proportions. Deliberate extermination of whole populations was contemplated and pursued by both Germany and Russia in the Eastern war. The hideous process of bombarding open cities from the air, once started by the Germans, was repaid twenty-fold by the ever-mounting power of the Allies, and found its culmination in the use of the atomic bombs which obliterated Hiroshima and Nagasaki.

We have at length emerged from a scene of material ruin and moral havoc the like of which had never darkened the imagination of former centuries. After all that we suffered and achieved we find ourselves still confronted with problems and perils not less but far more formidable than those through which we have so narrowly made our way.

It is my purpose, as one who lived and acted in these days, to show how easily the tragedy of the Second World War could have been prevented; how the malice of the wicked was reinforced by the weakness of the virtuous; how the structure and habits of democratic States, unless they are welded into larger organisms, lack those elements of persistence and conviction which can alone give security to humble masses; how, even in matters of self-preservation, no policy is pursued for even ten or fifteen years at a time. We shall see how the counsels of prudence and restraint may become the prime agents of mortal danger; how the middle course adopted from desires for safety and a quiet life may be found to lead direct to the bull's-eye of disaster. We shall see how absolute is the need of a broad path of international action pursued by many States in common across the years, irrespective of the ebb and flow of national politics.

It was a simple policy to keep Germany disarmed and the victors adequately armed for thirty years, and in the meanwhile, even if a reconciliation could not be made with Germany, to build ever more strongly a true League of Nations capable of making sure that treaties were kept, or changed only by discussion and agreement. When three or four powerful Governments acting together have demanded the most fearful sacrifices from their peoples, when these have been given freely for the common cause, and when the longed-for result has been attained, it would seem reasonable that concerted action should be preserved so that at least the essentials would not be cast away. But this modest requirement the might, civilisation, learning, knowledge, science, of the victors were unable to supply. They lived from hand to mouth and from day to day, and from one election to another, until, when scarcely twenty years were out, the dread signal of the Second World War was given, and we must write of the sons of those who had fought and died so faithfully and well:

> Shoulder to aching shoulder, side by side,
> They trudged away from life's broad wealds of light.*

* Siegfried Sassoon.

CHAPTER II

PEACE AT ITS ZENITH

1922–1931

Mr. Baldwin's Arrival – Fall of Lloyd George – The Revival of Protection – The First Socialist Government in Britain – Mr. Baldwin's Victory – I Become Chancellor of the Exchequer – War Debts and Reparations – Steady Progress at Home for All Classes – Hindenburg Elected President of Germany – The Conference at Locarno – Austen Chamberlain's Achievement – Peace at its Zenith – A Tranquil Europe – Revival of German Prosperity – The General Election of 1929 – My Differences with Mr. Baldwin – India – The Economic Blizzard – A Fine Hope Dies – Unemployment – Fall of Mr. MacDonald's Second Administration – My Political Exile from Office Begins – The British Financial Convulsion – The General Election of 1931.

DURING the year 1922 a new leader arose in Britain. Mr. Stanley Baldwin had been unknown or unnoticed in the world drama and played a modest part in domestic affairs. He had been Financial Secretary to the Treasury during the war, and was at this time President of the Board of Trade. He became the ruling force in British politics from October 1922, when he ousted Mr. Lloyd George, until May 1937, when, loaded with honours and enshrined in public esteem, he laid down his heavy task and retired in dignity and silence to his Worcestershire home. My relations with this statesman are a definite part of the tale I have to tell. Our differences at times were serious, but in all these years and later I never had an unpleasant personal interview or contact with him, and at no time did I feel we could not talk together in good faith and understanding as man to man.

* * * * *

The party stresses which the Irish Settlement had created inside Mr. Lloyd George's Coalition were growing with the approach of an inevitable General Election. The issue arose whether we should go to the country as a Coalition Government or break up beforehand. It seemed more in accordance with the public interest and the decencies of British politics that parties and Ministers who had come through so much together and bore a mass of joint responsibilities should present themselves unitedly to the nation. In order to make this easy for the Conservatives, who were by far the larger and stronger party, the Prime Minister and I had written earlier in the year offering to resign our offices and give our support from a private station to a new Government to be formed by Mr. Austen Chamberlain. The Conservative leaders, having considered this letter, replied firmly that they would not accept that sacrifice from us and that we must all stand or fall together. This chivalrous attitude was not endorsed by their followers in the party, which now felt itself strong enough to resume undivided power in the State.

By an overwhelming vote the Conservative Party determined to break with Lloyd George and end the National Coalition Government. The Prime Minister resigned that same afternoon. In the morning we had been friends and colleagues of all these people. By nightfall they were our party foes, intent on driving us from public life. With the solitary and unexpected exception of Lord Curzon, all the prominent Conservatives who had fought the war with us, and the majority of all the Ministers, adhered to Lloyd George. These included Arthur Balfour, Austen Chamberlain, Robert Horne, and Lord Birkenhead, the four ablest figures in the Conservative Party. At the crucial moment I was prostrated by a severe operation for appendicitis, and in the morning when I recovered consciousness I learned that the Lloyd George Government had resigned, and that I had lost not only my appendix but my office as Secretary of State for the Dominions and Colonies, in which I conceived myself to have had some Parliamentary and administrative success. Mr. Bonar Law, who had left us a year before for serious reasons of health, reluctantly became Prime Minister. He formed a Government of what one might call "the Second Eleven". Mr. Baldwin, the outstanding figure, was Chancellor of the Exchequer. The Prime Minister asked the King for a Dissolution. The people wanted a change. Mr. Bonar Law, with Mr. Baldwin at his side, and Lord Beaverbrook as his principal stimulant and mentor, gained a majority of seventy-three, with all the expectations of a five-year tenure of power. Early in the year 1923 Mr. Bonar Law resigned the Premiership and retired to die of his fell affliction. Mr. Baldwin succeeded him as Prime Minister, and Lord Curzon reconciled himself to the office of Foreign Secretary in the new Administration.

Thus began that period of fourteen years which may well be called "the Baldwin-MacDonald Régime". At first in alternation but eventually in political brotherhood, these two statesmen governed the country. Nominally the representatives of opposing parties, of contrary doctrines, of antagonistic interests, they proved in fact to be more nearly akin in outlook, temperament, and method than any other two men who had been Prime Ministers since that office was known to the Constitution. Curiously enough, the sympathies of each extended far into the territory of the other. Ramsay MacDonald nursed many of the sentiments of the old Tory. Stanley Baldwin, apart from a manufacturer's ingrained approval of Protection, was by disposition a truer representative of mild Socialism than many to be found in the Labour ranks.

* * * * *

Mr. Baldwin was by no means dazzled by his suddenly-acquired political eminence. "Give me your prayers," he said, when congratulations were offered. He was however soon disquieted by the fear that Mr. Lloyd George would rally, upon the cry of Protection, the numerous dissentient Conservative leaders who had gone out of office with the War Cabinet, and thus split the Government majority and even challenge the party leadership. He therefore resolved, in the autumn of 1923, to forestall his rivals by raising the Protectionist issue himself. He made a speech at Plymouth on October 25 which could only have the effect of bringing the newly-elected Parliament to an untimely end. He protested his innocence of any such design; but to accept this would be to underrate his profound knowledge of British party politics. Parliament was accordingly on his advice dissolved in October, and a second General Election was held within barely a twelvemonth.

The Liberal Party, rallying round the standard of Free Trade, to which I also adhered, gained a balancing position at the polls, and, though in a minority, might well have taken office had Mr. Asquith wished to do so. In view of his disinclination, Mr. Ramsay MacDonald, at the head of little more than two-fifths of the House, became the first Socialist Prime Minister of Great Britain, and lived in office for a year by the sufferance and on the quarrels of the two older parties. The nation was extremely restive under minority Socialist rule, and the political weather became so favourable that the two Oppositions—Liberal and Conservative—picked an occasion to defeat the Socialist Government on a major issue. There was another General Election—the third in less than two years. The Conservatives were returned by a majority of 222 over all other parties combined.* At the beginning of this election Mr. Baldwin's position was very weak, and he made no particular contribution to the result. He had however previously maintained himself as party leader, and as the results were declared it became certain that he would become again Prime Minister.

At this time I stood fairly high in Tory popularity. At the Westminster by-election six months later I proved my hold upon Conservative forces. Although I stood as an Independent Constitutionalist, great numbers of Tories worked and voted for me. In charge of each of my thirty-four committee rooms was a Conservative M.P. defying his leader Mr. Baldwin and the party machine. This was unprecedented. I was defeated only by forty-three votes out of over twenty thousand cast. At the General Election I was returned for Epping by a ten thousand majority, but as a "Constitutionalist". I would not at that time adopt the name "Conservative". I had had some friendly contacts with Mr. Baldwin in the interval; but I did not think he would survive to be Prime Minister. Now on the morrow of his victory I had no idea how he felt towards me. I was surprised, and the Conservative Party dumbfounded, when he invited me to become Chancellor of the Exchequer, the office which my father had once held. A year later, with the approval of my constituents, not having been pressed personally in any way, I formally rejoined the Conservative Party and the Carlton Club, which I had left twenty years before.

*   *   *   *   *

* Conservatives 413, Liberal 40, Labour 151.

My first question at the Treasury of an international character was our American debt. At the end of the war the European Allies owed the United States about ten thousand million dollars, of which four thousand million were owed by Britain. On the other hand, we were owed by the other Allies, principally by Russia, seven thousand million dollars. In 1920 Britain had proposed an all-round cancellation of war debts. This involved, on paper at least, a sacrifice by us of about seven hundred and fifty million pounds sterling. As the value of money has halved since then, the figures could in fact be doubled. No settlement was reached. On August 1, 1922, in Mr. Lloyd George's day, the Balfour Note had declared that Great Britain would collect no more from her debtors, Ally or former enemy, than the United States collected from her. This was a worthy statement. In December of 1922 a British delegation under Mr. Baldwin, Chancellor of the Exchequer in Mr. Bonar Law's Government, visited Washington; and as the result Britain agreed to pay the whole of her war debt to the United States at a rate of interest that had been reduced from 5 to 3½ per cent., irrespective of receipts from her debtors.

This agreement caused deep concern in many instructed quarters, and to no one more than the Prime Minister himself. It imposed upon Great Britain, much impoverished by the war, in which, as she was to do once again, she had fought from the first day to the last, the payment of thirty-five millions sterling a year for sixty-two years. The basis of this agreement was considered, not only in this Island, but by many disinterested financial authorities in America, to be a severe and improvident condition for both borrower and lender. "They hired the money, didn't they?" said President Coolidge. This laconic statement was true, but not exhaustive. Payments between countries which take the form of the transfer of goods and services, or still more of their fruitful exchange, are not only just but beneficial. Payments which are only the arbitrary, artificial transmission across the exchange of such very large sums as arise in war finance cannot fail to derange the whole process of world economy. This is equally true whether the payments are exacted from an ally who shared the victory and bore much of the brunt or from a defeated enemy nation. The enforcement of the Baldwin-Coolidge debt settlement is a recognisable factor in the economic collapse which was presently to overwhelm the world, to prevent its recovery and inflame its hatreds.

The service of the American debt was particularly difficult to render to a country which had newly raised its tariffs to even higher limits, and was soon to bury in its vaults nearly all the gold yet dug up. Similar but lighter settlements were imposed upon the other European Allies. The first result was that everyone put the screw on Germany. I was in full accord with the policy of the Balfour Note of 1922, and had argued for it at the time; and when I became Chancellor of the Exchequer I reiterated it, and acted accordingly. I thought that if Great Britain were thus made not only the debtor but the debt-collector of the United States the unwisdom of the debt collection would become apparent at Washington. However, no such reaction followed. Indeed, the argument was resented. The United States continued to insist upon its annual repayments from Great Britain, albeit at a reduced rate of interest.

It therefore fell to me to make settlements with all our Allies which, added to the German payments, which we had already scaled down, would enable us to produce the thirty-five millions annually for the American Treasury. Severest pressure was put upon Germany, and a vexatious *régime* of international control of German internal affairs was imposed. The United States received from England three payments in full, and these were extorted from Germany by indemnities on the modified Dawes scale.

* * * * *

For almost five years I lived next door to Mr. Baldwin at No. 11 Downing Street, and nearly every morning on my way through his house to the Treasury I looked in upon him for a few minutes' chat in the Cabinet Room. As I was one of his leading colleagues, I take my share of responsibility for all that happened. These five years were marked by very considerable recovery at home. This was a capable, sedate Government during a period in which marked improvement and recovery were gradually effected year by year. There was nothing sensational or controversial to boast about on the platforms, but, measured by every test, economic and financial, the mass of the people were definitely better off, and the state of the nation and of the world was easier and more fertile by the end of our term than at its beginning. Here is a modest but a solid claim.

It was in Europe that the distinction of the Administration was achieved.

★   ★   ★   ★   ★

Hindenburg now rose to power in Germany. At the end of February 1925 Friedrich Ebert, leader of the pre-war German Social-Democrat Party, and first President of the German Republic after the defeat, died. A new President had to be chosen. All Germans had long been brought up under paternal despotism, tempered by far-reaching customs of free speech and Parliamentary opposition. Defeat had brought them on its scaly wings democratic forms and liberties in an extreme degree. But the nation was rent and bewildered by all it had gone through, and many parties and groups contended for precedence and office. Out of the turmoil emerged a strong desire to turn to old Field-Marshal von Hindenburg, who was dwelling in dignified retirement. Hindenburg was faithful to the exiled Emperor, and favoured a restoration of the Imperial monarchy "on the English model". This of course was much the most sensible though least fashionable thing to do. When he was besought to stand as a candidate for the Presidency under the Weimar Constitution he was profoundly disturbed. "Leave me in peace," he said again and again.

However, the pressure was continuous, and only Grand-Admiral von Tirpitz at last was found capable of persuading him to abandon both his scruples and his inclinations at the call of duty, which he had always obeyed. Hindenburg's opponents were Marx of the Catholic Centre and Thaelmann the Communist. On Sunday, April 26, all Germany voted. The result was unexpectedly close: Hindenburg, 14,655,766; Marx, 13,751,615; Thaelmann, 1,931,151. Hindenburg, who towered above his opponents by being illustrious, reluctant, and disinterested, was elected by less than a million majority, and with no absolute majority on the total poll. He rebuked his son Oskar for waking him at seven to tell him the news: "Why did you want to wake me up an hour earlier? It would still have been true at eight." And with this he went to sleep again till his usual calling-time.

In France the election of Hindenburg was at first viewed as a renewal of the German challenge. In England there was an easier reaction. Always wishing as I did to see Germany recover her honour and self-respect and to let war-bitterness die, I was not at all distressed by the news. "He is a very sensible old man," said Lloyd George to me when we next met; and so indeed he proved as long as his faculties remained. Even some of his most bitter opponents were forced to admit "Better a Zero than a Nero".* However, he was seventy-seven, and his term of office was to be seven years. Few expected him to be returned again. He did his best to be impartial between the various parties, and certainly his tenure of the Presidency gave a sober strength and comfort to Germany without menace to her neighbours.

*   *   *   *   *

Meanwhile in February 1925 the German Government had addressed itself to M. Herriot, then French Premier. Their memorandum stated that Germany was willing to declare her acceptance of a pact by virtue of which the Powers interested in the Rhine, above all England, France, Italy, and Germany, would enter into a solemn obligation for a lengthy period towards the Government of the United States, as trustees, not to wage war against a contracting State. Furthermore, a pact expressly guaranteeing the existing territorial status on the Rhine would be acceptable to Germany. This was a remarkable event. The French Government undertook to consult their Allies. Mr. Austen Chamberlain made the news public in the House of Commons on March 5. Parliamentary crises in France and Germany delayed the process of negotiations, but after consultation between London and Paris a formal Note was handed to Herr Stresemann, the German Foreign Minister, by the French Ambassador in Berlin on June 16, 1925. The Note declared that no agreement could be reached unless as a prior condition Germany entered the League of Nations. There could be no suggestion in any proposed agreement of a modification of the conditions of the Peace Treaty. Belgium must be included among the contracting Powers, and finally the natural complement of a Rhineland Pact would be a Franco-German Arbitration Treaty.

The British attitude was debated in the House of Commons on

---

* Theodore Lessing (murdered by the Nazis, September 1933).

June 24. Mr. Chamberlain explained that British commitments under the Pact would be limited to the West. France would probably define her special relationships with Poland and Czechoslovakia; but Great Britain would not assume any obligations other than those specified in the Covenant of the League. The British Dominions were not enthusiastic about a Western Pact. General Smuts was anxious to avoid regional arrangements. The Canadians were lukewarm, and only New Zealand was unconditionally prepared to accept the view of the British Government. Nevertheless we persevered. To me the aim of ending the thousand-year strife between France and Germany seemed a supreme object. If we could only weave Gaul and Teuton so closely together economically, socially, and morally as to prevent the occasion of new quarrels, and make old antagonisms die in the realisation of mutual prosperity and interdependence, Europe would rise again. It seemed to me that the supreme interest of the British people in Europe lay in the assuagement of the Franco-German feud, and that they had no other interests comparable or contrary to that. This is still my view to-day.

Mr. Austen Chamberlain, as Foreign Secretary, had an outlook which was respected by all parties, and the whole Cabinet was united in his support. In July the Germans replied to the French Note, accepting the linking up of a Western Pact with the entry of Germany into the League of Nations, but stating the prior need for agreement upon general disarmament. M. Briand came to England, and prolonged discussions were held upon the Western Pact and its surroundings. In August the French, with the full agreement of Great Britain, replied officially to Germany. Germany must enter the League without reservations as the first and indispensable step. The German Government accepted this stipulation. This meant that the conditions of the Treaties were to continue in force unless or until modified by mutual arrangement, and that no specific pledge for a reduction of Allied armaments had been obtained. Further demands by the Germans, put forward under intense nationalistic pressure and excitement, for the eradication from the Peace Treaty of the War Guilt clause, for keeping open the issue of Alsace-Lorraine, and for the immediate evacuation of Cologne by Allied troops, were not pressed by the German Government, and would not have been conceded by the Allies.

On this basis the Conference at Locarno was formally opened on October 4. By the waters of this calm lake the delegates of Britain, France, Germany, Belgium, and Italy assembled. The Conference achieved: first, the Treaty of Mutual Guarantee between the five Powers; secondly, Arbitration Treaties between Germany and France, Germany and Belgium, Germany and Poland, Germany and Czechoslovakia; thirdly, special agreements between France and Poland, and France and Czechoslovakia, by which France undertook to afford them assistance if a breakdown of the Western Pact were followed by an unprovoked resort to arms. Thus did the Western European democracies agree to keep the peace among themselves in all circumstances, and to stand united against any one of their number who broke the contract and marched in aggression upon a brother land. As between France and Germany, Great Britain became solemnly pledged to come to the aid of whichever of these two States was the object of unprovoked aggression. This far-reaching military commitment was accepted by Parliament and endorsed warmly by the nation. The histories may be searched in vain for a parallel to such an undertaking.

The question whether there was any obligation on the part of France or Britain to disarm, or to disarm to any particular level, was not affected. I had been brought into these matters as Chancellor of the Exchequer at an early stage. My own view about this two-way guarantee was that while France remained armed and Germany disarmed Germany could not attack her; and that on the other hand France would never attack Germany if that automatically involved Britain becoming Germany's ally. Thus although the proposal seemed dangerous in theory—pledging us in fact to take part on one side or the other in any Franco-German war that might arise—there was little likelihood of such a disaster ever coming to pass; and this was the best means of preventing it. I was therefore always equally opposed to the disarmament of France and to the rearmament of Germany, because of the much greater danger this immediately brought on Great Britain. On the other hand, Britain and the League of Nations, which Germany joined as part of the agreement, offered a real protection to the German people. Thus there was a balance created in which Britain, whose major interest was the cessation of the quarrel between Germany and France, was to a large extent umpire and arbiter. One hoped that this equilibrium might have lasted twenty years, during which the Allied armaments would gradually and naturally have dwindled under the influence of a long peace, growing confidence, and financial burdens. It was evident that danger would arise if ever Germany became more or less equal with France, still more if she became stronger than France. But all this seemed excluded by solemn treaty obligations.

*   *   *   *   *

The pact of Locarno was concerned only with peace in the West, and it was hoped that what was called an "Eastern Locarno" might be its successor. We should have been very glad if the danger of some future war between Germany and Russia could have been controlled in the same spirit and by similar measures as the possibility of war between Germany and France. Even the Germany of Stresemann was however disinclined to close the door on German claims in the East, or to accept the territorial treaty position about Poland, Danzig, the Corridor, and Upper Silesia. Soviet Russia brooded in her isolation behind the *Cordon Sanitaire* of anti-Bolshevik States. Although our efforts were continued, no progress was made in the East. I did not at any time close my mind to an attempt to give Germany greater satisfaction on her eastern frontier. But no opportunity arose during these brief years of hope.

*   *   *   *   *

There were great rejoicings about the treaty which emerged at the end of 1925 from the Conference at Locarno. Mr. Baldwin was the first to sign it at the Foreign Office. The Foreign Secretary, having no official residence, asked me to lend my dining-room at No. 11 Downing Street for his intimate friendly luncheon with Herr Stresemann. We all met together in great amity, and thought what a wonderful future would await Europe if its greatest nation became truly united and felt themselves secure. After this memorable instrument had received the cordial assent of Parliament, Mr. Austen Chamberlain was given the Garter and the Nobel Peace Prize. His achievement was the high-water mark of Europe's restoration, and it inaugurated three years of peace and recovery. Although old antagonisms were but sleeping, and the drumbeat of new levies was already heard, we were justified in hoping that the ground thus solidly gained would open the road to a further forward march.

At the end of the second Baldwin Administration the state of Europe was tranquil, as it had not been for twenty years, and was not to be for at least another twenty. A friendly feeling existed towards Germany following upon our Treaty of Locarno, and the evacuation of the Rhineland by the French Army and Allied contingents at a much earlier date than had been prescribed at Versailles. The new Germany took her place in the truncated League of Nations. Under the genial influence of American and British loans Germany was reviving rapidly. Her new ocean liners gained the Blue Riband of the Atlantic. Her trade advanced by leaps and bounds, and internal prosperity ripened. France and her system of alliances also seemed secure in Europe. The disarmament clauses of the Treaty of Versailles were not openly violated. The German Navy was non-existent. The German Air Force was prohibited and still unborn. There were many influences in Germany strongly opposed, if only on grounds of prudence, to the idea of war, and the German High Command could not believe that the Allies would allow them to rearm. On the other hand, there lay before us what I later called the "Economic Blizzard". Knowledge of this was confined to rare financial circles, and these were cowed into silence by what they foresaw.

* * * * *

The General Election of May 1929 showed that the "swing of the pendulum" and the normal desire for change were powerful factors with the British electorate. The Socialists had a small majority over the Conservatives in the new House of Commons. The Liberals, with about sixty seats, held the balance, and it was plain that under Mr. Lloyd George's leadership they would, at the outset at least, be hostile to the Conservatives. Mr. Baldwin and I were in full agreement that we should not seek to hold office in a minority or on precarious Liberal support. Accordingly, although there were some differences of opinion in the Cabinet and the party about the course to be taken, Mr. Baldwin tendered his resignation to the King. We all went down to Windsor in a special train to give up our seals and offices; and on June 7 Mr. Ramsay MacDonald became for the second time Prime Minister at the head of a minority Government depending upon Liberal votes.

*　　*　　*　　*　　*

The Socialist Prime Minister wished his new Labour Government to distinguish itself by large concessions to Egypt, by a far-reaching constitutional change in India, and by a renewed effort for world, or at any rate British, disarmament. These were aims in which he could count upon Liberal aid, and for which he therefore commanded a Parliamentary majority. Here began my differences with Mr. Baldwin, and thereafter the relationship in which we had worked since he chose me for Chancellor of the Exchequer five years before became sensibly altered. We still of course remained in easy personal contact, but we knew we did not mean the same thing. My idea was that the Conservative Opposition should strongly confront the Labour Government on all great Imperial and national issues, should identify itself with the majesty of Britain as under Lord Beaconsfield and Lord Salisbury, and should not hesitate to face controversy, even though that might not immediately evoke a response from the nation. So far as I could see, Mr. Baldwin felt that the times were too far gone for any robust assertion of British Imperial greatness, and that the hope of the Conservative Party lay in accommodation with Liberal and Labour forces and in adroit, well-timed manœuvres to detach powerful moods of public opinion and large blocks of voters from them. He certainly was very successful. He was the greatest party manager the Conservatives had ever had. He fought, as their leader, five General Elections, of which he won three. History alone can judge these general issues.

It was on India that our definite breach occurred. The Prime Minister, strongly supported and even spurred by the Conservative Viceroy, Lord Irwin, afterwards Lord Halifax, pressed forward with his plan of Indian self-government. A portentous conference was held in London, of which Mr. Gandhi, lately released from commodious internment, was the central figure. There is no need to follow in these pages the details of the controversy which occupied the sessions of 1929 and 1930. On the release of Mr. Gandhi in order that he might become the envoy of Nationalist India to the London conference I reached the breaking-point in my relations with Mr. Baldwin. He seemed quite content with these developments, was in general accord with the Prime Minister and the Viceroy, and led the Conservative Opposition decidedly along this path. I felt sure we should lose India in the final result and that measureless disasters would come upon the Indian peoples. I therefore after a while resigned from the Shadow Cabinet upon this issue. On January 27, 1931 I wrote to Mr. Baldwin:

Now that our divergence of view upon Indian policy has become public, I feel that I ought not any longer to attend the meetings of your Business Committee, to which you have hitherto so kindly invited me. I need scarcely add that I will give you whatever aid is in my power in opposing the Socialist Government in the House of Commons, and I shall do my utmost to secure their defeat at the General Election.

* * * * *

The year 1929 reached almost the end of its third quarter under the promise and appearance of increasing prosperity, particularly in the United States. Extraordinary optimism sustained an orgy of speculation. Books were written to prove that economic crisis was a phase which expanding business organisation and science had at last mastered. "We are apparently finished and done with economic cycles as we have known them," said the President of the New York Stock Exchange in September. But in October a sudden and violent tempest swept over Wall Street. The intervention of the most powerful agencies failed to stem the tide of panic sales. A group of leading banks constituted a milliard-dollar pool to maintain and stabilise the market. All was vain.

The whole wealth so swiftly gathered in the paper values of previous years vanished. The prosperity of millions of American homes had grown upon a gigantic structure of inflated credit, now suddenly proved phantom. Apart from the nation-wide speculation in shares which even the most famous banks had encouraged by easy loans, a vast system of purchase by instalment of houses, furniture, cars, and numberless kinds of household conveniences and indulgences had grown up. All now fell together. The mighty production plants were thrown into confusion and paralysis. But yesterday there had been the urgent question of parking the motor-cars in which thousands of artisans and craftsmen were beginning to travel to their daily work. To-day the grievous pangs of falling wages and rising unemployment afflicted the whole community, engaged till this moment in the most active creation of all kinds of desirable articles for the enjoyment of millions. The American banking system was far less concentrated and solidly based than the British. Twenty thousand local banks suspended payment. The means of exchange of goods and services between man and man was smitten to the ground, and the crash on Wall Street reverberated in modest and rich households alike.

It should not however be supposed that the fair vision of far greater wealth and comfort ever more widely shared which had entranced the people of the United States had nothing behind it but delusion and market frenzy. Never before had such immense quantities of goods of all kinds been produced, shared, and exchanged in any society. There is in fact no limit to the benefits which human beings may bestow upon one another by the highest exertion of their diligence and skill. This splendid manifestation had been shattered and cast down by vain imaginative processes and greed of gain which far outstripped the great achievement itself. In the wake of the collapse of the stock market came during the years between 1929 and 1932 an unrelenting fall in prices and consequent cuts in production causing widespread unemployment.

The consequences of this dislocation of economic life became world-wide. A general contraction of trade in the face of unemployment and declining production followed. Tariff restrictions were imposed to protect the home markets. The general crisis brought with it acute monetary difficulties, and paralysed internal credit. This spread ruin and unemployment far and wide throughout the globe. Mr. MacDonald's Government, with all their promises behind them, saw unemployment during 1930 and 1931 bound up in their faces from one million to nearly three millions. It was said that in the United States ten million persons were without work. The entire banking system of the great Republic was thrown into confusion and temporary collapse. Consequential disasters fell upon Germany and other European countries. However, nobody starved in the English-speaking world.

*   *   *   *   *

It is always difficult for an Administration or party which is founded upon attacking capital to preserve the confidence and credit so important to the highly artificial economy of an island like Britain. Mr. MacDonald's Labour-Socialist Government were utterly unable to cope with the problems which confronted them. They could not command the party discipline or produce the vigour necessary even to balance the Budget. In such conditions a Government already in a minority and deprived of all financial confidence could not survive.

The failure of the Labour Party to face this tempest, the sudden collapse of British financial credit, and the break-up of the Liberal Party, with its unwholesome balancing power, led to a National Coalition. It seemed that only a Government of all parties was capable of coping with the crisis. Mr. MacDonald and his Chancellor of the Exchequer, on a strong patriotic emotion, attempted to carry the mass of the Labour Party into this combination. Mr. Baldwin, always content that others should have the function so long as he retained the power, was willing to serve under Mr. MacDonald. It was an attitude which, though deserving respect, did not correspond to the facts. Mr. Lloyd George was still recovering from an operation–serious at his age, and Sir Herbert Samuel led the bulk of the Liberals into the all-party combination.

I was not invited to take part in the Coalition Government. I was politically severed from Mr. Baldwin about India. I was an opponent of the policy of Mr. MacDonald's Labour Government. Like many others, I had felt the need of a national concentration. But I was neither surprised nor unhappy when I was left out of it. Indeed, I remained painting at Cannes while the political crisis lasted. What I should have done if I had been asked to join I cannot tell. It is superfluous to discuss doubtful temptations that have never existed. Certainly during the summer I had talked to MacDonald about a National Administration, and he had shown some interest. But I was awkwardly placed in the political scene. I had had fifteen years of Cabinet office, and was now busy with my life of Marlborough. Political dramas are very exciting at the time to those engaged in the clatter and whirlpool of politics, but I can truthfully affirm that I never felt resentment, still less pain, at being so decisively discarded in a moment of national stress. There was however an inconvenience. For all these years since 1905 I had sat on one or the other of the Front

Benches, and always had the advantage of speaking from the box, on which you can put your notes and pretend with more or less success to be making it up as you go along. Now I had to find with some difficulty a seat below the gangway on the Government side, where I had to hold my notes in my hand whenever I spoke, and take my chance in debate with other well-known ex-Cabinet Ministers. However, from time to time I got called.

★　　★　　★　　★　　★

The formation of the new Government did not end the financial crisis, and I returned from abroad to find everything unsettled in the advent of an inevitable General Election. The verdict of the electorate was worthy of the British nation. A National Government had been formed under Mr. Ramsay MacDonald, the founder of the Labour-Socialist Party. They proposed to the people a programme of severe austerity and sacrifice. It was an earlier version of "blood, toil, tears, and sweat", without the stimulus or the requirements of war and mortal peril. The sternest economy must be practised. Everyone would have his wages, salary, or income reduced. The mass of the people were asked to vote for a *régime* of self-denial. They responded as they always do when caught in the heroic temper. Although, contrary to their declarations, the Government abandoned the Gold Standard, and although Mr. Baldwin was obliged to suspend, as it proved for ever, those very payments on the American debt which he had forced on the Bonar Law Cabinet of 1923, confidence and credit were restored. There was an overwhelming majority for the new Administration. Mr. MacDonald as Prime Minister was only followed by seven or eight members of his own party; but barely fifty of his Labour opponents and former followers were returned to Parliament. His health and powers were failing fast, and he reigned in increasing decrepitude at the summit of the British system for nearly four fateful years. And very soon in these four years came Hitler.

CHAPTER III

LURKING DANGERS

My Reflections in 1928 – Annihilating Terrors of Future War – Some Technical Predictions – Allied Hatred of War and Militarism – "Ease Would Recant" – The German Army – The Hundred Thousand Volunteer Limit – General von Seeckt, His Work and Theme – "A Second Scharnhorst" – The Withdrawal of the Allied Mission of Control, January 1927 – German Aviation – Encroachment and Camouflage – The German Navy – Rathenau's Munitions Scheme – Convertible Factories – The "No Major War for Ten Years" Rule.

IN MY BOOK *The Aftermath* I have set down some reflections on the four years which elapsed between the Armistice and the change of Government in Britain at the end of 1922. Writing in 1928, I was deeply under the impression of a future catastrophe.

It was not until the dawn of the twentieth century of the Christian era that war began to enter into its kingdom as the potential destroyer of the human race. The organisation of mankind into great States and Empires, and the rise of nations to full collective consciousness, enabled enterprises of slaughter to be planned and executed upon a scale and with a perseverance never before imagined. All the noblest virtues of individuals were gathered to strengthen the destructive capacity of the mass. Good finances, the resources of world-wide credit and trade, the accumulation of large capital reserves, made it possible to divert for considerable periods the energies of whole peoples to the task of devastation. Democratic institutions gave expression to the will-power of millions. Education not only brought the course of the conflict within the comprehension of everyone, but rendered each person serviceable in a high degree for the purpose in hand. The Press afforded a means of unification and of mutual stimulation. Religion, having discreetly avoided conflict on the fundamental issues, offered its encouragements and consolations, through all its forms, impartially to all the combatants. Lastly, Science unfolded her treasures and her secrets to the desperate demands of men, and placed in their hands agencies and apparatus almost decisive in their character.

In consequence many novel features presented themselves. Instead of merely fortified towns being starved whole nations were methodically subjected, or sought to be subjected, to the process of reduction by famine. The entire population in one capacity or another took part in the war; all were equally the object of attack. The air opened paths along which death and terror could be carried far behind the lines of the actual armies, to women, children, the aged, the sick, who in earlier struggles would perforce have been left untouched. Marvellous organisations of railroads, steamships, and motor vehicles placed and maintained tens of millions of men continuously in action. Healing and surgery in their exquisite developments returned them again and again to the shambles. Nothing was wasted that could contribute to the process of waste. The last dying kick was brought into military utility.

But all that happened in the four years of the Great War was only a prelude to what was preparing for the fifth year. The campaign of the year 1919 would have witnessed an immense accession to the powers of destruction. Had the Germans retained the morale to make good their retreat to the Rhine, they would have been assaulted in the summer of 1919 with forces and by methods incomparably more prodigious than any yet employed. Thousands of aeroplanes would have shattered their cities. Scores of thousands of cannon would have blasted their front. Arrangements were being made to carry simultaneously a quarter of a million men, together with all their requirements, continuously forward across country in mechanical vehicles moving ten or fifteen miles each day. Poison gases of incredible malignity, against which only a secret mask (which the Germans could not obtain in time) was proof, would have stifled all resistance and paralysed all life on the hostile front subjected to attack. No doubt the Germans too had their plans. But the hour of wrath had passed. The signal of relief was given, and the horrors of 1919 remained buried in the archives of the great antagonists.

The war stopped as suddenly and as universally as it had begun. The world lifted its head, surveyed the scene of ruin, and victors and vanquished alike drew breath. In a hundred laboratories, in a thousand arsenals, factories, and bureaux, men pulled themselves up with a jerk, and turned from the task in which they had been absorbed. Their projects were put aside unfinished, unexecuted; but their knowledge was preserved; their data, calculations, and discoveries were hastily bundled together and docketed "for future reference" by the War Offices in every country. The campaign of 1919 was never fought; but its ideas go marching along. In every army they are being explored, elaborated, refined under the surface of peace, and should war come again to the world it is not with the weapons and agencies prepared for 1919 that it will be fought, but with developments and extensions of these which will be incomparably more formidable and fatal.

It is in these circumstances that we entered upon that period of exhaustion which has been described as Peace. It gives us, at any rate, an opportunity to consider the general situation. Certain sombre facts emerge, solid, inexorable, like the shapes of mountains from drifting mist. It is established that henceforward whole populations will take part in war, all doing their utmost, all subjected to the fury of the enemy. It is established that nations who believe their life is at stake will not be restrained from using any means to secure their existence. It is probable—nay, certain—that among the means which will next time be at their disposal will be agencies and processes of destruction wholesale, unlimited, and perhaps, once launched, uncontrollable.

Mankind has never been in this position before. Without having improved appreciably in virtue or enjoying wiser guidance, it has got into its hands for the first time the tools by which it can unfailingly accomplish its own extermination. That is the point in human destinies to which all the glories and toils of men have at last led them. They would do well to pause and ponder upon their new responsibilities. Death stands at attention, obedient, expectant, ready to serve, ready to shear away the peoples *en masse*; ready, if called on, to pulverise, without hope of repair, what is left of civilisation. He awaits only the word of command. He awaits it from a frail, bewildered being, long his victim, now—for one occasion only—his Master.

* * * * *

All this was published on January 1, 1929. Now, on another New Year's Day eighteen years later, I could not write it differently. All the words and actions for which I am accountable between the wars had as their object only the prevention of a second World War; and, of course, of making sure that if the worst happened we won, or at least survived. There can hardly ever have been a war more easy to prevent than this second Armageddon. I have always been ready to use force in order to defy tyranny or ward off ruin. But had our British, American, and Allied affairs been conducted with the ordinary consistency and common sense usual in decent households there was no need for Force to march unaccompanied by Law; and Strength, more-over, could have been used in righteous causes with little risk of bloodshed. In their loss of purpose, in their abandonment even of the themes they most sincerely espoused, Britain, France, and most of all, because of their immense power and impartiality, the United States, allowed conditions to be gradually built up which led to the very climax they dreaded most. They have only to repeat the same well-meaning, short-sighted behaviour towards the new problems which in singular resemblance confront us to-day to bring about a third convulsion from which none may live to tell the tale.

★　　★　　★　　★　　★

I had written even earlier, in 1925, some thoughts and queries of a technical character which it would be wrong to omit in these days:

May there not be methods of using explosive energy incomparably more intense than anything heretofore discovered? Might not a bomb no bigger than an orange be found to possess a secret power to destroy a whole block of buildings—nay, to concentrate the force of a thousand tons of cordite and blast a township at a stroke? Could not explosives even of the existing type be guided automatically in flying machines by wireless or other rays, without a human pilot, in ceaseless procession upon a hostile city, arsenal, camp, or dockyard?

As for poison gas and chemical warfare in all its forms, only the first chapter has been written of a terrible book. Certainly every one of these new avenues to destruction is being studied on both sides of the Rhine with all the science and patience of which man is capable. And why should it be supposed that these resources will be limited to inorganic chemistry? A study of disease—of pestilences methodically prepared and deliberately launched upon man and beast—is certainly being pursued in the laboratories of more than one great country. Blight to destroy crops, anthrax to slay horses and cattle, plague to poison not armies only but whole districts—such are the lines along which military science is remorselessly advancing.

All this is nearly a quarter of a century old.

★　　★　　★　　★　　★

It is natural that a proud people vanquished in war should strive to rearm themselves as soon as possible. They will not respect more than they can help treaties exacted from them under duress.

>   . . . Ease would recant
> Vows made in pain, as violent and void.

The responsibility therefore of imposing a continual state of
military disarmament upon a beaten foe rests upon the victors.
For this purpose they must pursue a twofold policy. First, while
remaining sufficiently armed themselves, they must enforce with
tireless vigilance and authority the clauses of the treaty which
forbid the revival of their late antagonist's military power.
Secondly, they should do all that is possible to reconcile the defeated nation to its lot by acts of benevolence designed to procure
the greatest amount of prosperity in the beaten country, and
labour by every means to create a basis of true friendship and of
common interests, so that the incentive to appeal again to arms
will be continually diminished. In these years I coined the maxim,
"The redress of the grievances of the vanquished should precede
the disarmament of the victors." As will be seen, the reverse
process was, to a large extent, followed by Britain, the United
States, and France. And thereby hangs this tale.

*   *   *   *   *

It is a prodigious task to make an army embodying the whole
manhood of a mighty nation. The victorious Allies had at
Mr. Lloyd George's suggestion limited the German Army to a
hundred thousand men, and conscription was forbidden. This
force therefore became the nucleus and the crucible out of which
an army of millions of men was if possible to be re-formed. The
hundred thousand men were a hundred thousand leaders. Once
the decision to expand was taken, the privates could become
sergeants, the sergeants officers. None the less, Mr. Lloyd
George's plan for preventing the re-creation of the German Army
was not ill-conceived. No foreign inspection could in times of
peace control the quality of the hundred thousand men allowed
to Germany. But the issue did not turn on this. Three or four
millions of trained soldiers were needed merely to hold the
German frontiers. To make a nation-wide army which could
compare with, still more surpass, the French Army required not
only the preparation of the leaders and the revival of the old
regiments and formations, but the national compulsory service of
each annual quota of men reaching the military age. Volunteer
corps, youth movements, extensions of the police and constabu-

lary forces, old-comrades associations, all kinds of non-official and indeed illegal organisations, might play their part in the interim period. But without universal national service the bones of the skeleton could never be clothed with flesh and sinew.

There was therefore no possibility of Germany creating an army which could face the French Army until conscription had been applied for several years. Here was a line which could not be transgressed without an obvious, flagrant breach of the Treaty of Versailles. Every kind of concealed, ingenious, elaborate preparation could be made beforehand, but the moment must come when the Rubicon would have to be crossed and the conquerors defied. Mr. Lloyd George's principle was thus sound. Had it been enforced with authority and prudence there could have been no new forging of the German war-machine. The class called up for each year, however well schooled beforehand, would also have to remain for at least two years in the regimental or other units, and it was only after this period of training that the reserves without which no modern army is possible could be gradually formed and accumulated. France, though her manhood had been depleted in a horrible degree by the previous war, had nevertheless maintained a regular uninterrupted routine of training annual quotas and of passing the trained soldiers into a reserve which comprised the whole fighting man-power of the nation. For fifteen years Germany was not allowed to build up a similar reserve. In all these years the German Army might nourish and cherish its military spirit and tradition, but it could not possibly even dream of entering the lists against the long-established unbroken developments of the armed, trained, organised manpower which flowed and gathered naturally from the French military system.

*　　*　　*　　*　　*

The creator of the nucleus and structure of the future German Army was General von Seeckt. As early as 1921 Seeckt was busy planning, in secret and on paper, a full-size German army, and arguing deferentially about his various activities with the Inter-Allied Military Commission of Control. His biographer, General von Rabenau, wrote in the triumphant days of 1940: "It would have been difficult to do the work of 1935–39 if from 1920–34 the centre of leadership had corresponded to the needs of the small army." For instance, the Treaty demanded a decrease in the Officer Corps from thirty-four thousand to four thousand. Every device was used to overcome this fatal barrier, and in spite of the efforts of the Allied Control Commission the process of planning for a revived German Army went forward. "The enemy," says Seeckt's biographer, "did his best to destroy the General Staff, and was supported by the political parties within Germany. The Inter-Allied Control had rightly, from its standpoint, tried for years to make the training in higher staffs so primitive that there could be no General Staff. They tried in the boldest ways to discover how General Staff officers were being trained, but we succeeded in giving nothing away, neither the system nor what was taught. Seeckt never gave in, for had the General Staff been destroyed it would have been difficult to recreate it. . . . Although the form had to be broken, the content was saved. . . ." In fact, under the pretence of being Departments of Reconstruction, Research, and Culture, several thousand staff officers in plain clothes and their assistants were held together in Berlin, thinking deeply about the past and the future.

Rabenau makes an illuminating comment: "Without Seeckt there would to-day [in 1940] be no General Staff in the German sense, for which generations are required and which cannot be achieved in a day, however gifted or industrious officers may be. Continuity of conception is imperative to safeguard leadership in the nervous trials of reality. Knowledge or capacity in individuals is not enough. In war the organically-developed capacity of a majority is necessary, and for this decades are needed. . . . In a small hundred-thousand army, if the generals were not also to be small, it was imperative to create a great theoretical framework. To this end large-scale practical exercises or war games were introduced . . . not so much to train the General Staff, but rather to create a class of higher commanders." These would be capable of thinking in full-scale military terms.

Seeckt insisted that false doctrines, springing from personal experiences of the Great War, should be avoided. All the lessons of that war were thoroughly and systematically studied. New principles of training and instructional courses of all kinds were introduced. All the existing manuals were rewritten, not for the hundred-thousand army, but for the armed might of the German Reich. In order to baffle the inquisitive Allies, whole sections of these manuals were printed in special type and made public.

Those for internal consumption were secret. The main principle inculcated was the need for the closest co-operation of *all* vital arms. Not only were the main services—infantry, motorised cavalry, and artillery—to be tactically interwoven, but machine-gun, trench mortar, and tommy-gun units, anti-tank weapons, army air squadrons, and much else were all to be blended. It is to this theme that the German war leaders attributed their tactical successes in the campaigns of 1939 and 1940. By 1924 Seeckt could feel that the strength of the German army was slowly increasing beyond the hundred-thousand limit. "The fruits of this," said his biographer, "were born only ten years later." In 1925 the old Field-Marshal von Mackensen congratulated Seeckt on his building up of the Reichswehr, and compared him, not unjustly, to the Scharnhorst who had secretly prepared the Prussian counter-stroke against Napoleon during the years of the French occupation of Germany after Jena. "The old fire burns still, and the Allied Control had not destroyed any of the lasting elements of German strength."

In the summer of 1926 Seeckt conducted his largest military exercise for commanders with staffs and signals. He had no troops, but practically all the generals, commanding officers, and General Staff officers of the army were introduced to the art of war and its innumerable technical problems on the scale of a German army which, when the time came, could raise the German nation to its former rank.

For several years short-service training of soldiers beyond the official establishments was practised on a small scale. These men were known as "black", *i.e.*, illegal. From 1925 onwards the whole sphere of "black" was centralised in the Reichswehr Ministry and sustained by national funds. The General Staff plan of 1925 for an extension and improvement of the army outside Treaty limits was to double and then to treble the existing legal seven infantry divisions. But Seeckt's ultimate aim was a minimum of sixty-three. From 1926 the main obstacle to this planning was the opposition of the Prussian Socialist Government. This was swept away in 1932. It was not till April 1933 that the establishment of the hundred-thousand army was officially exceeded, though its strength had for some time been rising steadily above that figure.

★　★　★　★　★

Amid the goodwill and hopes following Locarno a questionable, though by no means irremediable, decision was taken by the British and French Governments. The Inter-Allied Control Commission was to be withdrawn, and in substitution there should be an agreed scheme of investigation by the League of Nations ready to be put into operation when any of the parties desired. It was thought that some such arrangement might form a complement to the Locarno Treaty. This hope was not fulfilled. Marshal Foch reported that effective disarmament of Germany had taken place; but it had to be recognised that the disarmament of a nation of sixty-five millions could not be permanent, and that certain precautions were necessary. In January 1927 the Control Commission was nevertheless withdrawn from Germany. It was already known that the Germans were straining the interpretation of the Treaty in many covert and minor ways, and no doubt they were making paper plans to become a military nation once again. There were Boy Scouts, Cadet Corps, and many volunteer unarmed organisations both of youth and of veterans. But nothing could be done on a large scale in the Army or Navy which would not become obvious. The introduction of compulsory national service, the establishment of a Military Air Force, or the laying down of warships' beyond the Treaty limits would be an open breach of German obligations which could at any time have been raised in the League of Nations, of which Germany was now a member.

The Air was far less definable. The Treaty prohibited a German Military Air Force, and it was officially dissolved in May 1920. In his farewell order Seeckt said he hoped that it would again rise, and meanwhile its spirit would still live. He gave it every encouragement to do so. His first step had been to create within the Reichswehr Ministry a special group of experienced ex-Air Force officers, whose existence was hidden from the Allied Commission and protected against his own Government. This was gradually expanded until within the Ministry there were "air cells" in the various offices or inspectorates, and air personnel were gradually introduced throughout the cadres of the Army. The Civil Aviation Department was headed by an experienced war-time officer, a nominee of Seeckt's, who made sure that the control and development of civil aviation took place in harmony with military needs. This department, together with the German

Civil Air Transport, and various camouflaged military or naval air establishments, was to a great extent staffed by ex-flying officers without knowledge of commercial aviation.

Even before 1924 the beginnings of a system of airfields and civil aircraft factories and the training of pilots and instruction in passive air defence had come into existence throughout Germany. ·There was already much reasonable show of commercial flying, and very large numbers of Germans, both men and women, were encouraged to become "air-minded" by the institution of a network of gliding clubs. Severe limitations were observed, on paper, about the number of service personnel permitted to fly. But these rules, with so many others, were circumvented by Seeckt, who, with the connivance of the German Transport Ministry, succeeded in building up a sure foundation for an efficient industry and a future air arm. It was thought by the Allies, in the mood of 1926, derogatory to German national pride to go too far in curbing these German encroachments, and the victors rested on the line of principle which forbade a German Military Air Force. This proved a very vague and shadowy frontier.

In the naval sphere similar evasions were practised. By the Versailles Treaty Germany was allowed only to retain a small naval force with a maximum strength of fifteen thousand men. Subterfuges were used to increase this total. Naval organisations were covertly incorporated into civil Ministries. The coastal defences in Heligoland, although destroyed in accordance with the Treaty, were soon reconstructed. U-boats were illicitly built, and their officers and men trained in other countries. Everything possible was done to keep the Kaiser's Navy alive, and to prepare for the day when it could openly resume a place upon the sea.

Important progress was also made in another decisive direction. Herr Rathenau had, during his tenure of the Ministry of Reconstruction in 1919, set on foot on the broadest lines the reconstruction of German war industry. "They have destroyed your weapons," he had told the generals, in effect. "But these weapons would in any case have become obsolete before the next war. That war will be fought with brand-new ones, and the army which is least hampered with obsolete material will have a great advantage."

Nevertheless the struggle to preserve weapons from destruction was waged persistently by the German staffs throughout the years of control. Every form of deception and every obstacle baffled the Allied Commission. The work of evasion became thoroughly organised. The German police, which at first had interfered, presently became accessories of the Reichswehr in the amassing of arms. Under a civilian camouflage an organisation was set up to safeguard reserves of weapons and equipment. From 1926 this organisation had representatives all over Germany, and there was a network of depots of all kinds. Even more was ingenuity used to create machinery for future production of war material. Lathes which had been set up for war purposes and were capable of being reconverted to that use were retained for civil production in far greater numbers than were required for ordinary commercial work. State arsenals built for war were not closed down in accordance with the Treaty.

A general scheme had thus been put into action by which all the new factories, and many of the old, founded with American and British loans for reconstruction were designed from the outset for speedy conversion to war, and volumes could be written on the thoroughness and detail with which this was planned. Herr Rathenau had been brutally murdered in 1922 by anti-Semite and nascent Nazi secret societies, who fastened their hatred upon this Jew—Germany's faithful servant. When he came to power in 1929 Herr Bruening carried on the work with zeal and discretion. Thus, while the victors reposed on masses of obsolescent equipment an immense German potential of new munitions production was, year by year, coming into being.

* * * * *

It had been decided by the War Cabinet in 1919 that as part of the economy campaign the Service departments should frame their estimates on the assumption that "the British Empire will not be engaged in any great war during the next ten years, and that no Expeditionary Force will be required". In 1924, when I became Chancellor of the Exchequer, I asked the Committee of Imperial Defence to review this rule; but no recommendations were made for altering it. In 1927 the War Office suggested that the 1919 decision should be extended for the Army only to cover ten years "from the present date". This was approved by the Cabinet and Committee of Imperial Defence. The matter was next discussed on July 5, 1928, when I proposed with acceptance "that the basis of estimates for the Service departments should rest upon the statement that there would be no major war for a period of ten years, and that this basis should advance from day to day, but that the assumption should be reviewed every year by the Committee of Imperial Defence". It was left open for any Service department or Dominion Government to raise the issue at their discretion if they thought fit.

It has been contended that the acceptance of this principle lulled the fighting departments into a false sense of security, that research was neglected, and only short-term views prevailed, especially where expense was involved. Up till the time when I left office in 1929 I felt so hopeful that the peace of the world would be maintained that I saw no reason to take any new decision; nor in the event was I proved wrong. War did not break out till the autumn of 1939. Ten years is a long time in this fugitive world. The ten-year rule with its day-to-day advance remained in force until 1932, when, on March 23, Mr. MacDonald's Government rightly decided that its abandonment could be assumed.

All this time the Allies possessed the strength, and the right, to prevent any visible or tangible German rearmament, and Germany must have obeyed a strong united demand from Britain, France, and Italy to bring her actions into conformity with what the Peace Treaties had prescribed. In reviewing again the history of the eight years from 1930 to 1938 we can see how much time we had. Up till 1934 at least German rearmament could have been prevented without the loss of a single life. It was not time that was lacking.

CHAPTER IV

ADOLF HITLER

The Blinded Corporal – The Obscure Fuehrer – The Munich Putsch, 1923 – "Mein Kampf" – Hitler's Problems – Hitler and the Reichswehr – The Schleicher Intrigue – The Impact of the Economic Blizzard – Chancellor Bruening – A Constitutional Monarchy – Equality of Armaments – Schleicher Intervenes – The Fall of Bruening.

IN OCTOBER 1918 a German corporal had been temporarily blinded by mustard gas in a British attack near Comines. While he lay in hospital in Pomerania defeat and revolution swept over Germany. The son of an obscure Austrian customs official, he had nursed youthful dreams of becoming a great artist. Having failed to gain entry to the Academy of Art in Vienna, he had lived in poverty in that capital and later in Munich. Sometimes as a house-painter, often as a casual labourer, he suffered physical privations and bred a harsh though concealed resentment that the world had denied him success. These misfortunes did not lead him into Communist ranks. By an honourable inversion he cherished all the more an abnormal sense of racial loyalty and a fervent and mystic admiration for Germany and the German people. He sprang eagerly to arms at the outbreak of the war, and served for four years with a Bavarian regiment on the Western Front. Such were the early fortunes of Adolf Hitler.

As he lay sightless and helpless in hospital during the winter of 1918 his own personal failure seemed merged in the disaster of the whole German people. The shock of defeat, the collapse of law and order, the triumph of the French, caused this convalescent regimental orderly an agony which consumed his being, and generated those portentous and measureless forces of the spirit which may spell the rescue or the doom of mankind. The down-

47

fall of Germany seemed to him inexplicable by ordinary processes. Somewhere there had been a gigantic and monstrous betrayal. Lonely and pent within himself, the little soldier pondered and speculated upon the possible causes of the catastrophe, guided only by his narrow personal experiences. He had mingled in Vienna with extreme German Nationalist groups, and here he had heard stories of sinister, undermining activities of another race, foes and exploiters of the Nordic world—the Jews. His patriotic anger fused with his envy of the rich and successful into one overpowering hate.

When at length, as an unnoted patient, he was released from hospital, still wearing the uniform in which he had an almost school-boyish pride, what scenes met his newly unsealed eyes! Fearful are the convulsions of defeat. Around him in the atmosphere of despair and frenzy glared the lineaments of Red Revolution. Armoured cars dashed through the streets of Munich scattering leaflets or bullets upon the fugitive wayfarers. His own comrades, with defiant red arm-bands on their uniform, were shouting slogans of fury against all that he cared for on earth. As in a dream everything suddenly became clear. Germany had been stabbed in the back and clawed down by the Jews, by the profiteers and intriguers behind the Front, by the accursed Bolsheviks in their international conspiracy of Jewish intellectuals. Shining before him he saw his duty, to save Germany from these plagues, to avenge her wrongs, and lead the master race to its long-decreed destiny.

The officers of his regiment, deeply alarmed by the seditious and revolutionary temper of their men, were very glad to find one, at any rate, who seemed to have the root of the matter in him. Corporal Hitler desired to remain mobilised, and found employment as a "political education officer" or agent. In this guise he gathered information about mutinous and subversive designs. Presently he was told by the Security officer for whom he worked to attend meetings of the local political parties of all complexions. One evening in September 1919 the Corporal went to a rally of the German Workers' Party in a Munich brewery, and here he heard for the first time people talking in the style of his secret convictions against the Jews, the speculators, the "November Criminals" who had brought Germany into the abyss. On September 16 he joined this party, and shortly afterwards, in harmony with his military work, undertook its propaganda. In February 1920 the first mass meeting of the German Workers' Party was held in Munich, and here Adolf Hitler himself dominated the proceedings and in twenty-five points outlined the party programme. He had now become a politician. His campaign of national salvation had been opened. In April he was demobilised, and the expansion of the party absorbed his whole life. By the middle of the following year he had ousted the original leaders, and by his passion and genius forced upon the hypnotised company the acceptance of his personal control. Already he was "the Fuehrer". An unsuccessful newspaper, the *Voelkischer Beobachter*, was bought as the party organ.

The Communists were not long in recognising their foe. They tried to break up Hitler's meetings, and in the closing days of 1921 he organised his first units of storm-troopers. Up to this point all had moved in local circles in Bavaria. But in the tribulation of German life during these first post-war years many began here and there throughout the Reich to listen to the new gospel. The fierce anger of all Germany at the French occupation of the Ruhr in 1923 brought what was now called the National Socialist Party a broad wave of adherents. The collapse of the mark destroyed the basis of the German middle class, of whom many in their despair became recruits of the new party and found relief from their misery in hatred, vengeance, and patriotic fervour.

At the beginning Hitler had made it clear that the path to power lay through aggression and violence against a Weimar Republic born from the shame of defeat. By November 1923 "the Fuehrer" had a determined group around him, among whom Goering, Hess, Rosenberg, and Roehm were prominent. These men of action decided that the moment had come to attempt the seizure of authority in the State of Bavaria. General von Ludendorff lent the military prestige of his name to the venture, and marched forward in the *Putsch*. It used to be said before the war: "In Germany there will be no revolution, because in Germany all revolutions are strictly forbidden." This precept was revived on this occasion by the local authorities in Munich. The police troops fired, carefully avoiding the General, who marched straight forward into their ranks and was received with respect. About twenty of the demonstrators were killed. Hitler threw himself upon the ground, and presently escaped with other leaders from the scene. In April 1924 he was sentenced to four years' imprisonment.

Although the German authorities had maintained order, and the German court had inflicted punishment, the feeling was widespread throughout the land that they were striking at their own flesh and blood, and were playing the foreigners' game at the expense of Germany's most faithful sons. Hitler's sentence was reduced from four years to thirteen months. These months in the Landsberg fortress were however sufficient to enable him to complete in outline *Mein Kampf*, a treatise on his political philosophy inscribed to the dead of the recent *Putsch*. When eventually he came to power there was no book which deserved more careful study from the rulers, political and military, of the Allied Powers. All was there—the programme of German resurrection, the technique of party propaganda; the plan for combating Marxism; the concept of a National-Socialist State; the rightful position of Germany at the summit of the world. Here was the new Koran of faith and war: turgid, verbose, shapeless, but pregnant with its message.

The main thesis of *Mein Kampf* is simple. Man is a fighting animal; therefore the nation, being a community of fighters, is a fighting unit. Any living organism which ceases to fight for its existence is doomed to extinction. A country or race which ceases to fight is equally doomed. The fighting capacity of a race depends on its purity. Hence the need for ridding it of foreign defilements. The Jewish race, owing to its universality, is of necessity pacifist and internationalist. Pacifism is the deadliest sin, for it means the surrender of the race in the fight for existence. The first duty of every country is therefore to nationalise the masses. Intelligence in the case of the individual is not of first importance; will and determination are the prime qualities. The individual who is born to command is more valuable than countless thousands of subordinate natures. Only brute force can ensure the survival of the race; hence the necessity for military forms. The race must fight; a race that rests must rust and perish. Had the German race been united in good time it would have been already master of the globe. The new Reich must gather within its fold all the scattered German elements in Europe. A race which has suffered defeat can be rescued by restoring its self-confidence.

Above all things the army must be taught to believe in its own invincibility. To restore the German nation the people must be convinced that the recovery of freedom by force of arms is possible. The aristocratic principle is fundamentally sound. Intellectualism is undesirable. The ultimate aim of education is to produce a German who can be converted with the minimum of training into a soldier. The greatest upheavals in history would have been unthinkable had it not been for the driving force of fanatical and hysterical passions. Nothing could have been effected by the bourgeois virtues of peace and order. The world is now moving towards such an upheaval, and the new German State must see to it that the race is ready for the last and greatest decisions on this earth.

Foreign policy may be unscrupulous. It is not the task of diplomacy to allow a nation to founder heroically, but rather to see that it can prosper and survive. England and Italy are the only two possible allies for Germany. No country will enter into an alliance with a cowardly pacifist State run by democrats and Marxists. So long as Germany does not fend for herself, nobody will fend for her. Her lost provinces cannot be regained by solemn appeals to Heaven or by pious hopes in the League of Nations, but only by force of arms. Germany must not repeat the mistake of fighting all her enemies at once. She must single out the most dangerous and attack him with all her forces. The world will only cease to be anti-German when Germany recovers equality of rights and resumes her place in the sun. There must be no sentimentality about Germany's foreign policy. To attack France for purely sentimental reasons would be foolish. What Germany needs is increase of territory in Europe. Germany's pre-war colonial policy was a mistake and should be abandoned. Germany must look for expansion to Russia, and especially to the Baltic States. No alliance with Russia can be tolerated. To wage war together with Russia against the West would be criminal, for the aim of the Soviets is the triumph of international Judaism.

Such were the "granite pillars" of his policy.

*　　*　　*　　*　　*

The ceaseless struggles and gradual emergence of Adolf Hitler as a national figure were little noticed by the victors, oppressed and harassed as they were by their own troubles and party strife.

A long interval passed before National Socialism, or the "Nazi Party", as it came to be called, gained so strong a hold of the masses of the German people, of the armed forces, of the machinery of the State, and among industrialists not unreasonably terrified of Communism, as to become a power in German life of which world-wide notice had to be taken. When Hitler was released from prison at the end of 1924 he said that it would take him five years to reorganise his movement.

★　★　★　★　★

One of the democratic provisions of the Weimar Constitution prescribed elections to the Reichstag every four years. It was hoped by this provision to make sure that the masses of the German people should enjoy a complete and continuous control over their Parliament. In practice of course it only meant that they lived in a continual atmosphere of febrile political excitement and ceaseless electioneering. The progress of Hitler and his doctrines is thus registered with precision. In 1928 he had but twelve seats in the Reichstag. In 1930 this became 107; in 1932, 230. By that time the whole structure of Germany had been permeated by the agencies and discipline of the National Socialist Party, and intimidation of all kinds and insults and brutalities towards the Jews were rampant.

It is not necessary in this account to follow year by year this complex and formidable development, with all its passions and villainies and all its ups and downs. The pale sunlight of Locarno shone for a while upon the scene. The spending of the profuse American loans induced a sense of returning prosperity. Marshal Hindenburg presided over the German State, and Stresemann was his Foreign Minister. The stable decent majority of the German people, responding to their ingrained love of massive and majestic authority, clung to him till his dying gasp. But other powerful factors were also active in the distracted nation to which the Weimar Republic could offer no sense of security, and no satisfaction of national glory or revenge.

Behind the veneer of republican governments and democratic institutions, imposed by the victors and tainted with defeat, the real political power in Germany and the enduring structure of the nation in the post-war years had been the General Staff of the Reichswehr. They it was who made and unmade Presidents and

Cabinets. They had found in Marshal Hindenburg a symbol of their power and an agent of their will. But Hindenburg in 1930 was eighty-three years of age. From this time his character and mental grasp steadily declined. He became increasingly prejudiced, arbitrary, and senile. An enormous image had been made of him in the war, and patriots could show their admiration by paying for a nail to drive into it. This illustrates effectively what he had now become—"the Wooden Titan". It had for some time been clear to the generals that a satisfactory successor to the aged Marshal would have to be found. The search for the new man was however overtaken by the vehement growth and force of the National-Socialist movement. After the failure of the 1923 *Putsch* in Munich Hitler had professed a programme of strict legality within the framework of the Weimar Republic. Yet at the same time he had encouraged and planned the expansion of the military and para-military formations of the Nazi Party. From very small beginnings the S.A., the Storm Troops or Brownshirts, with their small disciplinary core, the S.S., grew in numbers and vigour to the point where the Reichswehr viewed their activities and potential strength with grave alarm.

At the head of the Storm Troop formations stood a German soldier of fortune, Ernst Roehm, the comrade and hitherto the close friend of Hitler through all the years of struggle. Roehm, Chief of the Staff of the S.A., was a man of proved ability and courage, but dominated by personal ambition and sexually perverted. His vices were no barrier to Hitler's collaboration with him along the hard and dangerous path to power. The Storm Troops had, as Bruening complains, absorbed most of the old German Nationalist formations, such as the Free Companies which had fought in the Baltic and Poland against the Bolsheviks in the 1920's, and also the Nationalist Veterans' Organisation of the Steel Helmets (Stahlhelm).

Pondering most carefully upon the tides that were flowing in the nation, the Reichswehr convinced themselves with much reluctance that, as a military caste and organisation in opposition to the Nazi movement, they could no longer maintain control of Germany. Both factions had in common the resolve to raise Germany from the abyss and avenge her defeat; but while the Reichswehr represented the ordered structure of the Kaiser's Empire, and gave shelter to the feudal, aristocratic, land-owning and well-to-do classes in German society, the S.A. had become to a large extent a revolutionary movement fanned by the discontents of temperamental or embittered subversives and the desperation of ruined men. They differed from the Bolsheviks whom they denounced no more than the North Pole does from the South.

For the Reichswehr to quarrel with the Nazi Party was to tear the defeated nation asunder. The Army chiefs in 1931 and 1932 felt they must, for their own sake and for that of the country, join forces with those to whom in domestic matters they were opposed with all the rigidity and severeness of the German mind. Hitler, for his part, although prepared to use any battering-ram to break into the citadels of power, had always before his eyes the leadership of the great and glittering Germany which had commanded the admiration and loyalty of his youthful years. The conditions for a compact between him and the Reichswehr were therefore present and natural on both sides. The Army chiefs had gradually realised that the strength of the Nazi Party was such that Hitler was the only possible successor to Hindenburg as head of the German nation. Hitler on his side knew that to carry out his programme of German resurrection an alliance with the governing *élite* of the Reichswehr was indispensable. A bargain was struck, and the German Army leaders began to persuade Hindenburg to look upon Hitler as eventual Chancellor of the Reich. Thus by agreeing to curtail the activities of the Brownshirts, to subordinate them to the General Staff, and ultimately, if unavoidable, to liquidate them, Hitler gained the allegiance of the controlling forces in Germany, official executive dominance, and the apparent reversion of the Headship of the German State. The Corporal had travelled far.

*   *   *   *   *

There was however an inner and separate complication. If the key to any master-combination of German internal forces was the General Staff of the Army, several hands were grasping for that key. General Kurt von Schleicher at this time exercised a subtle and on occasions a decisive influence. He was the political mentor of the reserved and potentially dominating military circle. He was viewed with a measure of distrust by all sections and factions, and regarded as an adroit and useful political agent possessed of much knowledge outside the General Staff manuals and not usually accessible to soldiers. Schleicher had long been convinced of the significance of the Nazi movement and of the need to stem and control it. On the other hand, he saw that in this terrific mob-thrust, with its ever-growing private army of S.A., there was a weapon which, if properly handled by his comrades of the General Staff, might reassert the greatness of Germany, and perhaps even establish his own. In this intention during the course of 1931 Schleicher began to plot secretly with Roehm, Chief of the Staff of the Nazi Storm Troopers. There was thus a major double process at work, the General Staff making their arrangements with Hitler, and Schleicher in their midst pursuing his personal conspiracy with Hitler's principal lieutenant and would-be rival, Roehm. Schleicher's contacts with the revolutionary element of the Nazi Party, and particularly with Roehm, lasted until both he and Roehm were shot by Hitler's orders three years later. This certainly simplified the political situation, and also that of the survivors.

★　　★　　★　　★　　★

Meanwhile the Economic Blizzard smote Germany in her turn. The United States banks, faced with increasing commitments at home, refused to increase their improvident loans to Germany. This reaction led to the widespread closing of factories and the sudden ruin of many enterprises on which the peaceful revival of Germany was based. Unemployment in Germany rose to 2,300,000 in the winter of 1930. At the same time Reparations entered a new phase. For the previous three years the American Agent-General, Mr. S. Parker Gilbert, had acted as Allied representative in the collection of the heavy payments demanded by the Allies, including the payments to Britain which I transmitted automatically to the United States Treasury. It was certain this system could not last. Already in the summer of 1929 Mr. Young, the American Commissioner, had framed, proposed, and negotiated in Paris an important scheme of mitigation, which not only put a final limit to the period of Reparation payments but freed both the Reichsbank and the German railways from Allied control, and abolished the Reparations Commission in favour of the Bank for International Settlements. Hitler and his National-Socialist movement joined forces with the business and commercial interests, which were represented, and to some extent led, by the truculent and transient figure of the commercial magnate Hugenberg. A vain but savage campaign was launched against this far-reaching and benevolent easement proffered by the Allies. The German Government succeeded by a dead-lift effort in procuring the assent of the Reichstag to the "Young Plan" by no more than 224 votes to 206. Stresemann, the Foreign Minister, who was now a dying man, gained his last success in the agreement for the complete evacuation of the Rhineland by the Allied armies, long before the Treaty required.

But the German masses were largely indifferent to the remarkable concessions of the victors. Earlier, or in happier circumstances, these would have been acclaimed as long steps upon the path of reconciliation and a return to true peace. But now the ever-present overshadowing fear of the German masses was unemployment. The middle classes had already been ruined and driven into violent courses by the flight from the mark. Stresemann's internal political position was undermined by the international economic stresses, and the vehement assaults of Hitler's Nazis and Hugenberg's capitalist magnates led to his overthrow. On March 28, 1930, Bruening, the leader of the Catholic Centre Party, became Chancellor.

*　*　*　*　*

Bruening was a Catholic from Westphalia and a patriot, seeking to re-create the former Germany in modern democratic guise. He pursued continuously the scheme of factory preparation for war which had been devised by Herr Rathenau before his murder. He had also to struggle towards financial stability amid mounting chaos. His programme of economy and reduction of Civil Service numbers and salaries was not popular. The tides of hatred flowed ever more turbulently. Supported by President Hindenburg, Bruening dissolved a hostile Reichstag, and the election of 1930 left him with a majority. He now made the last recognisable effort to rally what remained of the old Germany against the resurgent, violent, and debased nationalist agitation. For this purpose he had first to secure the re-election of Hindenburg as President. Chancellor Bruening looked to a new but obvious solution. He saw the peace, safety, and glory of Germany only in the restoration of an Emperor. Could he then induce the aged Marshal Hindenburg, if and when re-elected, to act for his last term of office as Regent for a restored monarchy to come into effect upon his death? This policy, if achieved, would have filled the void at the summit of the German nation towards which Hitler was now evidently making his way. In all the circumstances this was the right course. But how could Bruening lead Germany to it? The Conservative element, which was drifting to Hitler, might have been recalled by the return of Kaiser Wilhelm; but neither the Social Democrats nor the trade union forces would tolerate the return of the old Kaiser or the Crown Prince. Bruening's plan was not to recreate a Second Reich. He desired a constitutional monarchy on English lines. He hoped that one of the sons of the Crown Prince might be a suitable candidate.

In November 1931 he confided his plans to Hindenburg, on whom all depended. The aged Marshal's reaction was at once vehement and peculiar. He was astonished and hostile. He said that he regarded himself solely as trustee of the Kaiser. Any other solution was an insult to his military honour. The monarchical conception, to which he was devoted, could not be reconciled with picking and choosing among royal princes. Legitimacy must not be violated. Meanwhile, as Germany would not accept the return of the Kaiser, there was nothing left but he himself, Hindenburg. On this he rested. No compromise for him! "J'y suis, j'y reste." Bruening argued vehemently and perhaps over-long with the old veteran. The Chancellor had a strong case. Unless Hindenburg would accept this monarchical solution, albeit unorthodox, there must be a revolutionary Nazi dictatorship. No agreement was reached. But whether or not Bruening could convert Hindenburg, it was imperative to get him re-elected as President, in order at least to stave off an immediate political collapse of the German State. In its first stage Bruening's plan was successful. At the Presidential elections held in March 1932 Hindenburg was returned, after a second ballot, by a majority over his rivals, Hitler and the Communist Thaelmann. Both the economic position in Germany and her relations with Europe had now to be faced. The Disarmament Conference was sitting at Geneva, and Hitler throve upon a roaring campaign against the humiliation of Germany under Versailles.

In careful meditation Bruening drafted a far-reaching plan of

Treaty revision; and in April 1932 he went to Geneva and found an unexpectedly favourable reception. In conversations between him and MacDonald, Stimson, and Norman Davis it seemed that agreement could be reached. The extraordinary basis of this was the principle, subject to various reserved interpretations, of "equality of armaments" between Germany and France. It is indeed surprising, as future chapters will explain, that anyone in his senses should have imagined that peace could be built on such foundations. If this vital point were conceded by the victors, it might well pull Bruening out of his plight; and then the next step—and this one wise—would be the cancelling of Reparations for the sake of European revival. Such a settlement would of course have raised Bruening's personal position to one of triumph.

Norman Davis, the American Ambassador-at-Large, telephoned to the French Premier, Tardieu, to come immediately from Paris to Geneva. But, unfortunately for Bruening, Tardieu had other news. Schleicher had been busy in Berlin, and had just warned the French Ambassador not to negotiate with Bruening because his fall was imminent. It may well be also that Tardieu was concerned with the military position of France on the formula of "equality of armaments". At any rate, Tardieu did not come to Geneva, and on May 1 Bruening returned to Berlin. To arrive there empty-handed at such a moment was fatal to him. Drastic and even desperate measures were required to cope with the threatened economic collapse inside Germany. For these measures Bruening's unpopular Government had not the necessary strength. He struggled on through May, and meanwhile Tardieu, in the kaleidoscope of French Parliamentary politics, was replaced by M. Herriot.

The new French Premier declared himself ready to discuss the formula reached in the Geneva conversations. The American Ambassador in Berlin was instructed to urge the German Chancellor to go to Geneva without a moment's delay. This message was received by Bruening early on May 30. But meanwhile Schleicher's influence had prevailed. Hindenburg had already been persuaded to dismiss the Chancellor. In the course of that very morning, after the American invitation, with all its hope and imprudence, had reached Bruening, he learned that his fate was settled, and by midday he resigned to avoid actual dismissal. So ended the last Government in post-war Germany which might have led the German people into the enjoyment of a stable and civilised constitution and opened peaceful channels of intercourse with their neighbours. The offers which the Allies had made to Bruening would, but for Schleicher's intrigue and Tardieu's delay, certainly have saved him. These offers had presently to be discussed with a different System and a different man.

CHAPTER V

THE LOCUST YEARS*

1931–1935

The MacDonald-Baldwin Coalition – The Indian Collapse – All Germany Astir – Hindenburg and Hitler – Schleicher Fails as a Stopgap – Hitler Becomes Chancellor – The Burning of the Reichstag, February 27, 1933 – Hitler Wins a Majority at the Elections – The New Master – Qualitative Disarmament – 1932 in Germany – British Air Estimates of 1933 – Equality of Status in Armaments – "The MacDonald Plan" – "Thank God for the French Army" – Hitler Quits the League of Nations – A New York Adventure – Peace at Chartwell – Some Wise Friends – The Marlborough Battlefield – Putzi – The Attitude of the Conservative Party – Dangers in the Far East – Japan Attacks China – Accountability.

THE British Government which resulted from the General Election of 1931 was in appearance one of the strongest and in fact one of the weakest in British records. Mr. Ramsay MacDonald, the Prime Minister, had severed himself, with the utmost bitterness on both sides, from the Socialist Party which it had been his life's work to create. Henceforward he brooded supinely at the head of an Administration which, though nominally National, was in fact overwhelmingly Conservative. Mr. Baldwin preferred the substance to the form of power, and reigned placidly in the background. The Foreign Office was filled by Sir John Simon, one of the leaders of the Liberal contingent. The main work of the Administration at home was done by Mr. Neville Chamberlain, who soon succeeded Mr.

---

* Four years later Sir Thomas Inskip, Minister for Co-ordination of Defence, who was well versed in the Bible, used the expressive phrase about this dismal period, of which he was the heir: "The years that the locust hath eaten" (Joel, ii, 25).

Snowden as Chancellor of the Exchequer. The Labour Party, blamed for its failure in the financial crisis and sorely stricken at the polls, was led by the extreme pacifist, Mr. George Lansbury. During the period of four and a quarter years of this Administration, from August 1931 to November 1935, the entire situation on the Continent of Europe was reversed.

On the first return of the new Parliament the Government demanded a Vote of Confidence upon their Indian policy. To this I moved an Amendment as follows:

Provided that nothing in the said policy shall commit this House to the establishment in India of a Dominion Constitution as defined by the Statute of Westminster. . . . And that no question of self-government in India at this juncture shall impair the ultimate responsibility of Parliament for the peace, order, and good government of the Indian Empire.

On this occasion I spoke for as much as an hour and a half, and was heard with attention. But on this issue, as later upon Defence, nothing that one could say made the slightest difference. We have now along this subsidiary Eastern road also reached our horrible consummation in the slaughter of hundreds of thousands of poor people who only sought to earn their living under conditions of peace and justice. I ventured to tell the ignorant Members of all parties:

As the British authority passes for a time into collapse, the old hatreds between the Moslems and the Hindus revive and acquire new life and malignancy. We cannot easily conceive what these hatreds are. There are in India mobs of neighbours, people who have dwelt together in the closest propinquity all their lives, who when held and dominated by these passions will tear each other to pieces, men, women, and children, with their fingers. Not for a hundred years have the relations between Moslems and Hindus been so poisoned as they have been since England was deemed to be losing her grip, and was believed to be ready to quit the scene if told to go.

We mustered little more than forty in the Lobby against all the three parties in the House of Commons. This must be noted as a sad milestone on the downward path.

* * * * *

Meanwhile all Germany was astir and great events marched forward.

Much had happened in the year which followed the fall of the Bruening Cabinet in May 1932. Papen and the political general, Schleicher, had hitherto attempted to govern Germany by cleverness and intrigue. The time for these had now passed. Papen, who succeeded Bruening as Chancellor, hoped to rule with the support of the entourage of President Hindenburg and of the extreme Nationalist group in the Reichstag. On July 20 a decisive step was taken. The Socialist Government in Prussia was forcibly ousted from office. But Papen's rival was eager for power. In Schleicher's calculations the instrument lay in the dark, hidden forces storming into German politics behind the rising power and name of Adolf Hitler. He hoped to make the Hitler Movement a docile servant of the Reichswehr, and in so doing to gain the control of both himself. The contacts between Schleicher and Roehm, the leader of the Nazi Storm Troopers, which had begun in 1931, were extended in the following year to more precise relations between Schleicher and Hitler himself. The road to power for both men seemed to be obstructed only by Papen and by the confidence displayed by Hindenburg in him.

In August 1932 Hitler came to Berlin on a private summons from the President. The moment for a forward step seemed at hand. Thirteen million German voters stood behind the Fuehrer. A vital share of office must be his for the asking. He was now in somewhat the position of Mussolini on the eve of the march on Rome. But Papen did not care about recent Italian history. He had the support of Hindenburg and had no intention of resigning. The old Marshal saw Hitler. He was not impressed. "*That* man for Chancellor? I'll make him a postmaster and he can lick stamps with my head on them." In palace circles Hitler had not the influence of his competitors.

In the country the vast electorate was restless and adrift. In November 1932, for the fifth time in a year, elections were held throughout Germany. The Nazis lost ground and their 230 seats were reduced to 196, the Communists gaining the balance. The bargaining power of the Fuehrer was thus weakened. Perhaps General Schleicher would be able to do without him after all. The General gained favour in the circle of Hindenburg's advisers. On November 17 Papen resigned and Schleicher became Chancellor in his stead. But the new Chancellor was found to have been more apt at pulling wires behind the scenes than at the open summit of power. He had quarrelled with too many people. Hitler, together with Papen and the Nationalists, now ranged themselves against him; and the Communists, fighting the Nazis in the streets and the Government by their strikes, helped to make his rule impossible. Papen brought his personal influence to bear on President Hindenburg. Would not after all the best solution be to placate Hitler by thrusting upon him the responsibilities and burdens of office? Hindenburg at last reluctantly consented. On January 30, 1933, Adolf Hitler took office as Chancellor of Germany.

The hand of the Master was soon felt upon all who would or might oppose the New Order. On February 2 all meetings or demonstrations of the German Communist Party were forbidden, and throughout Germany a round-up of secret arms belonging to the Communists began. The climax came on the evening of February 27, 1933. The building of the Reichstag broke into flames. Brownshirts, Blackshirts, and their auxiliary formations were called out. Four thousand arrests, including the Central Committee of the Communist Party, were made overnight. These measures were entrusted to Goering, now Minister of the Interior of Prussia. They formed the preliminary to the forthcoming elections and secured the defeat of the Communists, the most formidable opponents of the new *régime*. The organising of the electoral campaign was the task of Goebbels, and he lacked neither skill nor zeal.

But there were still many forces in Germany reluctant, obstinate, or actively hostile to Hitlerism. The Communists, and many who in their perplexity and distress voted with them, obtained 81 seats, the Socialists 118, the Centre party 73, and the Nationalist allies of Hitler under Papen and Hugenberg 52. Thirty-three seats were allotted to minor Right Centre groups. The Nazis obtained a vote of 17,300,000, with 288 seats. These results gave Hitler and his Nationalist allies control of the Reichstag. Thus, and thus only, did Hitler obtain by hook and crook a majority vote from the German people. Under the ordinary processes of civilised Parliamentary government, so large a minority would have had great influence and due consideration in the State. But in the new Nazi Germany minorities were now to learn that they had no rights.

On March 21, 1933, Hitler opened, in the Garrison Church at Potsdam, hard by the tomb of Frederick the Great, the first Reichstag of the Third Reich. In the body of the church sat the representatives of the Reichswehr, the symbol of the continuity of German might, and the senior officers of the S.A. and S.S., the new figures of resurgent Germany. On March 24 the majority of the Reichstag, overbearing or overawing all opponents, confirmed by 441 votes to 94 complete emergency powers to Chancellor Hitler for four years. As the result was announced Hitler turned to the benches of the Socialists and cried, "And now I have no further need of you."

Amid the excitement of the election the exultant columns of the National Socialist Party filed past their leader in the pagan homage of a torchlight procession through the streets of Berlin. It had been a long struggle, difficult for foreigners, especially those who had not known the pangs of defeat, to comprehend. Adolf Hitler had at last arrived. But he was not alone. He had called from the depths of defeat the dark and savage furies latent in the most numerous, most serviceable, ruthless, contradictory, and ill-starred race in Europe. He had conjured up the fearful idol of an all-devouring Moloch of which he was the priest and incarnation. It is not within my scope to describe the inconceivable brutality and villainy by which this apparatus of hatred and tyranny had been fashioned and was now to be perfected. It is necessary, for the purpose of this account, only to present to the reader the new and fearful fact which had broken upon the still unwitting world: GERMANY UNDER HITLER, AND GERMANY ARMING.

While these deadly changes were taking place in Germany the MacDonald-Baldwin Government felt bound to enforce for some time the severe reductions and restrictions which the financial crisis had imposed upon our already modest armaments, and steadfastly closed their eyes and ears to the disquieting symptoms in Europe. In vehement efforts to procure a disarmament of the victors equal to that which had been enforced upon the vanquished by the Treaty of Versailles, Mr. MacDonald and his Conservative and Liberal colleagues pressed a series of proposals forward in the League of Nations and through every other channel that was open. The French, although their political affairs still remained in constant flux and in motion without particular significance, clung tenaciously to the French Army as the centre and prop of the life of France and of all her alliances. This attitude earned them rebukes both in Britain and in the United States. The opinions of the Press and public were in no way founded upon reality; but the adverse tide was strong.

When in May 1932 the virtues of disarmament were extolled in the House of Commons by all parties, the Foreign Secretary opened a new line in the classification of weapons which should be allowed or discouraged. He called this "qualitative disarmament". It was easier to expose the fallacy than to convince the Members. I said:

The Foreign Secretary told us that it was difficult to divide weapons into offensive and defensive categories. It certainly is, because almost every conceivable weapon may be used in defence or offence; either by an aggressor or by the innocent victim of his assault. To make it more difficult for the invader, heavy guns, tanks, and poison gas are to be relegated to the evil category of offensive weapons. The invasion of France by Germany in 1914 reached its climax without the employment of any of these weapons. The heavy gun is to be described as "an offensive weapon". It is all right in a fortress; there it is virtuous and pacific in its character; but bring it out into the field—and, of course, if it were needed, it would be brought out into the field—and it immediately becomes naughty, peccant, militaristic, and has to be placed under the ban of civilisation. Take the tank. The Germans, having invaded France, entrenched themselves; and in a couple of years they shot down 1,500,000 French and British soldiers who were trying to free the soil of France. The tank was invented to overcome the fire of the machine-guns with which the Germans were maintaining themselves in France, and it saved a lot of life in clearing the soil of the invader. Now, apparently, the machine-gun, which was the German weapon for holding on to thirteen provinces of France, is to be the virtuous, defensive machine-gun, and the tank, which was the means by which these Allied lives were saved, is to be placed under the censure and obloquy of all just and righteous men. . . .

A truer classification might be drawn in banning weapons which tend to be indiscriminate in their action and whose use entails death and wounds, not merely on the combatants in the fighting zones, but on the civil population, men, women, and children, far removed from those areas. There indeed, it seems to me, would be a direction in which the united nations assembled at Geneva might advance with hope. . . .

At the end I gave my first formal warning of approaching war:

I should very much regret to see any approximation in military strength between Germany and France. Those who speak of that as though it were right, or even a question of fair dealing, altogether underrate the gravity of the European situation. I would say to those who would like to see Germany and France on an equal footing in armaments: "Do you wish for war?" For my part, I earnestly hope that no such approximation will take place during my lifetime or that of my children. To say that is not in the least to imply any want of regard or admiration for the great qualities of the German people, but I am sure that the thesis that they should be placed in an equal military position with France is one which, if it ever emerged in fact, would bring us within practical distance of almost measureless calamity.

The British Air Estimates of March 1933 revealed a total lack of comprehension alike by the Government and the Oppositions, Labour and Liberal, of what was going on. I had to say (March 14, 1933):

I regretted to hear the Under-Secretary say that we were only the fifth air Power, and that the ten-year programme was suspended for another year. I was sorry to hear him boast that the Air Ministry had not laid down a single new unit this year. All these ideas are being increasingly stultified by the march of events, and we should be well advised to concentrate upon our air defences with greater vigour.

*   *   *   *   *

Under the so-called National Government British public opinion showed an increasing inclination to cast aside all care about Germany. In vain the French had pointed out correctly in a memorandum of July 21, 1931, that the general assurance given at Versailles that a universal limitation of armaments should follow the one-sided disarmament of Germany did not constitute a Treaty obligation. It certainly was not an obligation enforceable apart from time and circumstance. Yet when in 1932 the German delegation to the Disarmament Conference categorically demanded the removal of all restrictions upon their right to re-arm they found much support in the British Press. The *Times* spoke of "the timely redress of inequality", and the *New Statesman* of "the unqualified recognition of the principle of the equality of States". This meant that the seventy million Germans ought to be allowed to rearm and prepare for war without the victors in the late fearful struggle being entitled to make any objection. Equality of status between victors and vanquished; equality between a France of thirty-nine millions and a Germany of nearly double that number!

The German Government were emboldened by the British demeanour. They ascribed it to the fundamental weakness and inherent decadence imposed even upon a Nordic race by the democratic and Parliamentary form of society. With all Hitler's national drive behind them, they took a haughty line. In July their delegation gathered up its papers and quitted the Disarmament Conference. To coax them back then became the prime political objective of the victorious Allies. In November the French, under severe and constant British pressure, proposed what was somewhat unfairly called "the Herriot Plan". The essence of this was the reconstitution of all European defence forces as short-service armies with limited numbers, admitting equality of status but not necessarily accepting equality of strength. In fact and in principle, the admission of equality of status made it impossible ultimately not to accept equality of strength. This enabled the Allied Governments to offer to Germany "Equality of rights in a system which would provide security for all nations". Under certain safeguards of an illusory character the French were reduced to accepting this meaningless formula. On this the Germans consented to return to the Disarmament Conference. This was hailed as a notable victory for peace.

Fanned by the breeze of popularity, His Majesty's Government now produced on March 16, 1933, what was called after its author and inspirer "the MacDonald Plan". It accepted as its starting-point the adoption of the French conception of short-service armies—in this case of eight months' service—and proceeded to prescribe exact figures for the troops of each country. The French Army should be reduced from its peace-time establishment of 500,000 men to 200,000 and the Germans should increase to parity at that figure. By this time the German military forces, though not yet provided with the mass of trained reserves which only a succession of annual conscripted quotas could supply, may well have amounted to the equivalent of over a million ardent volunteers, partially equipped, and with many forms of the latest weapons coming along through the convertible and partially-converted factories to arm them.

At the end of the First World War, France, like Great Britain, had an enormous mass of heavy artillery, whereas the cannon of the German Army had in fact been blown to bits according to Treaty. Mr. MacDonald sought to remedy this evident inequality by proposing to limit the calibre of mobile artillery guns to 105 mm., or 4.2 inches. Existing guns up to 6 inches could be retained, but all replacements were to be limited to 4.2 inches. British interests, as distinct from those of France, were to be protected by the maintenance of the Treaty restrictions against German naval armaments until 1935, when it was proposed that a new Naval Conference should meet. Military aircraft were prohibited to Germany for the duration of the agreement; but the three Allied Powers were to reduce their own Air Forces to 500 planes apiece.

I viewed this attack upon the French armed forces and the attempt to establish equality between Germany and France with strong aversion; and on March 23, 1933, I had the opportunity of saying to Parliament:

I doubt the wisdom of pressing this plan upon France at the present time. I do not think the French will agree. They must be greatly concerned at what is taking place in Germany, as well as at the attitude of some others of their neighbours. I dare say that during this anxious month there are a good many people who have said to themselves, as I have been saying for several years: "Thank God for the French Army." When we read about Germany, when we watch with surprise and distress the tumultuous insurgence of ferocity and war spirit, the pitiless ill-treatment of minorities, the denial of the normal protections of civilised society, the persecution of large numbers of individuals solely on the ground of race—when we see all that occurring in one of the most gifted, learned, and scientific and formidable nations in the world, one cannot help feeling glad that the fierce passions that are raging in Germany have not yet found any other outlet but upon themselves. It seems to me that at a moment like this to ask France to halve her Army while Germany doubles hers, to ask France to halve her Air Force while the German Air Force remains whatever it is, is a proposal likely to be considered by the French Government, at present at any rate, as somewhat unseasonable. The figures that are given in the plan of the strength of armies and aeroplanes secure to France only as many aeroplanes as would be possessed by Italy, leaving any air-power possessed by Germany entirely out of consideration.

And again in April:

The Germans demand equality in weapons and equality in the organisation of armies and fleets, and we have been told, "You cannot keep so great a nation in an inferior position. What others have, they must have." I have never agreed. It is a most dangerous demand to make. Nothing in life is eternal, but as surely as Germany acquires full military equality with her neighbours while her own grievances are still unredressed and while she is in the temper which we have unhappily seen, so surely shall we see ourselves within a measurable distance of the renewal of general European war.

. . . One of the things which we were told after the Great War would be a security for us was that Germany would be a democracy with Parliamentary institutions. All that has been swept away. You have most grim dictatorship. You have militarism and appeals to every form of fighting spirit, from the reintroduction of duelling in the colleges to the Minister of Education advising the plentiful use of the cane in the elementary schools. You have these martial or pugnacious manifestations, and also this persecution of the Jews of which so many members have spoken. . . .

I will leave Germany and turn to France. France is not only the sole great surviving democracy in Europe; she is also the strongest military Power, I am glad to say, and she is the head of a system of States and nations. France is the guarantor and protector of the whole crescent of small States which runs right round from Belgium to Yugoslavia and Roumania. They all look to France. When any step is taken, by England or any other Power, to weaken the diplomatic or military security of France, all these small nations tremble with fear and anger. They fear that the central protective force will be weakened, and that then they will be at the mercy of the great Teutonic Power.

When one considers that the facts were hardly in dispute, the actions of a responsible Government of respectable men and the public opinion which so flocculently supported them are scarcely comprehensible. It was like being smothered by a feather-bed. I remember particularly the look of pain and aversion which I saw on the faces of Members in all parts of the House when I said, "Thank God for the French Army." Words were vain.

However, the French had the hardihood to insist that there should be a delay of four years before the destruction of their heavy war material. The British Government accepted this modification provided that the French agreement about the destruction of their artillery was specified in a document for immediate signature. France bowed to this, and on October 12, 1933, Sir John Simon, after complaining that Germany had shifted her ground in the course of the preceding weeks, brought these draft proposals before the Disarmament Conference. The result was unexpected. Hitler, now Chancellor and Master of all Germany, having already given orders on assuming power to drive ahead boldly on a nation-wide scale, both in the training-camps and the factories, felt himself in a strong position. He did not even trouble to accept the quixotic offers pressed upon him. With a gesture of disdain he directed the German Government to withdraw both from the Conference and from the League of Nations. Such was the fate of the MacDonald Plan.

*   *   *   *   *

It is difficult to find a parallel to the unwisdom of the British and weakness of the French Governments, who none the less reflected the opinion of their Parliaments in this disastrous period. Nor can the United States escape the censure of history. Absorbed in their own affairs and all the abounding interests, activities, and accidents of a free community, they simply gaped at the vast changes which were taking place in Europe, and imagined they were no concern of theirs. The considerable corps of highly competent, widely-trained professional American officers formed their own opinions, but these produced no noticeable effect upon the improvident aloofness of American foreign policy. If the influence of the United States had been exerted, it might have galvanised the French and British politicians into action. The League of Nations, battered though it had been, was still an august instrument which would have invested any challenge to the new Hitler war-menace with the sanctions of International Law. Under the strain the Americans merely shrugged their shoulders, so that in a few years they had to pour out the blood and treasures of the New World to save themselves from mortal danger.

Seven years later when at Tours I witnessed the French agony all this was in my mind, and that is why, even when proposals for a separate peace were mentioned, I spoke only words of comfort and reassurance, which I rejoice to feel have been made good.

*   *   *   *   *

I had arranged at the beginning of 1931 to undertake a considerable lecture tour in the United States, and travelled to New

York. Here I suffered a serious accident, which nearly cost me my life. On December 13, when on my way to visit Mr. Bernard Baruch, I got out of my car on the wrong side, and walked across Fifth Avenue without bearing in mind the opposite rule of the road which prevails in America, or the red lights, then unused in Britain. There was a shattering collision. For two months I was a wreck. I gradually regained at Nassau in the Bahamas enough strength to crawl around. In this condition I undertook a tour of forty lectures throughout the United States, living all day on my back in a railway compartment, and addressing in the evening large audiences. On the whole I consider this was the hardest time I have had in my life. I lay pretty low all through this year, but in time my strength returned.

Meanwhile at home our life flowed placidly downstream. At Westminster Mr. Baldwin adopted and espoused the main principles of Mr. MacDonald's India Bill, the conduct·of which in the Commons was entrusted to the new Secretary of State for India, Sir Samuel Hoare. The report of the Simon Commission was ignored, and no opportunity of debating it was given to Parliament. With about seventy other Conservatives I formed a group called "the India Defence League", which during the next four years resisted the Government's policy on India in so far as it went beyond the recommendations of the Commission. We fought the matter out at party conferences with a considerable measure of support, sometimes running very close, but always in a minority. The Labour Opposition voted in Parliament with the Government on the Indian issue, and it became, like Disarmament, a link between the two Front Benches. Their followers presented an overwhelming majority against our group, and derided us as "Die-hards". The rise of Hitler to power, the domination of the Nazi Party over all Germany, and the rapid, active growth of German armed power, led to further differences between me and the Government and the various political parties in the State.

The years from 1931 to 1935, apart from my anxiety on public affairs, were personally very pleasant to me. I earned my livelihood by dictating articles which had a wide circulation not only in Great Britain and the United States, but also, before Hitler's shadow fell upon them, in the most famous newspapers of sixteen

European countries. I lived in fact from mouth to hand. I produced in succession the various volumes of the life of Marlborough. I meditated constantly upon the European situation and the rearming of Germany. I lived mainly at Chartwell, where I had much to amuse me. I built with my own hands a large part of two cottages and extensive kitchen-garden walls, and made all kinds of rockeries and waterworks and a large swimming-pool which was filtered to limpidity and could be heated to supplement our fickle sunshine. Thus I never had a dull or idle moment from morning till midnight, and with my happy family around me dwelt at peace within my habitation.

During these years I saw a great deal of Frederick Lindemann, Professor of Experimental Philosophy at Oxford University. Lindemann was already an old friend of mine. I had met him first at the close of the previous war, in which he had distinguished himself by conducting in the air a number of experiments, hitherto reserved for daring pilots, to overcome the then almost mortal dangers of a "spin". We came much closer together from 1932 onwards, and he frequently motored over from Oxford to stay with me at Chartwell. Here we had many talks into the small hours of the morning about the dangers which seemed to be gathering upon us. Lindemann, "the Prof.", as he was called among his friends, became my chief adviser on the scientific aspects of modern war and particularly air defence, and also on questions involving statistics of all kinds. This pleasant and fertile association continued throughout the war.

Another of my close friends was Desmond Morton.* When in 1917 Field-Marshal Haig filled his personal staff with young officers fresh from the firing line, Desmond was recommended to him as the pick of the Artillery. He had commanded the most advanced field battery in Arras during the severe spring fighting of that year. To his Military Cross he added the unique distinction of having been shot through the heart, and living happily ever afterwards with the bullet in him. When I became Minister of Munitions in July 1917 I frequently visited the front as the Commander-in-Chief's guest, and he always sent his trusted aide-de-camp, Desmond Morton, with me. Together we visited many parts of the line. During these sometimes dangerous excursions, and at the Commander-in-Chief's house, I formed a

* Now Major Sir Desmond Morton, K.C.B., M.C.

great regard and friendship for this brilliant and gallant officer, and in 1919, when I became Secretary of State for War and Air, I appointed him to a key position in the Intelligence, which he held for many years. He was a neighbour of mine, dwelling only a mile away from Chartwell. He obtained from the Prime Minister, Mr. MacDonald, permission to talk freely to me and keep me well informed. He became, and continued during the war to be, one of my most intimate advisers till our final victory was won.

I had also formed a friendship with Ralph Wigram, then the rising star of the Foreign Office and in the centre of all its affairs. He had reached a level in that department which entitled him to express responsible opinions upon policy, and to use a wide discretion in his contacts, official and unofficial. He was a charming and fearless man, and his convictions, based upon profound knowledge and study, dominated his being. He saw as clearly as I did, but with more certain information, the awful peril which was closing in upon us. This drew us together. Often we met at his little house in North Street, and he and Mrs. Wigram came to stay with us at Chartwell. Like other officials of high rank, he spoke to me with complete confidence. All this helped me to form and fortify my opinion about the Hitler Movement. For my part, with the many connections which I now had in France, in Germany, and other countries, I had been able to send him a certain amount of information which we examined together.

From 1933 onwards Wigram became keenly distressed at the policy of the Government and the course of events. While his official chiefs formed every day a higher opinion of his capacity, and while his influence in the Foreign Office grew, his thoughts turned repeatedly to resignation. He had so much force and grace in his conversation that all who had grave business with him, and many others, gave ever-increasing importance to his views.

* * * * *

It was of great value to me, and it may be thought also to the country, that I should have the means of conducting searching and precise discussions for so many years in this very small circle. On my side however I gathered and contributed a great deal of information from foreign sources. I had confidential contacts with several of the French Ministers and with the successive chiefs of the French Government. Mr. Ian Colvin, the son of the famous leader-writer of the *Morning Post*, was the *News Chronicle* correspondent in Berlin. He plunged very deeply into German politics, and established contacts of a most secret character with some of the important German generals, and also with independent men of character and quality in Germany who saw in the Hitler Movement the approaching ruin of their native land. Several visitors of consequence came to me from Germany and poured their hearts out in their bitter distress. Most of these were executed by Hitler during the war. From other directions I was able to check and furnish information on the whole field of our air defence. In this way I became as well instructed as many Ministers of the Crown. All the facts I gathered from every source, including especially foreign connections, I reported to the Government from time to time. My personal relations with Ministers and also with many of their high officials were close and easy, and although I was often their critic we maintained a spirit of comradeship. Later on, as will be seen, I was made officially party to much of their most secret technical knowledge. From my own long experience in high office I was also possessed of the most precious secrets of the State. All this enabled me to form and maintain opinions which did not depend on what was published in the newspapers, though these brought many items to the discriminating eye.

* * * * *

At Westminster I pursued my two themes of India and the German menace, and went to Parliament from time to time to deliver warning speeches, which commanded attention, but did not, unhappily, wake to action the crowded, puzzled Houses which heard them. On the German danger, as on India, I found myself working in Parliament with a group of friends. It was composed differently from the India Defence League. Sir Austen Chamberlain, Sir Robert Horne, Sir Edward Grigg, Lord Winterton, Mr. Bracken, Sir Henry Croft, and several others formed our circle. We met regularly and to a large extent pooled our information. The Ministers eyed this significant but not unfriendly body of their own supporters and former colleagues or seniors with respect. We could at any time command the attention of Parliament and stage a full-dress debate.

* * * * *

The reader will pardon a personal digression in a lighter vein. In the summer of 1932 for the purposes of my life of Marlborough I visited his old battlefields in the Low Countries and Germany. Our family expedition, which included "the Prof.", journeyed agreeably along the line of Marlborough's celebrated march in 1705 from the Netherlands to the Danube, passing the Rhine at Coblenz. As we wended our way through these beautiful regions from one ancient, famous city to another, I naturally asked questions about the Hitler Movement, and found it the prime topic in every German mind. I sensed a Hitler atmosphere. After passing a day on the field of Blenheim, I drove into Munich, and spent the best part of a week there.

At the Regina Hotel a gentleman introduced himself to some of my party. He was Herr Hanfstaengl, and spoke a great deal about "the Fuehrer", with whom he appeared to be intimate. As he seemed to be a lively and talkative fellow, speaking excellent English, I asked him to dine. He gave a most interesting account of Hitler's activities and outlook. He spoke as one under the spell. He had probably been told to get in touch with me. He was evidently most anxious to please. After dinner he went to the piano and played and sang many tunes and songs in such remarkable style that we all enjoyed ourselves immensely. He seemed to know all the English tunes that I liked. He was a great entertainer, and at that time, as is known, a favourite of the Fuehrer. He said I ought to meet him, and that nothing would be easier to arrange. Herr Hitler came every day to the hotel about five o'clock, and would be very glad indeed to see me.

I had no national prejudices against Hitler at this time. I knew little of his doctrine or record and nothing of his character. I admire men who stand up for their country in defeat, even though I am on the other side. He had a perfect right to be a patriotic German if he chose. I always wanted England, Germany, and France to be friends. However, in the course of conversation with Hanfstaengl I happened to say, "Why is your chief so violent about the Jews? I can quite understand being angry with Jews who have done wrong or are against the country, and I understand resisting them if they try to monopolise power in any walk of life; but what is the sense of being against a man simply because of his birth? How can any man help how he is born?" He must have repeated this to Hitler, because about noon the next day he came round with rather a serious air and said that the appointment he had made with me to meet Hitler could not take place as the Fuehrer would not be coming to the hotel that afternoon. This was the last I saw of "Putzi"—for such was his pet name—although we stayed several more days at the hotel. Thus Hitler lost his only chance of meeting me. Later on, when he was all-powerful, I was to receive several invitations from him. But by that time a lot had happened, and I excused myself.

*   *   *   *   *

All this while the United States remained intensely preoccupied with its own vehement internal affairs and economic problems. Europe and far-off Japan watched with steady gaze the rise of German warlike power. Disquietude was increasingly expressed in Scandinavian countries and the States of the Little Entente and in some Balkan countries. Deep anxiety ruled in France, where a large amount of knowledge of Hitler's activities and of German preparations had come to hand. There was, I was told, a catalogue of breaches of the treaties of immense and formidable gravity, but when I asked my French friends why this matter was not raised in the League of Nations, and Germany invited, or even ultimately summoned, to explain her action and state precisely what she was doing, I was answered that the British Government would deprecate such an alarming step. Thus, while Mr. MacDonald, with Mr. Baldwin's full authority, preached disarmament to the French and practised it upon the British, the German might grew by leaps and bounds, and the time for overt action approached.

In justice to the Conservative Party it must be mentioned that at each of the Conferences of the National Union of Conservative Associations from 1932 onwards resolutions proposed by such worthies as Lord Lloyd and Sir Henry Croft in favour of an immediate strengthening of our armaments to meet the growing danger from abroad were carried almost unanimously. But the Parliamentary control by the Government Whips in the House of Commons was at this time so effective, and the three parties in the Government, as well as the Labour Opposition, so sunk in lethargy and blindness, that the warnings of their followers in the country were as ineffective as were the signs of the times and the evidence of the Secret Service. This was one of those awful periods which recur in our history, when the noble British nation seems to fall from its high estate, loses all trace of sense or purpose, and appears to cower from the menace of foreign peril, frothing pious platitudes while foemen forge their arms.

In this dark time the basest sentiments received acceptance or passed unchallenged by the responsible leaders of the political parties. In 1933 the students of the Oxford Union, under the inspiration of a Mr. Joad, passed their ever-shameful resolution, "That this House will in no circumstances fight for its King and Country." It was easy to laugh off such an episode in England, but in Germany, in Russia, in Italy, in Japan, the idea of a decadent, degenerate Britain took deep root and swayed many calculations. Little did the foolish boys who passed the resolution dream that they were destined quite soon to conquer or fall gloriously in the ensuing war, and prove themselves the finest generation ever bred in Britain. Less excuse can be found for their elders, who had no chance of self-redemption in action.*

* * * * *

In November 1933 we had another debate in the House of Commons. I returned to my main theme:

We read of large importations of scrap iron and nickel and war metals, quite out of the ordinary. We read all the news which accumulates of the military spirit which is rife throughout the country; we see that a philosophy of blood-lust is being inculcated into their youth to which no parallel can be found since the days of barbarism. We see all these forces on the move, and we must remember that this is the same mighty Germany which fought all the world and almost beat the world; it is the same mighty Germany which took two and a half lives for every German life that was taken.† No wonder, when you have these preparations, these doctrines, and these assertions openly made, that there is alarm throughout the whole circle of nations which surround Germany. . . .

* * * * *

* I cannot resist telling this story. I was asked to address the University Conservative Association in the Oxford Union. I declined to do so, but said I would give them an hour to ask me questions. One of the questions was, "Do you think Germany was guilty of making the last war?" I said, "Yes, of course." A young German Rhodes scholar rose from his place and said, "After this insult to my country I will not remain here." He then stalked out amid roars of applause. I thought him a spirited boy. Two years later it was found out in Germany that he had a Jewish ancestor. This ended his career in Germany.

† This excludes the Russian losses.

While this fearful transformation in the relative war-power of victors and vanquished was taking place in Europe, a complete lack of concert between the non-aggressive and peace-loving States had also developed in the Far East. This story forms a counterpart to the disastrous turn of events in Europe, and arose from the same paralysis of thought and action among the leaders of the former and future Allies.

The Economic Blizzard of 1929 to 1931 had affected Japan not less than the rest of the world. Since 1914 her population had grown from fifty to seventy millions. Her metallurgical factories had increased from fifty to one hundred and forty-eight. The cost of living had risen steadily. The production of rice was stationary, and its importation expensive. The need for raw material and for external markets was clamant. In the violent depression Britain and forty other countries felt increasingly compelled, as the years passed, to apply restrictions or tariffs against Japanese goods produced under labour conditions unrelated to European or American standards. China was more than ever Japan's principal export market for cotton and other manufactures, and almost her sole source of coal and iron. A new assertion of control over China became therefore the main theme of Japanese policy.

In September 1931, on a pretext of local disorders, the Japanese occupied Mukden and the zone of the Manchurian Railway. In January 1932 they demanded the dissolution of all Chinese associations of an anti-Japanese character. The Chinese Government refused, and on the 28th the Japanese landed to the north of the International Concession at Shanghai. The Chinese resisted with spirit, and, although without aeroplanes or anti-tank guns or any of the modern weapons, maintained their resistance for more than a month. At the end of February, after suffering very heavy losses, they were obliged to retire from their forts in the bay of Wu-Sung, and took up positions about twelve miles inland. Early in 1932 the Japanese created the puppet State of Manchukuo. A year later the Chinese province of Jehol was annexed to it, and Japanese troops, penetrating deeply into defenceless regions, had reached the Great Wall of China. This aggressive action corresponded to the growth of Japanese power in the Far East and her new naval position on the oceans.

From the first shot the outrage committed upon China aroused the strongest hostility in the United States. But the policy of Isolation cut both ways. Had the United States been a member of the League of Nations, she could undoubtedly have led that assembly into collective action against Japan, of which the United States would herself have been the principal mandatory. The British Government on their part showed no desire to act with the United States alone; nor did they wish to be drawn into antagonism with Japan further than their obligations under the League of Nations Charter required. There was a rueful feeling in some British circles at the loss of the Japanese Alliance and the consequential weakening of the British position with all its long-established interests in the Far East. His Majesty's Government could hardly be blamed if in their grave financial and growing European embarrassments they did not seek a prominent *rôle* at the side of the United States in the Far East without any hope of corresponding American support in Europe.

China however was a member of the League, and although she had not paid her subscription to that body she appealed to it for what was no more than justice. On September 30, 1931, the League called on Japan to remove her troops from Manchuria. In December a Commission was appointed to conduct an inquiry on the spot. The League of Nations entrusted the chairmanship of the Commission to the Earl of Lytton, the worthy descendant of a gifted line. He had had many years' experience in the East as Governor of Bengal and as acting Viceroy of India. The report, which was unanimous, was a remarkable document, and forms the basis of any serious study of the conflict between China and Japan. The whole background of the Manchurian affair was carefully presented. The conclusions drawn were plain: Manchu-kuo was the artificial creation of the Japanese General Staff, and the wishes of the population had played no part in the formation of this puppet State. Lord Lytton and his colleagues in their report not only analysed the situation, but put forward concrete proposals for an international solution. These were for the declaration of an autonomous Manchuria. It would still remain part of China, under the ægis of the League, and there would be a comprehensive treaty between China and Japan regulating their interests in Manchuria. The fact that the League could not follow up these proposals in no way detracts from the value of the Lytton report. The American Secretary of State, Stimson, wrote of the document: "It became at once and remains to-day the outstanding impartial authority upon the subject which it covers." In February 1933 the League of Nations declared that the State of Manchukuo could not be recognised. Although no sanctions were imposed upon Japan, nor any other action taken, Japan, on March 27, 1933, withdrew from the League of Nations. Germany and Japan had been on opposite sides in the war; they now looked towards each other in a different mood. The moral authority of the League was shown to be devoid of any physical support at a time when its activity and strength were most needed.

✶   ✶   ✶   ✶   ✶

We must regard as deeply blameworthy before history the conduct not only of the British National and mainly Conservative Government, but of the Labour-Socialist and Liberal Parties, both in and out of office, during this fatal period. Delight in smooth-sounding platitudes, refusal to face unpleasant facts, desire for popularity and electoral success irrespective of the vital interests of the State, genuine love of peace and pathetic belief that love can be its sole foundation, obvious lack of intellectual vigour in both leaders of the British Coalition Government, marked ignorance of Europe and aversion from its problems in Mr. Baldwin, the strong and violent pacifism which at this time dominated the Labour-Socialist Party, the utter devotion of the Liberals to sentiment apart from reality, the failure and worse than failure of Mr. Lloyd George, the erstwhile great war-time leader, to address himself to the continuity of his work, the whole supported by overwhelming majorities in both Houses of Parliament: all these constituted a picture of British fatuity and fecklessness which, though devoid of guile, was not devoid of guilt, and, though free from wickedness or evil design, played a definite part in the unleashing upon the world of horrors and miseries which, even so far as they have unfolded, are already beyond comparison in human experience.

CHAPTER VI

THE DARKENING SCENE

1934

Spring Warnings – The German Blood Purge of June 30 – The End of Disarmament – The Murder of Dr. Dollfuss, July 25 – The Death of Hindenburg – Hitler Head of the German State, August 2 – The Italian Dilemma – The Murder of King Alexander and M. Barthou at Marseilles, October 9 – M. Laval, French Foreign Minister, November – Italian-Abyssinian Clash at Wal-Wal, December – Franco-Italian Agreement, January 6, 1935 – The Saar Plebiscite, January 13, 1935.

HITLER'S accession to the Chancellorship in 1933 had not been regarded with enthusiasm in Rome. Nazism was viewed as a crude and brutalised version of the Fascist theme. The ambitions of a Greater Germany towards Austria and in South-eastern Europe were well known. Mussolini foresaw that in neither of these regions would Italian interests coincide with those of the new Germany. Nor had he long to wait for confirmation.

* * * * *

The acquisition of Austria by Germany was one of Hitler's most cherished ambitions. The first page of *Mein Kampf* contains the sentence, "German Austria must return to the great German Motherland." From the moment, therefore, of the acquisition of power in January 1933, the Nazi German Government cast its eyes upon Vienna. Hitler could not afford as yet to clash with Mussolini, whose interests in Austria had been loudly proclaimed. Even infiltration and underground activities had to be applied with caution by a Germany as yet militarily weak. Pressure on Austria however began in the first few months. Unceasing demands were made on the Austrian Government to force members of the satellite Austrian Nazi Party both into the Cabinet and into key posts in the Administration. Austrian Nazis were trained in an Austrian legion organised in Bavaria. Bomb outrages on the railways and at tourist centres, German aeroplanes showering leaflets over Salzburg and Innsbruck, disturbed the daily life of the republic. The Austrian Chancellor Dollfuss was equally opposed both by Socialist pressure within and external German designs against Austrian independence. Nor was this the only menace to the Austrian State. Following the evil example of their German neighbours, the Austrian Socialists had built up a private army with which to override the decision of the ballot-box. Both dangers loomed upon Dollfuss during 1933. The only quarter to which he could turn for protection and whence he had already received assurances of support was Fascist Italy. In August 1933 he met Mussolini at Riccione. A close personal and political understanding was reached between them. Dollfuss, who believed that Italy would hold the ring, felt strong enough to move against one set of his opponents—the Austrian Socialists.

In January 1934 Suvich, Mussolini's principal adviser on foreign affairs, visited Vienna as a gesture of warning to Germany. On January 21 he made the following public statement:

The importance of Austria, due to her position in the heart of Central Europe and in the Danube basin, far exceeds, as is well known, her territorial and numerical size. If she is to fulfil in the interests of all the mission accorded her by centuries-old tradition and geographical situation, the normal conditions of independence and peaceful life must first of all be secured. That is the standpoint which Italy has long maintained in regard to both political and economic conditions on the basis of unchangeable principles.

Three weeks later the Dollfuss Government took action against the Socialist organisations of Vienna. The Heimwehr, under Major Fey, belonging to Dollfuss's own party, received orders to disarm the equivalent and equally illegal body controlled by the Austrian Socialists. The latter resisted forcibly, and on February 12 street fighting broke out in the capital. Within a few hours the Socialist forces were broken. This event not only brought Dollfuss closer to Italy, but strengthened him in the next stage of his task against the Nazi penetration and conspiracy. On the other hand, many of the defeated Socialists or Communists swung over to the Nazi camp in their bitterness. In Austria as in Germany the Catholic-Socialist feud helped the Nazis.

* * * * *

Until the middle of 1934 the control of events was still largely in the hands of His Majesty's Government without the risk of war. They could at any time, in concert with France and through the agency of the League of Nations, have brought an overwhelming power to bear upon the Hitler Movement, about which Germany was profoundly divided. This would have involved no bloodshed. But this phase was passing. An armed Germany under Nazi control was approaching the threshold. And yet, incredible though it may seem, far into this cardinal year Mr. MacDonald, armed with Mr. Baldwin's political power, continued to work for the disarmament of France. I cannot but quote the unavailing protest which I made in Parliament on February 7:

What happens, for instance, if, after we have equalised and reduced the army of France to the level of that of Germany, and got an equality for Germany, and with all the reactions which will have followed in the sentiment of Europe upon such a change, Germany then proceeds to say, "How can you keep a great nation of seventy millions in a position in which it is not entitled to have a navy equal to the greatest of the fleets upon the seas?" You will say, "No; we do not agree. Armies—they belong to other people. Navies—that question affects Britain's interests and we are bound to say, 'No'." But what position shall we be in to say that "No"?

Wars come very suddenly. I have lived through a period when one looked forward, as we do now, with great anxiety and uncertainty to what would happen in the future. Suddenly something did happen— tremendous, swift, overpowering, irresistible. Let me remind the House of the sort of thing that happened in 1914. There was absolutely no quarrel between Germany and France. One July afternoon the German Ambassador drove down to the Quai d'Orsay and said to the French Prime Minister, "We have been forced to mobilise against Russia, and war will be declared. What is to be the position of France?" The French Premier made the answer which his Cabinet had agreed upon, that France would act in accordance with what she considered to be her own interests. The Ambassador said, "You have an alliance with Russia, have you not?" "Quite so," said the French Premier. And that was the process by which, in a few minutes, the area of the struggle, already serious in the East, was enormously widened and multiplied by the throwing in of the two great nations of the West on either side. But sometimes even a declaration of neutrality does not suffice. On this very occasion, as we now know, the German Ambassador was authorised by his Government, in case the French did not do their duty by their Russian ally, in case they showed any disposition to back out of the conflict which had been resolved on by Germany, to demand that the fortresses of Toul and Verdun should be handed over to German troops as a guarantee that the French, having declared neutrality, would not change their mind at a subsequent moment. . . .

We may ourselves, in the lifetime of those who are here, if we are not in a proper state of security, be confronted on some occasion with a visit from an Ambassador, and may have to give an answer, and if that answer is not satisfactory, within the next few hours the crash of bombs exploding in London and cataracts of masonry and fire and smoke will warn us of any inadequacy which has been permitted in our aerial defences. We are vulnerable as we have never been before. I often heard criticisms of the Liberal Government before the war. . . . A far graver case rests upon those who now hold power if by any chance, against our wishes and against our hopes, trouble should come.

Not one of the lessons of the past has been learned, not one of them has been applied, and the situation is incomparably more dangerous. Then we had the Navy and no air menace. Then the Navy was the "sure shield" of Britain. . . . We cannot say that now. This cursed, hellish invention and development of war from the air has revolutionised our position. We are not the same kind of country we used to be when we were an island, only twenty years ago.

I then asked for three definite decisions to be taken without delay. For the Army, the reorganisation of our civil factories, so that they could be turned over rapidly to war purposes, should be begun in Britain, as all over Europe. For the Navy we should regain freedom of design. We should get rid of this London Treaty which had crippled us in building the kind of ships we wanted, and had stopped the United States from building a great battleship which she probably needed, and to which we should not have had the slightest reason to object. We should be helped in doing this by the fact that another of the parties to that treaty* was resolved to regain her freedom too. Thirdly, the air. We ought to have an Air Force as strong as the Air Force of France or Germany, whichever was the stronger. The Government

* Japan.

commanded overwhelming majorities in both branches of the Legislature, and nothing would be denied to them. They had only to make their proposals with confidence and conviction for the safety of the country, and their countrymen would sustain them.

★   ★   ★   ★   ★

There was at this moment a flicker of European unity against the German menace. On February 17, 1934, the British, French, and Italian Governments made a common declaration upon the maintenance of Austrian independence. On March 14 I spoke again in Parliament:

The awful danger of our present foreign policy is that we go on perpetually asking the French to weaken themselves. And what do we say is the inducement? We say, "Weaken yourselves," and we always hold out the hope that if they do it and get into trouble we will then in some way or other go to their aid, although we have nothing with which to go to their aid. I cannot imagine a more dangerous policy. There is something to be said for isolation; there is something to be said for alliances. But there is nothing to be said for weakening the Power on the Continent with whom you would be in alliance, and then involving yourself more [deeply] in Continental tangles in order to make it up to them. In that way you have neither the one thing nor the other; you have the worst of both worlds.

The Romans had a maxim, "Shorten your weapons and lengthen your frontiers." But our maxim seems to be, "Diminish your weapons and increase your obligations." Aye, and diminish the weapons of your friends.

★   ★   ★   ★   ★

Italy now made a final attempt to carry out the aforesaid maxim. On March 17 Italy, Hungary, and Austria signed the so-called Rome Protocols, providing for mutual consultation in the event of a threat to any of the three parties. But Hitler was growing steadily stronger, and in May and June subversive activities increased throughout Austria. Dollfuss immediately sent reports on these terrorist acts to Suvich, with a note deploring their depressive effect upon Austrian trade and tourists.

It was with this dossier in his hand that Mussolini went to Venice on June 14 to meet Hitler for the first time. The German Chancellor stepped from his aeroplane in a brown mackintosh and Homburg hat into an array of sparkling Fascist uniforms, with a resplendent and portly Duce at their head. As Mussolini caught sight of his guest, he murmured to his aide, "*Non mi piace.*" ("I don't like the look of him.") At this strange meeting only a general exchange of ideas took place, with mutual lectures upon the virtues of dictatorship on the German and Italian models. Mussolini was clearly perplexed both by the personality and language of his guest. He summed up his final impression in these words, "A garrulous monk." He did however extract some assurances of relaxation of German pressure upon Dollfuss. Ciano told the journalists after the meeting, "You see. Nothing more will happen."

But the pause in German activities which followed was due not to Mussolini's appeal, but to Hitler's own internal preoccupations.

*   *   *   *   *

The acquisition of power had opened a deep divergence between the Fuehrer and many of those who had borne him forward. Under Roehm's leadership the S.A. increasingly represented the more revolutionary elements of the party. There were senior members of the party, such as Gregor Strasser, ardent for social revolution, who feared that Hitler in arriving at the first place would simply be taken over by the existing hierarchy, the Reichswehr, the bankers, and the industrialists. He would not have been the first revolutionary leader to kick down the ladder by which he had risen to exalted heights. To the rank and file of the S.A. ("Brownshirts") the triumph of January 1933 was meant to carry with it the freedom to pillage not only the Jews and profiteers, but also the well-to-do, established classes of society. Rumours of a great betrayal by their leader soon began to spread in certain circles of the party. Chief-of-Staff Roehm acted on this impulse with energy. In January 1933 the S.A. had been four hundred thousand strong. By the spring of 1934 he had recruited and organised nearly three million men. Hitler in his new situation was uneasy at the growth of this mammoth machine, which, while professing fervent loyalty to his name, and being for the most part deeply attached to him, was beginning to slip from his own personal control. Hitherto he had possessed a private army. Now he had the national Army. He did not intend to exchange the one for the other. He wanted both, and to use each, as events required, to control the other. He had now therefore to deal with Roehm. "I am resolved," he declared to the leaders of the S.A. in these days, "to repress severely any attempt to overturn the existing order. I will oppose with the sternest energy a second revolutionary wave, for it would bring with it inevitable chaos. Whoever raises his head against the established authority of the State will be severely treated, whatever his position."

In spite of his misgivings Hitler was not easily convinced of the disloyalty of his comrade of the Munich *Putsch*, who for the last seven years had been the Chief of Staff of his Brownshirt army. When, in December 1933, the unity of the party with the State had been proclaimed Roehm became a member of the German Cabinet. One of the consequences of the union of the party with the State was to be the merging of the Brownshirts with the Reichswehr. The rapid progress of national rearmament forced the issue of the status and control of all the German armed forces into the forefront of politics. In February 1934 Mr. Eden arrived in Berlin, and in the course of conversation Hitler agreed provisionally to give certain assurances about the non-military character of the S.A. Roehm was already in constant friction with General von Blomberg, the Chief of the General Staff. He now feared the sacrifice of the party army he had taken so many years to build, and, in spite of warnings of the gravity of his conduct, he published on April 18 an unmistakable challenge:

The Revolution we have made is not a national revolution, but a National *Socialist* Revolution. We would even underline this last word, "Socialist". The only rampart which exists against reaction is represented by our Assault Groups, for they are the absolute incarnation of the revolutionary idea. The militant in the Brown Shirt from the first day pledged himself to the path of revolution, and he will not deviate by a hairbreadth until our ultimate goal has been achieved.

He omitted on this occasion the "Heil Hitler!" which had been the invariable conclusion of Brownshirt harangues.

During the course of April and May Blomberg continually complained to Hitler about the insolence and activities of the S.A. The Fuehrer had to choose between the generals who hated him and the Brownshirt thugs to whom he owed so much. He chose the generals. At the beginning of June Hitler, in a five-hour conversation, made a last effort to conciliate and come to terms with Roehm. But with this abnormal fanatic, devoured by ambition, no compromise was possible. The mystic hierarchic Greater Germany of which Hitler dreamed and the Proletarian Republic of the People's Army desired by Roehm were separated by an impassable gulf.

Within the framework of the Brownshirts there had been formed a small and highly-trained *élite*, wearing black uniforms and known as the S.S., or later as Blackshirts. These units were intended for the personal protection of the Fuehrer and for special and confidential tasks. They were commanded by an unsuccessful ex-poultry-farmer, Heinrich Himmler. Foreseeing the impending clash between Hitler and the Army on the one hand and Roehm and the Brownshirts on the other, Himmler took care to carry the S.S. into Hitler's camp. On the other hand, Roehm had supporters of great influence within the party, who, like Gregor Strasser, saw their ferocious plans for Social Revolution being cast aside. The Reichswehr also had its rebels. Ex-Chancellor von Schleicher had never forgiven his disgrace in January 1933 and the failure of the Army chiefs to choose him as successor to Hindenburg. In a clash between Roehm and Hitler Schleicher saw an opportunity. He was imprudent enough to drop hints to the French Ambassador in Berlin that the fall of Hitler was not far off. This repeated the action he had taken in the case of Bruening. But the times had become more dangerous.

It will long be disputed in Germany whether Hitler was forced to strike by the imminence of the Roehm plot, or whether he and the generals, fearing what might be coming, resolved on a clean-cut liquidation while they had the power. Hitler's interest and that of the victorious faction was plainly to establish the case for a plot. It is improbable that Roehm and the Brownshirts had actually got as far as this. They were a menacing movement rather than a plot, but at any moment this line might have been crossed. It is certain they were drawing up their forces. It is also certain they were forestalled.

Events now moved rapidly. On June 25 the Reichswehr was confined to barracks, and ammunition was issued to the Blackshirts. On the opposite side the Brownshirts were ordered to stand in readiness, and Roehm with Hitler's consent called a meeting for June 30 of all their senior leaders to meet at Wiessee, in the Bavarian lakes. Hitler received warning of grave danger on the 29th. He flew to Godesberg, where he was joined by Goebbels, who brought alarming news of impending mutiny in Berlin. According to Goebbels, Roehm's adjutant, Karl Ernst, had been given orders to attempt a rising. This seems unlikely. Ernst was actually at Bremen, about to embark from that port on his honeymoon.

On this information, true or false, Hitler took instant decisions. He ordered Goering to take control in Berlin. He boarded his aeroplane for Munich, resolved to arrest his main opponents personally. In this life or death climax, as it had now become, he showed himself a terrible personality. Plunged in dark thought, he sat in the co-pilot's seat throughout the journey. The plane landed at an airfield near Munich at 4 o'clock in the morning of June 30. Hitler had with him, besides Goebbels, about a dozen of his personal bodyguard. He drove to the Brown House in Munich, summoned the leaders of the local S.A. to his presence, and placed them under arrest. At 6 o'clock, with Goebbels and his small escort only, he motored to Wiessee.

Roehm was ill in the summer of 1934 and had gone to Wiessee to take a cure. The establishment he had selected was a small chalet belonging to the doctor in charge of his case. No worse headquarters could have been chosen from which to organise an immediate revolt. The chalet stands at the end of a narrow cul-de-sac lane. All arrivals and departures could be easily noted. There was no room large enough to hold the alleged impending meeting of Brownshirt leaders. There was only one telephone. This ill accords with the theory of an imminent uprising. If Roehm and his followers were about to revolt, they were certainly careless.

At seven o'clock the Fuehrer's procession of cars arrived in front of Roehm's chalet. Alone and unarmed, Hitler mounted the stairs and entered Roehm's bedroom. What passed between the two men will never be known. Roehm was taken completely by surprise, and he and his personal staff were arrested without incident. The small party, with its prisoners, now left by road for Munich. It happened that they soon met a column of lorries of armed Brownshirts on their way to acclaim Roehm at the conference convened at Wiessee for noon. Hitler stepped out of his car, called for the commanding officer, and, with confident authority, ordered him to take his men home. He was instantly obeyed. If he had been an hour later, or they had been an hour earlier, great events would have taken a different course.

On arrival at Munich Roehm and his entourage were imprisoned in the same gaol where he and Hitler had been confined together ten years before. That afternoon the executions began. A revolver was placed in Roehm's cell, but as he disdained the invitation the cell door was opened within a few minutes and he was riddled with bullets. All the afternoon the executions proceeded in Munich at brief intervals. The firing parties of eight had to be relieved from time to time on account of the mental stress of the soldiers. But for several hours the recurrent volleys were heard every ten minutes or so.

Meanwhile in Berlin Goering, having heard from Hitler, followed a similar procedure. But here, in the capital, the killing spread beyond the hierarchy of the S.A. Schleicher and his wife, who threw herself in front of him, were shot in their house. Gregor Strasser was arrested and put to death. Papen's private secretary and immediate circle were also shot; but for some unknown reason he himself was spared. In the Lichterfelde barracks in Berlin Karl Ernst, clawed back from Bremen, met his fate; and here, as in Munich, the volleys of the executioners were heard all day. Throughout Germany, during these twenty-four hours, many men unconnected with the Roehm plot disappeared as the victims of private vengeance, sometimes for very old scores. Otto von Kahr, for instance, who as head of the Bavarian Government had broken the 1923 *Putsch*, was found dead in the woods near Munich. The total number of persons "liquidated" is variously estimated as between five and seven thousand.

Late in the afternoon of this bloody day Hitler returned by air to Berlin. It was time to put an end to the slaughter, which was spreading every moment. That evening a certain number of the S.S., who through excess of zeal had gone a little far in shooting prisoners, were themselves led out to execution. About one o'clock in the morning of July 1 the sounds of firing ceased. Later in the day the Fuehrer appeared on the balcony of the Chancellery to receive the acclamations of the Berlin crowds, many of whom thought that he had himself been a victim. Some say he looked haggard, others triumphant. He may well have been both. His promptitude and ruthlessness had saved his purpose and no doubt his life. In that "Night of the Long Knives", as it was called, the unity of National-Socialist Germany had been preserved to carry its curse throughout the world.

A fortnight later the Fuehrer addressed the Reichstag, who sat in loyalty or awe before him. In the course of two hours he delivered a reasoned defence of his action. The speech reveals his knowledge of the German mind and his own undoubted powers of argument. Its climax was:

The necessity for acting with lightning speed meant that in this decisive hour I had very few men with me. . . . Although only a few days before I had been prepared to exercise clemency, at this hour there was no place for any such consideration. Mutinies are suppressed in accordance with laws of iron which are eternally the same. If anyone reproaches me and asks why I did not resort to the regular Courts of Justice for conviction of the offenders, then all that I can say to him is this: In this hour I was responsible for the fate of the German people, and thereby I became the supreme Justiciar of the German people. . . . I did not wish to deliver up the Young Reich to the fate of the Old Reich. I gave the order to shoot those who were the ringleaders in this treason. . . .

Then followed this mixed but expressive metaphor:

And I further gave the order to burn out down to the raw flesh the ulcers of this poisoning of the wells in our domestic life, and of the poisoning of the outside world.

This massacre, however explicable by the hideous forces at work, showed that the new Master of Germany would stop at nothing, and that conditions in Germany bore no resemblance to those of a civilised State. A Dictatorship based upon terror and reeking with blood had confronted the world. Anti-Semitism was ferocious and brazen, and the concentration-camp system was already in full operation for all obnoxious or politically dissident classes. I was deeply affected by the episode, and the whole process of German rearmament, of which there was now overwhelming evidence, seemed to me invested with a ruthless, lurid tinge. It glittered and it glared.

* * * * *

We may now return for a moment to the House of Commons. In the course of June 1934 the Standing Committee of the Disarmament Conference at Geneva was adjourned indefinitely. On July 13 I said:

I am very glad that the Disarmament Conference is passing out of life into history. It is the greatest mistake to mix up disarmament with peace. When you have peace you will have disarmament. But there has been during these recent years a steady deterioration in the relations between different countries, a steady growth of ill-will, and a steady, indeed a rapid increase in armaments that has gone on through all these years in spite of the endless flow of oratory, of perorations, of well-meaning sentiments, of banquets, which have marked this epoch.

Europe will be secure when nations no longer feel themselves in great danger, as many of them do now. Then the pressure and the burden of armaments will fall away automatically, as they ought to have done in a long peace; and it might be quite easy to seal a movement of that character by some general agreement. I hope, indeed, that we have now also reached the end of the period of the Government pressing France—this peaceful France with no militarism—to weaken her armed forces. I rejoice that the French have not taken the advice which has been offered to them so freely from various quarters, and which the Leader of the Opposition [Mr. Lansbury] no doubt would strongly endorse.

This is not the only Germany which we shall live to see, but we have to consider that at present two or three men, in what may well be a desperate position, have the whole of that mighty country in their grip, have that wonderful scientific, intelligent, docile, valiant people in their grip, a population of seventy millions; that there is no dynastic interest such as the monarchy brings as a restraint upon policy, because it looks long ahead and has much to lose; and that there is no public opinion except what is manufactured by those new and terrible engines—broadcasting and a controlled Press. Politics in Germany are not as they are over here. There you do not leave office to go into Opposition. You do not leave the Front Bench to sit below the Gangway. You may well leave your high office at a quarter of an hour's notice to drive to the police station, and you may be conducted thereafter very rapidly to an even graver ordeal.

It seems to me that men in that position might very easily be tempted to do what even a military dictatorship would not do, because a military dictatorship, with all its many faults, at any rate is one that is based on a very accurate study of the real facts; and there is more danger in this kind of dictatorship than there would be in a military dictatorship, because you have men who, to relieve themselves from the great peril which confronts them at home, might easily plunge into a foreign adventure of the most dangerous and catastrophic character to the whole world.

*   *   *   *   *

The first temptation to such an adventure was soon to be revealed.

During the early part of July 1934 there was much coming and going over the mountain paths leading from Bavaria into Austrian territory. At the end of July a German courier fell into the hands of the Austrian frontier police. He carried documents, including cipher keys, which showed that a complete plan of revolt was reaching fruition. The organiser of the *coup d'état* was to be Anton von Rintelen, at that time Austrian Minister to Italy. Dollfuss and his Ministers were slow to respond to the warnings of an impending crisis, and to the signs of imminent revolt which became apparent in the early hours of July 25. The Nazi adherents in Vienna mobilised during the morning. Just before one o'clock in the afternoon a party of armed rebels entered the Chancellery, and Dollfuss, hit by two revolver bullets, was left to bleed slowly to death. Another detachment of Nazis seized the broadcasting station and announced the resignation of the Dollfuss Government and the assumption of office by Rintelen.

But the other members of the Dollfuss Cabinet reacted with firmness and energy. President Dr. Miklas issued a formal command to restore order at all costs. Dr. Schuschnigg assumed the administration. The majority of the Austrian Army and police rallied to his Government, and besieged the Chancellery building, where, surrounded by a small party of rebels, Dollfuss was dying. The revolt had also broken out in the provinces, and parties from the Austrian legion in Bavaria crossed the frontier. Mussolini had by now heard the news. He telegraphed at once to Prince Starhemberg, the head of the Austrian Heimwehr, promising Italian support for Austrian independence. Flying specially to Venice, the Duce received the widow of Dr. Dollfuss with every circumstance of sympathy. At the same time three Italian divisions were dispatched to the Brenner Pass. On this Hitler, who knew the limits of his strength, recoiled. The German Minister in Vienna, Rieth, and other German officials implicated in the rising, were recalled or dismissed. The attempt had failed. A longer process was needed. Papen, newly spared from the blood-bath, was appointed as German Minister to Vienna, with instructions to work by more subtle means.

Papen had been appointed German Minister to Vienna for the explicit purpose of organising the overthrow of the Austrian

Republic. He had a double task: the encouragement of the underground Austrian Nazi Party, which received henceforth a monthly subsidy of 200,000 marks; and the undermining or winning over of leading personalities in Austrian politics. In the early days of his appointment he expressed himself with frankness verging upon indiscretion to his American colleagues in Vienna. "In the boldest and most cynical manner," says the American Minister, "Papen proceeded to tell me that all South-eastern Europe to the borders of Turkey was Germany's natural hinterland, and that he had been charged with the mission of effecting German economic and political control over the whole of this region. He blandly and directly said that getting control of Austria was to be the first step. He intended to use his reputation as a good Catholic to gain influence with Austrians like Cardinal Innitzer. The German Government was determined to gain control of South-eastern Europe. There was nothing to stop them. The policy of the United States, like that of France and England, was not 'realistic'."

Amid these tragedies and alarms the aged Marshal Hindenburg, who had for some months been almost completely senile, and so more than ever a tool of the Reichswehr, expired. Hitler became the head of the German State while retaining the office of Chancellor. He was now the Sovereign of Germany. His bargain with the Reichswehr had been sealed and kept by the blood-purge. The Brownshirts had been reduced to obedience and reaffirmed their loyalty to the Fuehrer. All foes and potential rivals had been extirpated from their ranks. Henceforward they lost their influence and became a kind of special constabulary for ceremonial occasions. The Blackshirts, on the other hand, increased in numbers, and, strengthened by privileges and discipline, became under Himmler a Prætorian Guard for the person of the Fuehrer, a counterpoise to the Army leaders and military caste, and also political troops to arm with considerable military force the activities of the expanding Secret Police or Gestapo. It was only necessary to invest these powers with the formal sanction of a managed plebiscite to make Hitler's dictatorship absolute and perfect.

* * * * *

Events in Austria drew France and Italy together, and the shock of the Dollfuss assassination led to General Staff contacts. The menace to Austrian independence promoted a revision of Franco-Italian relations, and this had to comprise not only the balance of power in the Mediterranean and North Africa, but the relative positions of France and Italy in South-eastern Europe. But Mussolini was anxious not only to safeguard Italy's position in Europe against the potential German threat, but also to secure her imperial future in Africa. Against Germany, close relations with France and Great Britain would be useful; but in the Mediterranean and Africa disagreements with both these Powers might be inevitable. The Duce wondered whether the common need for security felt by Italy, France, and Great Britain might not induce the two former allies of Italy to accept the Italian Imperialist programme in Africa. At any rate this seemed a hopeful course for Italian policy.

✷   ✷   ✷   ✷   ✷

In France, after the Stavisky scandal and the riots of February, M. Daladier had been succeeded as Premier by a Government of the Right Centre under M. Doumergue, with M. Barthou as Foreign Minister. Ever since the signature of the Locarno treaties France had been anxious to reach formal agreement on security measures in the East. British reluctance to undertake commitments beyond the Rhine, the German refusal to make binding agreements with Poland and Czechoslovakia, the fears of the Little Entente as to Russian intentions, Russian suspicion of the capitalist West, all united to thwart such a programme. In September 1934, however, Louis Barthou determined to go forward. His original plan was to propose an Eastern Pact, grouping together Germany, Russia, Poland, Czechoslovakia, and the Baltic States on the basis of a guarantee by France of the European frontiers of Russia, and by Russia of the eastern borders of Germany. Both Germany and Poland were opposed to an Eastern Pact; but Barthou succeeded in obtaining the entry of Russia into the League of Nations on September 18, 1934. This was an important step. Litvinov, who represented the Soviet Government, was versed in every aspect of foreign affairs. He adapted himself to the atmosphere of the League of Nations and spoke its moral language with so much success that he soon became an outstanding figure.

In her search for allies against the new Germany that had been allowed to grow up, it was natural that France should turn her eyes to Russia and try to re-create the balance of power which had existed before the war. But in October a tragedy occurred. In pursuance of French policy in the Balkans King Alexander of Yugoslavia had been invited to pay an official visit to Paris. He landed at Marseilles, was met by M. Barthou, and drove with him and General Georges through the welcoming crowds who thronged the streets, gay with flags and flowers. Once again from the dark recesses of the Serbian and Croat underworld a hideous murder-plot sprang upon the European stage, and, as at Sarajevo in 1914, a band of assassins, ready to give their lives, were at hand. The French police arrangements were loose and casual. A figure darted from the cheering crowds, mounted the running-board of the car, and discharged his automatic pistol into the King and its other occupants, all of whom were stricken. The murderer was immediately cut down and killed by the mounted Republican guardsman behind whom he had slipped. A scene of wild confusion occurred. King Alexander expired almost immediately. General Georges and M. Barthou stepped out of the carriage streaming with blood. The General was too weak to move, but soon received medical aid. The Minister wandered off into the crowd. It was twenty minutes before he was attended to. He was made to walk upstairs to the Prefect's office before he could be given medical care; the doctor then applied the tourniquet *below* the wound. He had already lost much blood; he was seventy-two, and he died in a few hours. This was a heavy blow to French foreign policy, which under him was beginning to take a coherent form. He was succeeded as Foreign Secretary by Pierre Laval.

Laval's later shameful record and fate must not obscure the fact of his personal force and capacity. He had a clear and intense view. He believed that France must at all costs avoid war, and he hoped to secure this by arrangements with the Dictators of Italy and Germany, against whose systems he entertained no prejudice. He distrusted Soviet Russia. Despite his occasional protestations of friendship, he disliked England and thought her a worthless ally. At that time indeed British repute did not stand very high in France. Laval's first object was to reach a definite understanding with Italy, and he deemed the moment ripe. The French Government was obsessed by the German danger, and was prepared to make solid concessions to gain Italy. In January 1935 M. Laval went to Rome and signed a series of agreements with the object of removing the main obstacles between the two countries. Both Governments were united upon the illegality of German rearmament. They agreed to consult each other in the event of future threats to the independence of Austria. In the colonial sphere France undertook to make administrative concessions about the status of Italians in Tunisia, and handed over to Italy certain tracts of territory on the borders both of Libya and of Somaliland, together with a 20 per cent. share in the Jibuti—Addis-Ababa railway. These conversations were designed to lay the foundations for more formal discussions between France, Italy, and Great Britain about a common front against the growing German menace. Across them all there cut in the ensuing months the fact of Italian aggression in Abyssinia.

* * * * *

In December 1934 a clash took place between Italian and Abyssinian soldiers at the wells of Wal Wal, on the borders of Abyssinia and Italian Somaliland. This was to be the pretext for the ultimate presentation before the world of Italian claims upon the Ethiopian kingdom. Thus the problem of containing Germany in Europe was henceforth confused and distorted by the fate of Abyssinia.

* * * * *

There is one more incident at this juncture which should be mentioned. Under the terms of the Treaty of Versailles, the Saar Valley, a small strip of German territory, possessing rich coal-mines and important iron-works, was to decide at the end of fifteen years by a plebiscite whether the population wished to return to Germany or not. The date fixed for this event was in January 1935. There could be no doubt of the outcome. The majority would certainly vote for reincorporation into the German Fatherland; and, to make assurance doubly sure, the valley, though nominally governed by a League of Nations Commission, was in fact under the control of the local Nazi Party centre. Barthou realised that ultimately the Saar was bound to return to Germany, but was inclined to insist upon some guarantees to those who might vote against immediate incorporation with Germany. His assassination changed the tone of French policy. On December 3, 1934, Laval made a direct bargain with the Germans over the coal-mines, and three days later announced publicly before the League Council that France would not oppose the return of the Saar to Germany. The actual plebiscite was held on January 13, 1935, under international supervision, in which a British brigade took part; and this little enclave—except Danzig, the only territorial embodiment of League sovereignty—voted by 90.3 per cent. for return to Germany. This moral triumph for National Socialism, although the result of a normal and inevitable procedure, added to Hitler's prestige, and seemed to crown his authority with an honest sample of the will of the German people. He was not at all conciliated, still less impressed, by the proof of the League's impartiality or fair play. No doubt it confirmed his view that the Allies were decadent fools. For his own part he proceeded to concentrate on his main objective, the expansion of the German forces.

CHAPTER VII

AIR PARITY LOST

1934–1935

The German Short Cut – The East Fulham Election, October 25, 1933 – Debate of February 7, 1934 – Mr. Baldwin's Pledge of Air Parity – The Labour Vote of Censure against Air Increases – Liberal Hostility – My Precise Warning, November 28, 1934 – Mr. Baldwin's Contradiction – Hitler Claims that Germany Has Air Parity, March 1935 – Mr. MacDonald's Alarm – Mr. Baldwin's Confession, May 22 – The Labour and Liberal Attitudes – The Air Ministry View – Lord Londonderry Presently Succeeded by Sir Philip Cunliffe-Lister.

*T*HE German General Staff did not believe that the German Army could be formed and matured on a scale greater than that of France, and suitably provided with arsenals and equipment, before 1943. The German Navy, except for U-boats, could not be rebuilt in its old state under twelve or fifteen years, and in the process would compete heavily with all other plans. But owing to the unlucky discovery by an immature civilisation of the internal combustion engine and the art of flying, a new weapon of national rivalry had leapt upon the scene capable of altering much more rapidly the relative war-power of States. Granted a share in the ever-accumulating knowledge of mankind and in the march of Science, only four or five years might be required by a nation of the first magnitude, devoting itself to the task, to create a powerful, and perhaps a supreme, Air Force. This period would of course be shortened by any preliminary work and thought.

As in the case of the German Army, the re-creation of the German air power was long and carefully prepared in secret. As early as 1923 Seeckt had decided that the future German Air

Force must be a part of the German war-machine. For the time being he was content to build inside the "airforceless army" a well-articulated Air Force skeleton which could not be discerned, or at any rate was not discerned in its early years, from without. Air-power is the most difficult of all forms of military force to measure, or even to express in precise terms. The extent to which the factories and training-grounds of civil aviation have acquired a military value and significance at any given moment cannot easily be judged and still less exactly defined. The opportunities for concealment, camouflage, and treaty-evasion are numerous and varied. The air, and the air alone, offered Hitler the chance of a short cut, first to equality and next to predominance, in a vital military arm over France and Britain. But what would France and Britain do?

By the autumn of 1933 it was plain that neither by precept nor still less by example would the British effort for disarmament succeed. The pacifism of the Labour and Liberal Parties was not affected even by the grave event of the German withdrawal from the League of Nations. Both continued in the name of Peace to urge British disarmament, and anyone who differed was called "warmonger" and "scaremonger". It appeared that their feeling was endorsed by the people, who of course did not understand what was unfolding. At a by-election which occurred in East Fulham on October 25 a wave of pacifist emotion increased the Socialist vote by nearly 9,000, and the Conservative vote fell by over 10,000. The successful candidate, Mr. Wilmot, said after the poll that "British people demand . . . that the British Government shall give a lead to the whole world by initiating immediately a policy of general disarmament." And Mr. Lansbury, then Leader of the Labour Party, said that all nations must "disarm to the level of Germany as a preliminary to total disarmament". This election left a deep impression upon Mr. Baldwin, and he referred to it in a remarkable speech three years later. In November came the Reichstag election, at which no candidates except those endorsed by Hitler were tolerated, and the Nazis obtained 95 per cent. of the votes polled.

It would be wrong in judging the policy of the British Government not to remember the passionate desire for peace which animated the uninformed, misinformed majority of the British people, and seemed to threaten with political extinction any party or politician who dared to take any other line. This, of course, is no excuse for political leaders who fall short of their duty. It is much better for parties or politicians to be turned out of office than to imperil the life of the nation. Moreover, there is no record in our history of any Government asking Parliament and the people for the necessary measures for defence and being refused. Nevertheless, those who scared the timid MacDonald-Baldwin Government from their path should at least keep silent.

The Air Estimates of March 1934 totalled only twenty millions, and contained provision for four new squadrons, or an increase in our first-line air strength from 850 to 890. The financial cost involved in the first year was £130,000.

On this I said:

We are, it is admitted, the fifth air Power only—if that. We are but half the strength of France, our nearest neighbour. Germany is arming fast and no one is going to stop her. That seems quite clear. No one proposes a preventive war to stop Germany breaking the Treaty of Versailles. She is going to arm; she is doing it; she has been doing it. I have no knowledge of the details, but it is well known that those very gifted people, with their science and with their factories— with what they call their "Air-Sport"—are capable of developing with great rapidity the most powerful Air Force for all purposes, offensive and defensive, within a very short period of time.

I dread the day when the means of threatening the heart of the British Empire should pass into the hands of the present rulers of Germany. We should be in a position which would be odious to every man who values freedom of action and independence, and also in a position of the utmost peril for our crowded, peaceful population engaged in their daily toil. I dread that day, but it is not perhaps far distant. It is perhaps only a year, or perhaps eighteen months distant. It has not come yet—at least, so I believe or I hope and pray; but it is not far distant. There is time for us to take the neces- sary measures, but it is the measures we want. We want the measures to achieve parity. No nation playing the part we play and aspire to play in the world has a right to be in a position where it can be blackmailed. . . .

None of the grievances between the victors and the vanquished have been redressed. The spirit of aggressive Nationalism was never more rife in Europe and in the world. Far away are the days of Locarno, when we nourished bright hopes of the reunion of the European family. . . .

I called upon Mr. Baldwin, as the man who possessed the power, for action. His was the power, and his the responsibility. In the course of his reply Mr. Baldwin said:

If all our efforts for an agreement fail, and if it is not possible to obtain this equality in such matters as I have indicated, then any Government of this country—a National Government more than any, and *this* Government—will see to it that in air strength and air power this country shall no longer be in a position inferior to any country within striking distance of its shores.

Here was a most solemn and definite pledge, given at a time when it could almost certainly have been made good by vigorous action on a large scale.

★　　★　　★　　★　　★

Although Germany had not yet openly violated the clauses of the Treaty which forbade her a military Air Force, civil aviation and an immense development of gliding had now reached a point where they could very rapidly reinforce and extend the secret and illegal military Air Force already formed. The blatant denunciations of Communism and Bolshevism by Hitler had not prevented the clandestine sending by Germany of arms to Russia. On the other hand, from 1927 onwards a number of German pilots had been trained by the Soviets for military purposes. There were fluctuations, but in 1932 the British Ambassador in Berlin reported that the Reichswehr had close technical liaison with the Red Army. Just as the Fascist dictator of Italy had, almost from his accession to power, been the first to make a trade agreement with Soviet Russia, so now the relations between Nazi Germany and the vast Soviet State appeared to be unprejudiced by public ideological controversy.

★　　★　　★　　★　　★

Nevertheless, when on July 20, 1934, the Government brought forward some belated and inadequate proposals for strengthening the Royal Air Force by 41 squadrons, or about 820 machines, *only to be completed in five years,* the Labour Party, supported by the Liberals, moved a Vote of Censure upon them in the House of Commons.

The motion regretted that

His Majesty's Government should enter upon a policy of rearmament neither necessitated by any new commitment nor calculated to add to the security of the nation, but certain to jeopardise the prospects of international disarmament and to encourage a revival of dangerous and wasteful competition in preparation for war.

In support of this complete refusal by the Opposition to take any measures to strengthen our air-power Mr. Attlee, speaking in their name, said: "We deny the need for increased air armaments. . . . We deny the proposition that an increased British Air Force will make for the peace of the world, and we reject altogether the claim to parity." The Liberal Party supported this Censure Motion, although they would have preferred their own, which ran as follows:

That this House views with grave concern the tendency among the nations of the world to resume the competitive race of armaments which has always proved a precursor of war; it will not approve any expansion of our own armaments unless it is clear that the Disarmament Conference has failed and unless a definite case is established; and these conditions not being present as regards the proposed additional expenditure of £20,000,000 upon air armaments, the House declines its assent.

In his speech the Liberal Leader, Sir Herbert Samuel, said: "What is the case in regard to Germany? Nothing we have so far seen or heard would suggest that our present Air Force is not adequate to meet any peril at the present time from this quarter."

When we remember that this was language used after careful deliberation by the responsible heads of parties, the danger of our country becomes apparent. This was the formative time when by extreme exertions we could have preserved the air strength on which our independence of action was founded. If Great Britain and France had each maintained quantitative parity with Germany they would together have been double as strong, and Hitler's career of violence might have been nipped in the bud without the loss of a single life. Thereafter it was too late. We cannot doubt the sincerity of the Leaders of the Socialist and Liberal Parties. They were completely wrong and mistaken, and they bear their share of the burden before history. It is indeed astonishing that the Socialist Party should have endeavoured in after years to claim superior foresight and should have reproached their opponents with failing to provide for national safety.

* * * * *

I now enjoyed for once the advantage of being able to urge rearmament in the guise of a defender of the Government. I therefore received an unusually friendly hearing from the Conservative Party.

One would have thought that the character of His Majesty's Government and the record of its principal Ministers would have induced the Opposition to view the request for an increase in the national defence with some confidence and some consideration. I do not suppose there has ever been such a pacifist-minded Government. There is the Prime Minister, who in the war proved in the most extreme manner and with very great courage his convictions and the sacrifices he would make for what he believed was the cause of pacifism. The Lord President of the Council is chiefly associated in the public mind with the repetition of the prayer "Give peace in our time". One would have supposed that when Ministers like these come forward and say that they feel it their duty to ask for some small increase in the means they have of guaranteeing the public safety, it would weigh with the Opposition and would be considered as a proof of the reality of the danger from which they seek to protect us.

Then look at the apologies which the Government have made. No one could have put forward a proposal in more extremely inoffensive terms. Meekness has characterised every word which they have spoken since this subject was first mooted. We are told that we can see for ourselves how small is the proposal. We are assured that it can be stopped at any minute if Geneva succeeds. And we are also assured that the steps we are taking, although they may to some lower minds have associated with them some idea of national self-defence, are really only associated with the great principle of collective security.

But all these apologies and soothing procedures are most curtly repulsed by the Opposition. Their only answer to these efforts to conciliate them is a Vote of Censure, which is to be decided to-night. It seems to me that we have got very nearly to the end of the period when it is worth while endeavouring to conciliate some classes of opinion upon this subject. We are in the presence of an attempt to establish a kind of tyranny of opinion, and if its reign could be perpetuated the effect might be profoundly injurious to the stability and security of this country. We are a rich and easy prey. No country is so vulnerable, and no country would better repay pillage than our own. . . . *With our enormous metropolis here, the greatest target in the world, a kind of tremendous, fat, valuable cow tied up to attract the beast of prey*, we are in a position in which we have never been before, and in which no other country is at the present time.

Let us remember this: our weakness does not only involve ourselves; our weakness involves also the stability of Europe.

I then proceeded to argue that Germany was already approaching air parity with Britain:

I first assert that Germany has already, in violation of the Treaty, created *a military Air Force which is now nearly two-thirds as strong as our present home defence Air Force.* That is the first statement which I put before the Government for their consideration. The second is that Germany is rapidly increasing this Air Force, not only by large sums of money which figure in her estimates, but also by public subscriptions—very often almost forced subscriptions—which are in progress and have been in progress for some time all over Germany. *By the end of* 1935 *the German Air Force will be nearly equal in numbers and efficiency to our home defence Air Force at that date even if the Government's present proposals are carried out.*

The third statement is that if Germany continues this expansion and if we continue to carry out our scheme, then some time in 1936 Germany will be definitely and substantially stronger in the air than Great Britain. Fourthly, and this is the point which is causing anxiety, once they have got that lead we may never be able to overtake them. If these assertions cannot be contradicted, then there is cause for the anxiety which exists in all parts of the House, not only because of the physical strength of the German Air Force, but I am bound to say also because of the character of the present German dictatorship. *If the Government have to admit at any time in the next few years that the German air forces are stronger than our own, then they will be held, and I think rightly held, to have failed in their prime duty to the country.*

I ended as follows:

The Opposition are very free-spoken, as most of us are in this country, on the conduct of the German Nazi Government. No one has been more severe in criticism than the Labour Party or that section of the Liberal Party which I see opposite. And their great newspapers, now united in the common cause, have been the most forward in the severity of their strictures. But these criticisms are fiercely resented by the powerful men who have Germany in their hands. So that we are to disarm our friends, we are to have no allies, we are to affront powerful nations, and we are to neglect our own defences entirely. That is a miserable and perilous situation. Indeed, the position to which they seek to reduce us by the course which they have pursued and by the vote which they ask us to take is one of terrible jeopardy, and in voting against them to-night we shall hope that a better path for national safety will be found than that along which they would conduct us.

The Labour Party's Vote of Censure was of course defeated by a large majority, and I have no doubt that the nation, had it been appealed to with proper preparation on these issues, would equally have sustained the measures necessary for national safety.

*   *   *   *   *

It is not possible to tell this story without recording the milestones which we passed on our long journey from security to the jaws of Death. Looking back, I am astonished at the length of time that was granted to us. It would have been possible in 1933, or even in 1934, for Britain to have created an air-power which would have imposed the necessary restraints upon Hitler's ambition, or would perhaps have enabled the military leaders of Germany to control his violent acts. More than five whole years had yet to run before we were to be confronted with the supreme ordeal. Had we acted even now with reasonable prudence and healthy energy, it might never have come to pass. Based upon superior air-power, Britain and France could safely have invoked the aid of the League of Nations, and all the States of Europe would have gathered behind them. For the first time the League would have had an Instrument of Authority.

When the winter session opened on November 28, 1934, I moved in the name of some of my friends* an Amendment to the Address, declaring that "the strength of our national defences and especially of our air defences is no longer adequate to secure the peace, safety, and freedom of Your Majesty's faithful subjects". The House was packed and very ready to listen. After using all the arguments which emphasised the heavy danger to us and to the world, I came to precise facts:

I assert, first, that Germany already, at this moment, has a military Air Force—that is to say, military squadrons, with the necessary ground services, and the necessary reserves of trained personnel and material—which only awaits an order to assemble in full open combination; and that this illegal Air Force is rapidly approaching equality with our own. Secondly, by this time next year, if Germany executes her existing programme without acceleration, and if we execute our existing

* The amendment stood in the names of Mr. Churchill, Sir Robert Horne, Mr. Amery, Captain F. E. Guest, Lord Winterton, and Mr. Boothby.

programme on the basis which now lies before us without slowing down, and carry out the increases announced to Parliament in July last, the German military Air Force will this time next year be in fact at least as strong as our own, and it may be even stronger. Thirdly, on the same basis—that is to say, both sides continuing with their existing programmes as at present arranged—by the end of 1936, that is, one year further on, and two years from now, the German military Air Force will be nearly 50 per cent. stronger, and in 1937 nearly double. All this is on the assumption, as I say, that there is no acceleration on the part of Germany, and no slowing down on our part.

Mr. Baldwin, who followed me at once, faced this issue squarely, and, on the case made out by his Air Ministry advisers, met me with direct contradiction:

It is not the case that Germany is rapidly approaching equality with us. I pointed out that the German figures are total figures, not first-line strength figures, and I have given our own first-line figures and said they are only first-line figures, with a considerably larger reserve at our disposal behind them, even if we confine the comparison to the German air strength and the strength of the Royal Air Force immediately available in Europe. Germany is actively engaged in the production of service aircraft, but her real strength is not 50 per cent. of our strength in Europe to-day. As for the position this time next year, if she continues to execute her air programme without acceleration, and if we continue to carry out at the present approved rate the expansion announced to Parliament in July, *so far from the German military Air Force being at least as strong as, and probably stronger than, our own, we estimate that we shall still have a margin in Europe alone of nearly 50 per cent.* I cannot look farther forward than the next two years. Mr. Churchill speaks of what may happen in 1937. Such investigations as I have been able to make lead me to believe that his figures are considerably exaggerated.

* * * * *

This sweeping assurance from the virtual Prime Minister soothed most of the alarmed, and silenced many of the critics. Everyone was glad to learn that my precise statements had been denied upon unimpeachable authority. I was not at all convinced. I believed that Mr. Baldwin was not being told the truth by his advisers, and anyhow that he did not know the facts.

* * * * *

Thus the winter months slipped away, and it was not till the spring that I again had the opportunity of raising the issue. I gave full and precise notice.

Mr. Churchill to Mr. Baldwin   17.III.35

On the Air Estimates on Tuesday I propose to renew our discussion of last November and to analyse as far as I can your figures of British and German air strength for home defence at the various dates in question, viz.: then, now, at the end of the year 1935 calendar and financial, etc. I believe that the Germans are already as strong as we are and possibly stronger, and that if we carry out our new programme as prescribed Germany will be 50 per cent. stronger than we by the end of 1935 or the beginning of 1936. This, as you will see, runs counter to your statement of November, that we should have a 50 per cent. superiority at that date. I shall of course refer to your undertaking of March 1934 that "this country shall no longer be in a position inferior to any country within striking distance of its shores", and I shall argue that according to such knowledge as I have been able to acquire this is not being made good, as will rapidly be proved by events.

I thought it would be convenient to you if I let you know beforehand as I did on the last occasion what my general line will be, and if whoever speaks for the Government is able to prove the contrary no one will be better pleased than I.

On March 19 the Air Estimates were presented to the House. I reiterated my statement of November, and again directly challenged the assurances which Mr. Baldwin had then given. A very confident reply was made by the Under-Secretary for Air. However, at the end of March the Foreign Secretary and Mr. Eden paid a visit to Herr Hitler in Germany, and in the course of an important conversation, the text of which is on record, they were told personally by him that the German Air Force had already reached parity with Great Britain. This fact was made public by the Government on April 3. At the beginning of May the Prime Minister wrote an article in his own organ, *The Newsletter*, in which he emphasised the dangers of German rearmament in terms akin to those which I had so often expressed since 1932. He used the revealing word "ambush", which must have sprung from the anxiety of his heart. We had indeed fallen into an ambush. Mr. MacDonald himself opened the debate. After referring to the declared German intention to build a Navy beyond the Treaty and submarines in breach of it, he came to the air position:

In the debate last November certain estimates were put forward on the basis of our then estimates as to the strength of the German Air Force, and the assurance was given by the Lord President, on behalf of the Government, that in no circumstances would we accept any position of inferiority with regard to whatever Air Force might be raised in Germany in the future. If it were not so, that would put us in an impossible position, of which the Government and the Air Ministry are fully aware. In the course of the visit which the Foreign Secretary and the Lord Privy Seal paid to Berlin at the end of March, the German Chancellor stated, as the House was informed on April 3, that Germany had reached parity with Great Britain in the air. Whatever may be the exact interpretation of this phrase in terms of air strength, it undoubtedly indicated that the German force has been expanded to a point considerably in excess of the estimates which we were able to place before the House last year. That is a grave fact, with regard to which both the Government and the Air Ministry have taken immediate notice.

When in due course I was called I said:

Even now we are not taking the measures which would be in true proportion to our needs. The Government have proposed these increases. They must face the storm. They will have to encounter every form of unfair attack. Their motives will be misrepresented. They will be calumniated and called warmongers. Every kind of attack will be made upon them by many powerful, numerous, and extremely vocal forces in this country. They are going to get it anyway. Why, then, not fight for something that will give us safety? Why, then, not insist that the provision for the Air Force should be adequate, and then, however severe may be the censure and however strident the abuse which they have to face, at any rate there will be this satisfactory result—that His Majesty's Government will be able to feel that in this, of all matters the prime responsibility of a Government, they have done their duty.

Although the House listened to me with close attention, I felt a sensation of despair. To be so entirely convinced and vindicated in a matter of life and death to one's country, and not to be able to make Parliament and the nation heed the warning, or bow to the proof by taking action, was an experience most painful. I went on:

I confess that words fail me. In the year 1708 Mr. Secretary St. John, by a calculated Ministerial indiscretion, revealed to the House the fact that the Battle of Almanza had been lost in the previous summer because only 8,000 English troops were actually in Spain out of the 29,000 that had been voted by the House of Commons for this service. When a month later this revelation was confirmed by the Government, it is recorded that the House sat in silence for half an hour, no Member caring to speak or wishing to make a comment upon so staggering an announcement. And yet how incomparably small that event was to what we have now to face. That was merely a frustration of policy. Nothing that could happen to Spain in that war could possibly have contained in it any form of danger which was potentially mortal. . . .

There is a wide measure of agreement in the House to-night upon our foreign policy. We are bound to act in concert with France and Italy and other Powers, great and small, who are anxious to preserve peace. I would not refuse the co-operation of any Government which plainly conformed to that test, so long as it was willing to work under the authority and sanction of the League of Nations. Such a policy does not close the door upon a revision of the treaties, but it procures a sense of stability, and an adequate gathering together of all reasonable Powers for self-defence, before any inquiry of that character [i.e., Treaty revision] can be entered upon. In this august association for collective security we must build up defence forces of all kinds and combine our action with that of friendly Powers, so that we may be allowed to live in quiet ourselves and retrieve the woeful miscalculations of which we are at present the dupes, and of which, unless we take warning in time, we may some day be the victims.

There lay in my memory at this time some lines from an unknown writer about a railway accident. I had learnt them from a volume of *Punch* cartoons which I used to pore over when I was eight or nine years old at school at Brighton.

> Who is in charge of the clattering train?
> The axles creak and the couplings strain,
> And the pace is hot, and the points are near,
> And Sleep has deadened the driver's ear;
> And the signals flash through the night in vain,
> For Death is in charge of the clattering train.

However, I did not repeat them.

*   *   *   *   *

It was not until May 22 that Mr. Baldwin made his celebrated confession. I am forced to cite it:

> First of all, with regard to the figure I gave in November of German aeroplanes, nothing has come to my knowledge since that makes me think that figure was wrong. I believed at that time it was right. *Where I was wrong was in my estimate of the future. There I was completely wrong. We were completely misled on that subject.* . . .
>
> I would repeat here that there is no occasion, in my view, in what we are doing, for panic. But I will say this deliberately, with all the knowledge I have of the situation, that I would not remain for one moment in any Government which took less determined steps than we are taking to-day. I think it is only due to say that there has been a great deal of criticism, both in the Press and verbally, about the Air Ministry, as though they were responsible for possibly an inadequate programme, for not having gone ahead faster, and for many other things. I only want to repeat that whatever responsibility there may be—and we are perfectly ready to meet criticism—*that responsibility is not that of any single Minister; it is the responsibility of the Government as a whole, and we are all responsible, and we are all to blame.*

I hoped that this shocking confession would be a decisive event, and that at the least a Parliamentary Committee of all parties would be set up to report upon the facts and upon our safety. The House of Commons had a different reaction. The Labour and Liberal Oppositions, having nine months earlier moved or supported a Vote of Censure even upon the modest steps the Government had taken, were ineffectual and undecided. They were looking forward to an election against "Tory armaments". Neither the Labour nor the Liberal spokesmen had prepared themselves for Mr. Baldwin's disclosures and admission, and they did not attempt to adapt their speeches to this outstanding episode. Mr. Attlee said:

> As a party we do not stand for unilateral disarmament. . . . We stand for Collective Security through the League of Nations. We reject the use of force as an instrument of policy. We stand for the reduction of armaments and pooled security. . . . We have stated that this country must be prepared to make its contribution to collective security. Our policy is not one of seeking security through rearmament, but through disarmament. Our aim is the reduction of armaments, and then the complete abolition of all national armaments and the creation of an International Police Force under the League.

What was to happen if this spacious policy could not be immediately achieved or till it was achieved he did not say. He complained that the White Paper on defence justified increases in the Navy by reference to the United States, and increases in our Air Force by references to the Air Forces of Russia, Japan, and the United States. "All that was old-fashioned talk and right outside the collective system." He recognised that the fact of German rearmament had become dominating, but "The measure of the counterweight to any particular armed forces is not the forces of this country or of France, but the combined force of all loyal Powers in the League of Nations. An aggressor must be made to realise that if he challenges the world he will be met by the co-ordinated forces of the world, not by a number of disjointed national forces." The only way was to concentrate all air-power in the hands of the League, which must be united and become a reality. Meanwhile he and his party voted against the measure proposed.

For the Liberals, Sir Archibald Sinclair asked the Government to summon "a fresh economic conference, and to bring Germany not only within the political comity of nations but also into active co-operation with ourselves in all the works of civilisation and in raising the standards of life of both peoples. . . . Let the Government table detailed and definite proposals for the abolition of military air forces and the control of civil aviation. If the proposals are resisted let the responsibility be clearly and properly fixed."

Nevertheless [said he], while disarmament ought vigorously to be pursued as the chief objective of the Government, a situation in which a great country not a member of the League of Nations possesses the most powerful Army and perhaps the most powerful Air Force in Western Europe, with probably a greater coefficient of expansion than any other Air Force . . . cannot be allowed to endure. . . . The Liberal Party will feel bound to support measures of national defence when clear proof is afforded of their necessity. . . . I cannot therefore agree that to increase our national armaments is necessarily inconsistent with our obligations under the collective peace system.

He then proceeded to deal at length with "the question of private profits being made out of the means of death", and quoted a recent speech by Lord Halifax, President of the Board of Education, who had said that the British people were "disposed to regard the preparation of instruments of war as too high and too grave a thing to be entrusted to any hands less responsible than those of the State itself". Sir Archibald Sinclair thought that there ought to be national factories for dealing with the rapid expansion in air armaments, for which expansion, he said, a case had been made out.

The existence of private armament firms had long been a bugbear to Labour and Liberal minds, and it lent itself readily to the making of popular speeches. It was of course absurd to suppose that at this time our air expansion, recognised as necessary, could be achieved through national factories only. A large part of the private industry of the country was urgently required for immediate adaptation and to reinforce our existing sources of manufacture. Nothing in the speeches of the Opposition leaders was in the slightest degree related to the emergency in which they admitted we stood, or to the far graver facts which we now know lay behind it.

The Government majority for their part appeared captivated by Mr. Baldwin's candour. His admission of having been utterly wrong, with all his sources of knowledge, upon a vital matter for which he was responsible was held to be redeemed by the frankness with which he declared his error and shouldered the blame. There was even a strange wave of enthusiasm for a Minister who did not hesitate to say that he was wrong. Indeed, many Conservative Members seemed angry with me for having brought their trusted leader to a plight from which only his native manliness and honesty had extricated him; but not, alas, his country.

* * * * *

My kinsman, Lord Londonderry, a friend from childhood days, the direct descendant of the famous Castlereagh of Napoleonic times, was a man of unquestionable loyalty and patriotism. He had presided over the Air Ministry since the formation of the Coalition. In this period the grave changes which have been described had overshadowed our affairs, and the Air Ministry had become one of the most important offices in the State. During the years of retrenchment and disarmament he and his Ministry had tried to keep and get as much as they could from a severe and arbitrary Chancellor of the Exchequer. They were overjoyed when in the summer of 1934 an air programme of forty-one additional squadrons was conceded to them by the Cabinet. But in British politics the hot fits very quickly succeed the cold. When the Foreign Secretary returned from Berlin, profoundly startled by Hitler's assertion that his Air Force was equal to that of Britain, the whole Cabinet became deeply concerned. Mr. Baldwin had to face, in the light of what was now generally accepted as a new situation, his assertions of November, when he had contradicted me. The Cabinet had no idea they had been overtaken in the air, and turned, as is usually the case, inquisitorial looks upon the department involved and its Minister.

The Air Ministry did not realise that a new inheritance awaited them. The Treasury's fetters were broken. They had but to ask for more. Instead of this they reacted strongly against Hitler's claim to air parity. Londonderry, who was their spokesman, even rested upon the statement that "*when Simon and Eden went to Berlin there was only one German operation squadron in being. From their training establishments they hoped to form fifteen to twenty squadron formations by the end of the month.*"* All this is a matter of nomenclature. It is of course very difficult to classify Air Forces, because of the absence of any common "yardstick" and all the variations in defining "first-line air strength" and "operational units". The Air Ministry now led its chief into an elaborate vindication of their own past conduct, and in consequence were entirely out of harmony with the new mood of a genuinely alarmed Government and public. The experts and officials at the Air Ministry had given Mr. Baldwin the figures and forecasts with which he had answered me in November. They wished him to go into action in defence of these statements; but this was no longer practical politics. There seems no doubt that these experts and officials of the Air Ministry at this time were themselves misled and misled their chief. A great air Power, at least the equal of our own, long pent up, had at last sprung into daylight in Germany.

It was an odd and painful experience for Londonderry, as his book describes, after having gone through several years of asking for more, to be suddenly turned out for not asking enough. But apart from all this his political standing was not sufficient to enable him to head a department now at the very centre and

* The Marquess of Londonderry, *Wings of Destiny*, 1943, p. 128.

almost at the summit of our affairs. Besides, everyone could see that in such times the Air Minister must be in the House of Commons. Accordingly Mr. Ramsay MacDonald's vacation of the Premiership later in the year became also the occasion for the appointment of Sir Philip Cunliffe-Lister, then Secretary of State for the Colonies, as Air Minister, as part of a new policy for vigorous air expansion. Lord Londonderry with much reluctance became Lord Privy Seal and Leader of the House of Lords; but after the General Election Mr. Baldwin dispensed with his services in both these capacities. The great achievement of his period in office was the designing and promotion of the ever-famous Hurricane and Spitfire fighters. The first prototypes of these flew in November 1935 and March 1936 respectively. Londonderry does not mention this in his defence, but he might well have done so, since he took the blame for so much that he had not done. The new Secretary of State, wafted by favourable breezes and fresh tides, ordered immediate large-scale production of these types, and they were ready in some numbers none too soon. Cunliffe-Lister was a much more potent political figure than his predecessor, and had a better chance and a more inspiriting task. He brought an altogether more powerful force to bear upon our air policy and administration, and set himself actively to work to make up for the time lost by the Cabinet from 1932 to 1934. He however made the serious mistake of quitting the House of Commons for the House of Lords in November 1935, thus stultifying one of the arguments for his transfer to the Secretaryship of State for Air. This was to cost him his office a few years later.

* * * * *

A disaster of the first magnitude had fallen upon us. Hitler had already obtained parity with Great Britain. Henceforward he had merely to drive his factories and training-schools at full speed not only to keep his lead in the air, but steadily to improve it. Henceforward all the unknown, immeasurable threats which overhung London from air attack would be a definite and compelling factor in all our decisions. Moreover, we could never catch up; or at any rate the Government never did catch up. Credit is due to them and to the Air Ministry for the high efficiency of the Royal Air Force. But the pledge that air parity would be maintained was irretrievably broken. It is true that the immediate further expansion of the German Air Force did not proceed at the same rate as in the period when they gained parity. No doubt a supreme effort had been made by them to achieve at a bound this commanding position and to assist and exploit it in their diplomacy. It gave Hitler the foundation for the successive acts of aggression which he had planned and which were now soon to take place. Very considerable efforts were made by the British Government in the next four years, and there is no doubt that we excelled in air quality; but quantity was henceforth beyond us. The outbreak of the war found us with barely half the German numbers.

CHAPTER VIII

CHALLENGE AND RESPONSE

1935

Hitler Decrees Conscription, March 16, 1935 – Two Years' Military Service in France, March 16 – Sir John Simon and Mr. Eden in Berlin, March 24 – The Stresa Conference – The Franco-Soviet Pact, May 2 – Mr. Baldwin Becomes Prime Minister, June 7 – Sir Samuel Hoare, Foreign Secretary – Mr. Eden Appointed Minister for League of Nations Affairs – The Anglo-German Naval Agreement – Its Dangers – Far-reaching Effects in Europe – The Foreign Secretary's Defence – The Growth of the German Army – French and German Man-power.

*T*HE years of underground burrowings, of secret or disguised preparations, were now over, and Hitler at length felt himself strong enough to make his first open challenge. On March 9, 1935, the official constitution of the German Air Force was announced, and on the 16th it was declared that the German Army would henceforth be based on national compulsory service. The laws to implement these decisions were soon promulgated, and action had already begun in anticipation. The French Government, who were well informed of what was coming, had actually declared the consequential extension of their own military service to two years a few hours earlier on the same momentous day. The German action was an open formal affront to the treaties of peace upon which the League of Nations was founded. As long as the breaches had taken the form of evasions or calling things by other names, it was easy for the responsible victorious Powers, obsessed by pacifism and preoccupied with domestic politics, to avoid the responsibility of declaring that the Peace Treaty was being broken or repudiated. Now the issue came with blunt and brutal force. Almost on the same day the Ethiopian Government

117

appealed to the League of Nations against the threatening demands of Italy. When, on March 24, against this background, Sir John Simon with the Lord Privy Seal, Mr. Eden, visited Berlin at Hitler's invitation, the French Government thought the occasion ill-chosen. They had now themselves at once to face, not the reduction of their Army, so eagerly pressed upon them by Mr. MacDonald the year before, but the extension of compulsory military service from one year to two. In the prevailing state of public opinion this was a heavy task. Not only the Communists but the Socialists had voted against the measure. When M. Léon Blum said, "The workers of France will rise to resist Hitlerite aggression," Thorez replied, amid the applause of his Soviet-bound faction, "We will not tolerate the working classes being drawn into a so-called war in defence of Democracy against Fascism."

The United States had washed their hands of all concern with Europe, apart from wishing well to everybody, and were sure they would never have to be bothered with it again. But France, Great Britain, and also—decidedly—Italy, in spite of their discordances, felt bound to challenge this definite act of treaty-violation by Hitler. A Conference of the former principal Allies was summoned under the League of Nations at Stresa, and all these matters were brought to debate.

*   *   *   *   *

Anthony Eden had for nearly ten years devoted himself almost entirely to the study of foreign affairs. Taken from Eton at eighteen to the World War, he had served for four years with distinction in the 60th Rifles through many of the bloodiest battles, and risen to the position of Brigade-Major, with the Military Cross. Shortly after entering the House of Commons in 1925, he became Parliamentary Private Secretary to Austen Chamberlain at the Foreign Office during Mr. Baldwin's second Administration. In the MacDonald-Baldwin Coalition of 1931 he was appointed Under-Secretary of State and served under the new Foreign Secretary, Sir John Simon. The duties of an Under-Secretary are often changed, but his responsibilities are always limited. He has to serve his chief in carrying out the policy settled in the Cabinet, of which he is not a member and to which he has no access. Only in an extreme case where conscience and honour are involved is he justified in carrying any difference about foreign policy to the point of public controversy or resignation.

Eden had however during all these years obtained a wide view of the foreign scene, and he was intimately acquainted with the life and thought of the great department upon which so much depends. Sir John Simon's conduct of foreign affairs was not in 1935 viewed with favour either by the Opposition or in influential circles of the Conservative Party. Eden, with all his knowledge and exceptional gifts, began therefore to acquire prominence. For this reason, after becoming Lord Privy Seal at the end of 1934, he had retained by the desire of the Cabinet an informal but close association with the Foreign Office, and thus had been invited to accompany his former chief, Sir John Simon, on the inopportune, but not unfruitful, visit to Berlin. The Foreign Secretary returned to London after the interview with Hitler, bringing with him the important news, already mentioned, that, according to Hitler, Germany *had now gained air parity with Britain*. Eden was sent on to Moscow, where he established contacts with Stalin which were to be revived with advantage after some years. On the homeward journey his aeroplane ran into a severe and prolonged storm, and when after a dangerous flight they landed he was almost in a state of collapse. The doctors declared that he was not fit to go with Simon to the Stresa Conference, and indeed for several months he was an invalid. In these circumstances the Prime Minister decided himself to accompany the Foreign Secretary, although at this time his own health, eyesight, and mental powers were evidently failing. Great Britain was therefore weakly represented at this all-important meeting, which MM. Flandin and Laval attended on behalf of France, and Signors Mussolini and Suvich on behalf of Italy.

There was general agreement that open violation of solemn treaties, for the making of which millions of men had died, could not be borne. But the British representatives made it clear at the outset that they would not consider the possibility of sanctions in the event of treaty-violation. This naturally confined the Conference to the region of words. A resolution was passed unanimously to the effect that "unilateral"—by which they meant one-sided—breaches of treaties could not be accepted, and the Executive Council of the League of Nations was invited to pronounce upon the situation disclosed. On the second afternoon of the Conference Mussolini strongly supported this action, and was outspoken against aggression by one Power upon another. The final declaration was as follows:

The three Powers, the object of whose policy is the collective maintenance of peace within the framework of the League of Nations, find themselves in complete agreement in opposing, by all practicable means, any unilateral repudiation of treaties which may endanger the peace of Europe, and will act in close and cordial collaboration for this purpose.

The Italian Dictator in his speech had stressed the words *"peace of Europe"*, and paused after "Europe" in a noticeable manner. This emphasis on Europe at once struck the attention of the British Foreign Office representatives. They pricked up their ears, and well understood that while Mussolini would work with France and Britain to prevent Germany from rearming he reserved for himself any excursion in Africa against Abyssinia on which he might later resolve. Should this point be raised or not? Discussions were held that night among the Foreign Office officials. Everyone was so anxious for Mussolini's support in dealing with Germany that it was felt undesirable at that moment to warn him off Abyssinia, which would obviously have very much annoyed him. Therefore the question was not raised; it passed by default, and Mussolini felt, and in a sense had reason to feel, that the Allies had acquiesced in his statement and would give him a free hand against Abyssinia. The French remained mute on the point, and the Conference separated.

In due course, on April 15–17, the Council of the League of Nations examined the alleged breach of the Treaty of Versailles committed by Germany in decreeing universal compulsory military service. The following Powers were represented on the Council: the Argentine Republic, Australia, Great Britain, Chile, Czechoslovakia, Denmark, France, Italy, Mexico, Poland, Portugal, Spain, Turkey, and the U.S.S.R. All these Powers voted for the principle that treaties should not be broken by "unilateral" action, and referred the issue to the Plenary Assembly of the League. At the same time the Foreign Ministers of the three Scandinavian countries, Sweden, Norway, and Denmark, being deeply concerned about the naval balance in the Baltic, also met together in general support. In all nineteen countries formally protested. But how vain was all their voting without the readiness of any single Power or any group of Powers to contemplate the use of FORCE, even in the last resort!

★ ★ ★ ★ ★

Laval was not disposed to approach Russia in the firm spirit of Barthou. But in France there was now an urgent need. It seemed, above all, necessary to those concerned with the life of France to obtain national unity on the two years' military service which had been approved by a narrow majority in March. Only the Soviet Government could give permission to the important section of Frenchmen whose allegiance they commanded. Besides this, there was a general desire in France for a revival of the old alliance of 1895, or something like it. On May 2 the French Government put their signature to a Franco-Soviet pact. This was a nebulous document guaranteeing mutual assistance in the face of aggression over a period of five years.

To obtain tangible results in the French political field M. Laval now went on a three days' visit to Moscow, where he was welcomed by Stalin. There were lengthy discussions, of which a fragment not hitherto published may be recorded. Stalin and Molotov were of course anxious to know above all else what was to be the strength of the French Army on the Western Front: how many divisions? what period of service? After this field had been explored Laval said: "Can't you do something to encourage religion and the Catholics in Russia? It would help me so much with the Pope." "Oho!" said Stalin. "The Pope! How many divisions has *he* got?" Laval's answer was not reported to me; but he might certainly have mentioned a number of legions not always visible on parade. Laval had never intended to commit France to any of the specific obligations which it is the habit of the Soviet to demand. Nevertheless he obtained a public declaration from Stalin on May 15 approving the policy of national defence carried out by France in order to maintain her armed forces at the level of security. On these instructions the French Communists immediately turned about and gave vociferous support to the defence programme and the two years' service. As a factor in European security the Franco-Soviet Pact, which contained no engagements binding on either party in the event of German aggression, had only limited advantages. No real confederacy was achieved with Russia. Moreover, on his return journey the French Foreign Minister stopped at Cracow to attend the funeral of Marshal Pilsudski. Here he met Goering, with whom he talked with much cordiality. His expressions of distrust and dislike of the Soviets were duly reported through German channels to Moscow.

Mr. MacDonald's health and capacity had declined to a point which made his continuance as Prime Minister impossible. He had never been popular with the Conservative Party, who regarded him, on account of his political and war records and Socialist faith, with long-bred prejudice, softened in later years by pity. No man was more hated, or with better reason, by the Labour-Socialist Party, which he had so largely created and then laid low by what they viewed as his treacherous desertion in 1931. In the massive majority of the Government he had but seven party followers. The disarmament policy to which he had given his utmost personal efforts had now proved a disastrous failure. A General Election could not be far distant, in which he could play no helpful part. In these circumstances there was no surprise when, on June 7, it was announced that he and Mr. Baldwin had changed places and offices and that Mr. Baldwin had become Prime Minister for the third time. The Foreign Office also passed to another hand. Sir Samuel Hoare's labours at the India Office had been crowned by the passing of the Government of India Bill, and he was now free to turn to a more immediately important sphere. For some time past Sir John Simon had been bitterly attacked for his foreign policy by influential Conservatives closely associated with the Government. He now moved to the Home Office, with which he was well acquainted, and Sir Samuel Hoare became Secretary of State for Foreign Affairs.

At the same time Mr. Baldwin adopted a novel expedient. He appointed Mr. Eden, whose prestige was steadily growing and whose health was now restored, to be Minister for League of Nations Affairs. Mr. Eden was to work in the Foreign Office with equal status to the Foreign Secretary and with full access to the dispatches and the departmental staff. Mr. Baldwin's object was no doubt to conciliate the strong tide of public opinion associated with the League of Nations Union by showing the importance which he attached to the League and to the conduct of our affairs at Geneva. When about a month later I had the opportunity of commenting on what I described as "the new plan of having two equal Foreign Secretaries", I drew attention to its defects:

I was very glad indeed that the Prime Minister said yesterday that this was only a temporary experiment. I cannot feel that it will last long or ever be renewed. . . . We need the integral thought of a single man responsible for Foreign Affairs, ranging over the entire field and making every factor and every incident contribute to the general purpose upon which Parliament has agreed. The Foreign Secretary, whoever he is, whichever he is, must be supreme in his department, and everyone in that great office ought to look to him, and to him alone. I remember that we had a discussion in the war about unity of command, and that Mr. Lloyd George said, "It is not a question of one General being better than another, but of one General being better than two." There is no reason why a strong Cabinet Committee should not sit with the Foreign Secretary every day in these difficult times, or why the Prime Minister should not see him or his officials at any time; but when the topic is so complicated and vast, when it is in such continued flux, it seems to me that confusion will only be made worse confounded by dual allegiances and equal dual responsibilities.

All this was certainly borne out by events.

* * ÷ * *

While men and matters were in this posture a most surprising act was committed by the British Government. Some at least of its impulse came from the Admiralty. It is always dangerous for soldiers, sailors, or airmen to play at politics. They enter a sphere in which the values are quite different from those to which they have hitherto been accustomed. Of course they were following the inclination or even the direction of the First Lord and the Cabinet, who alone bore the responsibility. But there was a strong favourable Admiralty breeze. There had been for some time conversations between the British and German Admiralties about the proportions of the two Navies. By the Treaty of Versailles the Germans were not entitled to build more than six armoured ships of 10,000 tons, in addition to six light cruisers not exceeding 6,000 tons. The British Admiralty had recently found out that the last two pocket-battleships being constructed, the *Scharnhorst* and the *Gneisenau*, were of a far larger size than the Treaty allowed, and of a quite different type. In fact, they turned out to be 26,000-ton light battle-cruisers, or commerce-destroyers of the highest class.

In the face of this brazen and fraudulent violation of the Peace Treaty, carefully planned and begun at least two years earlier (1933), the Admiralty actually thought it was worth while making an Anglo-German Naval Agreement. His Majesty's Government did this without consulting their French ally or informing the League of Nations. At the very time when they themselves were appealing to the League and enlisting the support of its members to protest against Hitler's violation of the military clauses of the Treaty they proceeded by a private agreement to sweep away the naval clauses of the same Treaty.

The main feature of the agreement was that the German Navy should not exceed one-third of the British. This greatly attracted the Admiralty, who looked back to the days before the Great War when we had been content with a ratio of sixteen to ten. For the sake of that prospect, taking German assurances at their face value, they proceeded to concede to Germany the right to build U-boats, explicitly denied to her in the Peace Treaty. Germany might build 60 per cent. of the British submarine strength, and if she decided that the circumstances were exceptional she might build to 100 per cent. The Germans, of course, gave assurances that their U-boats would never be used against merchant ships. Why, then, were they needed? For, clearly, if the rest of the agreement was kept, they could not influence the naval decision, so far as warships were concerned.

The limitation of the German Fleet to a third of the British allowed Germany a programme of new construction which would set her yards to work at maximum activity for at least ten years. There was therefore no practical limitation or restraint of any kind imposed upon German naval expansion. They could build as fast as was physically possible. The quota of ships assigned to Germany by the British project was, in fact, far more lavish than Germany found it expedient to use, having regard partly no doubt to the competition for armour-plate arising between warship and tank construction. They were authorised to build five capital ships, two aircraft-carriers, twenty-one cruisers, and sixty-four destroyers. In fact however all they had ready or approaching completion by the outbreak of war was two capital ships, no aircraft-carriers, eleven cruisers, and twenty-five destroyers, or considerably less than half what we had so complacently accorded them. By concentrating their available resources on cruisers and destroyers at the expense of battleships, they could have put themselves in a more advantageous position for a war with Britain in 1939 or 1940. Hitler, as we now know, informed Admiral Raeder that war with England would not be likely till 1944-45. The development of the German Navy was therefore planned on a long-term basis. In U-boats alone did they build to the full paper limits allowed. As soon as they were able to pass the 60 per cent. limit they invoked the provision allowing them to build to 100 per cent., and fifty-seven were actually constructed when war began.

In the design of new battleships the Germans had the further advantage of not being parties to the provisions of the Washington Naval Agreement or the London Conference. They immediately laid down the *Bismarck* and *Tirpitz*, and, while Britain, France, and the United States were all bound by the 35,000 tons limitation, these two great vessels were being designed with a displacement of over 45,000 tons, which made them, when completed, certainly the strongest vessels afloat in the world.

It was also at this moment a great diplomatic advantage to Hitler to divide the Allies, to have one of them ready to condone breaches of the Treaty of Versailles, and to invest the regaining of full freedom to rearm with the sanction of agreement with Britain. The effect of the announcement was another blow at the League of Nations. The French had every right to complain that their vital interests were affected by the permission accorded by Great Britain for the building of U-boats. Mussolini saw in this episode evidence that Great Britain was not acting in good faith with her other allies, and that, so long as her special naval interests were secured, she would apparently go to any length in accommodation with Germany, regardless of the detriment to friendly Powers menaced by the growth of the German land forces. He was encouraged by what seemed the cynical and selfish attitude of Great Britain to press on with his plans against Abyssinia. The Scandinavian Powers, who only a fortnight before had courageously sustained the protest against Hitler's introduction of compulsory service in the German Army, now found that Great Britain had behind the scenes agreed to a German Navy which, though only a third of the British, would within this limit be master of the Baltic.

Great play was made by British Ministers with the German offer to co-operate with us in abolishing the submarine. Considering that the condition attached to it was that all other countries should agree at the same time, and that it was well known there was not the slightest chance of other countries agreeing, this was a very safe offer for the Germans to make. This also applied to the German agreement to restrict the use of submarines so as to strip submarine warfare against commerce of inhumanity. Who could suppose that the Germans, possessing a great fleet of U-boats and watching their women and children being starved by a British blockade, would abstain from the fullest use of that arm? I described this view as "the acme of gullibility".

Far from being a step towards disarmament, the agreement, had it been carried out over a period of years, would inevitably have provoked a world-wide development of new warship-building. The French Navy, except for its latest vessels, would require reconstruction. This again would react upon Italy. For ourselves, it was evident that we should have to rebuild the British Fleet on a very large scale in order to maintain our three to one superiority in modern ships. It may be that the idea of the German Navy being one-third of the British also presented itself to our Admiralty as the British Navy being three times the German. This perhaps might clear the path to a reasonable and overdue rebuilding of our Fleet. But where were the statesmen?

This agreement was announced to Parliament by the First Lord of the Admiralty, Sir Bolton Eyres-Monsell, on June 21, 1935. On the first opportunity, July 11, and again on July 22, I condemned it:

I do not believe that this isolated action by Great Britain will be found to work for the cause of peace. The immediate reaction is that every day the German Fleet approaches a tonnage which gives it absolute command of the Baltic, and very soon one of the deterrents of a European war will gradually fade away. So far as the position in the Mediterranean is concerned, it seems to me that we are in for very great difficulties. Certainly a large addition of new shipbuilding must come when the French have to modernise their Fleet to meet German construction and the Italians follow suit, and we shall have pressure upon us to rebuild from that point of view, or else our position in the

Mediterranean will be affected. But worst of all is the effect upon our position at the other end of the world, in China and in the Far East. What a windfall this has been to Japan! Observe what the consequences are. The First Lord said, "Face the facts." The British Fleet, when this programme is completed, will be largely anchored to the North Sea. That means to say that the whole position in the Far East has been very gravely altered, to the detriment of the United States and of Great Britain and to the detriment of China. . . .

I regret that we are not dealing with this problem of the resuscitation of German naval power with the Concert of Europe on our side, and in conjunction with many other nations whose fortunes are affected and whose fears are aroused equally with our own by the enormous developments of German armaments. What those developments are no one can accurately measure. We have seen that powerful vessels, much more powerful than we expected, can be constructed unknown even to the Admiralty. We have seen what has been done in the air. I believe that if the figures of the expenditure of Germany during the current financial year could be ascertained the House and the country would be staggered and appalled by the enormous expenditure upon war preparations which is being poured out all over that country, converting the whole mighty nation and empire of Germany into an arsenal virtually on the threshold of mobilisation.

★ ★ ★ ★ ★

It is only right to state here the contrary argument as put forward by Sir Samuel Hoare in his first speech as Foreign Secretary on July 11, 1935, in response to many domestic and European criticisms:

The Anglo-German Naval Agreement is in no sense a selfish agreement. On no account could we have made an agreement that was not manifestly in our view to the advantage of the other naval Powers. On no account could we have made an agreement that we did not think, so far from hindering general agreement, would actually further it. The question of naval disarmament has always been treated distinctively from the question of land and air disarmament. The naval question has always been treated apart, and it was always the intention, so far as I know, of the naval Powers to treat it apart.

Apart however from the juridical position, there seemed to us to be, in the interests of peace—which is the main objective of the British Government—overwhelming reasons why we should conclude the agreement. In the opinion of our naval experts, we were advised to accept the agreement as a safe agreement for the British Empire. Here again we saw a chance that might not recur of eliminating one of the causes that chiefly led to the embitterment before the Great War—the race of German naval armaments. Incidentally, out of that discussion arose the very important statement of the German Government that henceforth, so far as they were concerned, they would eliminate one of the causes that made the war so terrible, namely, the unrestricted use of submarines against merchant ships. Thirdly, we came definitely to the view that there was a chance of making an agreement that seemed on naval grounds manifestly to the advantage of other naval Powers, including France. . . . With the French Fleet at approximately its present level as compared with our own Fleet, the agreement gives France a permanent superiority over the German Fleet of 43 per cent., as compared with an inferiority of about 30 per cent. before the war. . . . I am therefore bold enough to believe that, when the world looks more dispassionately at these results, the overwhelming majority of those who stand for peace and a restriction of armaments will say that the British Government took not only a wise course but the only course that in the circumstances was open to them.

What had in fact been done was to authorise Germany to build to her utmost capacity for five or six years to come.

* * * * *

Meanwhile in the military sphere the formal establishment of conscription in Germany on March 16, 1935, marked the fundamental challenge to Versailles. But the steps by which the German Army was now magnified and reorganised are not of technical interest only. The whole function of the Army in the National-Socialist State required definition. The purpose of the Law of May 21, 1935, was to expand the technical *élite* of secretly-trained specialists into the armed expression of the whole nation. The name Reichswehr was changed to that of Wehrmacht. The Army was to be subordinated to the supreme leadership of the Fuehrer. Every soldier took the oath, not, as formerly, to the Constitution, but to the person of Adolf Hitler. The War Ministry was directly subordinated to the orders of the Fuehrer. Military service was an essential civic duty, and it was the responsibility of the Army to educate and to unify once and for all the population of the Reich. The second clause of the Law reads: "The Wehrmacht is the armed force and the school of military education of the German people."

Here indeed was the formal and legal embodiment of Hitler's words in *Mein Kampf*:

The coming National-Socialist State should not fall into the error of the past and assign to the Army a task which it does not and should not have. The German Army is not to be a school for the maintenance of tribal peculiarities, but rather a school for the mutual understanding and adjustment of all Germans. Whatever may have a disruptive effect in national life should be given a unifying effect through the Army. It should furthermore raise the individual youth above the narrow horizon of his little countryside and place him in the German nation. He must learn to respect, not the boundaries of his birthplace, but the boundaries of his Fatherland; for it is these which he too must some day defend.

Upon these ideological bases the Law also established a new territorial organisation. The Army was now organised in three commands, with headquarters at Berlin, Cassel, and Dresden, subdivided into ten (later twelve) Wehrkreise (military districts). Each Wehrkreis contained an Army Corps of three divisions. In addition a new kind of formation was planned—the Armoured Division, of which three were soon in being.

Detailed arrangements were also made regarding military service. The regimentation of German youth was the prime task of the new *régime*. Starting in the ranks of the Hitler Youth, the boyhood of Germany passed at the age of eighteen on a voluntary basis into the S.A. for two years. By a Law of June 26, 1935, service in the Work Battalions or Arbeitsdienst became a compulsory duty on every male German reaching the age of twenty. For six months he would have to serve his country, constructing roads, building barracks, or draining marshes, thus fitting him physically and morally for the crowning duty of a German citizen, service with the armed forces. In the Work Battalions the emphasis lay upon the abolition of class and the stressing of the social unity of the German people; in the Army it was put upon discipline and the territorial unity of the nation.

The gigantic task of training the new body and of expanding the cadres prescribed by the technical conception of Seeckt now began. On October 15, 1935, again in defiance of the clauses of Versailles, the German Staff College was reopened with formal ceremony by Hitler, accompanied by the chiefs of the armed services. Here was the apex of the pyramid, whose base was now already constituted by the myriad formations of the Work Battalions. On November 7, 1935, the first class, born in 1914, was called up for service: 596,000 young men to be trained in the profession of arms. Thus at one stroke, on paper at least, the German Army was raised to nearly 700,000 effectives.

With the task of training came the problems of financing rearmament and expanding German industry to meet the needs of the new national Army. By secret decrees Dr. Schacht had been made virtual Economic Dictator of Germany. Seeckt's pioneer work was now put to its supreme test. The two major difficulties were, first, the expansion of the Officers Corps, and, secondly, the organisation of the specialised units, the Artillery, the Engineers, and the Signals. By October 1935 ten Army Corps were forming. Two more followed a year later, and a thirteenth in October 1937. The police formations were also incorporated in the armed forces.

It was realised that after the first call-up of the 1914 class, in Germany as in France, the succeeding years would bring a diminishing number of recruits, owing to the decline in births during the period of the World War. Therefore in August 1936 the period of active military service in Germany was raised to two years. The 1915 class numbered 464,000, and with the retention of the 1914 class for another year the number of Germans under regular military training in 1936 was 1,511,000 men, excluding the para-military formations of the party and the Work Battalions. The effective strength of the French Army, apart from reserves, in the same year was 623,000 men, of whom only 407,000 were in France.

The following figures, which actuaries could foresee with some precision, tell their tale:

TABLE OF THE COMPARATIVE FRENCH AND GERMAN FIGURES FOR THE CLASSES BORN FROM 1914 TO 1920, AND CALLED UP FROM 1934 TO 1940

Class				German		French
1914	..	..	..	596,000 men	..	279,000 men
1915	..	..	..	464,000 ,,	..	184,000 ,,
1916	..	..	..	351,000 ,,	..	165,000 ,,
1917	..	..	..	314,000 ,,	..	171,000 ,,
1918	..	..	..	326,000 ,,	..	197,000 ,,
1919	..	..	..	485,000 ,,	..	218,000 ,,
1920	..	..	..	636,000 ,,	..	360,000 ,,
				3,172,000 men		1,574,000 men

Until these figures became facts as the years unfolded they were still but warning shadows. All that was done up to 1935 fell far short of the strength and power of the French Army and its vast reserves, apart from its numerous and vigorous allies. Even at this time a resolute decision upon the authority, which could easily have been obtained, of the League of Nations might have arrested the whole process. Germany could either have been brought to the bar at Geneva and invited to give a full explanation and allow inter-Allied missions of inquiry to examine the state of her armaments and military formations in breach of the Treaty, or, in the event of refusal, the Rhine bridgeheads could have been reoccupied until compliance with the Treaty had been secured, without there being any possibility of effective resistance or much likelihood of bloodshed. In this way the Second World War could have been at least delayed indefinitely. Many of the facts and their whole general tendency were well known to the French and British Staffs, and were to a lesser extent realised by the Governments. The French Government, which was in ceaseless flux in the fascinating game of party politics, and the British Government, which arrived at the same vices by the opposite process of general agreement to keep things quiet, were equally incapable of any drastic or clear-cut action, however justifiable both by treaty and by common prudence. The French Government had not accepted all the reductions of their own forces pressed upon them by their ally; but, like their British colleagues, they lacked the quality to resist in any effective manner what Seeckt in his day had called "the Resurrection of German Military Power".

CHAPTER IX

PROBLEMS OF AIR AND SEA

1935-1939

*A Technical Interlude – German Power to Blackmail – Approaches to
Mr. Baldwin and the Prime Minister – The Earth v. the Air – Mr.
Baldwin's Invitation – The Air Defence Research Committee – Some
General Principles – Progress of Our Work – The Development of
Radar – Professor Watson-Watt and Radio Echoes – The Tizard
Report – The Chain of Coastal Stations – Air-Marshal Dowding's
Network of Telephonic Communications – The Graf Zeppelin Flies
Up Our East Coast: Spring of 1939 – I.F.F. – A Visit to Martlesham,
1939 – My Admiralty Contacts – The Fleet Air Arm – The Question
of Building New Battleships – Calibre of Guns – Weight of Broad-
sides – Number of Turrets – My Letter to Sir Samuel Hoare of
August 1, 1936 – The Admiralty Case – Quadruple Turrets – An
Unfortunate Sequel – A Visit to Portland: the "Asdics".*

TECHNICAL decisions of high consequence affecting our
future safety now require to be mentioned, and it will be
convenient in this chapter to cover the whole four years
which lay between us and the outbreak of war.

After the loss of air parity we were liable to be blackmailed by
Hitler. If we had taken steps betimes to create an Air Force half
as strong again, or twice as strong, as any that Germany could
produce in breach of her treaty we should have kept control of
the future. But even air parity, which no one could say was
aggressive, would have given us a solid measure of defensive
confidence in these critical years, and a broad basis from which
to conduct our diplomacy or expand our Air Force. But we had
lost air parity. And such attempts as were made to recover it
were vain. We had entered a period when the weapon which

had played a considerable part in the previous war had become obsessive in men's minds, and also a prime military factor. Ministers had to imagine the most frightful scenes of ruin and slaughter in London if we quarrelled with the German Dictator. Although these considerations were not special to Great Britain they affected our policy, and by consequence all the world.

During the summer of 1934 Professor Lindemann wrote to the *Times* newspaper pointing out the possibility of decisive scientific results being obtained in air defence research. In August we tried to bring the subject to the attention not merely of the officials at the Air Ministry, who were already on the move, but of their masters in the Government. In September we journeyed from Cannes to Aix-les-Bains, and had an agreeable conversation with Mr. Baldwin, who appeared deeply interested. Our request was for an inquiry on a high level. When we came back to London departmental difficulties arose, and the matter hung in suspense. Early in 1935 an Air Ministry committ.e composed of scientists was set up and instructed to explore the future. We remembered that it was upon the advice of the Air Ministry that Mr. Baldwin had made the speech which produced so great an impression in 1932, when he said that there was really no defence. "The bomber will always get through." We had therefore no confidence in any Air Ministry departmental committee, and thought the subject should be transferred from the Air Ministry to the Committee of Imperial Defence, where the heads of the Government, the most powerful politicians in the country, would be able to supervise and superintend its actions and also to make sure that the necessary funds were not denied. At this stage we were joined by Sir Austen Chamberlain, and we continued at intervals to address Ministers on the subject.

In February we were received by Mr. MacDonald personally, and we laid our case before him. No difference of principle at all existed between us. The Prime Minister was most sympathetic when I pointed out the peace aspect of the argument. Nothing, I said, could lessen the terrors and anxieties which overclouded the world so much as the removal of the idea of surprise attacks upon the civil populations. Mr. MacDonald seemed at this time greatly troubled with his eyesight. He gazed blankly out of the windows on to Palace Yard, and assured us he was hardening his heart to overcome departmental resistance. The Air Ministry, for their part, resented the idea of any outside or superior body interfering in their special affairs, and for a while nothing happened.

I therefore raised the matter in the House on June 7, 1935.

The point [I said] is limited, and largely scientific in its character. It is concerned with the methods which can be invented or adopted or discovered to enable the Earth to control the Air, to enable defence from the ground to exercise control—indeed domination—upon aeroplanes high above its surface. . . . My experience is that in these matters, when the need is fully explained by military and political authorities, science is always able to provide something. We were told that it was impossible to grapple with submarines, but methods were found which enabled us to strangle the submarines below the surface of the water, a problem not necessarily harder than that of clawing down marauding aeroplanes. Many things were adopted in the war which we were told were technically impossible, but patience, perseverance, and above all the spur of necessity under war conditions, made men's brains act with greater vigour, and science responded to the demands. . . .

It is only in the twentieth century that this hateful conception of inducing nations to surrender by terrorising the helpless civil population by massacring the women and children has gained acceptance and countenance among men. This is not the cause of any one nation. Every country would feel safer if once it were found that the bombing aeroplane was at the mercy of appliances directed from the earth, and the haunting fears and suspicions which are leading nations nearer and nearer to another catastrophe would be abated. . . . We have not only to fear attacks upon our civil population in our great cities, in respect of which we are more vulnerable than any other country in the world, but also attacks upon the dockyards and other technical establishments without which our Fleet, still an essential factor in our defence, might be paralysed or even destroyed. Therefore it is not only for the sake of a world effort to eliminate one of the worst causes of suspicion and of war, but as a means of restoring to us here in Great Britain the old security of our Island, that this matter should receive and command the most vigorous thought of the greatest men in our country and our Government, and should be pressed forward by every resource that the science of Britain can apply and the wealth of the country can liberate.

On the very next day the Ministerial changes recorded in the previous chapter took place, and Mr. Baldwin became Prime Minister. Sir Philip Cunliffe-Lister—Lord Swinton as he soon afterwards became—succeeded Lord Londonderry as Air Minister. One afternoon a month later I was in the smoking-room of the

House of Commons when Mr. Baldwin came in. He sat down next to me and said at once: "I have a proposal to make to you. Philip is very anxious that you should join the newly-formed Committee of Imperial Defence on Air Defence Research, and I hope you will." I said I was a critic of our air preparations and must reserve my freedom of action. He said: "That is quite understood. Of course you will be perfectly free except upon the secret matters you learn only at the Committee."

I made it a condition that Professor Lindemann should at least be a member of the Technical Sub-Committee, because I depended upon his aid. A few days later the Prime Minister wrote:

July 8, 1935

I am glad you have seen Hankey, and I take your letter as an expression of your willingness to serve on that Committee.

I am glad, and I think you may be of real help in a most important investigation.

Of course, you are free as air (the correct expression in this case!) to debate the general issues of policy, programmes, and all else connected with the Air Services.

My invitation was not intended as a muzzle, but as a gesture of friendliness to an old colleague.

Accordingly for the next four years I attended these meetings, and thus obtained a full view of this vital sphere of our air defence, and built up my ideas upon it year by year in close and constant discussion with Lindemann. I immediately prepared a memorandum for the Committee which embodied the thought and knowledge I had already gathered, without official information, in my talks and studies with Lindemann and from my own military conceptions. This paper is of interest because of the light which it throws on the position in July 1935. No one at that time had considered the use of radio beams for guiding bombers. The difficulties of training large numbers of individual pilots were obvious, and it was generally held that at night large fleets of aircraft would be led by a few master-bombers. Great advances into new fields were made in the four years which were to pass before the life of the nation was to be at stake; and meanwhile the adoption of bombing guided by radio beams caused profound tactical changes. Hence much that was written then was superseded, but a good deal was tried by me when I had power—not all with success.

July 23, 1935

The following notes are submitted with much diffidence, and in haste on account of our early meeting, in the hopes that they may be a contribution to our combined thought.

General tactical conceptions and what is technically feasible act and react upon one another. Thus the scientist should be told what facilities the Air Force would like to have, and aeroplane design be made to fit into and implement a definite scheme of warfare.

At this stage we must assume a reasonable war hypothesis, namely, that Great Britain, France, and Belgium are allies attacked by Germany.

After the outbreak of such a war the dominating event will be the mobilisation of the great Continental armies. This will take at least a fortnight, diversified and hampered by mechanised and motorised inroads. The French and German General Staff's minds will be riveted upon the assembly and deployment of the armies. Neither could afford to be markedly behindhand at the first main shock. It may be hoped that Germany will not be ready for a war in which the Army and Navy are to play an important part for two or three years. Their Navy is at the moment exiguous; they have not yet obtained the command of the Baltic; and it would appear that their heavy artillery is still inadequate. To build a navy and to produce heavy artillery and train the men will take a time measured in years rather than in months.

A large part of German munitions production is concentrated in the Ruhr, which is easily accessible to enemy bombing. She must realise that she would be cut off from foreign supplies of many essential war materials (copper, tungsten, cobalt, vanadium, petrol, rubber, wool, etc.), and *even her iron supply will be reduced unless she dominates the Baltic*, so that she is scarcely yet in a position to undertake a war of long duration. Great efforts are of course being made to overcome these handicaps, such as the removal of certain factories from the frontier to Central Germany, the synthetic production of substances such as petrol and rubber, and the accumulation of large stocks. But it seems unlikely that Germany will be in a position before 1937 or 1938 to begin with any hope of success a war of the three Services, which might last for years, and in which she would have scarcely any allies.

It would appear in such a war the first task of the Anglo-French Air Force should be the breaking down of enemy communications, their railways, motor roads, Rhine bridges, viaducts, etc., and the maximum disturbance of their assembly zones and munition dumps. Next in priority come the most accessible factories for their war industry in all its forms. It seems fairly certain that if our efforts from zero hour were concentrated on these vital targets *we should impose a similar policy on the enemy*. Otherwise the French would have an unobstructed mobilisation and command the initiative in the great land battle. Thus any German aircraft used to commit acts of terror upon the British and French civil populations will be grudged and sparingly diverted.

Nevertheless we must expect that even in a three-Service war attempts will be made to burn down London, or other great cities within easy reach, in order to test the resisting will-power of the Government and people under these terrible ordeals. Secondly, the Port of London, and the dockyards upon which the life of the Fleet depends, are also military targets of the highest possible consequence.

There is however always the ugly possibility that those in authority in Germany may believe that it would be possible to beat a nation to its knees in a very few months, or even weeks, by violent aerial mass attack. The conception of psychological shock tactics has a great attraction for the German mind. Whether they are right or wrong is beside the point. If the German Government believes that it can force a country to sue for peace by destroying its great cities and slaughtering the civilian population from the air before the Allied armies have mobilised and advanced materially, this might well lead it to commence hostilities with the air arm alone. It need scarcely be added that England, if she could be separated from France, would be a particularly apt victim for this form of aggression. For her main form of counter-attack, apart from aerial reprisals, namely, a naval blockade, only makes itself felt after a considerable time.

If the aerial bombardment of our cities can be restricted or prevented, the chance (which may in any case be illusory) that our morale could be broken by "frightfulness" will vanish, and the decision will remain in the long run with the armies and navies. The more our defences are respected the greater will be the deterrent upon a purely air war.

*   *   *   *   *

I had some ideas to contribute. It must be remembered that in 1935 we had still more than four years to run before any radio-detection method came into play.

*   *   *   *   *

The Committee worked in secret, and no statement was ever made of my association with the Government, whom I continued to criticise and attack with increasing severity in other parts of the field. It is often possible in England for experienced politicians to reconcile functions of this kind, in the same way as the sharpest political differences are sometimes found not incompatible with personal friendship. Scientists are however a far more jealous society. In 1937 a considerable difference on the Technical Sub-Committee grew between them and Professor Lindemann. His colleagues resented the fact that he was in constant touch with me, and that I pressed his points on the main Committee, to which they considered Sir Henry Tizard, Rector of the Imperial College of Science and Technology, should alone explain their collective view. Lindemann was therefore asked to retire. He was perfectly right in arming me with the facts on which to argue; indeed this was the basis on which we had both joined in the work. Nevertheless, in the public interest, in spite of his departure, I continued, with his full agreement, to remain a member; and in 1938, as will presently be described, I was able to procure his reinstatement.

★　★　★　★　★

The possibility of using radio waves scattered back from aircraft and other metal objects seems to have occurred to a very large number of people in England, America, Germany, and France in the nineteen-thirties. We talked of them as R.D.F. (Radio Direction Finding), or later as "Radar". The practical aim was to discern the approach of hostile aircraft, not by human senses, by eye or ear, but by the echo which they sent back from radio waves. About seventy miles up there is a reflecting canopy (ionosphere), the existence of which prevents ordinary wireless waves from wandering off into space, and thus makes long-range wireless communication possible. The technique of sending up very short pulses and observing their echo had been actively developed for some years by our scientists, notably by Professor Appleton.

In February 1935 a Government research scientist, Professor Watson-Watt, had first explained to the Technical Sub-Committee that the detection of aircraft by radio echoes might be feasible and had proposed that it should be tested. The Committee was impressed. It was assumed that it would take five years to detect aircraft up to a range of fifty miles. On July 25, 1935, at the fourth meeting of the Air Defence Research Committee, and the first which I attended, Tizard made his report upon radio-location. The preliminary experiments were held to justify further executive action. The Service departments were invited to formulate plans. A special organisation was set up, and a chain of stations established in the Dover-Orfordness area for experimental purposes. The possibility of radio-location of ships was also to be explored.

By March 1936 stations were being erected and equipped along the south coast, and it was hoped to carry out experimental exercises in the autumn. During the summer there were considerable delays in construction, and the problem of hostile jamming appeared. In July 1937 plans were brought forward by the Air Ministry, and approved by the Air Defence Research Committee, to create a chain of twenty stations from the Isle of Wight to the Tees by the end of 1939 at the cost of over a million pounds. Experiments were now tried for finding hostile aircraft after they had come inland. By the end of the year we could track them up to a distance of thirty-five miles at ten thousand feet. Progress was also being made about ships. It had been proved possible to fix vessels from the air at a range of nine miles. Two ships of the Home Fleet were already equipped with apparatus for aircraft detection, and experiments were taking place for range-finding on aircraft, for fire control of anti-aircraft guns, and for the direction of searchlights. Work proceeded. By December 1938 fourteen of the twenty new stations planned were operating with temporary equipment. Location of ships from the air was now possible at thirty miles.

By 1939 the Air Ministry, using comparatively long-wave radio (ten metres), had constructed the so-called coastal chain, which enabled us to detect aircraft approaching over the sea at distances up to about sixty miles. An elaborate network of telephonic communication had been installed under Air-Marshal Dowding, of Fighter Command, linking all these stations with a central command station at Uxbridge, where the movements of all aircraft observed could be plotted on large maps and thus the control in action of all our own air forces be maintained. Apparatus called I.F.F.* had also been devised which enabled our coastal chain Radar stations to distinguish British aircraft which carried it from enemy aircraft. It was found that these long-wave stations did not detect aircraft approaching at low heights over the sea, and as a counter to this danger a supplementary set of

* Identification Friend or Foe.

stations called C.H.L.* was constructed, using much shorter waves (one and a half metres), but only effective over a shorter range.

To follow enemy aircraft once they had come inland we had meanwhile to rely upon the Royal Observer Corps, which only operated by ear and eye, but which when linked up with all the telephone exchanges proved of high value, and in the early part of the Battle of Britain was our main foundation. It was not enough to detect approaching enemy aircraft over the sea, though that gave at least fifteen to twenty minutes' warning. We must seek to guide our own aircraft towards the attackers and intercept them over the land. For this purpose a number of stations with what were called G.C.I.† were being erected. But all this was still embryonic at the outbreak of war.

*   *   *   *   *

The Germans were also busy, and in the spring of 1939 the "Graf Zeppelin" flew up the East Coast of Britain. General Martini, Director-General of Signals in the Luftwaffe, had arranged that she carried special listening equipment to discover the existence of British Radar transmissions, if any. The attempt failed, but had her listening equipment been working properly the "Graf Zeppelin" ought certainly to have been able to carry back to Germany the information that we had Radar, for our Radar stations were not only operating at the time, but also detected her movements and divined her intention. The Germans would not have been surprised to hear our Radar pulses, for they had developed a technically efficient Radar system which was in some respects ahead of our own. What would have surprised them however was the extent to which we had turned our discoveries to practical effect, and woven all into our general air defence system. In this we led the world, and it was operational efficiency rather than novelty of equipment that was the British achievement.

The final meeting of the Air Defence Research Committee took place on July 11, 1939. Twenty Radar stations were at that time in existence between Portsmouth and Scapa Flow, able to detect aircraft flying above ten thousand feet, with ranges varying

* Chain Stations Home Service Low Cover.
† Ground Control of Interception.

from fifty to a hundred and twenty miles. A satisfactory anti-jamming device and a simplified method of I.F.F. were now actually in production. Flight trials were taking place with experimental sets in aircraft to try to "home" on enemy machines. The experimental sets for the location of ships from the air had proved too bulky for air-service purposes, and were passed to the Admiralty for possible use by ships.

★　★　★　★　★

I add a final note. In June 1939 Sir Henry Tizard, at the desire of the Secretary of State, conducted me in a rather disreputable aeroplane to see the establishments which had been developed on the East Coast. We flew around all day. I sent my impressions to the Air Minister, and I print them here because they give a glimpse of where we were in this Radar field on the eve of the task.

Mr. Churchill to Sir Kingsley Wood
. . . I found my visit to Martlesham and Bawdsey under Tizard's guidance profoundly interesting, and also encouraging. It may be useful if I put down a few points which rest in my mind.

These vital R.D.F. stations require immediate protection. We thought at first of erecting dummy duplicates and triplicates of them at little expense; but on reflection it seems to me that here is a case for using the smoke-cloud. . . .

A weak point in this wonderful development is of course that when the raider crosses the coast it leaves the R.D.F., and we become dependent upon the Observer Corps. This would seem transition from the middle of the twentieth century to the Early Stone Age. Although I hear that good results are obtained from the Observer Corps, we must regard following the raider inland by some application of R.D.F. as most urgently needed. It will be some time before the R.D.F. stations can look back inland, and then only upon a crowded and confused air theatre. . . .

The progress in R.D.F. especially applied to range-finding must surely be of high consequence to the Navy. It would give power to engage an enemy irrespective of visibility. How different would have been the fate of the German battle-cruisers when they attacked Scarborough and Hartlepool in 1914 if we could have pierced the mist! I cannot conceive why the Admiralty are not now hot upon this trail. Tizard also pointed out the enormous value to destroyers and submarines of directing torpedoes accurately, irrespective of visibility by night or day. I should have thought this was one of the biggest things that had happened for a long time, and all for our benefit.

The method of discrimination between friend and foe is also of the highest consequence to the Navy, and should entirely supersede recognition signals, with all their peril. I presume the Admiralty knows all about it.

Finally, let me congratulate you upon the progress that has been made. We are on the threshold of immense securities for our Island. Unfortunately we want to go farther than the threshold, and time is short.

I shall in a later volume explain the way in which, by these and other processes known only to a very small circle, the German attack on Great Britain was to a large extent parried in the autumn and winter of 1940. There is no doubt that the work of the Air Ministry and the Air Defence Research Committee, both under Lord Swinton and his successor, played the decisive part in procuring this precious reinforcement to our fighter aircraft. When in 1940 the chief responsibility fell upon me and our national survival depended upon victory in the air, I had the advantage of a layman's insight into the problems of air warfare resulting from four long years of study and thought based upon the fullest official and technical information. Although I have never tried to be learned in technical matters, this mental field was well lit for me. I knew the various pieces and the moves on the board, and could understand anything I was told about the game.

* * * * *

My contacts with the Admiralty during these years were also constant and intimate. In the summer of 1936 Sir Samuel Hoare became First Lord, and he authorised his officers to discuss Admiralty matters freely with me; and as I took a keen interest in the Navy I availed myself fully of these opportunities. I had known the First Sea Lord, Admiral Chatfield, from the Beatty days of 1914, and my correspondence with him on naval problems began in 1936. I also had a long-standing acquaintance with Admiral Henderson, the Controller of the Navy and Third Sea Lord, who dealt with all questions of construction and design. He was one of our finest gunnery experts in 1912, and as I used when First Lord often to go out and see the initial firings of battleships before their gun-mountings were accepted from the contractors I was able to form a very high opinion of his work. Both these officers at the summit of their careers treated me with the utmost confidence, and although I differed from them and criticised severely much that was done or not done, no complaint or personal reproaches ever disturbed our association.

The question of whether the Fleet Air Arm should be under the Admiralty or the Air Ministry was hotly disputed between the two departments and services. I took the Navy view, and my advocacy of it in Parliament drew a cordial letter of thanks from the First Sea Lord, in which he entered upon the whole question of naval policy. Sir Thomas Inskip came down to see me at Chartwell, and asked for my advice on this nicely-balanced issue. I drew up for him a memorandum which was eventually adopted almost word for word by His Majesty's Government.*

*　　*　　*　　*　　*

When at last it was decided to begin building battleships again, the question of their design caused me great concern. Up to this moment practically all the capital ships of the Royal Navy had been built or designed during my administration of the Admiralty from 1911 to 1915. Only the *Nelson* and the *Rodney* were created after the First World War. I have in *The World Crisis* described all the process of rebuilding the Navy and the designing of the *Queen Elizabeth* class of fast battleships in my first tenure of the Admiralty, when I had at my disposal so much of the genius and inspiration of Lord Fisher. To this I was always able to apply my own thought, gathered from many other naval expert sources, and I still held strong opinions.

As soon as I heard that a battleship programme had been agreed to by the Cabinet, I was at once sure that our new ships should continue to mount the 16-inch gun, and that this could be achieved within 35,000 tons displacement—the Treaty limit, which we alone rigidly respected—by three triple 16-inch-gun turrets. I had several talks and some correspondence with Sir Samuel Hoare, and as I was not convinced by the arguments I heard I began to ask questions in the House about the relative weight of broadsides from 14-inch- and 16-inch-gunned ships. For my private information the following figures were given:

14-inch 9-gun broadside	..	..	..	6.38 tons
16-inch 9-gun broadside	..	..	..	9.55 tons

The figure for the 16-inch gun is based, not on the existing 16-inch

* See Appendix B.

gun of H.M.S. *Nelson*, but on a hypothetical 16-inch gun of the type which the Americans have in mind for their new capital ships.

I was deeply impressed by the superior weight of the 16-inch broadside. I therefore wrote to Sir Samuel Hoare:

Sir,   1.VIII.36

It is very civil of you to attach any importance to my opinion, and *prima facie* there is a case. I cannot answer the argument about the long delay involved. Once again we alone are injured by treaties. I cannot doubt that a far stronger ship could be built with three triple 16-inch-gun turrets in a 35,000-ton hull, than any combination of 14-inch. Not only would she be a better ship, but she would be rated a better ship and a more powerful token of naval power by everyone, including those who serve in her. Remember the Germans get far better results out of their guns per calibre than we do. They throw a heavier shell farther and more accurately. The answer is a big punch. Not only is there an enormous increase in the weight of broadside, but in addition the explosive charge of a 16-inch shell must be far larger than that of a 14-inch. If you can get through the armour it is worth while doing something inside with the explosion.

Another aspect is the number of turrets. What a waste to have four turrets, which I suppose weigh 2,000 tons each, when three will give a bigger punch! With three turrets the centralisation of armour against gunfire and torpedoes can be much more intense, and the decks all the more clear for the anti-aircraft batteries. If you ask your people to give you a legend for a 16-inch-gun ship, I am persuaded they would show you decidedly better proportions than could be achieved at 14-inch. Of course, there may be an argument about gunnery control, the spread of shot, etc., with which I am not familiar. Still, I should have thought that the optimum gunnery effect could be reached with salvos of four and five alternately.

Nothing would induce me to succumb to 14-inch if I were in your shoes. The Admiralty will look rather silly if they are committed to two 14-inch-gun ships and both Japan and the United States go in for 16-inch a few months later. I should have thought it was quite possible to lie back and save six months in construction. It is terrible deliberately to build British battleships costing £7,000,000 apiece that are not the strongest in the world! As old Fisher used to say, "The British Navy always travels first class."

However, these are only vaticinations! I went through all this in bygone years, or I would not venture to obtrude it on you. I will get in touch with Chatfield as you suggest.

The First Lord in no way resented my arguments and a considerable correspondence took place between us; and I also had several conversations with him and the First Sea Lord. Before leaving the Admiralty at the end of May 1937 Sir Samuel Hoare sent me two memoranda prepared by the Naval Staff, one dealing with battleships and the other with cruisers. The Admiralty case about battleship design was that since the Washington Treaty Great Britain had continually pressed for a reduction in displacement and size of guns on grounds of economy. It had not been possible when the new British battleships were at last sanctioned in 1936 to throw over the Treaty limitations of the 14-inch gun or the 35,000-ton ship. The design of the battleships of the *King George V* class had to be started before it could become known whether other Powers would accept these limits as governing the immediate future. The turrets of the *King George V* class had in fact been ordered in May 1936. Had the Admiralty delayed decision upon design until April 1937 only two ships would be available by 1941, instead of five. Should foreign countries go beyond the Washington limits the designs for the 1938 programme ships, which would be complete in 1942, could take a larger scope.

If however we should eventually be forced to go to fully-balanced 16-inch-gun ships and not sacrifice any of the structural strength and other characteristics of the *King George V* class, there would be considerable increase in displacement. The resultant vessels could not pass through the Panama Canal, and we should have to enlarge our docks as well as add to the cost of each ship. The Admiralty concurred with my preference for a ship of nine 16-inch guns in three turrets, rather than one with ten 14-inch guns in four turrets. All their battleship designs were of ships having three "multi-gun turrets".

After studying this long and massive paper I recognised that we could not face the delay involved in putting larger guns in the first five battleships. The decision was irrevocable. I urged however that the designs for the larger guns and turrets should be completed as a precaution, and that the tools and appliances necessary to adapt the gun-plants, etc., to the larger calibre should actually be made, even at considerable expense.

In my discussions with the Admiralty about battleship design I had not appreciated the fact that they had designed and were in process of drawing out quadruple turrets for the 14-inch gun, thus achieving a total of twelve guns. Had I realised this I should have been forced to reconsider my view. The expression "multi-gun turrets" led to this misunderstanding on my part. Three quadruple turrets would have avoided many of the evils which I saw in a four-turret ship, and twelve 14-inch guns, though not the equal of nine 16-inch, were a considerable improvement in weight of metal.

However, the sequel of the Admiralty policy was unfortunate. Serious delays took place in the designing of the entirely novel quadruple turret for the 14-inch gun. No sooner had work been started upon this than the Admiralty Board decided to change the third turret superposed forward for a two-gun turret. This of course meant redesigning the two or three thousand parts which composed these amazing pieces of mechanism, and a further delay of at least a year in the completion of the *King George V* and *Prince of Wales* was caused by this change of plan. Moreover, our new ships were now reduced to ten guns, and all my arguments about the inferiority of their broadsides compared with 16-inch-gun ships resumed their force. Meanwhile the Americans got round the problem of putting three triple 16-inch turrets into a 35,000-ton hull. The French and the Germans chose the 15-inch gun, the French mounting eight guns in two quadruple turrets, and the Germans eight in four twin turrets. The Germans, however, like the Japanese, had no intention of being bound by any treaty limitations, and the *Bismarck's* displacement exceeded 45,000 tons, with all the advantages which thus accrued. We alone, having after all these years at last decided to build five battleships, on which the life of the Navy and the maintenance of sea-power were judged to depend, went back from the 16-inch gun to the 14-inch, while others increased their calibres. We therefore produced a series of vessels, each taking five years to build, which might well have carried heavier gun-power.

*   *   *   *   *

On June 15, 1938, the First Sea Lord took me down to Portland to show me the Asdics. This was the name which described the system of groping for submarines below the surface by means of sound waves through the water which echoed back from any steel structure they met. From this echo the position of the submarine could be fixed with some accuracy. We were on the threshold of this development at the end of the First World War.

We slept on board the flagship, and had a long talk with Sir Charles Forbes, the Commander-in-Chief. All the morning was spent at the Anti-Submarine School, and in about four hours I received a very full account. We then went to sea in a destroyer, and during the afternoon and evening an exercise of great interest was conducted for my benefit. A number of submarines were scattered about in the offing. Standing on the bridge of the destroyer which was using the Asdic, with another destroyer half a mile away, in constant intercourse, I could see and hear the whole process, which was the Sacred Treasure of the Admiralty, and in the culture of which for a whole generation they had faithfully persevered. Often I had criticised their policy. No doubt on this occasion I overrated, as they did, the magnitude of their achievement, and forgot for a moment how broad are the seas. Nevertheless, if this twenty years' study had not been pursued with large annual expenditure and thousands of highly-skilled officers and men employed and trained with nothing to show for it—all quite unmentionable—our problem in dealing with the U-boat, grievous though it proved, might well have found no answer but defeat.

To Chatfield I wrote:

I have reflected constantly on all that you showed me, and I am sure the nation owes the Admiralty, and those who have guided it, an inestimable debt for the faithful effort sustained over so many years, which has, as I feel convinced, relieved us of one of our great dangers.

What surprised me was the clarity and force of the [Asdic] indications. I had imagined something almost imperceptible, certainly vague and doubtful. I never imagined that I should hear one of those creatures asking to be destroyed. It is a marvellous system and achievement.

The Asdics did not conquer the U-boat; but without the Asdics the U-boat would not have been conquered.

CHAPTER X

SANCTIONS AGAINST ITALY

1935

A Second Heavy Stroke – Adowa Memories – A Time of Caution – A Talk at the Foreign Office – The Peace Ballot – British Naval Strength in the Mediterranean – Sir Samuel Hoare's Speech at Geneva and British Naval Movements – My Speech to the City Carlton Club – Mussolini Invades Abyssinia – Strong Reaction in Britain; ·Mr. Lansbury Resigns the Leadership of the Labour Parliamentary Party – Sham Sanctions – Mr. Baldwin Resolved on Peace – The Conservative Party Conference – His Conduct of the Election – His Great Majority – The Hoare-Laval Agreement – The Parliamentary Convulsion – I Stay Abroad – The Effect upon Europe of Mussolini's Conquest of Abyssinia.

WORLD peace now suffered its second heavy stroke. The loss by Britain of air parity was followed by the transference of Italy to the German side. The two events combined enabled Hitler to advance along his predetermined deadly course. We have seen how helpful Mussolini had been in the protection of Austrian independence, with all that it implied in Central and South-eastern Europe. Now he was to march over to the opposite camp. Nazi Germany was no longer to be alone. One of the principal Allies of the First World War would soon join her. The gravity of this downward turn in the balance of safety oppressed my mind.

Mussolini's designs upon Abyssinia were unsuited to the ethics of the twentieth century. They belonged to those dark ages when white men felt themselves entitled to conquer yellow, brown, black, or red men, and subjugate them by their superior strength and weapons. In our enlightened days, when crimes and cruelties have been committed from which the savages of former times would have recoiled, or of which they would at least have been incapable, such conduct was at once obsolete and reprehensible. Moreover, Abyssinia was a member of the League of Nations. By a curious inversion it was Italy who had in 1923 pressed for her inclusion, and Britain who had opposed it. The British view was that the character of the Ethiopian Government and the conditions prevailing in that wild land of tyranny, slavery, and tribal war were not consonant with membership of the League. But the Italians had had their way, and Abyssinia was a member of the League, with all its rights and such securities as it could offer. Here indeed was a testing case for the instrument of world government upon which the hopes of all good men were founded.

The Italian Dictator was not actuated solely by desire for territorial gains. His rule, his safety, depended upon prestige. The humiliating defeat which Italy had suffered forty years before at Adowa, and the mockery of the world when an Italian army had not only been destroyed or captured but shamefully mutilated, rankled in the minds of all Italians. They had seen how Britain had after the passage of years avenged both Khartoum and Majuba. To proclaim their manhood by avenging Adowa meant almost as much in Italy as the recovery of Alsace-Lorraine in France. There seemed no way in which Mussolini could more easily or at less risk and cost consolidate his own power or, as he saw it, raise the authority of Italy in Europe than by wiping out the stain of bygone years and adding Abyssinia to the recently-built Italian Empire. All such thoug'ts were wrong and evil, but since it is always wise to try to understand another country's point of view they may be recorded.

In the fearful struggle against rearming Nazi Germany which I could feel approaching with inexorable strides, I was most reluctant to see Italy estranged, and even driven into the opposite camp. There was no doubt that the attack by one member of the League of Nations upon another at this juncture, if not resented, would be finally destructive of the League as a factor for welding together the forces which could alone control the might of resurgent Germany and the awful Hitler menace. More could perhaps be got out of the vindicated majesty of the League than Italy could ever give, withhold, or transfer. If therefore the League were prepared to use the united strength of all its members to curb

Mussolini's policy, it was our bounden duty to take our share and play a faithful part. There seemed in all the circumstances no obligation upon Britain to take the lead herself. She had a duty to take account of her own weakness caused by the loss of air parity, and even more of the military position of France, in the face of German rearmament. One thing was clear and certain. Half-measures were useless for the League, and pernicious to Britain if she assumed its leadership. If we thought it right and necessary for the law and welfare of Europe to quarrel mortally with Mussolini's Italy, we must also strike him down. The fall of the lesser Dictator might combine and bring into action all the forces—and they were still overwhelming—which would enable us to restrain the greater Dictator, and thus prevent a second German war.

These general reflections are a prelude to the narrative of this chapter.

★　　★　　★　　★　　★

Ever since the Stresa Conference Mussolini's preparations for the conquest of Abyssinia had been apparent. It was evident that British opinion would be hostile to such an act of Italian aggression. Those of us who saw in Hitler's Germany a danger not only to peace but to survival dreaded this movement of a first-class Power, as Italy was then rated, from our side to the other. I remember a dinner at which Sir Robert Vansittart and Mr. Duff Cooper, then only an Under-Secretary, were present, at which this adverse change in the balance of Europe was clearly foreseen. The project was mooted of some of us going out to see Mussolini in order to explain to him the inevitable effects which would be produced in Great Britain. Nothing came of this; nor would it have been of any good. Mussolini, like Hitler, regarded Britannia as a frightened, flabby old woman, who at the worst would only bluster, and was anyhow incapable of making war. Lord Lloyd, who was on friendly terms with him, noted how he had been struck by the Joad resolution of the Oxford undergraduates in 1933 refusing to "fight for King and Country".

★　　★　　★　　★　　★

In Parliament I expressed my misgivings on July 11:

We seem to have allowed the impression to be created that we were ourselves coming forward as a sort of bell-wether or fugleman to lead opinion in Europe against Italy's Abyssinian designs. It was even suggested that we would act individually and independently. I am glad to hear from the Foreign Secretary that there is no foundation for that. We must do our duty, but we must do it with other nations only in accordance with the obligations which others recognise as well. We are not strong enough to be the law-giver and the spokesman of the world. We will do our part, but we cannot be asked to do more than our part in these matters. . . .

As we stand to-day there is no doubt that a cloud has come over the old friendship between Great Britain and Italy, a cloud which, it seems to me, may very easily not pass away, although undoubtedly it is everyone's desire that it should. It is an old friendship, and we must not forget, what is a little-known fact, that at the time Italy entered into the Triple Alliance in the last century she stipulated particularly that in no circumstances would the obligations under the Alliance bring her into armed conflict with Great Britain.

* * * * *

In August the Foreign Secretary invited me and also the Opposition Party leaders to visit him separately at the Foreign Office, and the fact of these consultations was made public by the Government. Sir Samuel Hoare told me of this growing anxiety about Italian aggression against Abyssinia, and asked me how far I should be prepared to go against it. Wishing to know more about the internal and personal situation at the Foreign Office under diarchy before replying, I asked about Eden's view. "I will get him to come," said Hoare, and in a few minutes Anthony arrived, smiling and in the best of tempers. We had an easy talk. I said I thought the Foreign Secretary was *justified in going as far with the League of Nations against Italy as he could carry France*; but I added that he ought not to put any pressure upon France, because of her military convention with Italy and her German preoccupation; and that in the circumstances I did not expect France would go very far. I then spoke of the Italian divisions on the Brenner Pass, of the unguarded southern front of France, and other military aspects.

Generally I strongly advised the Ministers not to try to take a leading part or to put themselves forward too prominently. In this I was of course oppressed by my German fears and the conditions to which our defences had been reduced.

* * * * *

In the early months of 1935 there was organised a Peace Ballot for collective security and for upholding the Covenant of the League of Nations. This scheme received the blessing of the League of Nations Union, but was sponsored by a separate organisation largely supported by the Labour and Liberal Parties. The following were the questions put:

THE PEACE BALLOT

1. Should Great Britain remain a member of the League of Nations?
2. Are you in favour of an all-round reduction of armaments by international agreement?
3. Are you in favour of the all-round abolition of national military and naval aircraft by international agreement?
4. Should the manufacture and sale of armaments for private profit be prohibited by international agreement?
5. Do you consider that if a nation insists on attacking another the other nations should combine to compel it to stop by:
   (*a*) economic and non-military measures,
   (*b*) if necessary military measures?

It was announced on June 27 that over eleven million persons had subscribed their names affirmatively to this. The Peace Ballot seemed at first to be misunderstood by Ministers. Its name overshadowed its purpose. It of course combined the contradictory propositions of reduction of armaments and forcible resistance to aggression. It was regarded in many quarters as a part of the Pacifist campaign. On the contrary, Clause 5 affirmed a positive and courageous policy which could at this time have been followed with an overwhelming measure of national support. Lord Cecil and other leaders of the League of Nations Union were, as this clause declared, and as events soon showed, willing, and indeed resolved, to go to war in a righteous cause, provided that all necessary action was taken under the auspices of the League of Nations. Their evaluation of the facts underwent considerable changes in the next few months. Indeed, within a year I was working with them in harmony upon the policy which I described as "Arms and the Covenant".

*   *   *   *   *

As the summer drew on the movement of Italian troopships through the Suez Canal was continuous, and considerable forces and supplies were assembled along the eastern Abyssinian frontier. Suddenly an extraordinary and to me, after my talks at the Foreign Office, a quite unexpected event occurred. On August 24 the Cabinet resolved and declared that Britain would uphold its obligation under its treaties and under the Covenant of the League. This produced an immediate crisis in the Mediterranean, and I thought it right, since I had been so recently consulted, to ask the Foreign Secretary to reassure me about the naval situation.

Mr. Churchill to Sir Samuel Hoare   *August 25, 1935*
I am sure you will be on your guard against the capital fault of letting diplomacy get ahead of naval preparedness. We took care about this in 1914.

Where are the fleets? Are they in good order? Are they adequate? Are they capable of rapid and complete concentration? Are they safe? Have they been formally warned to take precautions? Remember you are putting extreme pressure upon a Dictator who may get into desperate straits. He may well measure your corn by his bushel. He may at any moment in the next fortnight credit you with designs far beyond what the Cabinet at present harbour. While you are talking judicious, nicely-graded formulas he may act with violence. Far better put temptation out of his way.

I see by the newspapers that the Mediterranean Fleet is leaving Malta for the Levant. Certainly it is wise [for the Fleet] to quit Malta, which, I understand, is totally unprovided with anti-aircraft defence. The Mediterranean Fleet based at Alexandria, etc., is on paper—that is all we are justified in going by—far weaker than the Italian Navy. I spent some time to-day looking up the cruiser and flotilla construction of the two countries since the war. It seems to me that you have not half the strength of Italy in modern cruisers and destroyers, and still less in modern submarines. Therefore it seems to me that very searching questions should be asked of the Admiralty *now* as to the position of this British Fleet in the Levant. It is enough to do us grievous loss. Is it enough to defend itself? It is more than three thousand miles from reinforcement by the Atlantic and Home Fleets. Much might happen before these could effect a junction. I do not, indeed I dare not, doubt but that the Admiralty have studied the dispositions with vigilance. I hope you will satisfy yourself that their answers to these suggestions are adequate.

I heard some time ago talk about a plan for evacuating the Mediterranean in the event of a war with Italy and holding only the Straits of Gibraltar and the Red Sea. The movement of the Mediterranean Fleet to the Levant looks like a piece of this policy. If so I hope it has been thought out. If we abandon the Mediterranean while in a state of war or quasi-war with Italy there is nothing to prevent Mussolini landing in Egypt in force and seizing the Canal. Nothing but France. Is the Admiralty sure of France in such a contingency?

George Lloyd, who is with me, thinks I ought to send you this letter in view of the hazards of the situation. I do not ask you for a detailed answer; but we should like your assurance that you have been satisfied with the Admiralty dispositions.

The Foreign Secretary replied on August 27:

You may rest assured that all the points you have mentioned have been, and are being, actively discussed. I am fully alive to the kind of risks that you mention, and I will do my best to see that they are not ignored. Please have no hesitation in sending me any suggestions or warnings that you think necessary. You know as well as anyone the risks of a situation such as this, and you also know as well as anyone, at least outside the Government, the present state of our Imperial defences.

*   *   *   *   *

Mr. Eden, Minister for League of Nations Affairs and almost co-equal of the Foreign Secretary, had already been for some weeks at Geneva, where he had rallied the Assembly to a policy of "Sanctions" against Italy if she invaded Abyssinia. The peculiar office to which he had been appointed made him by its very nature concentrate upon the Abyssinian question with an emphasis which outweighed other aspects. "Sanctions" meant the cutting off from Italy of all financial aid and of economic supplies, and the giving of all such assistance to Abyssinia. To a country like Italy, dependent for so many commodities needed in war upon unhampered imports from overseas, this was indeed a formidable deterrent. Eden's zeal and address and the principles which he proclaimed dominated the Assembly. On September 11 the Foreign Secretary, Sir Samuel Hoare, having arrived at Geneva, himself addressed them:

I will begin by reaffirming the support of the League by the Government I represent and the interest of the British people in collective security. . . . The ideas enshrined in the Covenant and in particular the aspiration to establish the rule of law in international affairs have become a part of our national conscience. It is to the principles of the League and not to any particular manifestation that the British nation has demonstrated its adherence. Any other view is at once an under-estimation of our good faith and an imputation upon our sincerity. In conformity with its precise and explicit obligations the League stands, and my country stands with it, for the collective maintenance of the Covenant in its entirety, and particularly for steady and collective resistance to all acts of unprovoked aggression.

In spite of my anxieties about Germany, and little as I liked the way our affairs were handled, I remember being stirred by this speech when I read it in Riviera sunshine. It aroused everyone, and reverberated throughout the United States. It united all those forces in Britain which stood for a fearless combination of righteousness and strength. Here at least was a policy. If only the orator had realised what tremendous powers he held unleashed in his hand at that moment he might indeed for a while have led the world.

These declarations gathered their validity from the fact that they had behind them, like many causes which in the past have proved vital to human progress and freedom, the British Navy. For the first and the last time the League of Nations seemed to have at its disposal a secular arm. Here was the international police force upon the ultimate authority of which all kinds of diplomatic and economic pressures and persuasion could be employed. When on September 12, the very next day, the battle-cruisers *Hood* and *Renown*, accompanied by the Second Cruiser Squadron and a destroyer flotilla, arrived at Gibraltar, it was assumed on all sides that Britain would back her words with deeds. Policy and action alike gained immediate and overwhelming support at home. It was taken for granted, not unnaturally, that neither the declaration nor the movement of warships would have been made without careful expert calculation by the Admiralty of the fleet or fleets required in the Mediterranean to make our undertakings good.

At the end of September I had to make a speech at the City Carlton Club, an orthodox body of some influence. I tried to convey a warning to Mussolini, which I believe he read:

To cast an army of nearly a quarter of a million men, embodying the flower of Italian manhood, upon a barren shore two thousand miles from home, against the goodwill of the whole world and without command of the sea, and then in this position embark upon what may well be a series of campaigns against a people and in regions which no conqueror in four thousand years ever thought it worth while to subdue, is to give hostages to fortune unparalleled in all history.*

Sir Austen Chamberlain wrote to me agreeing with this speech, and I replied:

October 1, 1935

I am glad you approve the line I took about Abyssinia; but I am very unhappy. It would be a terrible deed to smash up Italy, and it will cost us dear. How strange it is that after all these years of begging France to make it up with Italy we are now forcing her to choose between Italy and ourselves! I do not think we ought to have taken the lead in such a vehement way. If we had felt so strongly on the subject we should have warned Mussolini two months before. The sensible course would have been gradually to strengthen the Fleet in the Mediterranean during the early summer, and so let him see how grave the matter was. Now what can he do? I expect a very serious rise of temperature when the fighting [in Abyssinia] begins.

*   *   *   *   *

In October Mussolini, undeterred by belated British naval movements, launched the Italian armies upon the invasion of Abyssinia. On the 10th, by the votes of fifty sovereign States to one, the Assembly of the League resolved to take collective measures against Italy, and a Committee of Eighteen was appointed to make further efforts for a peaceful solution. Mussolini, thus confronted, made a clear-cut statement marked by deep shrewdness. Instead of saying, "Italy will meet sanctions with war," he said, "Italy will meet them with discipline, with frugality, and with sacrifice." At the same time however he intimated that *he would not tolerate the imposition of any sanctions which hampered his invasion of Abyssinia.* If that enterprise were endangered he would go to war with whoever stood in his path. "Fifty nations!" he said. "Fifty nations, led by one!" Such was the position in the weeks which preceded the Dissolution of Parliament in Britain and the General Election, which was now constitutionally due.

*   *   *   *   *

Bloodshed in Abyssinia, hatred of Fascism, the invocation of Sanctions by the League, produced a convulsion within the British Labour Party. Trade unionists, among whom Mr. Ernest Bevin

* See also my conversation with Count Grandi, Appendix A.

was outstanding, were by no means pacifist by temperament. A very strong desire to fight the Italian Dictator, to enforce Sanctions of a decisive character, and to use the British Fleet, if need be, surged through the sturdy wage-earners. Rough and harsh words were spoken at excited meetings. On one occasion Mr. Bevin complained that "he was tired of having George Lansbury's conscience carted about from conference to conference". Many members of the Parliamentary Labour Party shared the trade union mood. In a far wider sphere, all the leaders of the League of Nations Union felt themselves bound to the cause of the League. Clause 5 of their Peace Ballot was plainly involved. Here were principles in obedience to which lifelong humanitarians were ready to die, and if to die, also to kill. On October 8 Mr. Lansbury resigned his leadership of the Labour Parliamentary Party, and Major Attlee, who had a fine war record, reigned in his stead.

*　*　*　*　*

But this national awakening was not in accord with Mr. Baldwin's outlook or intentions. It was not till several months after the election that I began to understand the principles upon which Sanctions were founded. The Prime Minister had declared that Sanctions meant war; secondly, he was resolved there must be no war; and, thirdly, he decided upon Sanctions. It was evidently impossible to reconcile these three conditions. Under the guidance of Britain and the pressures of Laval the League of Nations Committee, charged with devising Sanctions, kept clear of any that would provoke war. A large number of commodities, some of which were war materials, were prohibited from entering Italy, and an imposing schedule was drawn up. But oil, without which the campaign in Abyssinia could not have been maintained, continued to enter freely, because it was understood that to stop it meant war. Here the attitude of the United States, not a member of the League of Nations, and the world's main oil supplier, though benevolent, was uncertain. Moreover, to stop it to Italy involved also stopping it to Germany. The export of aluminium to Italy was strictly forbidden; but aluminium was almost the only metal that Italy produced in quantities beyond her own needs. The importation of scrap iron and iron ore into Italy was sternly vetoed in the name of public justice. But as the Italian metallurgical industry made but little use of them, and as steel billets and pig iron were not interfered with, Italy suffered no hindrance. Thus the measures pressed with so great a parade were not real sanctions to paralyse the aggressor, but merely such half-hearted sanctions as the aggressor would tolerate, because in fact, though onerous, they stimulated Italian war spirit. The League of Nations therefore proceeded to the rescue of Abyssinia on the basis that nothing must be done to hamper the invading Italian armies. These facts were not known to the British public at the time of the election. They earnestly supported the policy of Sanctions, and believed that this was a sure way of bringing the Italian assault upon Abyssinia to an end.

Still less did His Majesty's Government contemplate the use of the Fleet. All kinds of tales were told of Italian suicide squadrons of dive-bombers which would hurl themselves upon the decks of our ships and blow them to pieces. The British fleet which was lying at Alexandria had now been reinforced. It could by a gesture have turned back Italian transports from the Suez Canal, and would as a consequence have had to offer battle to the Italian Navy. We were told that it was not capable of meeting such an antagonist. I had raised the question at the outset, but had been reassured. Our battleships of course were old, and it now appeared that we had no aircraft cover and very little anti-aircraft ammunition. It transpired however that the Admiral commanding resented the suggestion attributed to him that he was not strong enough to fight a fleet action. It would seem that before taking their first decision to oppose the Italian aggression His Majesty's Government should carefully have examined ways and means, and also made up their minds.

There is no doubt, on our present knowledge, that a bold decision would have cut the Italian communications with Ethiopia, and that we should have been successful in any naval battle which might have followed. I was never in favour of isolated action by Great Britain, but having gone so far it was a grievous deed to recoil. Moreover, Mussolini would never have dared to come to grips with a resolute British Government. Nearly the whole of the world was against him, and he would have had to risk his *régime* upon a single-handed war with Britain, in which a fleet action in the Mediterranean would be the early and decisive test. How could Italy have fought this war? Apart from a limited advantage in modern light cruisers, her Navy was but a fourth the size of the British. Her numerous conscript Army, which was vaunted in millions, could not come into action. Her air-power was in quantity and quality far below even our modest establishments. She would instantly have been blockaded. The Italian armies in Abyssinia would have famished for supplies and ammunition. Germany could as yet give no effective help. If ever there was an opportunity of striking a decisive blow in a generous cause with the minimum of risk, it was here and now. The fact that the nerve of the British Government was not equal to the occasion can be excused only by their sincere love of peace. Actually it played a part in leading to an infinitely more terrible war. Mussolini's bluff succeeded, and an important spectator drew far-reaching conclusions from the fact. Hitler had long resolved on war for German aggrandisement. He now formed a view of Great Britain's degeneracy, which was only to be changed too late for peace and too late for him. In Japan also there were pensive spectators.

*　　*　　*　　*　　*

The two opposite processes of gathering national unity on the burning issue of the hour and the clash of party interests inseparable from a General Election moved forward together. This was greatly to the advantage of Mr. Baldwin and his supporters. "The League of Nations would remain as heretofore the keystone of British foreign policy": so ran the Government's election manifesto. "The prevention of war and the establishment of peace in the world must always be the most vital interest of the British people, and the League is the instrument which has been framed and to which we look for the attainment of these objects. We shall therefore continue to do all in our power to uphold the Covenant and to maintain and increase the efficiency of the League. In the present unhappy dispute between Italy and Abyssinia *there will be no wavering in the policy we have hitherto pursued.*"

The Labour Party, on the other hand, was much divided. The majority was pacifist, but Mr. Bevin's active campaign commanded many supporters among the masses. The official leaders therefore tried to give general satisfaction by pointing opposite ways at once. On the one hand they clamoured for decisive action against the Italian Dictator; on the other they denounced the policy of rearmament. Thus Mr. Attlee in the House of Commons on October 22: "We want effective sanctions, effectively applied.

We support economic sanctions. We support the League system." But then, later in the same speech: "We are not persuaded that the way to safety is by piling up armaments. We do not believe that in this [time] there is such a thing as national defence. We think that you have to go forward to disarmament and not to the piling up of armaments." Neither side usually has much to be proud of at election times. The Prime Minister himself was no doubt conscious of the growing strength behind the Government's foreign policy. He was however determined not to be drawn into war on any account. It seemed to me, viewing the proceedings from outside, that he was anxious to gather as much support as possible and use it to begin British rearmament on a modest scale.

★ ★ ★ ★ ★

The Conservative Party Conference was held at Bournemouth on the very day when Mussolini began his attack on Abyssinia and his bombs were falling on Adowa. In view of this, and not less of the now imminent General Election, we all closed our ranks as party men.

I supported a resolution which was carried unanimously:

(1) To repair the serious deficiencies in the defence forces of the Crown, and in particular, first, to organise our industry for speedy conversion to defence purposes, if need be.

(2) To make a renewed effort to establish equality in the air with the strongest foreign Air Force within striking distance of our shores.

(3) To rebuild the British Fleet and strengthen the Royal Navy, so as to safeguard our food and livelihood and preserve the coherence of the British Empire.

Hitherto in these years I had not desired office, having had so much of it, and being opposed to the Government on their Indian policy. But with the passage of the India Bill, which was to take some years to come into force, this barrier had fallen away. The growing German menace made me anxious to lay my hands upon our military machine. I could now feel very keenly what was coming. Distracted France and timid, peace-loving Britain would soon be confronted with the challenge of the European Dictators. I was in sympathy with the changing temper of the Labour Party. Here was the chance of a true National Government. It was understood that the Admiralty would be vacant, and I wished very much to go there should the Conservatives be returned to power. I was of course well aware that this desire was not shared by several of Mr. Baldwin's principal colleagues. I represented a policy, and it was known that I should strive for it whether from without or from within. If they could do without me they would certainly be very glad. To some extent this depended upon their majority.

* * * * *

At the General Election the Prime Minister spoke in strong terms of the need for rearmament, and his principal speech was devoted to the unsatisfactory condition of the Navy. However, having gained all that there was in sight upon a programme of Sanctions and rearmament, he became very anxious to comfort the professional peace-loving elements in the nation and allay any fears in their breasts which his talk about naval requirements might have caused. On October 1, six weeks before the poll, he made a speech to the Peace Society at the Guildhall. In the course of this he said: "I give you my word there will be no great armaments." In the light of the knowledge which the Government had of strenuous German preparations, this was a singular promise. Thus the votes both of those who sought to see the nation prepare itself against the dangers of the future and of those who believed that peace could be preserved by praising its virtues were gained.

* * * * *

I fought my contest in the Epping Division upon the need for rearmament and upon a severe and *bona fide* policy of Sanctions. Generally speaking I supported the Government, and although many of my Conservative friends had been offended by my almost ceaseless criticism of Government measures I was returned by an ample majority. Upon the declaration of the poll I thought it right to safeguard my own position. "I take it from your vote, in view of the speeches I have made, that you desire me to exercise my independent judgment as a Member of Parliament, and, in accordance with the highest traditions of that House, to give the fruits of my knowledge and experience freely and without fear." The result of the General Election was a triumph for Mr. Baldwin. The electors accorded him a majority of two hundred and forty-seven over all other parties combined, and after five years of office he reached a position of personal power unequalled by any Prime Minister since the close of the Great War. All who had opposed him, whether on India or on the neglect of our defences, were stultified by this renewed vote of confidence, which he had gained by his skilful and fortunate tactics in home politics and by the esteem so widely felt for his personal character. Thus an Administration more disastrous than any in our history saw all its errors and shortcomings acclaimed by the nation. There was however a bill to be paid, and it took the new House of Commons nearly ten years to pay it.

* * * * *

It had been widely bruited that I should join the Government as First Lord of the Admiralty. But after the figures of his victory had been proclaimed Mr. Baldwin lost no time in announcing through the Central Office that there was no intention to include me in the Government. In this way he paid some of his debt to the pacifist deputation which he had received in the last days of the election. There was much mocking in the Press about my exclusion. But now one can see how lucky I was. Over me beat the invisible wings.

And I had agreeable consolations. I set out with my paint-box for more genial climes without waiting for the meeting of Parliament.

* * * * *

There was an awkward sequel to Mr. Baldwin's triumph, for the sake of which we may sacrifice chronology. His Foreign Secretary, Sir Samuel Hoare, travelling through Paris to Switzerland on a well-earned skating holiday, had a talk with M. Laval, still French Foreign Minister. The result of this was the Hoare-Laval pact of December 9. It is worth while to look a little into the background of this celebrated incident.

The idea of Britain leading the League of Nations against Mussolini's Fascist invasion of Abyssinia had carried the nation in one of its big swings. But once the election was over and the Ministers found themselves in possession of a majority which might give them for five years the guidance of the State many tiresome consequences had to be considered. At the root of them all lay Mr. Baldwin's "There must be no war", and also "There must be no large rearmaments". This remarkable Party Manager, having won the election on world leadership against aggression, was profoundly convinced that we must keep peace at any price.

Moreover, now from the Foreign Office came a very powerful thrust. Sir Robert Vansittart never removed his eyes for one moment from the Hitler peril. He and I were of one mind on that point. And now British policy had forced Mussolini to change sides. Germany was no longer isolated. The four Western Powers were divided two against two instead of three against one. This marked deterioration in our affairs aggravated the anxiety in France. The French Government had already made the Franco-Italian agreement of January. Following thereupon had come the military convention with Italy. It was calculated that this convention saved eighteen French divisions from the Italian front for transference to the front against Germany. In his negotiations it is certain that M. Laval had given more than a hint to Mussolini that France would not trouble herself about anything that might happen to Abyssinia. The French had a considerable case to argue with British Ministers. First, for several years we had tried to make them reduce their army, which was all they had to live upon. Secondly, the British had had a very good run in the leadership of the League of Nations against Mussolini. They had even won an election upon it; and in democracies elections are very important. Thirdly, we had made a naval agreement, supposed to be very good for ourselves, which made us quite comfortable upon the seas, apart from submarine warfare.

But what about the French front? How was it to be manned against the ever-growing German military power? Two divisions, to be sent only under many reservations, was all the British could offer for the first six months, so really they should not talk too much. Now the British Government, in a fine flow of martial, moral, and world sentiment, "fifty nations led by one", were making a mortal feud with Italy. France had much to worry about, and only very silly people, of whom there are extremely large numbers in every country, could ignore all this. If Britain had used her naval power, closed the Suez Canal, and defeated the Italian Navy in a general engagement, she would have had the right to call the tune in Europe. But, on the contrary, she had definitely declared that whatever happened she would not go to war over Abyssinia. Honest Mr. Baldwin: a triumphant vote in the constituencies; a solid Tory majority for five more years;

every aspect of righteous indignation, but no war, no war! The French therefore felt very strongly that they should not be drawn into permanent estrangement from Italy because of all the strong feeling which had suddenly surged up in England against Mussolini. Especially did they feel this when they remembered that Britain had bowed before the Italian naval challenge in the Mediterranean, and when two divisions of troops were all we could send at the outset to help France if she were invaded by Germany. One can certainly understand Monsieur Laval's point of view at this time.

Now in December a new set of arguments marched upon the scene. Mussolini, hard pressed by Sanctions, and under the very heavy threat of "fifty nations led by one", would, it was whispered, welcome a compromise on Abyssinia. Poison gas, though effective against the native Ethiopians, would certainly not elevate the name of Italy in the world. The Abyssinians were being defeated. They were not, it was said, prepared to make large concessions and wide surrenders of territory. Could not a peace be made which gave Italy what she had aggressively demanded and left Abyssinia four-fifths of her entire empire? Vansittart, who happened to be in Paris at the time the Foreign Secretary passed through, and was thus drawn into the affair, should not be misjudged because he thought continuously of the German threat and wished to have Britain and France organised at their strongest to face this major danger, with Italy in their rear a friend and not a foe.

But the British nation from time to time gives way to waves of crusading sentiment. More than any other country in the world, it is at rare intervals ready to fight for a cause or a theme, just because it is convinced in its heart and soul that it will not get any material advantage out of the conflict. Baldwin and his Ministers had given a great uplift to Britain in their resistance to Mussolini at Geneva. They had gone so far that their only salvation before history was to go all lengths. Unless they were prepared to back words and gestures by action, it might have been better to keep out of it all, like the United States, and let things rip and see what happened. Here was an arguable plan. But it was not the plan they had adopted. They had appealed to the millions, and the unarmed, and hitherto unconcerned, millions had answered with a loud shout, overpowering all other cries,

"Yes, we will march against evil, and we will march now. Give us the weapons."

The new House of Commons was a spirited body. With all that lay before them in the next ten years, they had need to be. It was therefore with a horrible shock that, while tingling from the election, they received the news that a compromise had been made between Sir Samuel Hoare and M. Laval about Abyssinia. This crisis nearly cost Mr. Baldwin his political life. It shook Parliament and the nation to its base. Mr. Baldwin fell almost overnight from his pinnacle of acclaimed national leadership to a depth where he was derided and despised. His position in the House during these days was pitiful. He had never understood why people should worry about all these bothersome foreign affairs. They had a Conservative majority and no war. What more could they want? But the experienced pilot felt and measured the full force of the storm.

The Cabinet, on December 9, had approved the Hoare-Laval plan to partition Abyssinia between Italy and the Emperor. On the 13th the full text of the Hoare-Laval proposals was laid before the League. On the 18th the Cabinet abandoned the Hoare-Laval proposals, thus entailing the resignation of Sir Samuel Hoare. In the debate on the 19th Mr. Baldwin said:

I felt that these proposals went too far. I was not at all surprised at the expression of feeling in that direction. I was not expecting that deeper feeling that was manifest in many parts of the country on what I may call the grounds of conscience and of honour. The moment I am confronted with that, I know that something has happened that has appealed to the deepest feelings of our countrymen, that some note has been struck that brings back from them a response from the depths. I examined again all that I had done, and I felt that . . . there could not be support in this country behind those proposals even as terms of negotiation. It is perfectly obvious now that the proposals are absolutely and completely dead. This Government is certainly going to make no attempt to resurrect them. If there arose a storm when I knew I was in the right, I would let it break on me, and I would either survive it or break. If I felt after examination of myself that there was in that storm something which showed me that I had done something that was not wise or right, then I would bow to it.

The House accepted this apologia. The crisis passed. On his return from Geneva Mr. Eden was summoned to 10 Downing

Street by the Prime Minister to discuss the situation following Sir Samuel Hoare's resignation. Mr. Eden at once suggested that Sir Austen Chamberlain should be invited to take over the Foreign Office, and added that if desired he was prepared to serve under him in any capacity. Mr. Baldwin replied that he had already considered this and had informed Sir Austen himself that he did not feel able to offer the Foreign Office to him. This may have been due to Sir Austen's health. On December 22 Mr. Eden became Foreign Secretary.

*　　*　　*　　*　　*

My wife and I passed this exciting week at Barcelona. Several of my best friends advised me not to return. They said I should only do myself harm if I were mixed up in this violent conflict. Our comfortable Barcelona hotel was the rendezvous of the Spanish Left. In the excellent restaurant where we lunched and dined were always several groups of eager-faced, black-coated young men purring together with glistening eyes about Spanish politics, in which quite soon a million Spaniards were to die. Looking back, I think I ought to have come home. I might have brought an element of decision and combination to the anti-Government gatherings which would have ended the Baldwin *régime*. Perhaps a Government under Sir Austen Chamberlain might have been established at this moment. On the other hand, my friends cried, "Better stay away. Your return will only be regarded as a personal challenge to the Government." I did not relish the advice, which was certainly not flattering; but I yielded to the impression that I could do no good, and stayed on at Barcelona daubing canvases in the sunshine. Thereafter Frederick Lindemann joined me, and we cruised in a nice steamship around the eastern coasts of Spain and landed at Tangier. Here I found Lord Rothermere with a pleasant circle. He told me that Mr. Lloyd George was at Marrakesh, where the weather was lovely. We all motored thither. I lingered painting in delightful Morocco, and did not return till the sudden death of King George V on January 20.

*　　*　　*　　*　　*

The collapse of Abyssinian resistance and the annexation of the whole country by Italy produced unhelpful effects in German public opinion. Even those elements which did not approve of

166

Mussolini's policy or action admired the swift, efficient, and ruthless manner in which, as it seemed, the campaign had been conducted. The general view was that Great Britain had emerged thoroughly weakened. She had earned the undying hatred of Italy; she had wrecked the Stresa front once and for all; and her loss of prestige in the world contrasted agreeably with the growing strength and repute of the new Germany. "I am impressed," wrote one of our representatives in Bavaria, "by the note of contempt in references to Great Britain in many quarters. . . . It is to be feared that Germany's attitude in the negotiations for a settlement in Western Europe and for a more general settlement of European and extra-European questions will be found to have stiffened."

An article in the *Muenchener Zeitung* (May 16, 1936) contains some illuminating passages:

The English like a comfortable life compared with our German standards. This does not indeed mean that they are incapable of sustained efforts, but they avoid them so far as they can, without impairing their personal and national security. They also control means and wealth which have enabled them, in contrast with us, for a century or so, to increase their capital more or less automatically. . . . After the war, in which the English after some preliminary hesitation showed certainly an amazing energy, the British masters of the world thought they had at last earned a little rest. They disarmed along the whole line—in civil life even more than on land and sea. They reconciled themselves to abandoning the two-Power [naval] standard and accepted parity with America. . . . How about the Army? How about the Air Force? . . . For the land and air defence forces England needs above all men, not merely money, but also the lives of her citizens for Empire defence. Indeed, of the eleven thousand men needed for the new air programme seven thousand are lacking. Again, the small Regular Army shows a large deficiency, about one whole division, and the Territorial Army (a sort of Sunday school for amateur soldiers) is so far below its authorised numbers that it cannot in any way be considered an effective combatant force. Mr. Baldwin himself said a short time ago that he had no intention of changing the system of recruiting by the introduction of conscription.

A policy which seeks to achieve success by postponing decisions can to-day hardly hope to resist the whirlwind which is shaking Europe, and indeed the whole world. Few are the men who, upon national and not upon party grounds, rage against the spineless and ambiguous attitude of the Government, and hold them responsible for the dangers into which the Empire is being driven all unaware. The masses seem to agree with the Government that the situation will improve by marking time, and that by means of small adjustments and carefully-thought-out manœuvres the balance can once again be rectified. . . .

To-day all Abyssinia is irrevocably, fully, and finally Italian alone. This being so, neither Geneva nor London can have any doubt that only the use of extraordinary force can drive the Italians out of Abyssinia. But neither the power nor the courage to use force is at hand.

All this was only too true. His Majesty's Government had imprudently advanced to champion a great world cause. They had led fifty nations forward with much brave language. Confronted with brute facts Mr. Baldwin had recoiled. Their policy had for a long time been designed to give satisfaction to powerful elements of opinion at home rather than to seek the realities of the European situation. By estranging Italy they had upset the whole balance of Europe and gained nothing for Abyssinia. They had led the League of Nations into an utter fiasco, most damaging if not fatally injurious to its effective life as an institution.

CHAPTER XI

HITLER STRIKES

1936

*A New Atmosphere in Britain – Hitler Free to Strike – Ratification of
the Franco-Soviet Pact – The Rhineland and the Treaties of Versailles
and Locarno – Hitler Reoccupies the Rhineland, March 7 – French
Hesitation – Flandin's Visit to London – British Pacifism – Flandin
and Baldwin – Ralph Wigram's Grief – Hitler's Vindication and
Triumph – A Minister of Co-ordination of Defence – Sir Thomas
Inskip Chosen – A Blessing in Disguise – My Hopes of the League –
Eden Insists on Staff Conversations with France – German Fortifica-
tion of the Rhineland – My Warnings in Parliament – Mr. Bullitt's
Post-War Revelations – Hitler's Pledge to Austria, July 11.*

WHEN I returned at the end of January 1936 I was conscious of a new atmosphere in England. Mussolini's
conquest of Ethiopia and the brutal methods by which it
had been accomplished, the shock of the Hoare-Laval negotiations, the discomfiture of the League of Nations, the obvious
breakdown of "Collective Security", had altered the mood not
only of the Labour and Liberal Parties, but of that great body of
well-meaning but hitherto futile opinion represented by the
eleven million votes cast in the Peace Ballot only seven months
before. All these forces were now prepared to contemplate war
against Fascist or Nazi tyranny. Far from being excluded from
lawful thought, the use of force gradually became a decisive
point in the minds of a vast mass of peace-loving people, and
even of many who had hitherto been proud to be called pacifists.
But force, according to the principles which they served, could
only be used on the initiative and under the authority of the
League of Nations. Although both the Opposition parties con-

tinued to oppose all measures of rearmament, there was an immense measure of agreement open, and had His Majesty's Government risen to the occasion they could have led a united people forward into the whole business of preparation in an emergency spirit.

The Government adhered to their policy of moderation, half-measures, and keeping things quiet. It was astonishing to me that they did not seek to utilise all the growing harmonies that now existed in the nation. By this means they would enormously have strengthened themselves and have gained the power to strengthen the country. Mr. Baldwin had no such inclinations. He was ageing fast. He rested upon the great majority which the election had given him, and the Conservative Party lay tranquil in his hand.

★ ★ ★ ★ ★

Once Hitler's Germany had been allowed to rearm without active interference by the Allies and former associated Powers, a second World War was almost certain. The longer a decisive trial of strength was put off the worse would be our chances, at first of stopping Hitler without serious fighting, and as a second stage of being victorious after a terrible ordeal. In the summer of 1935 Germany had reinstituted conscription in breach of the treaties. Great Britain had condoned this, and by a separate agreement her rebuilding of a Navy, if desired with U-boats on the British scale. Nazi Germany had secretly and unlawfully created a military Air Force which, by the spring of 1935, openly claimed to be equal to the British. She was now in the second year of active munitions production after long covert preparations. Great Britain and all Europe, and what was then thought distant America, were faced with the organised might and will-to-war of seventy millions of the most efficient race in Europe, longing to regain their national glory, and driven, in case they faltered, by a merciless military, social, and party régime.

Hitler was now free to strike. The successive steps which he took encountered no effective resistance from the two liberal democracies of Europe, and, apart from their far-seeing President, only gradually excited the attention of the United States. The battle for peace which could during 1935 have been won was now almost lost. Mussolini had triumphed in Abyssinia, and had successfully defied the League of Nations and especially Great

Britain. He was now bitterly estranged from us, and had joined hands with Hitler. The Berlin-Rome Axis was in being. There was now, as it turned out, little hope of averting war, or of postponing it by a trial of strength equivalent to war. Almost all that remained open to France and Britain was to await the moment of the challenge and do the best they could.

There was, perhaps, still time for an assertion of Collective Security, based upon the avowed readiness of all members concerned to enforce the decisions of the League of Nations by the sword. The democracies and their dependent States were still actually and potentially far stronger than the dictatorships, but their position relatively to their opponents was less than half as good as it had been twelve months before. Virtuous motives, trammelled by inertia and timidity, are no match for armed and resolute wickedness. A sincere love of peace is no excuse for muddling hundreds of millions of humble folk into total war. The cheers of weak, well-meaning assemblies soon cease to echo, and their votes soon cease to count. Doom marches on.

* * * * *

Germany had, during the course of 1935, repulsed and sabotaged the efforts of the Western Powers to negotiate an Eastern Locarno. The new Reich at this moment declared itself a bulwark against Bolshevism, and for them, they said, there could be no question of working with the Soviets. Hitler told the Polish Ambassador in Berlin on December 18 that "he was resolutely opposed to any co-operation of the West with Russia". It was in this mood that he sought to hinder and undermine the French attempts to reach direct agreement with Moscow. The Franco-Soviet Pact had been signed in May, but not ratified by either party. It became a major object of German diplomacy to prevent such a ratification. Laval was warned from Berlin that if this move took place there could be no hope of any further Franco-German rapprochement. His reluctance to persevere thereafter became marked, but did not affect the event.

In January 1936, M. Flandin, the new French Foreign Minister, came to London for the funeral of King George V. On the evening of his visit he dined at Downing Street with Mr. Baldwin and Mr. Eden. The conversation turned to the future attitude of France and Britain in the event of a violation of the Locarno

Treaty by Germany. Such a step by Hitler was considered probable, as the French Government now intended to proceed with the ratification of the Franco-Soviet Pact. Flandin undertook to seek the official views of the French Cabinet and General Staff. In February at Geneva, according to his account, he informed Mr. Eden that the armed forces of France would be put at the disposal of the League in the event of a Treaty violation by Germany, and asked the British Minister for the eventual assistance of Great Britain in conformity with the clauses of Locarno.

On February 27 the French Chamber ratified the Franco-Soviet Pact, and the following day the French Ambassador in Berlin was instructed to approach the German Government and inquire upon what basis general negotiations for a Franco-German understanding could be initiated. Hitler, in reply, asked for a few days in which to reflect. At 10 o'clock on the morning of March 7 Herr von Neurath, the German Foreign Minister, summoned the British, French, Belgian, and Italian Ambassadors to the Wilhelmstrasse to announce to them a proposal for a twenty-five-year pact, a demilitarisation on both sides of the Rhine frontier, a pact limiting air forces, and non-aggression pacts to be negotiated with Eastern and Western neighbours.

*　*　*　*　*

The "demilitarised zone" in the Rhineland had been established by Articles 42, 43, and 44 of the Treaty of Versailles. These articles declared that Germany should not have or establish fortifications on the left bank of the Rhine or within fifty kilometres of its right bank. Neither should Germany have in this zone any military forces, nor hold at any time any military manœuvres, nor maintain any facilities for military mobilisation. On top of this lay the Treaty of Locarno, freely negotiated by both sides. In this treaty the signatory Powers guaranteed individually and collectively the permanence of the frontiers of Germany and Belgium and of Germany and France. Article 2 of the Treaty of Locarno promised that Germany, France, and Belgium would never invade or attack across these frontiers. Should, however, Articles 42 or 43 of the Treaty of Versailles be infringed, such a violation would constitute "an unprovoked act of aggression", and immediate action would be required from the offended signatories because of the assembling of armed forces in the demilitarised zone. Such a violation should be brought at once before the League of Nations, and the League, having established the fact of violation, must then advise the signatory Powers that they were bound to give their military aid to the Power against whom the offence had been perpetrated.

★　　★　　★　　★　　★

At noon on this same March 7, 1936, two hours after his proposal for a twenty-five-years pact, Hitler announced to the Reichstag that he intended to reoccupy the Rhineland, and even while he spoke German columns streamed across the boundary and entered all the main German towns. They were everywhere received with rejoicing, tempered by the fear of Allied action. Simultaneously, in order to baffle British and American public opinion, Hitler declared that the occupation was purely symbolic. The German Ambassador in London handed Mr. Eden similar proposals to those which Neurath in Berlin had given to the Ambassadors of the other Locarno Powers in the morning. This provided comfort for everyone on both sides of the Atlantic who wished to be humbugged. Mr. Eden made a stern reply to the Ambassador. We now know of course that Hitler was merely using these conciliatory proposals as part of his design and as a cover for the violent act he had committed, the success of which was vital to his prestige and thus to the next step in his programme.

It was not only a breach of an obligation exacted by force of arms in war, and of the Treaty of Locarno, signed freely in full peace, but the taking advantage of the friendly evacuation by the Allies of the Rhineland several years before it was due. This news caused a world-wide sensation. The French Government, under M. Sarraut, in which M. Flandin was Foreign Minister, uprose in vociferous wrath and appealed to all its allies and to the League. At this time France commanded the loyalty of the "Little Entente", namely, Czechoslovakia, Yugoslavia, and Roumania. The Baltic States and Poland were also associated with the French system. Above all France also had a right to look to Great Britain, having regard to the guarantee we had given for the French frontier against German aggression, and the pressure we had put upon France for the earlier evacuation of the Rhineland. Here if ever was the violation, not only of the Peace Treaty, but of the

Treaty of Locarno, and an obligation binding upon all the Powers concerned.

* * * * *

In France there was a hideous shock. MM. Sarraut and Flandin had the impulse to act at once by general mobilisation. If they had been equal to their task they would have done so, and thus compelled all others to come into line. It was a vital issue for France. But they appeared unable to move without the concurrence of Britain. This is an explanation, but no excuse. The issue was vital to France, and any French Government worthy of the name should have made up its own mind and trusted to the Treaty obligations. More than once in these fluid years French Ministers in their ever-changing Governments were content to find in British pacifism an excuse for their own. Be this as it may, they did not meet with any encouragement to resist the German aggression from the British. On the contrary, if they hesitated to act, their British allies did not hesitate to dissuade them. During the whole of Sunday there were agitated telephonic conversations between London and Paris. His Majesty's Government exhorted the French to wait in order that both countries might act jointly and after full consideration. A velvet carpet for retreat!

The unofficial responses from London were chilling. Mr. Lloyd George hastened to say, "In my judgment Herr Hitler's greatest crime was not the breach of a treaty, because there was provocation." He added that "he hoped we should keep our heads". The provocation was presumably the failure of the Allies to disarm themselves more than they had done. Lord Snowden concentrated upon the proposed Non-Aggression Pact, and said that Hitler's previous peace overtures had been ignored, but the peoples would not permit *this* peace offer to be neglected. These utterances may have expressed misguided British public opinion at the moment, but will not be deemed creditable to their authors. The British Cabinet, seeking the line of least resistance, felt that the easiest way out was to press France into another appeal to the League of Nations.

* * * * *

There was also great division in France. On the whole it was the politicians who wished to mobilise the Army and send an ultimatum to Hitler, and the generals who, like their German counterparts, pleaded for calm, patience, and delay. We now know of the conflicts of opinion which arose at this time between Hitler and the German High Command. If the French Government had mobilised the French Army, with nearly a hundred divisions, and its Air Force (then still falsely believed to be the strongest in Europe), there is no doubt that Hitler would have been compelled by his own General Staff to withdraw, and a check would have been given to his pretensions which might well have proved fatal to his rule. It must be remembered that France alone was at this time quite strong enough to drive the Germans out of the Rhineland, even without the aid which her own action, once begun, and the invocation of the Locarno Treaty would certainly have drawn from Great Britain. In fact she remained completely inert and paralysed, and thus lost irretrievably the last chance of arresting Hitler's ambitions without a serious war. Instead, the French Government were urged by Britain to cast their burden upon the League of Nations, already weakened and disheartened by the fiasco of Sanctions and the Anglo-German Naval Agreement of the previous year.

On Monday, March 9, Mr. Eden went to Paris, accompanied by Lord Halifax and Ralph Wigram. The first plan had been to convene a meeting of the League in Paris, but presently Wigram, on Eden's authority, was sent to tell Flandin to come to London to have the meeting of the League in England, as he would thus get more effective support from Britain. This was an unwelcome mission for the faithful official. Immediately on his return to London on March 11 he came to see me, and told me the story. Flandin himself arrived late the same night, and at about 8.30 on Thursday morning he came to my flat in Morpeth Mansions. He told me that he proposed to demand from the British Government simultaneous mobilisation of the land, sea, and air forces of both countries, and that he had received assurances of support from all the nations of the "Little Entente" and from other States. He read out an impressive list of the replies received. There was no doubt that superior strength still lay with the Allies of the former war. They had only to act to win. Although we did not know what was passing between Hitler and his generals, it was evident that overwhelming force lay on our side. There was little I could do in my detached private position, but I wished our visitor all success in bringing matters to a head and promised any assistance that was in my power. I gathered my principal associates at dinner that night to hear M. Flandin's exhortations.

Mr. Chamberlain was at this time, as Chancellor of the Exchequer, the most effective Member of the Government. His able biographer, Mr. Keith Feiling, gives the following extract from his diary: "March 12, talked to Flandin, emphasising that public opinion would not support us in sanctions of any kind. His view is that if a firm front is maintained Germany will yield without war. We cannot accept this as a reliable estimate of a mad Dictator's reaction." When Flandin urged at least an economic boycott Chamberlain replied by suggesting an international force during negotiations, agreed to a pact for mutual assistance, and declared that if by giving up a colony we could secure lasting peace he would consider it.*

Meanwhile most of the British Press, with the *Times* and the *Daily Herald* in the van, expressed their belief in the sincerity of Hitler's offers of a non-aggression pact. Austen Chamberlain, in a speech at Cambridge, proclaimed the opposite view. Wigram thought it was within the compass of his duty to bring Flandin into touch with everyone he could think of from the City, from the Press, and from the Government, and also with Lord Lothian. To all whom Flandin met at the Wigrams' he spoke in the following terms: "The whole world and especially the small nations to-day turn their eyes towards England. If England will act now she can lead Europe. You will have a policy, all the world will follow you, and thus you will prevent war. It is your last chance. If you do not stop Germany now, all is over. France cannot guarantee Czechoslovakia any more, because that will become geographically impossible. If you do not maintain the Treaty of Locarno all that will remain to you is to await a re-armament by Germany, against which France can do nothing. If you do not stop Germany by force to-day war is inevitable, even if you make a temporary friendship with Germany. As for myself, I do not believe that friendship is possible between France and Germany; the two countries will always be in tension, Nevertheless, if you abandon Locarno I shall change my policy, for there will be nothing else to do." These were brave words; but action would have spoken louder.

Lord Lothian's contribution was: "After all, they are only going

* Keith Feiling, *Life of Neville Chamberlain.*

into their own back-garden." This was a representative British view.

*   *   *   *   *

When I heard how ill things were going, and after a talk with Wigram, I advised M. Flandin to demand an interview with Mr. Baldwin before he left. This took place at Downing Street. The Prime Minister received M. Flandin with the utmost courtesy. Mr. Baldwin explained that although he knew little of foreign affairs he was able to interpret accurately the feelings of the British people. And they wanted peace. M. Flandin says that he rejoined that the only way to ensure this was to stop Hitlerite aggression while such action was still possible. France had no wish to drag Great Britain into war; she asked for no practical aid, and she would herself undertake what would be a simple police operation, as, according to French information, the German troops in the Rhineland had orders to withdraw if opposed in a forcible manner. Flandin asserts that he said that all that France asked of her Ally was a free hand. This is certainly not true. How could Britain have restrained France from action to which, under the Locarno Treaty, she was legally entitled? The British Prime Minister repeated that his country could not accept the risk of war. He asked what the French Government had resolved to do. To this no plain answer was returned. According to Flandin,* Mr. Baldwin then said: "You may be right, but if there is *even one chance in a hundred* that war would follow from your police operation I have not the right to commit England." And after a pause he added: "England is not in a state to go to war." There is no confirmation of this. M. Flandin returned to France convinced, first that his own divided country could not be united except in the presence of a strong will-power in Britain, and secondly that, so far from this being forthcoming, no strong impulse could be expected from her. Quite wrongly he plunged into the dismal conclusion that the only hope for France was in an arrangement with an ever more aggressive Germany.

Nevertheless, in view of what I saw of Flandin's attitude during these anxious days, I felt it my duty, in spite of his subsequent lapses, to come to his aid, so far as I was able, in later years. I used my power in the winter of 1943–44 to protect him when he was

* Pierre-Étienne Flandin, *Politique Française*, 1919–40, pp. 207–8.

arrested in Algeria by the de Gaulle Administration. In this I invoked and received active help from President Roosevelt. When after the war Flandin was brought to trial, my son Randolph, who had seen much of Flandin during the African campaign, was summoned as a witness, and I am glad to think that his advocacy, and also a letter which I wrote for Flandin to use in his defence, were not without influence in procuring the acquittal which he received from the French tribunal. Weakness is not treason, though it may be equally disastrous. Nothing however can relieve the French Government of their prime responsibility. Clemenceau or Poincaré would have left Mr. Baldwin no option.

★　　★　　★　　★　　★

The British and French submission to the violations of the Treaties of Versailles and Locarno involved in Hitler's seizure of the Rhineland was a mortal blow to Wigram. "After the French delegation had left," wrote his wife to me, "Ralph came back, and sat down in a corner of the room where he had never sat before, and said to me, 'War is now *inevitable*, and it will be the most terrible war there has ever been. I don't think I shall see it, but you will. Wait now for bombs on this little house.'★ I was frightened at his words, and he went on, 'All my work these many years has been no use. I am a failure. I have failed to make the people here realise what is at stake. I am not strong enough, I suppose. I have not been able to make them understand. Winston has always, always understood, and he is strong and will go on to the end.'"

My friend never seemed to recover from this shock. He took it too much to heart. After all, one can always go on doing what one believes to be his duty, and running ever greater risks till knocked out. Wigram's profound comprehension reacted on his sensitive nature unduly. His untimely death in December 1936 was an irreparable loss to the Foreign Office, and played its part in the miserable decline of our fortunes.

★　　★　　★　　★　　★

When Hitler met his generals after the successful reoccupation of the Rhineland he was able to confront them with the falsity of their fears and prove to them how superior his judgment o₁

★ It was actually smitten.

"intuition" was to that of ordinary military men. The generals bowed. As good Germans they were glad to see their country gaining ground so rapidly in Europe and its former adversaries so divided and tame. Undoubtedly Hitler's prestige and authority in the supreme circle of German power was sufficiently enhanced by this episode to encourage and enable him to march forward to greater tests. To the world he said: "All Germany's territorial ambitions have now been satisfied."

France was thrown into incoherency, amid which fear of war, and relief that it had been avoided, predominated. The simple English were taught by their simple Press to comfort themselves with the reflection, "After all, the Germans are only going back to their own country. How should we feel if we had been kept out of, say, Yorkshire for ten or fifteen years?" No one stopped to note that the detrainment points from which the German Army could invade France had been advanced by one hundred miles. No one worried about the proof given to all the Powers of the "Little Entente" and to Europe that France would not fight, and that England would hold her back even if she would. This episode confirmed Hitler's power over the Reich, and stultified, in a manner ignominious and slurring upon their patriotism, the generals who had hitherto sought to restrain him.

★　★　★　★　★

During this exciting period my own personal fortunes were, it now appears, discussed in high quarters. The Prime Minister, under constant pressure, had decided at last to create a new Ministry—not of Defence, but of the Co-ordination of Defence. Neville Chamberlain's biographer has given some account of this. Austen Chamberlain, whose influence with the Government stood high, thought and said that it was an "immense mistake" to exclude me. Sir Samuel Hoare had returned from convalescence, and in view of the docility with which he had accepted his dismissal after the Hoare-Laval crisis he evidently had strong claims for re-employment. The Prime Minister thought it would be best for Neville Chamberlain to take the new office, and for Austen to go back to the Exchequer. Neville, who was certain to succeed Baldwin in the immediate future, declined this proposal. "The party," says Mr. Feiling, "would not have the immediate return of Hoare. If the new Ministry went to

Churchill it would alarm those Liberal and Central elements who had taken his exclusion as a pledge against militarism,* it would be against the advice of those responsible for interpreting the party's general will, and would it not when Baldwin disappeared raise a disputed succession?" For a whole month, we are told, "these niceties and gravities were well weighed".

I was naturally aware that this process was going on. In the debate of March 9 I was careful not to derogate in the slightest degree from my attitude of severe though friendly criticism of Government policy, and I was held to have made a successful speech. I did not consider the constitution of the new office and its powers satisfactory. But I would gladly have accepted the post, being confident that knowledge and experience would prevail. Apparently (according to Mr. Feiling) the German entry into the Rhineland on March 7 was decisive against my appointment. It was certainly obvious that Hitler would not like it. On the 9th Mr. Baldwin selected Sir Thomas Inskip, an able lawyer, who had the advantages of being little known himself and knowing nothing about military subjects. The Prime Minister's choice was received with astonishment by Press and public. To me this definite and as it seemed final exclusion from all share in our preparations for defence was a heavy blow.

I had to be very careful not to lose my poise in the great discussions and debates which crowded upon us, and in which I was often prominent. I had to control my feelings and appear serene, indifferent, detached. In this endeavour continuous recurrence to the safety of the country was a good and simple rule. In order to steady and absorb my mind I planned in outline a history of what had happened since the Treaty of Versailles down to the date we had reached. I even began the opening chapter, and part of what I wrote then finds its place without the need of alteration in this present book. I did not however carry this project very far because of the press of events, and also of the current literary work by which I earned my pleasant life at Chartwell. Moreover, by the end of 1936 I became absorbed in my *History of the English-Speaking Peoples*, which I actually finished before the outbreak of war and which will some day be published. Writing a long and substantial book is like having a friend and companion at your side,

---

* This was the reverse of the truth at this time. The signers of the Peace Ballot were at one with me upon armed collective security.

to whom you can always turn for comfort and amusement, and whose society becomes more attractive as a new and widening field of interest is lighted in the mind.

Mr. Baldwin certainly had good reason to use the last flickers of his power against one who had exposed his mistakes so severely and so often. Moreover, as a profoundly astute party manager, thinking in majorities and aiming at a quiet life between elections, he did not wish to have my disturbing aid. He thought, no doubt, that he had dealt me a politically fatal stroke, and I felt he might well be right. How little can we foresee the consequences either of wise or unwise action, of virtue or of malice! Without this measureless and perpetual uncertainty the drama of human life would be destroyed. Mr. Baldwin knew no more than I how great was the service he was doing me in preventing me from becoming involved in all the Cabinet compromises and shortcomings of the next three years, and from having, if I had remained a Minister, to enter upon a war bearing direct responsibility for conditions of national defence bound to prove fearfully inadequate.

This was not the first time—or indeed the last—that I have received a blessing in what was at the time a very effective disguise.

*　*　*　*　*

I still had the hope that the appeal which France had made to the League of Nations would result in bringing into being an international pressure upon Germany to carry out the decisions of the League.

France [I wrote on March 13, 1936] has taken her case before the Court, and she asks for justice there. If the Court finds that her case is just but is unable to offer her any satisfaction, the Covenant of the League of Nations will have been proved a fraud, and Collective Security a sham. If no means of lawful redress can be offered to the aggrieved party, the whole doctrine of international law and co-operation upon which the hopes of the future are based would lapse ignominiously. It would be replaced immediately by a system of alliances and groups of nations deprived of all guarantees but their own right arm. On the other hand, if the League of Nations were able to enforce its decree upon one of the most powerful countries in the world found to be an aggressor, then the authority of the League would be set upon so majestic a pedestal that it must henceforth be the accepted sovereign authority by which all the quarrels of people can be determined and controlled. Thus we might upon this occasion reach by one single bound the realisation of our most cherished dreams.

But the risk! No one must ignore it. How can it be minimised? There is a simple method: the assembly of an overwhelming force, moral and physical, in support of international law. If the relative strengths are narrowly balanced war may break out in a few weeks, and no one can measure what the course of war may be, or who will be drawn into its whirlpools, or how, if ever, they will emerge. But if the forces at the disposal of the League of Nations are four or five times as strong as those which the aggressor can as yet command the chances of a peaceful and friendly solution are very good. Therefore every nation, great or small, should play its part according to the Covenant of the League.

Upon what force can the League of Nations count at this cardinal moment? Has she sheriffs and constables with whom to sustain her judgments, or is she left alone, impotent, a hollow mockery amid the lip-serving platitudes of irresolute or cynical devotees? Strangely enough for the destiny of the world, there was never a moment or occasion when the League of Nations could command such overwhelming force. The Constabulary of the world is at hand. On every side of Geneva stand great nations, armed and ready, whose interests as well as whose obligations bind them to uphold, and in the last resort enforce, the public law. This may never come to pass again. The fateful moment has arrived for choice between the New Age and the Old.

All this language was agreeable to the Liberal and Labour forces with whom I and several of my Conservative friends were at this time working. It united Conservatives alarmed about national safety with trade unionists, with Liberals, and with the immense body of peace-minded men and women who had signed the Peace Ballot of a year before. There is no doubt that had His Majesty's Government chosen to act with firmness and resolve through the League of Nations they could have led a united Britain forward on a final quest to avert war.

*　*　*　*　*

The violation of the Rhineland was not debated till March 26. The interval was partly filled by a meeting of the Council of the League of Nations in London. As the result Germany was invited to submit to the Hague Court her case against the Franco-Soviet Pact, about which Hitler had complained, and to under-

take not to increase her troops in the Rhineland pending further
negotiations. If Germany refused this latter request, the British
and Italian Governments undertook to carry out the steps entailed
by their obligations under the Treaty of Locarno. Not much
value could be assigned to the Italian promise. Mussolini was
already in close contact with Hitler. Germany felt strong enough
to decline any conditions limiting her forces in the Rhineland.
Mr. Eden therefore insisted that Staff conversations should take
place between Great Britain, France, and Belgium to enable any
joint action which might at some future time become necessary
under the Treaty of Locarno to be studied and prepared in advance. The youthful Foreign Secretary made a courageous speech,
and carried the House with him. Austen Chamberlain and I both
spoke at length in his support. The Cabinet was lukewarm, and it
was no easy task for Eden even to procure the institution of Staff
conversations. Usually such conversations do not play any part
as diplomatic counters, and take place secretly or even informally.
Now they were the only practical outcome of three weeks' parleyings and protestations, and the only Allied reply to Hitler's
breach of the Treaty and solid gain of the Rhineland.

In the course of my speech I said:

We cannot look back with much pleasure on our foreign policy in
the last five years. They certainly have been disastrous years. God
forbid that I should lay on the Government of my own country the
charge of responsibility for the evils which have come upon the
world in that period. . . . But certainly we have seen the most depressing and alarming change in the outlook of mankind which has ever
taken place in so short a period. Five years ago all felt safe; five years
ago all were looking forward to peace, to a period in which mankind
would rejoice in the treasures which science can spread to all classes if
conditions of peace and justice prevail. Five years ago to talk of war
would have been regarded not only as a folly and a crime, but almost
as a sign of lunacy. . . .

The violation of the Rhineland is serious because of the menace to
which it exposes Holland, Belgium, and France. I listened with apprehension to what the Secretary of State said about the Germans declining even to refrain from entrenching themselves during the period
of negotiations. When there is a line of fortifications, as I suppose
there will be in a very short time, it will produce reactions on the
European situation. *It will be a barrier across Germany's front door which
will leave her free to sally out eastwards and southwards by the other doors.*

The far-reaching consequences of the fortification of the Rhine-land were only gradually comprehended in Britain and the United States. On April 6, when the Government asked for a Vote of Confidence in their foreign policy, I recurred to this subject:

Herr Hitler has torn up the treaties and has garrisoned the Rhineland. His troops are there, and there they are going to stay. All this means that the Nazi *régime* has gained a new prestige in Germany and in all the neighbouring countries. But more than that, Germany is now fortifying the Rhine zone or is about to fortify it. No doubt it will take some time. We are told that in the first instance only field entrenchments will be erected, but those who know to what perfection the Germans can carry field entrenchments, like the Hindenburg Line, with all the masses of concrete and the underground chambers there included, will realise that field entrenchments differ only in degree from permanent fortifications, and work steadily up from the first cutting of the sods to their final and perfect form.

I do not doubt that the whole of the German frontier opposite to France is to be fortified as strongly and as speedily as possible. Three, four, or six months will certainly see a barrier of enormous strength. What will be the diplomatic and strategic consequences of that? . . . *The creation of a line of forts opposite to the French frontier will enable the German troops to be economised on that line, and will enable the main forces to swing round through Belgium and Holland.* . . . Then look East. There the consequences of the Rhineland fortification may be more immediate. That is to us a less direct danger, but it is a more imminent danger. The moment those fortifications are completed, and in proportion as they are completed, the whole aspect of Middle Europe is changed. *The Baltic States, Poland, and Czechoslovakia, with which must be associated Yugoslavia, Roumania, Austria, and some other countries, are all affected very decisively the moment that this great work of construction has been completed.*

Every word of this warning was successively and swiftly proved true.

*   *   *   *   *

After the occupation of the Rhineland and the development of the line of fortifications against France, the incorporation of Austria in the German Reich was evidently to be the next step. The story that had opened with the murder of Chancellor Doll-fuss in July 1934 had soon another and a consequential chapter to unfold. With illuminating candour, as we now know, the German Foreign Minister Neurath told the American Ambassador in Moscow, Mr. Bullitt, on May 18, 1936, that it was the policy of the German Government to do nothing active in foreign affairs until the Rhineland had been digested. He explained that *until the German defences had been built on the French and Belgian frontiers* the German Government would do everything to prevent rather than encourage an outbreak by the Nazis in Austria, and that they would pursue a quiet line with regard to Czechoslovakia. *"As soon as our fortifications are constructed," he said, "and the countries in Central Europe realise that France cannot enter German territory, all these countries will begin to feel very differently about their foreign policies, and a new constellation will develop."* Neurath further informed Mr. Bullitt that the youth of Austria was turning more and more towards the Nazis, and the dominance of the Nazi Party in Austria was inevitable and only a question of time. *But the governing factor was the completion of the German fortifications on the French frontier,* for otherwise a German quarrel with Italy might lead to a French attack on Germany.

On May 21, 1936, Hitler in a speech to the Reichstag declared that "Germany neither intends nor wishes to interfere in the internal affairs of Austria, to annex Austria, or to conclude an Anschluss". On July 11, 1936, he signed a pact with the Austrian Government agreeing not to influence in any way the internal affairs of Austria, and especially not to give any active support to the Austrian National-Socialist movement. Within five days of this agreement secret instructions were sent to the National-Socialist Party in Austria to extend and intensify their activities. Meanwhile the German General Staff, under Hitler's orders, were set to draw up military plans for the occupation of Austria when the hour should strike.

CHAPTER XII

THE LOADED PAUSE. SPAIN

1936–1937

The Foreign Policy of England – The New Dominator – The League of Nations – Two Years' Interlude – My Memorandum on Supply Organisation, June 6, 1936 – The Civil War in Spain – Non-Intervention – The Anti-Comintern Pact – Mr. Baldwin's "Frankness" Speech – Arms and the Covenant – The Albert Hall Meeting – The Abdication of King Edward VIII – Mr. Baldwin's Wisdom – The Coronation of King George VI – A Letter from the King – Mr. Baldwin's Retirement – Mr. Chamberlain Prime Minister – Ministerial Changes – Baldwin and Chamberlain – A Talk with Ribbentrop.

HERE is the place to set forth the principles of British policy towards Europe which I had followed for many years and follow still. I cannot better express them than in the words which I used to the Conservative Members' Committee on Foreign Affairs, who invited me to address them in private at the end of March 1936.

"For four hundred years the foreign policy of England has been to oppose the strongest, most aggressive, most dominating Power on the Continent, and particularly to prevent the Low Countries falling into the hands of such a Power. Viewed in the light of history, these four centuries of consistent purpose amid so many changes of names and facts, of circumstances and conditions, must rank as one of the most remarkable episodes which the records of any race, nation, State, or people can show. Moreover, on all occasions England took the more difficult course. Faced by Philip II of Spain, against Louis XIV under William III and Marlborough, against Napoleon, against William II of Germany, it would have been easy and must have been very tempting to join with the stronger and share the fruits of his conquest. However, we always took the harder course, joined with the less strong Powers, made a combination among them, and thus defeated and frustrated the Continental military tyrant, whoever he was, whatever nation he led. Thus we preserved the liberties of Europe, protected the growth of its vivacious and varied society, and emerged after four terrible struggles with an ever-growing fame and widening Empire, and with the Low Countries safely protected in their independence. Here is the wonderful unconscious tradition of British Foreign Policy. All our thoughts rest in that tradition to-day. I know of nothing which has occurred to alter or weaken the justice, wisdom, valour, and prudence upon which our ancestors acted. I know of nothing that has happened to human nature which in the slightest degree alters the validity of their conclusions. I know of nothing in military, political, economic, or scientific fact which makes me feel that we are less capable. I know of nothing which makes me feel that we might not, or cannot, march along the same road. I venture to put this very general proposition before you because it seems to me that if it is accepted everything else becomes much more simple.

"Observe that the policy of England takes no account of which nation it is that seeks the overlordship of Europe. The question is not whether it is Spain, or the French Monarchy, or the French Empire, or the German Empire, or the Hitler *régime*. It has nothing to do with rulers or nations; it is concerned solely with whoever is the strongest or the potentially dominating tyrant. Therefore we should not be afraid of being accused of being pro-French or anti-German. If the circumstances were reversed, we could equally be pro-German and anti-French. It is a law of public policy which we are following, and not a mere expedient dictated by accidental circumstances, or likes and dislikes, or any other sentiment.

"The question therefore arises which is to-day the Power in Europe which is the strongest, and which seeks in a dangerous and oppressive sense to dominate. To-day, for this year, probably for part of 1937, the French Army is the strongest in Europe. But no one is afraid of France. Everyone knows that France wants to be let alone, and that with her it is only a case of self-preservation. Everyone knows that the French are peaceful and overhung by fear. They are at once brave, resolute, peace-loving, and weighed down by anxiety. They are a liberal nation with free Parliamentary institutions.

"Germany, on the other hand, fears no one. She is arming in a manner which has never been seen in German history. She is led by a handful of triumphant desperadoes. The money is running short, discontents are arising beneath these despotic rulers. Very soon they will have to choose on the one hand between economic and financial collapse or internal upheaval, and on the other a war which could have no other object, and which if successful can have no other result, than a Germanised Europe under Nazi control. Therefore it seems to me that all the old conditions present themselves again, and that our national salvation depends upon our gathering once again all the forces of Europe to contain, to restrain, and if necessary to frustrate German domination. For, believe me, if any of those other Powers, Spain, Louis XIV, Napoleon, Kaiser Wilhelm II, had with our aid become the absolute masters of Europe, they could have despoiled us, reduced us to insignificance and penury on the morrow of their victory. We ought to set the life and endurance of the British Empire and the greatness of this Island very high in our duty, and not be led astray by illusions about an ideal world, which only means that other and worse controls will step into our place, and that the future direction will belong to them.

"It is at this stage that the spacious conception and extremely vital organisation of the League of Nations presents itself as a prime factor. The League of Nations is, in a practical sense, a British conception, and it harmonises perfectly with all our past methods and actions. Moreover, it harmonises with those broad ideas of right and wrong, and of peace based upon controlling the major aggressor, which we have always followed. We wish for the reign of law and freedom among nations and within nations, and it was for that, and nothing less than that, that those bygone architects of our repute, magnitude, and civilisation fought, toiled, and won. The dream of a reign of International Law and of the settlement of disputes by patient discussion, but still in accordance with what is lawful and just, is very dear to the British people. You must not underrate the force which these ideals exert upon the modern British democracy. One does not know how these seeds are planted by the winds of the centuries in the hearts of the working people. They are there, and just as

strong as their love of liberty. We should not neglect them, because they are the essence of the genius of this Island. Therefore
we believe that in the fostering and fortifying of the League of
Nations will be found the best means of defending our island
security, as well as maintaining grand universal causes with which
we have very often found our own interests in natural accord.

"My three main propositions are: first, that we must oppose the
would-be dominator or potential aggressor; secondly, that Germany, under its present Nazi *régime*, and with its prodigious
armaments, so swiftly developing, fills unmistakably that part;
thirdly, that the League of Nations rallies many countries, and
unites our own people here at home in the most effective way to
control the would-be aggressor. I venture most respectfully to
submit these main themes to your consideration. Everything else
will follow from them.

"It is always more easy to discover and proclaim general principles than to apply them. First, we ought to count our effective
association with France. That does not mean that we should
develop a needlessly hostile mood against Germany. It is a part
of our duty and our interest to keep the temperature low between
these two countries. We shall not have any difficulty in this so
far as France is concerned. Like us, they are a Parliamentary
democracy with tremendous inhibitions against war, and, like
us, under considerable drawbacks in preparing their defence.
Therefore I say we ought to regard our defensive association with
France as fundamental. Everything else must be viewed in
proper subordination now that the times have become so sharp
and perilous. Those who are possessed of a definite body of
doctrine and of deeply-rooted convictions upon it will be in a
much better position to deal with the shifts and surprises of daily
affairs than those who are merely taking short views, and indulging their natural impulses as they are evoked by what they read
from day to day. The first thing is to decide where you want to
go. For myself, I am for the armed League of all Nations, or as
many as you can get, against the potential aggressor, with England
and France as the core of it. Let us neglect nothing in our power
to establish the great international framework. If that should
prove to be beyond our strength, or if it breaks down through
the weakness or wrongdoing of others, then at least let us make
sure that England and France, the two surviving free great

countries of Europe, can together ride out any storm that may blow with good and reasonable hopes of once again coming safely into port."

If we add the United States to Britain and France; if we change the name of the potential aggressor; if we substitute the United Nations Organisation for the League of Nations, the Atlantic Ocean for the English Channel, and the world for Europe, the argument is not necessarily without its application to-day.

*   *   *   *   *

Two whole years passed between Hitler's seizure of the Rhineland in March 1936 and his rape of Austria in March 1938. This was a longer interval than I had expected. Everything happened in the order foreseen and stated, but the spacing between the successive blows was longer. During this period no time was wasted by Germany. The fortification of the Rhineland, or "the West Wall", proceeded apace, and an immense line of permanent and semi-permanent fortifications grew continually. The German Army, now on the full methodical basis of compulsory service and reinforced by ardent volunteering, grew stronger month by month, both in numbers and in the maturity and quality of its formations. The German Air Force held and steadily improved the lead it had obtained over Great Britain. The German munitions plants were working at high pressure. The wheels revolved and the hammers descended day and night in Germany, making its whole industry an arsenal, and welding all its population into one disciplined war machine. At home in the autumn of 1936 Hitler inaugurated a Four Years' Plan to reorganise German economy for greater self-sufficiency in war. Abroad he obtained that "strong alliance" which he had stated in *Mein Kampf* would be necessary for Germany's foreign policy. He came to terms with Mussolini, and the Rome-Berlin Axis was formed.

Up till the middle of 1936 Hitler's aggressive policy and treaty-breaking had rested, not upon Germany's strength, but upon the disunion and timidity of France and Britain and the isolation of the United States. Each of his preliminary steps had been gambles in which he knew he could not afford to be seriously challenged. The seizure of the Rhineland and its subsequent fortification was the greatest gamble of all. It had succeeded brilliantly. His opponents were too irresolute to call his bluff. When next he moved in 1938 his bluff was bluff no more. Aggression was backed by force, and it might well be by superior force. When the Governments of France and Britain realised the terrible transformation which had taken place it was too late.

*   *   *   *   *

I continued to give the closest attention to our military prepara-tions. My relations with Sir Thomas Inskip, Minister for Co-ordination of Defence, were friendly, and I did my best to help him privately. At his request I wrote and sent him a memoran-dum about the much-needed Ministry of Supply, which is dated June 6, 1936. See Appendix C. No effective action was however taken to create a Ministry of Supply until the spring of 1939, nearly three years later, nor was any attempt made to introduce emergency conditions into our munitions production.

*   *   *   *   *

At the end of July 1936 the increasing degeneration of the Parliamentary *régime* in Spain, and the growing strength of the movements for a Communist, or alternatively an anarchist revolution, led to a military revolt which had long been prepar-ing. It is part of the Communist doctrine and drill-book, laid down by Lenin himself, that Communists should aid all move-ments towards the Left and help into office weak Constitutional, Radical, or Socialist Governments. These they should undermine, and from their falling hands snatch absolute power, and found the Marxist State. In fact, a perfect reproduction of the Kerensky period in Russia was taking place in Spain. But the strength of Spain had not been shattered by foreign war. The Army still maintained a measure of cohesion. Side by side with the Com-munist conspiracy there was elaborated in secret a deep military counterplot. Neither side could claim with justice the title-deeds of legality, and Spaniards of all classes were bound to consider the life of Spain.

Many of the ordinary guarantees of civilised society had been already liquidated by the Communist pervasion of the decayed Parliamentary Government. Murders began on both sides, and the Communist pestilence had reached a point where it could take political opponents in the streets or from their beds and kill them.

Already a large number of these assassinations had taken place in and around Madrid. The climax was the murder of Señor Sotelo, the Conservative leader, who corresponded somewhat to the type of Sir Edward Carson in British politics before the 1914 war. This crime was the signal for the generals of the Army to act. General Franco had a month before written a letter to the Spanish War Minister, making it clear that if the Spanish Government could not maintain the normal securities of law in daily life the Army would have to intervene. Spain had seen many *pronunciamientos* by military chiefs in the past. When, after General Sanjurjo had perished in an air crash, General Franco raised the standard of revolt, he was supported by the Army, including the rank and file. The Church, with the noteworthy exception of the Dominicans, and nearly all the elements of the Right and Centre adhered to him, and he became immediately the master of several important provinces. The Spanish sailors killed their officers and joined what soon became the Communist side. In the collapse of civilised government the Communist sect obtained control, and acted in accordance with their drill. Bitter civil war now began. Wholesale cold-blooded massacres of their political opponents, and of the well-to-do, were perpetrated by the Communists who had seized power. These were repaid with interest by the forces under Franco. All Spaniards went to their deaths with remarkable composure, and great numbers on both sides were shot. The military cadets defended their college at the Alcazar in Toledo with the utmost tenacity, and Franco's troops, forcing their way up from the south, leaving a trail of vengeance behind them in every Communist village, presently achieved their relief. This episode deserves the notice of historians.

In this quarrel I was neutral. Naturally I was not in favour of the Communists. How could I be, when if I had been a Spaniard they would have murdered me and my family and friends? I was sure however that with all the rest they had on their hands the British Government were right to keep out of Spain. France proposed a plan of Non-Intervention, whereby both sides would be left to fight it out without any external aid. The British, German, Italian, and Russian Governments subscribed to this. In consequence the Spanish Government, now in the hands of the most extreme revolutionaries, found itself deprived of the right even to buy the arms ordered with the gold it physically possessed. It would have been more reasonable to follow the normal course and to have recognised the belligerency of both sides, as was done in the American Civil War of 1861–65. Instead, however, the policy of Non-Intervention was adopted and formally agreed to by all the Great Powers. This agreement was strictly observed by Great Britain; but Italy and Germany on the one side, and Soviet Russia on the other, broke their engagement constantly and threw their weight into the struggle one against the other. Germany in particular used her air-power to commit such experimental horrors as the bombing of the defenceless little township of Guernica.

The Government of M. Léon Blum, which had succeeded the Ministry of M. Albert Sarraut on June 4, was under pressure from its Communist supporters in the Chamber to support the Spanish Government with war material. The Air Minister, M. Cot, without too much regard for the strength of the French Air Force, then in a state of decay, was secretly delivering planes and equipment to the Republican armies. I was perturbed at such developments, and on July 31, 1936, I wrote to M. Corbin, the French Ambassador:

One of the greatest difficulties I meet with in trying to hold on to the old position is the German talk that the anti-Communist countries should stand together. I am sure if France sent aeroplanes, etc., to the present Madrid Government, and the Germans and Italians pushed in from the other angle, the dominant forces here would be pleased with Germany and Italy, and estranged from France. I hope you will not mind my writing this, which I do of course entirely on my own account. I do not like to hear people talking of England, Germany, and Italy forming up against European Communism. It is too easy to be good.

I am sure that an absolutely rigid neutrality, with the strongest protest against any breach of it, is the only correct and safe course at the present time. A day may come, if there is a stalemate, when the League of Nations may intervene to wind up the horrors. But even that is very doubtful.

*   *   *   *   *

There is another event which must be recorded here. On November 25, 1936, the Ambassadors of all the Powers represented in Berlin were summoned to the Foreign Office, where Herr von Neurath disclosed the details of the Anti-Comintern

Pact, which had been negotiated with the Japanese Government. The purpose of the pact was to take common action against the international activities of the Comintern, either within the boundaries of the contracting States or beyond them.

*   *   *   *   *

During the whole of 1936 the anxiety of the nation and Parliament continued to mount, and was concentrated in particular upon our air defences. In the Debate on the Address on November 12 I severely reproached Mr. Baldwin for having failed to keep his pledge that "any Government of this country—a National Government more than any, and this Government—will see to it that in air strength and air power this country shall no longer be in a position inferior to any country within striking distance of its shores". I said: "The Government simply cannot make up their minds, or they cannot get the Prime Minister to make up his mind. So they go on in strange paradox, decided only to be undecided, resolved to be irresolute, adamant for drift, solid for fluidity, all-powerful to be impotent. So we go on preparing more months and years—precious, perhaps vital, to the greatness of Britain—for the locusts to eat."

Mr. Baldwin replied to me in a remarkable speech, in which he said:

I want to speak to the House with the utmost frankness. . . . The difference of opinion between Mr. Churchill and myself is in the years 1933 onwards. In 1931–32, although it is not admitted by the Opposition, there was a period of financial crisis. But there was another reason. I would remind the House that not once but on many occasions in speeches and in various places, when I have been speaking and advocating as far as I am able the democratic principle, I have stated that *a democracy is always two years behind the dictator*. I believe that to be true. It has been true in this case. I put before the whole House my own views with an appalling frankness. You will remember at that time the Disarmament Conference was sitting in Geneva. You will remember at that time there was probably a stronger pacifist feeling running through this country than at any time since the war. You will remember *the election at Fulham in the autumn of* 1933, *when a seat which the National Government held was lost by about* 7,000 *votes on no issue but the pacifist.* . . . My position as the leader of a great party was not altogether a comfortable one. I asked myself what chance was there—when that feeling that was given expression to in Fulham was common throughout the country—what chance was there within the next year or two of that feeling being so changed that the country would give a mandate for rearmament? Supposing I had gone to the country and said that Germany was rearming, and that we must rearm, does anybody think that this pacific democracy would have rallied to that cry at that moment? *I cannot think of anything that would have made the loss of the election from my point of view more certain.*

This was indeed appalling frankness. It carried naked truth about his motives into indecency. That a Prime Minister should avow that he had not done his duty in regard to national safety because he was afraid of losing the election was an incident without parallel in our Parliamentary history. Mr. Baldwin was of course not moved by any ignoble wish to remain in office. He was in fact in 1936 earnestly desirous of retiring. His policy was dictated by the fear that if the Socialists came into power even less would be done than his Government intended. All their declarations and votes against defence measures are upon record. But this was no complete defence, and less than justice to the spirit of the British people. The success which had attended the naïve confession of miscalculation in air parity the previous year was not repeated on this occasion. The House was shocked. Indeed, the impression produced was so painful that it might well have been fatal to Mr. Baldwin, who was also at that time in failing health, had not the unexpected intervened.

★   ★   ★   ★   ★

At this time there was a great drawing together of men and women of all parties in England who saw the perils of the future, and were resolute upon practical measures to secure our safety and the cause of freedom, equally menaced by both the totalitarian impulsions and our Government's complacency. Our plan was the most rapid large-scale rearmament of Britain, combined with the complete acceptance and employment of the authority of the League of Nations. I called this policy "Arms and the Covenant". Mr. Baldwin's performance in the House of Commons was viewed among us all with disdain. The culmination of this campaign was to be a meeting at the Albert Hall. Here on December 3 we gathered many of the leading men in all the parties—strong Tories of the Right Wing earnestly convinced of the national peril; the leaders of the League of Nations Peace

Ballot; the representatives of many great trade unions, including in the chair my old opponent of the General Strike, Sir Walter Citrine; the Liberal Party and its leader, Sir Archibald Sinclair. We had the feeling that we were upon the threshold of not only gaining respect for our views, but of making them dominant. It was at this moment that the King's passion to marry the woman he loved caused the casting of all else into the background. The Abdication crisis was at hand.

Before I replied to the Vote of Thanks there was a cry, "God Save the King," and this excited prolonged cheering. I explained therefore on the spur of the moment my personal position.

There is another grave matter which overshadows our minds to-night. In a few minutes we are going to sing "God Save the King". I shall sing it with more heartfelt fervour than I have ever sung it in my life. I hope and pray that no irrevocable decision will be taken in haste, but that time and public opinion will be allowed to play their part, and that a cherished and unique personality may not be incontinently severed from the people he loves so well. I hope that Parliament will be allowed to discharge its function in these high constitutional questions. I trust that our King may be guided by the opinions that are now for the first time being expressed by the British nation and the British Empire, and that the British people will not in their turn be found wanting in generous consideration for the occupant of the Throne.

It is not relevant to this account to describe the brief but intensely violent controversy that followed. I had known King Edward VIII since he was a child, and had in 1910 as Home Secretary read out to a wonderful assembly the Proclamation creating him Prince of Wales at Carnarvon Castle. I felt bound to place my personal loyalty to him upon the highest plane. Although during the summer I had been made fully aware of what was going forward, I in no way interfered or communicated with him at any time. However, presently in his distress he asked the Prime Minister for permission to consult me. Mr. Baldwin gave formal consent, and on this being conveyed to me I went to the King at Fort Belvedere. I remained in contact with him till his abdication, and did my utmost to plead both to the King and to the public for patience and delay. I have never repented of this—indeed, I could do no other.

The Prime Minister proved himself to be a shrewd judge of

British national feeling. Undoubtedly he perceived and expressed the profound will of the nation. His deft and skilful handling of the Abdication issue raised him in a fortnight from the depths to the pinnacle. There were several moments when I seemed to be entirely alone against a wrathful House of Commons. I am not, when in action, unduly affected by hostile currents of feeling; but it was on more than one occasion almost physically impossible to make myself heard. All the forces I had gathered together on "Arms and the Covenant", of which I conceived myself to be the mainspring, were estranged or dissolved, and I was myself so smitten in public opinion that it was the almost universal view that my political life was at last ended. How strange it is that this very House of Commons which had regarded me with so much hostility should have been the same instrument which hearkened to my guidance and upheld me through the long adverse years of war till victory over all our foes was gained! What a proof is here offered that the only wise and safe course is to act from day to day in accordance with what one's own conscience seems to decree!

From the Abdication of one King we passed to the Coronation of another, and until the end of May 1937 the ceremonial and pageantry of a solemn national act of allegiance and the consecration of British loyalties at home and throughout the Empire to the new Sovereign filled all minds. Foreign affairs and the state of our defences lost all claim upon the public mood. Our Island might have been ten thousand miles away from Europe. However, I am permitted to record that on May 18, 1937, on the morrow of the Coronation, I received from the new King, His present Majesty, a letter in his own handwriting:

THE ROYAL LODGE,
THE GREAT PARK,
WINDSOR, BERKS.
18.V.37

My dear Mr. Churchill,

I am writing to thank you for your very nice letter to me. I know how devoted you have been, and still are, to my dear brother, and I feel touched beyond words by your sympathy and understanding in the very difficult problems that have arisen since he left us in December. I fully realise the great responsibilities and cares that I have taken on as King, and I feel most encouraged to receive your good wishes, as one of our great statesmen, and from one who has served his country so faithfully. I can only hope and trust that the good feeling and hope that exists in the Country and Empire now will prove a good example to other Nations in the world.

Believe me,

Yours very sincerely,

GEORGE R.I.

This gesture of magnanimity towards one whose influence at that time had fallen to zero will ever be a cherished experience in my life.

★　　★　　★　　★　　★

On May 28, 1937, after King George VI had been crowned, Mr. Baldwin retired. His long public services were suitably rewarded by an Earldom and the Garter. He laid down the wide authority he had gathered and carefully maintained, but had used as little as possible. He departed in a glow of public gratitude and esteem. There was no doubt who his successor should be. Mr. Neville Chamberlain had, as Chancellor of the Exchequer, not only done the main work of the Government for five years past, but was the ablest and most forceful Minister, with high abilities and an historic name. I had described him a year earlier at Birmingham in Shakespeare's words as the "packhorse in our great affairs", and he had accepted this description as a compliment. I had no expectation that he would wish to work with me, nor would he have been wise to do so at such a time. His ideas were far different from mine on the treatment of the dominant issues of the day. But I welcomed the accession to power of a live, competent, executive figure. While still Chancellor of the Exchequer he had involved himself in a fiscal proposal for a small-scale national defence contribution which had been ill-received by the Conservative Party and was of course criticised by the Opposition. I was able, in the first days of his Premiership, to make a speech upon this subject which helped him to withdraw, without any loss of dignity, from a position which had become untenable. Our relations continued to be cool, easy, and polite both in public and in private.

Mr. Chamberlain made few changes in the Government. He had had disagreements with Mr. Duff Cooper about War Office administration, and much surprised him by offering him advancement to the great key office of the Admiralty. The Prime

Minister evidently did not know the eyes through which his new First Lord, whose early career had been in the Foreign Office, viewed the European scene. In my turn I was astonished that Sir Samuel Hoare, who had just secured a large expansion of the naval programme, should wish to leave the Admiralty for the Home Office. Hoare seems to have believed that prison reform in a broad humanitarian sense would become the prevailing topic in the immediate future; and since his family was connected with the famous Elizabeth Fry, he had a strong personal sentiment about it.

*　　*　　*　　*　　*

I may here set down a comparative appreciation of these two Prime Ministers, Baldwin and Chamberlain, whom I had known so long and under whom I had served or was to serve. Stanley Baldwin was the wiser, more comprehending personality, but without detailed executive capacity. He was largely detached from foreign and military affairs. He knew little of Europe, and disliked what he knew. He had a deep knowledge of British party politics, and represented in a broad way some of the strengths and many of the infirmities of our Island race. He had fought five General Elections as leader of the Conservative Party and had won three of them. He had a genius for waiting upon events and an imperturbability under adverse criticism. He was singularly adroit in letting events work for him, and capable of seizing the ripe moment when it came. He seemed to me to revive the impressions history gives us of Sir Robert Walpole, without of course the eighteenth-century corruption, and he was master of British politics for nearly as long.

Neville Chamberlain, on the other hand, was alert, businesslike, opinionated and self-confident in a very high degree. Unlike Baldwin, he conceived himself able to comprehend the whole field of Europe, and indeed the world. Instead of a vague but none the less deep-seated intuition, we had now a narrow, sharp-edged efficiency within the limits of the policy in which he believed. Both as Chancellor of the Exchequer and as Prime Minister he kept the tightest and most rigid control upon military expenditure. He was throughout this period the masterful opponent of all emergency measures. He had formed decided judgments about all the political figures of the day, both at home and abroad, and felt himself capable of dealing with them. His all-pervading hope was to go down to history as the great Peace-maker, and for this he was prepared to strive continually in the teeth of facts, and face great risks for himself and his country. Unhappily he ran into tides the force of which he could not measure, and met hurricanes from which he did not flinch, but with which he could not cope. In these closing years before the war I should have found it easier to work with Baldwin, as I knew him, than with Chamberlain; but neither of them had any wish to work with me except in the last resort.

★　★　★　★　★

One day in 1937 I had a meeting with Herr von Ribbentrop, German Ambassador to Britain. In one of my fortnightly articles I had noted that he had been misrepresented in some speech he had made. I had of course met him several times in society. He now asked me whether I would come to see him and have a talk. He received me in the large upstairs room at the German Embassy. We had a conversation lasting for more than two hours. Ribben-trop was most polite, and we ranged over the European scene, both in respect of armaments and policy. The gist of his state-ment to me was that Germany sought the friendship of England (on the Continent we are still often called "England"). He said he could have been Foreign Minister of Germany, but he had asked Hitler to let him come over to London in order to make the full case for an Anglo-German entente or even alliance. Germany would stand guard for the British Empire in all its greatness and extent. They might ask for the return of the German colonies, but this was evidently not cardinal. What was required was that Britain should give Germany a free hand in the East of Europe. She must have her Lebensraum, or living-space, for her increas-ing population. Therefore Poland and the Danzig Corridor must be absorbed. White Russia and the Ukraine were indispensable to the future life of the German Reich of some seventy million souls. Nothing less would suffice. All that was asked of the British Commonwealth and Empire was not to interfere. There was a large map on the wall, and the Ambassador several times led me to it to illustrate his projects.

After hearing all this I said at once that I was sure the British Government would not agree to give Germany a free hand in Eastern Europe. It was true we were on bad terms with Soviet

Russia and that we hated Communism as much as, Hitler did, but he might be sure that even if France were safeguarded Great Britain would never disinterest herself in the fortunes of the Continent to an extent which would enable Germany to gain the domination of Central and Eastern Europe. We were actually standing before the map when I said this. Ribbentrop turned abruptly away. He then said, "In that case, war is inevitable. There is no way out. The Fuehrer is resolved. Nothing will stop him and nothing will stop us." We then returned to our chairs. I was only a private Member of Parliament, but of some prominence. I thought it right to say to the German Ambassador —in fact, I remember the words well, "When you talk of war, which no doubt would be general war, you must not underrate England. She is a curious country, and few foreigners can understand her mind. Do not judge by the attitude of the present Administration. Once a great cause is presented to the people all kinds of unexpected actions might be taken by this very Government and by the British nation." And I repeated, "Do not underrate England. She is very clever. If you plunge us all into another Great War she will bring the whole world against you, like last time." At this the Ambassador rose in heat and said, "Ah, England may be very clever, but this time she will not bring the world against Germany." We turned the conversation on to easier lines, and nothing more of note occurred. The incident however remains in my memory, and as I reported it at the time to the Foreign Office I feel it right to put it on record.

When he was on his trial for his life by the conquerors Ribbentrop gave a distorted version of this conversation and claimed that I should be summoned as a witness. What I have set down about it is what I should have said had I been called.

CHAPTER XIII

GERMANY ARMED

1936-1938

The "Overall Strategic Objective" – German Expenditure on Armaments – Independent Inquiries – The Conservative Deputation to the Prime Minister, July 28, 1936 – My Statement of the Case – General Conclusions – My Fear – Our Second Meeting, November 23, 1936 – Lord Swinton Leaves the Air Ministry, May 12, 1938 – Debate in Parliament – Lindemann Rejoins the Air Defence Research Committee – My Correspondence with M. Daladier – The French Estimate of German Air Strength, 1938 – My Estimate of the German Army, June 1938 – M. Daladier Concurs – The Decay of the French Air Force – The Careless Islanders.

ADVANTAGE is gained in war and also in foreign policy and other things by selecting from many attractive or unpleasant alternatives the dominating point. American military thought has coined the expression "Overall Strategic Objective". When our officers first heard this they laughed; but later on its wisdom became apparent and accepted. Evidently this should be the rule, and other great business be set in subordinate relationship to it. Failure to adhere to this simple principle produces confusion and futility of action, and nearly always makes things much worse later on.

Personally I had no difficulty in conforming to the rule long before I heard it proclaimed. My mind was obsessed by the impression of the terrific Germany I had seen and felt in action during the years of 1914 to 1918 suddenly becoming again possessed of all her martial power, while the Allies, who had so narrowly survived, gaped idle and bewildered. Therefore I continued by every means and on every occasion to use what influence I had with the House of Commons and also with individual Ministers to urge forward our military preparations and to procure Allies and associates for what would before long become again the Common Cause.

One day a friend of mine in a high confidential position under the Government came over to Chartwell to swim with me in my pool when the sun shone bright and the water was fairly warm. We talked of nothing but the coming war, of the certainty of which he was not entirely convinced. As I saw him off he suddenly on an impulse turned and said to me, "The Germans are spending a thousand million pounds sterling a year on their armaments." I thought Parliament and the British public ought to know the facts. I therefore set to work to examine German finance. Budgets were produced and still published every year in Germany; but from their wealth of figures it was very difficult to tell what was happening. However, in April 1936 I privately instituted two separate lines of scrutiny. The first rested upon two German refugees of high ability and inflexible purpose. They understood all the details of the presentation of German budgets, the value of the mark, and so forth. At the same time I asked my friend Sir Henry Strakosch whether he could not find out what was actually happening. Strakosch was the head of the firm called Union Corporation, with great resources, and a highly-skilled, devoted personnel. The brains of this City company were turned for several weeks on to the problem. Presently they reported with precise and lengthy detail that the German war expenditure was certainly round about a thousand million pounds sterling a year. At the same time the German refugees, by a totally different series of arguments, arrived independently at the same conclusion. One thousand million pounds sterling per annum at the money values of 1936!

I had therefore two separate structures of fact on which to base a public assertion. So I accosted Mr. Neville Chamberlain, still Chancellor of the Exchequer, in the Lobby the day before a debate and said to him, "To-morrow I shall ask you whether it is not a fact that the Germans are spending a thousand million pounds a year on warlike preparations, and I shall ask you to confirm or deny." Chamberlain said, "I cannot deny it, and if you put the point I shall confirm it." I must quote my words:

Taking the figures from German official sources, the expenditure on capital account from the end of March 1933 to the end of June 1935 has been as follows: in 1933 nearly five milliards of marks; in 1934 nearly eight milliards; and in 1935 nearly eleven milliards—a total of twenty-four milliards, or roughly £2,000,000,000. Look at these figures, 5, 8, and 11 for the three years. They give you exactly the kind of progression which a properly-developing munitions industry would make.

Specifically I asked the Chancellor:

Whether he is aware that the expenditure by Germany upon purposes directly and indirectly concerned with military preparations, including strategic roads, may well have amounted to the equivalent of £800 millions during the calendar year 1935; and whether this rate of expenditure seems to be continuing in the current calendar year.

Mr. Chamberlain: The Government have no official figures, but from such information as they have I see no reason to think that the figure mentioned in my right hon. friend's question is necessarily excessive as applied to either year, although, as he himself would agree, there are elements of conjecture.

I substituted the figure of £800 millions for £1,000 millions to cover my secret information, and also to be on the safe side.

* * * * *

I sought by several means to bring the relative state of British and German armaments to a clear-cut issue. I asked for a debate in Secret Session. This was refused. "It would cause needless alarm." I got little support. All Secret Sessions are unpopular with the Press. Then on July 20, 1936, I asked the Prime Minister whether he would receive a deputation of Privy Councillors and a few others who would lay before him the facts so far as they knew them. Lord Salisbury requested that a similar deputation from the House of Lords should also come. This was agreed. Although I made personal appeals both to Mr. Attlee and Sir Archibald Sinclair, the Labour and Liberal Parties declined to be represented. Accordingly, on July 28 we were received in the Prime Minister's House of Commons room by Mr. Baldwin, Lord Halifax, and Sir Thomas Inskip. The following Conservative and non-party notables came with me. Sir Austen Chamberlain introduced us.

THE DEPUTATION

House of Commons	House of Lords
Sir Austen Chamberlain.	The Marquess of Salisbury.
Mr. Churchill.	Viscount Fitzalan.
Sir Robert Horne.	Viscount Trenchard.
Mr. Amery.	Lord Lloyd.
Sir John Gilmour.	Lord Milne.
Captain Guest.	
Admiral Sir Roger Keyes.	
Earl Winterton.	
Sir Henry Croft.	
Sir Edward Grigg.	
Viscount Wolmer.	
Lieut.-Col. Moore-Brabazon.	
Sir Hugh O'Neill.	

This was a great occasion. I cannot recall anything like it in what I have seen of British public life. The group of eminent men, with no thought of personal advantage, but whose lives had been centred upon public affairs, represented a weight of Conservative opinion which could not easily be disregarded. If the leaders of the Labour and Liberal Oppositions had come with us there might have been a political situation so tense as to enforce remedial action. The proceedings occupied three or four hours on each of two successive days. I have always said Mr. Baldwin was a good listener. He certainly seemed to listen with the greatest interest and attention. With him were various members of the staff of the Committee of Imperial Defence. On the first day I opened the case in a statement of an hour and a quarter, of which some extracts, given in Appendix D, throw a true light on the scene.

I ended as follows:

First, we are facing the greatest danger and emergency of our history. Second, we have no hope of solving our problem except in conjunction with the French Republic. The union of the British Fleet and the French Army, together with their combined Air Forces operating from close behind the French and Belgian frontiers, together with all that Britain and France stand for, constitutes a deterrent in which salvation may reside. Anyhow it is the best hope. Coming down to detail, we must lay aside every impediment in raising our own strength. We cannot possibly provide against all possible dangers.

We must concentrate upon what is vital and take our punishment elsewhere. Coming to still more definite propositions, we must increase the development of our air-power in priority over every other consideration. At all costs we must draw the flower of our youth into piloting aeroplanes. Never mind what inducements must be offered; we must draw from every source, by every means. We must accelerate and simplify our aeroplane production and push it to the largest scale, and not hesitate to make contracts with the United States and elsewhere for the largest possible quantities of aviation material and equipment of all kinds. We are in danger, as we have never been in danger before—no, not even at the height of the submarine campaign [1917].

This thought preys upon me: *The months slip by rapidly. If we delay too long in repairing our defences we may be forbidden by superior power to complete the process.*

★　　★　　★　　★　　★

We were much disappointed that the Chancellor of the Exchequer could not be present. It was evident that Mr. Baldwin's health was failing, and it was well known that he would soon seek rest from his burdens. There could be no doubt who would be his successor. Unhappily, Mr. Neville Chamberlain was absent upon a well-deserved holiday, and did not have the opportunity of this direct confrontation with the facts from members of the Conservative Party, who included his brother and so many of his most valued personal friends.

Most earnest consideration was given by Ministers to our formidable representations, but it was not till after the Recess, on November 23, 1936, that we were all invited by Mr. Baldwin to receive a more fully considered statement on the whole position. Sir Thomas Inskip then gave a frank and able account, in which he did not conceal from us the gravity of the plight into which we had come. In substance this was to the effect that our estimates, and in particular my statements, took a too gloomy view of our prospects; that great efforts were being made (as indeed they were) to recover the lost ground; but that no case existed which would justify the Government in adopting emergency measures; that these would necessarily be of a character to upset the whole industrial life of this country, would cause widespread alarm, and advertise any deficiencies that existed, and that within these limits everything possible was being done. On this Sir Austen Chamberlain recorded our general impression that our anxieties were not relieved and that we were by no means satisfied. Thus we took our leave.

I cannot contend that at this date, the end of 1936, the position could have been retrieved. Much more however could and ought to have been done by an intense convulsive effort. And of course the fact and proof of this effort must have had its immeasurable effect on Germany, if not on Hitler. But the paramount fact remained that the Germans had the lead of us in the air, and also over the whole field of munitions production, even making allowance for our smaller military needs, and for the fact that we had a right also to count upon France and the French Army and Air Force. It was no longer in our power to forestall Hitler or to regain air parity. Nothing could now prevent the German Army and the German Air Force from becoming the strongest in Europe. By extraordinary and disturbing exertions we could improve our position. We could not cure it.

These sombre conclusions, which were not seriously disputed by the Government, no doubt influenced their foreign policy; and full account must be taken of them when we try to form a judgment upon the decision which Mr. Chamberlain, when he became Prime Minister, took before and during the Munich crisis. I was at this time only a private Member of Parliament, and I bore no official responsibility. I strove my utmost to galvanise the Government into vehemence and extraordinary preparation, even at the cost of world alarm. In these endeavours no doubt I painted the picture even darker than it was. The emphasis which I had put upon the two years' lag which afflicted us may well be judged inconsistent with my desire to come to grips with Hitler in October 1938. I remain convinced however that it was right to spur the Government by every means, and that it would have been better in all the circumstances, which will presently be described, to fight Hitler in 1938 than it was when we finally had to do so in September 1939. Of this more later.

★ ★ ★ ★ ★

Presently Mr. Baldwin, as we have seen, gave place to Mr. Neville Chamberlain; and we must now move on to 1938. Lord Swinton was a very keen and efficient Air Minister, and for a long time had great influence in the Cabinet in procuring the necessary facilities and funds. The anxiety about our air defences continued to grow, and reached its climax in May. The many great and valuable expansions and improvements which Lord Swinton had made could not become apparent quickly, and in any case the whole policy of the Government lacked both magnitude and urgency. I continued to press for an inquiry into the state of our air programme, and found increasing support. Swinton had made the mistake of accepting a peerage. He was not therefore able to defend himself and his department in the House of Commons. The spokesman who was chosen from the Government Front Bench was utterly unable to stem the rising tide of alarm and dissatisfaction. After one most unfortunate debate it became obvious that the Air Minister should be in the House of Commons.

One morning (May 12) at the Air Defence Research Committee we were all busily engaged—scientists, politicians, and officials—on technical problems, when a note was brought in to the Air Minister asking him to go to Downing Street. He desired us to continue our discussions, and left at once. He never returned. He had been dismissed by Mr. Chamberlain.

In the agitated debate which followed on the 25th I tried to distinguish between the exertions and capacity of the fallen Minister and the general complaint against the Government:

The credit of Government statements has been compromised by what has occurred. The House has been consistently misled about the air position. The Prime Minister himself has been misled. He was misled right up to the last moment, apparently. Look at the statement which he made in March, when he spoke about our armaments:

"The sight of this enormous, this almost terrifying, power which Britain is building up has a sobering effect, a steadying effect, on the opinion of the world."

I have often warned the House that the air programmes were falling into arrear. But I have never attacked Lord Swinton. I have never thought that he was the one to blame—certainly not the only one to blame. It is usual for the critics of a Government to discover hitherto unnoticed virtues in any Minister who is forced to resign. But perhaps I may quote what I said three months ago: "It would be unfair to throw the blame on any one Minister, or upon Lord Swinton, for our deficiency. He certainly represents an extremely able and whole-hearted effort to do the best he possibly could to expand our air-power, and the results which he achieved would be bright if they were not darkened by the time-table, and if they were not outshone by other relative facts occurring elsewhere." . . .

The hard responsibility for the failure to fulfil the promises made to us rests upon those who have governed and guided this Island for the last five years, that is to say, from the date when German rearmament in real earnest became apparent and known. I certainly did not attempt to join in a man-hunt of Lord Swinton. I was very glad to-day to hear the Prime Minister's tribute to him. Certainly he deserves our sympathy. He had the confidence and friendship of the Prime Minister, he had the support of an enormous Parliamentary majority; yet he has been taken from his post at what, I think, is the worst moment in the story of air expansion. It may be that in a few months there will be a considerable flow of aircraft arriving; yet he has had to answer for his record at this particularly dark moment for him. I was reading the other day a letter of the great Duke of Marlborough, in which he said: "To remove a General in the midst of a campaign—that is the mortal stroke."

I turned to other aspects of our defences:

We are now in the third year of openly avowed rearmament. Why is it, if all is going well, there are so many deficiencies? Why, for instance, are the Guards drilling with flags instead of machine-guns? Why is it that our small Territorial Army is in a rudimentary condition? Is that all according to schedule? Why, when you consider how small are our forces, should it be impossible to equip the Territorial Army simultaneously with the Regular Army? It would have been a paltry task for British industry, which is more flexible and more fertile than German industry in every sphere except munitions. . . .

The other day the Secretary of State for War was asked about the anti-aircraft artillery. The old 3-inch guns of the Great War, he said, had been modernised, and deliveries of the newer guns—and there is more than one type of newer gun—were proceeding "in advance of schedule". But what is the schedule? If your schedule prescribes a delivery of half a dozen, ten, a dozen, twenty, or whatever it may be, guns per month, no doubt that may easily be up to schedule, and easily be in advance of it. But what is the adequacy of such a schedule to our needs? A year ago I reminded the House of the published progress of Germany in anti-aircraft artillery—thirty regiments of twelve batteries each of mobile artillery alone, aggregating something between twelve and thirteen hundred guns, in addition to three or four thousand guns in fixed positions. These are all modern guns, not guns of 1915, but all guns made since the year 1933.

Does not that give the House an idea of the tremendous scale of these transactions? We do not need to have a gigantic army like Continental countries; but in the matter of anti-aircraft defence we are on equal terms. We are just as vulnerable, and perhaps more vulnerable. Here is the Government thinking of anti-aircraft artillery in terms of hundreds where the Germans have it to-day in terms of thousands. . . .

We are thinking at the present time in terms of production for three separate armed forces. In fact and in truth, the supply of arms for all fighting forces resolves itself into a common problem of the provision and distribution of skilled labour, raw materials, plant, machinery, and technical appliances. That problem can only be dealt with comprehensively, harmoniously, and economically through one central dominating control. At the present time there is inefficiency and over-lapping, and there is certainly waste. Why is it that this skilful aircraft industry of Britain requires ninety thousand men, and that it produces only one-half to one-third of what is being produced by about one hundred and ten thousand men in Germany? Is that not an extra-ordinary fact? It is incredible that we have not been able to produce a greater supply of aeroplanes at this time. Given a plain office table, an empty field, money, and labour, we should receive a flow of aero-planes by eighteen months; yet this is the thirty-fourth month since Lord Baldwin decided that the Air Force must be tripled.

*   *   *   *   *

The new Secretary of State for Air, Sir Kingsley Wood, invited me to remain on the Air Defence Research Committee. The skies had now grown much darker, and I felt keenly the need of Lindemann's interpretation of the technical aspects and of his advice and aid. I therefore wrote to him saying that unless he was associated with me I would not continue. After some tussling behind the scenes Lindemann was placed on the main Committee, and we resumed our joint work.

*   *   *   *   *

Always, up till the armistice of June 1940, whether in peace or war, in a private station or as head of the Government, I enjoyed confidential relations with the often-changing Premiers of the French Republic and with many of its leading Ministers. I was most anxious to find out the truth about German rearmament and to cross-check my own calculations by theirs.

I therefore wrote to M. Daladier, with whom I was personally acquainted.

Mr. Churchill to M. Daladier   *May* 3, 1938

Your predecessors, MM. Blum and Flandin, were both kind enough to give me the French estimates of the German air strength at particular periods in recent years. I should be much obliged if you could let me know what your view is now. I have several sources of information which have proved accurate in the past, but am anxious to have a cross-check from an independent source.

I am so glad that your visit here was so successful, and I hope now that all those staff arrangements will be made, the need for which I have pressed upon our Ministers.

In response M. Daladier sent me a document of seventeen pages dated May 11, 1938, which "had been deeply thought out by the French Air Staff". I showed this important paper to my friends in the British department concerned, who examined it searchingly and reported that "it agreed in every essential with the independent opinions formed by the British Air Staff on the basis of their own information". The French estimate of the size of the German Air Force was slightly higher than the British. Early in June I was in a position to write to M. Daladier with a considerable amount of authoritative opinion behind me.

Mr. Churchill to M. Daladier   *June* 6, 1938

I am very much obliged to you for the invaluable information which I have received through the French Military Attaché. You may be sure I shall use it only with the greatest discretion, and in our common interests.

The general estimate of the German Air Force at the present time agrees with the private views I have been able to form. I am inclined to think however that the German aircraft industry is turning out aircraft at a somewhat higher rate than is allowed, and that the figure given is that for the actual deliveries of aircraft of military types to the German Air Force, excluding deliveries for export, and to General Franco. It is probable that the German Air Force will consist of three hundred squadrons by April 1, 1939, and four hundred squadrons by April 1, 1940.

I was also most anxious to cross-check my own estimates of the German Army with those which I had been able to form from English sources. Accordingly I added the following:

I venture to enclose a very short note of the information I have been able to gather from various sources about the present and prospective strength of the German Army. It would be a convenience to me to know whether this agrees broadly with your estimates. It would be quite sufficient if the figures, as you understand them, could be pencilled in in any case where you think I am in error.

NOTE

The German Army at this date, June 1, consists of 36 regular divisions, and 4 armoured divisions, the whole at full war-strength. The non-armoured divisions are rapidly acquiring the power to triple themselves, and can at the present time be doubled. The artillery beyond 70 divisions is markedly incomplete. The Officer Corps is thin over the whole force. Nevertheless by October 1, 1938, we cannot expect less than 56 plus 4 armoured, equals 60 fully equipped and armed divisional formations. Behind these will stand a reservoir of trained men equal in man-power to about another 36 divisions, for which skeleton formations have been devised and for which armaments, small arms, and a very low complement of artillery would be available if a lower standard were accepted for part of the active army. This takes no account of the man-power of Austria, which at the extreme computation could provide 12 divisions without arms but ready to draw on the general pool of German munitions industry. In addition there are a number of men and formations of an unbrigaded nature—frontier defence force, Landwehr divisions, and so on, who are relatively unarmed.

On June 18, 1938, M. Daladier wrote:

I am particularly pleased to learn that the information enclosed in my letter of May 16 corresponds to yours.

I am entirely in accord with you in the facts relating to the German Army contained in the note annexed to your letter of June 6. It should be pointed out however that of the 36 ordinary divisions of which Germany actually disposes 4 are entirely motorised and 2 are in the course of becoming so soon.

In fact, according to our post-war information from German sources, this epitome of the German Army in the summer of 1938 was remarkably accurate, considering that it was produced by a private person. It shows that in my long series of campaigns for British rearmament I was by no means ill-informed.

* * * * *

References have been made at various points in this tale to the French air-power. At one time it was double our own and

Germany was not supposed to have an Air Force at all. Until 1933 France had held a high place among the air fleets of Europe. But in the very year in which Hitler came into power a fateful lack of interest and support began to be displayed. Money was grudged; the productive capacity of the factories was allowed to dwindle; modern types were not developed. The French forty-hour week could not rival the output of a Germany working harsh hours under war-time conditions. All this happened about the same time as the loss of air parity in Britain which has been so fully described. In fact the Western Allies, who had the right to create whatever Air Forces they thought necessary for their safety, neglected this vital weapon, while the Germans, who were prohibited by treaty from touching it, made it the spear-point of their diplomacy and eventual attack.

The French "Popular Front" Government of 1936 and later took many substantial measures to prepare the French Army and Navy for war. No corresponding exertion was made in the air. There is an ugly graph in Appendix E which shows in a decisive fashion the downward streak of French air-power and its inter-section in 1935 by the line of ever-rising German achievement. It was not until January 1938, when M. Guy La Chambre became Air Minister, that vigorous steps were taken to revive the French Air Force. But then only eighteen months remained. Nothing that the French could do could prevent the German Army grow-ing and ripening as each year passed and thus overtaking their own Army. But it is astonishing that their air-power should have been allowed to fall by the wayside. It is not for me to apportion responsibility and blame to the Ministers of friendly and Allied foreign countries, but when in France they are looking out for "guilty men" it would seem that here is a field which might well be searchingly explored.

*　*　*　*　*

The spirit of the British nation and of the Parliament they had newly elected gradually rose as consciousness of the German, and soon of the German-Italian, menace slowly and fitfully dawned upon them. They became willing, and even eager, for all kinds of steps which, taken two or three years earlier, would have prevented their troubles. But as their mood improved the power of their opponents and also the difficulty of their task increased.

Many say that nothing except war could have stopped Hitler after we had submitted to the seizure of the Rhineland. This may indeed be the verdict of future generations. Much however could have been done to make us better prepared and thus lessen our hazards. And who shall say what could not have happened?

CHAPTER XIV

MR. EDEN AT THE FOREIGN OFFICE.
HIS RESIGNATION

Foreign Secretary and Prime Minister – Eden and Chamberlain – Sir Robert Vansittart – My Contacts with the Foreign Secretary about Spain – The Nyon Conference – Our Correspondence – A British Success – Divergence between Prime Minister and Foreign Secretary – Lord Halifax Visits Germany and Hitler – I Decline an Invitation – Eden Feels Isolated – President Roosevelt's Overture – The Prime Minister's Reply – The President Rebuffed and Discouraged – Mr. Chamberlain's Grave Responsibility – Final Breach between Eden and Chamberlain about Conversations in Rome – A Sleepless Night at Chartwell.

THE Foreign Secretary has a special position in a British Cabinet. He is treated with marked respect in his high and responsible office, but he usually conducts his affairs under the continuous scrutiny, if not of the whole Cabinet, at least of its principal members. He is under an obligation to keep them informed. He circulates to his colleagues, as a matter of custom and routine, all his executive telegrams, the reports from our Embassies abroad, the records of his interviews with foreign Ambassadors or other notables. At least this has been the case during my experience of Cabinet life. This supervision is of course especially maintained by the Prime Minister, who personally or through his Cabinet is responsible for controlling, and has the power to control, the main course of foreign policy. From him at least there must be no secrets. No Foreign Secretary can do his work unless he is supported constantly by his chief. To make things go smoothly, there must not only be agreement between them on fundamentals, but also a harmony of outlook and even to some extent of temperament. This is all the more important if the Prime Minister himself devotes special attention to foreign affairs.

Eden was the Foreign Secretary of Mr. Baldwin, who, apart from his main well-known desire for peace and a quiet life, took no active share in foreign policy. Mr. Chamberlain, on the other hand, sought to exercise a masterful control in many departments. He had strong views about foreign affairs, and from the beginning asserted his undoubted right to discuss them with foreign Ambassadors. His assumption of the Premiership therefore implied a delicate but perceptible change in the position of the Foreign Secretary.

To this was added a profound, though at first latent, difference of spirit and opinion. The Prime Minister wished to get on good terms with the two European Dictators, and believed that conciliation and the avoidance of anything likely to offend them was the best method. Eden, on the other hand, had won his reputation at Geneva by rallying the nations of Europe against one Dictator; and, left to himself, might well have carried Sanctions to the verge of war, and perhaps beyond. He was a devoted adherent of the French Entente. He had just insisted upon "Staff conversations". He was anxious to have more intimate relations with Soviet Russia. He felt and feared the Hitler peril. He was alarmed by the weakness of our armaments, and its reaction on foreign affairs. It might almost be said that there was not much difference of view between him and me, except of course that he was in harness. It seemed therefore to me from the beginning that differences would be likely to arise between these two leading Ministerial figures as the world situation became more acute.

Moreover, in Lord Halifax the Prime Minister had a colleague who seemed to share his views on foreign affairs with sympathy and conviction. My long and intimate associations with Edward Halifax dated from 1922, when, in the days of Lloyd George, he became my Under-Secretary at the Dominions and Colonial Office. Political differences—even as serious and prolonged as those which arose between us about his policy as Viceroy of India—had never destroyed our personal relations. I thought I knew him very well, and I was sure that there was a gulf between us. I felt also that this same gulf, or one like it, was open between him and Anthony Eden. It would have been wiser, on the whole, for Mr. Chamberlain to have made Lord Halifax his Foreign

Secretary when he formed his Government. Eden would have been far more happily placed in the War Office or the Admiralty, and the Prime Minister would have had a kindred spirit and his own man at the Foreign Office. This inauspicious situation developed steadily during the year that Eden and Chamberlain worked together.

* * * * *

Up to this time and during many anxious years Sir Robert Vansittart had been the official head of the Foreign Office. His fortuitous connection with the Hoare-Laval pact had affected his position both with the new Foreign Secretary, Mr. Eden, and in wide political circles. The Prime Minister, who leaned more and more upon his Chief Industrial Adviser, Sir Horace Wilson, and consulted him a great deal on matters entirely outside his province or compass, regarded Vansittart as hostile to Germany. This was indeed true, for no one more clearly realised or foresaw the growth of the German danger or was more ready to subordinate other considerations to meeting it. The Foreign Secretary felt he could work more easily with Sir Alexander Cadogan, a Foreign Office official also of the highest character and ability. Therefore, at the end of 1937 Vansittart was apprised of his impending dismissal, and on January 1, 1938, was appointed to the special post of "Chief Diplomatic Adviser to His Majesty's Government". This was represented to the public as promotion, and might well indeed appear to be so. In fact however the whole responsibility for managing the Foreign Office passed out of his hands. He kept his old traditional room, but he saw the Foreign Office telegrams only after they had reached the Foreign Secretary with the minutes of the department upon them. Vansittart, who refused the Embassy in Paris, continued in this detached position for some time.

* * * * *

Between the summer of 1937 and the end of that year divergence, both in method and aim, grew between the Prime Minister and his Foreign Secretary. The sequence of events which led to Mr. Eden's resignation in February 1938 followed a logical course.

The original points of difference arose about our relations with Germany and Italy. Mr. Chamberlain was determined to press his suit with the two Dictators. In July 1937 he invited the Italian

Ambassador, Count Grandi, to Downing Street. The conversation took place with the knowledge but not in the presence of Mr. Eden. Mr. Chamberlain spoke of his desire for an improvement in Anglo-Italian relations. Count Grandi suggested to him that as a preliminary move it might be well if the Prime Minister were to write a personal appeal to Mussolini. Mr. Chamberlain sat down and wrote such a letter during the interview. It was dispatched without reference to the Foreign Secretary, who was in the Foreign Office a few yards away. The letter produced no apparent results, and our relations with Italy, because of the increasing Italian intervention in Spain, got steadily worse.

Mr. Chamberlain was imbued with a sense of a special and personal mission to come to friendly terms with the Dictators of Italy and Germany, and he conceived himself capable of achieving this relationship. To Mussolini he wished to accord recognition of the Italian conquest of Abyssinia as a prelude to a general settlement of differences. To Hitler he was prepared to offer colonial concessions. At the same time he was disinclined to consider in a conspicuous manner the improvements of British armaments or the necessity of close collaboration with France, both on the Staff and political levels. Mr. Eden, on the other hand, was convinced that any arrangement with Italy must be part of a general Mediterranean settlement, which must include Spain, and be reached in close understanding with France. In the negotiation of such a settlement our recognition of Italy's position in Abyssinia would clearly be an important bargaining counter. To throw this away in the prelude and appear eager to initiate negotiations was, in the Foreign Secretary's view, unwise.

During the autumn of 1937 these differences became more severe. Mr. Chamberlain considered that the Foreign Office was obstructing his attempts to open discussions with Germany and Italy, and Mr. Eden felt that his chief was displaying immoderate haste in approaching the Dictators, particularly while British armaments were so weak. There was in fact a profound practical and psychological divergence of view.

*　　*　　*　　*　　*

In spite of my differences with the Government, I was in close sympathy with their Foreign Secretary. He seemed to me the most resolute and courageous figure in the Administration, and although as a Private Secretary and later as an Under-Secretary of State in the Foreign Office he had had to adapt himself to many things I had attacked and still condemn, I felt sure his heart was in the right place and that he had the root of the matter in him. For his part, he made a point of inviting me to Foreign Office functions, and we corresponded freely. There was of course no impropriety in this practice, and Mr. Eden held to the well-established precedent whereby the Foreign Secretary is accustomed to keep in contact with the prominent political figures of the day on all broad international issues.

On August 7, 1937, I wrote to him:

This Spanish business cuts across my thoughts. It seems to me most important to make Blum stay with us strictly neutral, even if Germany and Italy continue to back the rebels and Russia sends money to the Government. If the French Government takes sides against the rebels it will be a godsend to the Germans and pro-Germans. In case you have a spare moment look at my article in the *Evening Standard* on Monday.

In this article I had written:

The worst quarrels only arise when both sides are equally in the right and in the wrong. Here, on the one hand, the passions of a poverty-stricken and backward proletariat demand the overthrow of Church, State, and property, and the inauguration of a Communist *régime*. On the other hand, the patriotic, religious, and *bourgeois* forces, under the leadership of the Army, and sustained by the countryside in many provinces, are marching to re-establish order by setting up a military dictatorship. The cruelties and ruthless executions extorted by the desperation of both sides, the appalling hatreds unloosed, the clash of creed and interest, make it only too probable that victory will be followed by the merciless extermination of the active elements of the vanquished and by a prolonged period of iron rule.

In the autumn of 1937 Eden and I had reached, though by somewhat different paths, a similar standpoint against active Axis intervention in the Spanish Civil War. I always supported him in the House when he took resolute action, even though it was upon a very limited scale. I knew well what his difficulties were with some of his senior colleagues in the Cabinet and with his chief, and that he would act more boldly if he were not enmeshed. We saw a good deal of each other at the end of August at Cannes, and one day I gave him and Mr. Lloyd George luncheon at a restaurant half-way between Cannes and Nice. Our conversation ran over the whole field—the Spanish struggle, Mussolini's persistent bad faith and intervention in Spain, and finally of course the dark background of ever-growing German power. I thought we were all three pretty well agreed. The Foreign Secretary was naturally most guarded about his relations with his chief and colleagues, and no reference was made to this delicate topic. Nothing could have been more correct than his demeanour. Nevertheless I was sure he was not a happy man in his great office.

★ ★ ★ ★ ★

Soon in the Mediterranean a crisis arose which he handled with firmness and skill, and which was accordingly solved in a manner reflecting a gleam of credit upon our course. A number of merchant ships had been sunk by so-called Spanish submarines. Actually there was no doubt that they were not Spanish but Italian. This was sheer piracy, and it stirred all who knew about it to action. A Conference of the Mediterranean Powers was convened at Nyon for September 10. To this the Foreign Secretary, accompanied by Vansittart and Lord Chatfield, the First Sea Lord, proceeded.

Mr. Churchill to Mr. Eden 9.IX.37

In your last letter you said that you would be very glad to see Lloyd George and me before you left for Geneva. We have met to-day, and I venture to let you know our views.

This is the moment to rally Italy to her international duty. Submarine piracy in the Mediterranean and the sinking of ships of many countries without any care for the lives of their crews must be suppressed. For this purpose all Mediterranean Powers should agree to keep their own submarines away from certain defined routes for commerce. In these routes the French and British Navies should search for all submarines, and any found by the detector apparatus should be pursued and sunk as pirates. Italy should be asked in the most courteous manner to participate in this. If however she will not do so, she should be told "that is what we are going to do".

At the same time, as it is very important to have the friendly concurrence of Italy, France should say that unless this concurrence is obtained she will open the Pyrenees frontier to the import of munitions of all kinds. Thus, on the one hand, Italy would be faced by the fact that the sea routes through the Mediterranean are going to be cleared of pirate submarines whatever happens, while at the same time she will gain nothing by not joining in, because the French frontier will be open. This point we consider essential. This combination of pressure upon Italy to join with the other Mediterranean Powers, coupled with the fact that she would risk much and gain nothing by standing out, would almost certainly be effective provided Mussolini knows that France and England are in earnest.

It is not believed that Germany is ready for a major war this year, and if it is hoped to have good relations with Italy in the future matters should be brought to a head now. The danger from which we suffer is that Mussolini thinks all can be carried off by bluff and bullying, and that in the end we shall only blether and withdraw. It is in the interests of European peace that a firm front should be shown now, and if you feel able to act in this sense we wish to assure you of our support upon such a policy in the House of Commons and in the country however matters may turn.

Speaking personally, I feel that this is as important a moment for you as when you insisted upon the Staff conversations with France after the violation of the Rhineland. The bold path is the path of safety.

Pray make any use of this letter privately or publicly that you may consider helpful to British interests and to the interests of peace.

P.S.—I have read this letter to Mr. Lloyd George, who declares himself in full agreement with it.

The Conference at Nyon was brief and successful. It was agreed to establish British and French anti-submarine patrols, with orders which left no doubt as to the fate of any submarine encountered. This was acquiesced in by Italy, and the outrages stopped at once.

Mr. Eden to Mr. Churchill   14.IX.37

You will now have seen the line which we have taken at Nyon, which in part at least coincides with that suggested in your letter. I hope you will agree that the results of the Conference are satisfactory. They seem so as viewed from here. The really important political fact is that we have emphasised that co-operation between Britain and France can be effective, and that the two Western Democracies can still play a decisive part in European affairs. The programme upon which we eventually agreed was worked out jointly by the French and ourselves. I must say that they could not have co-operated more sincerely, and we have been surprised at the extent of the naval co-operation which they have been ready to offer. It is fair to say that if we include their help in the air we shall be working on a fifty-fifty basis.

I agree that what we have done here only deals with one aspect of the Spanish problem. But it has much increased our authority among the nations at a time when we needed such an increase badly. The attitude of the smaller Powers of the Mediterranean was no less satisfactory. They played up well under the almost effusively friendly lead of Turkey. Chatfield has been a great success with everyone, and I feel that the Nyon Conference, by its brevity and success, has done something to put us on the map again. I hope that this may be your feeling too.

At least it has heartened the French and ourselves to tackle our immensely formidable task together.

Mr. Churchill to Mr. Eden   20.IX.37

It was very good of you, when so busy, to write to me. Indeed I congratulate you on a very considerable achievement. It is only rarely that an opportunity comes when stern and effective measures can be brought to bear upon an evildoer without incurring the risk of war. I have no doubt that the House of Commons will be very much pleased with the result.

I was very glad to see that Neville has been backing you up, and not, as represented by the Popular Press, holding you back by the coat-tails. My hope is that the advantages you have gained will be firmly held on to. Mussolini only understands superior force, such as he is now confronted with in the Mediterranean. The whole naval position there is transformed from the moment that the French bases are at our disposal. Italy cannot resist an effective Anglo-French combination. I hope therefore that Mussolini will be left to find his own way out of the diplomatic ditch into which he has blundered. The crystallisation against him for an unassailable purpose which has taken place in the Mediterranean is the one thing above all that he should have laboured to avoid. He has brought it about. I hope that the Anglo-French naval co-operation which has now begun will be continued indefinitely, and that both Navies and Air Forces will continue to use each other's facilities. This will be needed to prevent trouble arising about the Balearic Islands. The continued fortification of the Mediterranean by Italy against us will have to be dealt with in the future, as it is a capital danger to the British Empire. The more permanent the present arrangement becomes the less loaded with danger will this situation be.

Bernard Baruch telegraphs he is writing the results of his interview with the President [after our talks in London]. I have little doubt that the President's speech against dictatorships has been largely influenced by our talk, and I trust that the ground on the tariff and currency side is also being explored.

Mr. Eden to Mr. Churchill   25.IX.37

Thank you so much for your letter of September 20, and for the generous things you have written about Nyon, which I much appreciate I thought your summing up of the position at Nyon, "It is only rarely that an opportunity comes when stern and effective measures can be brought to bear upon an evildoer without incurring the risk of war," effectively described the position. Mussolini has been unwise enough to overstep the limits, and he has had to pay the penalty. There is no doubt that the spectacle of eighty Anglo-French destroyers patrolling the Mediterranean assisted by a considerable force of aircraft has made a profound impression on opinion in Europe. From reports which I have received, Germany herself has not been slow to take note of this fact. It was a great relief, both to Delbos and me, to be able to assert the position of our respective countries in this way in the autumn of a year in which we have inevitably had to be so much on the defensive. There is plenty of trouble ahead, and we are not yet of course anything like as strong in the military sense as I would wish, but Nyon has enabled us to improve our position and to gain more time.

I also cordially agree with you on the importance of the Anglo-French co-operation which we have now created in the Mediterranean. The whole French attitude was of course fundamentally different from that which prevailed when Laval was in command. The French Naval Staff could not have been more helpful, and they really made a great effort to make an important contribution to the joint force. Our Admiralty were, I am sure, impressed. Moreover, the mutual advantages to which you refer in respect of the use of each other's bases are very valuable. Nor will Italian participation, whatever its ultimate form, be able to affect the realities of the situation.

The Nyon Conference, although an incident, is a proof of how powerful the combined influence of Britain and France, if expressed with conviction and a readiness to use force, would have been upon the mood and policy of the Dictators. That such a policy would have prevented war at this stage cannot be asserted. It might easily have delayed it. It is the fact that whereas "appeasement" in all its forms only encouraged their aggression and gave the Dictators more power with their own peoples, any sign of a positive counter-offensive by the Western Democracies immediately produced an abatement of tension. This rule prevailed during the whole of 1937. After that the scene and conditions were different.

*   *   *   *   *

Early in October 1937 I was invited to a dinner at the Foreign
Office for the Yugoslav Premier, M. Stoyadinovitch. Afterwards, when we were all standing about and I was talking to
Eden, Lord Halifax came up and said in a genial way that Goering
had invited him to Germany on a sports visit, and the hope was
held out that he would certainly be able to see Hitler. He said
that he had spoken about it to the Prime Minister, who thought
it would be a very good thing, and therefore he had accepted.
I had the impression that Eden was surprised and did not like it;
but everything passed off pleasantly. Halifax therefore visited
Germany in his capacity as a Master of Foxhounds. The Nazi
Press welcomed him as *Lord Halalifax*, *"Halali!"* being a Continental hunting-cry, and after some sporting entertainment he
was in fact bidden to Berchtesgaden and had an informal and
none too ceremonious interview with the Fuehrer. This did not
go very well. One could hardly conceive two personalities less
able to comprehend one another. This High Church Yorkshire
aristocrat and ardent peace-lover, reared in all the smiling goodwill of former English life, who had taken his part in the war as
a good officer, met on the other side the demon-genius sprung
from the abyss of poverty, inflamed by defeat, devoured by
hatred and revenge, and convulsed by his design to make the
German race masters of Europe or maybe the world. Nothing
came of all this but chatter and bewilderment.

*   *   *   *   *

I may mention here that Ribbentrop twice tendered me an
invitation to visit Herr Hitler. Long before, as Colonial Under-
Secretary and a major in the Oxfordshire Yeomanry, I had been
the guest of the Kaiser at the German Manœuvres in 1907 and
in 1909. But now there was a different tune. Mortal quarrels
were afoot, and I had my station in them. I would gladly have
met Hitler with the authority of Britain behind me. But as a
private individual I should have placed myself and my country
at a disadvantage. If I had agreed with the Dictator-host I should
have misled him. If I had disagreed he would have been offended,
and I should have been accused of spoiling Anglo-German relations. Therefore I declined, or rather let lapse, both invitations.
All those Englishmen who visited the German Fuehrer in these
years were embarrassed or compromised. No one was more

completely misled than Mr. Lloyd George, whose rapturous accounts of his conversations make odd reading to-day. There is no doubt that Hitler had a power of fascinating men, and the sense of force and authority is apt to assert itself unduly upon the tourist. Unless the terms are equal it is better to keep away.

*　　*　　*　　*　　*

During these November days Eden became increasingly concerned about our slow rearmament. On the 11th he had an interview with the Prime Minister and tried to convey his misgivings. Mr. Neville Chamberlain after a while refused to listen to him. He advised him to "go home and take an aspirin". When Halifax returned from Berlin he reported that Hitler had told him the colonial question was the only outstanding issue between Britain and Germany. He believed the Germans were in no hurry. There was no immediate prospect of a peace deal. His conclusions were negative and his mood passive.

In February 1938 the Foreign Secretary conceived himself to be almost isolated in the Cabinet. The Prime Minister had strong support against him and his outlook. A whole band of important Ministers thought the Foreign Office policy dangerous and even provocative. On the other hand, a number of the younger Ministers were very ready to understand his point of view. Some of them later complained that he did not take them into his confidence. He did not however contemplate anything like forming a group against his leader. The Chiefs of Staff could give him no help. Indeed, they enjoined caution and dwelt upon the dangers of the situation. They were reluctant to draw too close to the French lest we should enter into engagements beyond our power to fulfil. They took a gloomy view of Russian military strength after the purge. They believed it necessary to deal with our problems as though we had three enemies—Germany, Italy, and Japan—who might all attack us together, and few to help us. We might ask for air bases in France, but we were not able to send an army in the first instance. Even this modest suggestion encountered strong resistance in the Cabinet.

*　　*　　*　　*　　*

But the actual breach came over a new and separate issue. On the evening of January 11, 1938, Mr. Sumner Welles, the American Under-Secretary of State, called upon the British Ambassador in Washington. He was the bearer of a secret and confidential message from President Roosevelt to Mr. Chamberlain. The President was deeply anxious at the deterioration of the international situation, and proposed to take the initiative by inviting the representatives of certain Governments to Washington to discuss the underlying causes of present difficulties. Before taking this step however he wished to consult the British Government on their view of such a plan, and stipulated that no other Government should be informed either of the nature or the existence of such a proposal. He asked that not later than January 17 he should be given a reply to his message, and intimated that only if his suggestion met with "the cordial approval and whole-hearted support of His Majesty's Government" would he then approach the Governments of France, Germany, and Italy. Here was a formidable and measureless step.

In forwarding this most secret message to London the British Ambassador, Sir Ronald Lindsay, commented that in his view the President's plan was a genuine effort to relax international tension, and that if His Majesty's Government withheld their support the progress which had been made in Anglo-American co-operation during the previous two years would be destroyed. He urged in the most earnest manner acceptance of the proposal by the British Government. The Foreign Office received the Washington telegram on January 12, and copies were sent to the Prime Minister in the country that evening. On the following morning he came to London, and on his instructions a reply was sent to the President's message. Mr. Eden was at this time on a brief holiday in the South of France. Mr. Chamberlain's reply was to the effect that he appreciated the confidence of President Roosevelt in consulting him in this fashion upon his proposed plan to alleviate the existing tension in Europe, but he wished to explain the position of his own efforts to reach agreement with Germany and Italy, particularly in the case of the latter. "His Majesty's Government would be prepared, for their part, if possible with the authority of the League of Nations, to recognise *de jure* the Italian occupation of Abyssinia, if they found that the Italian Government on their side were ready to give evidence of their desire to contribute to the restoration of confidence and friendly relations." The Prime Minister mentioned these facts, the message continued, so that the President might consider whether his present proposal might not cut across the British efforts. Would it not therefore be wiser to postpone the launching of the American plan?

This reply was received by the President with some disappointment. He intimated that he would reply by letter to Mr. Chamberlain on January 17. On the evening of January 15 the Foreign Secretary returned to England. He had been urged to come back, not by his chief, who was content to act without him, but by his devoted officials at the Foreign Office. The vigilant Alexander Cadogan awaited him upon the pier at Dover. Mr. Eden, who had worked long and hard to improve Anglo-American relations, was deeply perturbed. He immediately sent a telegram to Sir Ronald Lindsay attempting to minimise the effects of Mr. Chamberlain's chilling answer. The President's letter reached London on the morning of January 18. In it he agreed to postpone making his proposal in view of the fact that the British Government were contemplating direct negotiations, but he added that he was gravely concerned at the suggestion that His Majesty's Government might accord recognition to the Italian position in Abyssinia. He thought that this would have a most harmful effect upon Japanese policy in the Far East and upon American public opinion. Mr. Cordell Hull, in delivering this letter to the British Ambassador in Washington, expressed himself even more emphatically. He said that such a recognition would "rouse a feeling of disgust, would revive and multiply all fears of pulling the chestnuts out of the fire; it would be represented as a corrupt bargain completed in Europe at the expense of interests in the Far East in which America was intimately concerned".

The President's letter was considered at a series of meetings of the Foreign Affairs Committee of the Cabinet. Mr. Eden succeeded in procuring a considerable modification of the previous attitude. Most of the Ministers thought he was satisfied. He did not make it clear to them that he was not. Following these discussions two messages were sent to Washington on the evening of January 21. The substance of these replies was that the Prime Minister warmly welcomed the President's initiative, but was not anxious to bear any responsibility for its failure if American overtures were badly received. Mr. Chamberlain wished to point out that we did not accept in an unqualified manner the

President's suggested procedure, which would clearly irritate both the Dictators and Japan. Nor did His Majesty's Government feel that the President had fully understood our position in regard to *de jure* recognition. The second message was in fact an explanation of our attitude in this matter. We intended to accord such recognition only as part of a general settlement with Italy.

The British Ambassador reported his conversation with Mr. Sumner Welles when he handed these messages to the President on January 22. He stated that Mr. Welles told him that "the President regarded recognition as an unpleasant pill which we should both have to swallow, and he wished that we should both swallow it together".

Thus it was that President Roosevelt's proposal to use American influence for the purpose of bringing together the leading European Powers to discuss the chances of a general settlement, this of course involving however tentatively the mighty power of the United States, was rebuffed by Mr. Chamberlain. This attitude defined in a decisive manner the difference of view between the British Prime Minister and his Foreign Secretary. Their disagreements were still confined to the circle of the Cabinet for a little time longer; but the split was fundamental. The comments of Mr. Chamberlain's biographer, Professor Feiling, upon this episode, are not without interest. "While Chamberlain feared the Dictators would pay no heed or else *would use this line-up of the Democracies as a pretext for a break*, it was found on Eden's return that he would rather risk that calamity than the loss of American goodwill. There was the first breath of resignation. But a compromise was beaten out. . . ." Poor England! Leading her free, careless life from day to day, amid endless good-tempered Parliamentary babble, she followed, wondering, along the downward path which led to all she wanted to avoid. She was continually reassured by the leading articles of the most influential newspapers, with some honourable exceptions, and behaved as if all the world were as easy, uncalculating, and well-meaning as herself.

★ ★ ★ ★ ★

It was plain that no resignation by the Foreign Secretary could be founded upon the rebuff administered by Mr. Chamberlain to the President's overture. Mr. Roosevelt was indeed running great risks in his own domestic politics by deliberately involving the United States in the darkening European scene. All the forces of Isolationism would have been aroused if any part of these interchanges had transpired. On the other hand, no event could have been more likely to stave off, or even prevent, war than the arrival of the United States in the circle of European hates and fears. To Britain it was a matter almost of life and death. No one can measure in retrospect its effect upon the course of events in Austria and later at Munich. We must regard its rejection—for such it was—as the loss of the last frail chance to save the world from tyranny otherwise than by war. That Mr. Chamberlain, with his limited outlook and inexperience of the European scene, should have possessed the self-sufficiency to wave away the proffered hand stretched out across the Atlantic leaves one, even at this date, breathless with amazement. The lack of all sense of proportion, and even of self-preservation, which this episode reveals in an upright, competent, well-meaning man, charged with the destinies of our country and all who depended upon it, is appalling. One cannot to-day even reconstruct the state of mind which would render such gestures possible.

* * * * *

I have yet to unfold the story of the treatment of the Russian offers of collaboration in the advent of Munich. If only the British people could have known and realised that, having neglected our defences and sought to diminish the defences of France, we were now disengaging ourselves, one after the other, from the two mighty nations whose extreme efforts were needed to save our lives and their own, history might have taken a different turn. But all seemed so easy from day to day. Now ten years later let the lessons of the past be a guide.

* * * * *

It must have been with declining confidence in the future that Mr. Eden went to Paris on January 25 to consult with the French. Everything now turned upon the success of the approach to Italy, of which we had made such a point in our replies to the President. The French Ministers impressed upon Mr. Eden the necessity of the inclusion of Spain in any general settlement with the Italians; on this he needed little convincing. On February 10 the Prime Minister and the Foreign Secretary met Count Grandi, who declared that the Italians were ready in principle to open the conversations.

On February 15 the news came of the submission of the Austrian Chancellor, Schuschnigg, to the German demand for the introduction into the Austrian Cabinet of the chief Nazi agent, Seyss-Inquart, as Minister of the Interior and Head of the Austrian Police. This grave event did not avert the personal crisis between Mr. Chamberlain and Mr. Eden. On February 18 they saw Count Grandi again. This was the last business they conducted together. The Ambassador refused either to discuss the Italian position towards Austria, or to consider the British plan for the withdrawal of volunteers, or so-called volunteers—in this case five divisions of the regular Italian Army—from Spain. Grandi asked however for general conversations to be opened in Rome. The Prime Minister wished for these, and the Foreign Secretary was strongly opposed to such a step.

There were prolonged parleyings and Cabinet meetings. Of these the only authoritative account yet disclosed is in Mr. Chamberlain's biography. Professor Feiling says that the Prime Minister "let the Cabinet see that the alternative to Eden's resignation might be his own". He quotes from some diary or private letter, to which he was given access, the following statement by the Prime Minister: "I thought it necessary to say clearly that I could not accept any decision in the opposite sense." "The Cabinet," says Mr. Feiling, "were unanimous, though with a few reserves." We have no knowledge of how and when these statements were made during the protracted discussions. But at the end Mr. Eden briefly tendered his resignation on the issue of the Italian conversations taking place at this stage and in these circumstances. At this his colleagues were astonished. Mr. Feiling says they were "much shaken". They had not realised that the differences between the Foreign Secretary and the Prime Minister had reached breaking-point. Evidently if Mr. Eden's resignation was involved a new question raising larger and more general issues was raised. However, they had all committed themselves on the merits of the matter in dispute. The rest of the long day was spent in efforts to induce the Foreign Secretary to change his mind. Mr. Chamberlain was impressed by the distress of the Cabinet. "Seeing how my colleagues had been taken aback, I proposed an adjournment until next day." But Eden saw no use in continuing a search for formulas, and by midnight, on the 20th, his resignation became final. "Greatly to his credit, as I see it," noted the Prime Minister. Lord Halifax was at once appointed Foreign Secretary in his place.[*]

It had of course become known that there were serious differences in the Cabinet, though the cause was obscure. I had heard something of this, but carefully abstained from any communication with Mr. Eden. I hoped that he would not on any account resign without building up his case beforehand, and giving his many friends in Parliament a chance to draw out the issues. But the Government at this time was so powerful and aloof that the struggle was fought out inside the Ministerial conclave, and mainly between the two men.

★   ★   ★   ★   ★

Late in the night of February 20 a telephone message reached me as I sat in my old room at Chartwell (as I often sit now) that Eden had resigned. I must confess that my heart sank, and for a while the dark waters of despair overwhelmed me. In a long life I have had many ups and downs. During all the war soon to come and in its darkest times I never had any trouble in sleeping. In the crisis of 1940, when so much responsibility lay upon me, and also at many very anxious, awkward moments in the following five years, I could always flop into bed and go to sleep after the day's work was done—subject of course to any emergency call. I slept sound and awoke refreshed, and had no feelings except appetite to grapple with whatever the morning's boxes might bring. But now on this night of February 20, 1938, and on this occasion only, sleep deserted me. From midnight till dawn I lay in my bed consumed by emotions of sorrow and fear. There seemed one strong young figure standing up against long, dismal, drawling tides of drift and surrender, of wrong measurements and feeble impulses. My conduct of affairs would have been different from his in various ways; but he seemed to me at this moment to embody the life-hope of the British nation, the grand old British race that had done so much for men, and had yet some more to give. Now he was gone. I watched the daylight slowly creep in through the windows, and saw before me in mental gaze the vision of Death.

[*] Feiling, *op. cit.*, p. 338.

CHAPTER XV

THE RAPE OF AUSTRIA

February 1938

"Case Otto" – Hitler Assumes Supreme Command – The Austrian Chancellor Summoned to Berchtesgaden – His Ordeal – Schuschnigg's Collapse – Hitler's Speech to the Reichstag, February 20 – Debate on Mr. Eden's Resignation – Hitler and Mussolini in Combination – The Austrian Plebiscite – The Invasion of Austria – Hitler's Debt to Mussolini – The Triumphal Entry into Vienna and its Background – A Farewell Luncheon to Ribbentrop – The Debate of March 12 – Consequences of the Fall of Vienna – Danger to Czechoslovakia – Mr. Chamberlain and the Soviet Overture – A Side Blow – Negotiation with Mr. de Valera – Surrender of the Irish Ports – A Major Injury to Britain – Irish Neutrality – My Vain Protest.

USUALLY in modern times when States have been defeated in war they have preserved their structure, their identity, and the secrecy of their archives. On this occasion, the war being fought to an utter finish, we have come into full possession of the inside story of the enemy. From this we can check up with some exactness our own information and performances. We have seen how in July 1936 Hitler had instructed the German General Staff to draw up military plans for the occupation of Austria when the hour should strike. This operation was labelled "Case Otto". Now, a year later, on June 24, 1937, he crystallised these plans by a special directive. On November 5 he unfolded his future designs to the chiefs of his armed forces. Germany must have more "living space". This could best be found in Eastern Europe—Poland, White Russia, and the Ukraine. To obtain this would involve a major war, and incidentally the extermination of the people then living in those parts. Germany would have to reckon with her two "hateful enemies", England and France, to whom "a German Colossus in the centre of Europe would be intolerable". In order to profit by the lead she had gained in munitions production and by the patriotic fervour aroused and represented by the Nazi Party, she must therefore make war at the first promising opportunity, and deal with her two obvious opponents before they were ready to fight.

Neurath, Fritsch, and even Blomberg, all of them influenced by the views of the German Foreign Office, General Staff, and Officer Corps, were alarmed by this policy. They thought that the risks to be run were too high. They recognised that by the audacity of the Fuehrer they were definitely ahead of the Allies in every form of rearmament. The Army was maturing month by month; the internal decay of France and the lack of will-power in Britain were favourable factors which might well run their full course. What was a year or two when all was moving so well? They must have time to complete the war machine, and a conciliatory speech now and again from the Fuehrer would keep these futile and degenerate democracies chattering. But Hitler was not sure of this. His genius taught him that victory would not be achieved by processes of certainty. Risks had to be run. The leap had to be made. He was flushed with his successes, first in rearmament, second in conscription, third in the Rhineland, fourth by the accession of Mussolini's Italy. To wait till everything was ready was probably to wait till all was too late. It is very easy for historians and other people, who do not have to live and act from day to day, to say that he would have had the whole fortunes of the world in his hand if he had gone on growing in strength for another two or three years before striking. However, this does not follow. There are no certainties in human life or in the life of States. Hitler was resolved to hurry, and have the war while he was in his prime.

Blomberg, weakened with the Officer Corps by an inappropriate marriage, was first removed; and then, on February 4, 1938, Hitler dismissed Fritsch, and himself assumed supreme command of the armed forces. So far as it is possible for one man, however gifted and powerful, however terrible the penalties he can inflict, to make his will effective over spheres so vast, the Fuehrer assumed direct control, not only of the policy of the

State, but of the military machine. He had at this time something like the power of Napoleon after Austerlitz and Jena, without of course the glory of winning great battles by his personal direction on horseback, but with triumphs in the political and diplomatic field which all his circle and followers knew were due alone to him and to his judgment and daring.

★ ★ ★ ★ ★

Apart from his resolve, so plainly described in *Mein Kampf,* to bring all Teutonic races into the Reich, Hitler had two reasons for wishing to absorb the Austrian Republic. It opened to Germany both the door of Czechoslovakia and the more spacious portals of South-eastern Europe. Since the murder of Chancellor Dollfuss in July 1934 by the Austrian section of the Nazi Party the process of subverting the independent Austrian Government by money, intrigue, and force had never ceased. The Nazi movement in Austria grew with every success that Hitler reaped elsewhere, whether inside Germany or against the Allies. It had been necessary to proceed step by step. Officially Papen was instructed to maintain the most cordial relations with the Austrian Government, and to procure the official recognition by them of the Austrian Nazi Party as a legal body. At that time the attitude of Mussolini had imposed restraint. After the murder of Dr. Dollfuss the Italian Dictator had flown to Venice to receive and comfort the widow, who had taken refuge there, and considerable Italian forces had been concentrated on the southern frontier of Austria. But now in the dawn of 1938 decisive changes in European groupings and values had taken place. The Siegfried Line confronted France with a growing barrier of steel and concrete, requiring as it seemed an enormous sacrifice of French manhood to pierce. The door from the West was shut. Mussolini had been driven into the German system by sanctions so ineffectual that they had angered him without weakening his power. He might well have pondered with relish on Machiavelli's celebrated remark, "Men avenge slight injuries, but not grave ones." Above all the Western Democracies had seemed to give repeated proofs that they would bow to violence so long as they were not themselves directly assailed. Papen was working skilfully inside the Austrian political structure. Many Austrian notables had yielded to his pressure and intrigues. The tourist trade, so important to Vienna, was impeded by the prevailing uncertainty. In the background terrorist activity and bomb outrages shook the frail life of the Austrian Republic.

It was thought that the hour had now come to obtain control of Austrian policy by procuring the entry into the Vienna Cabinet of the leaders of the lately legalised Austrian Nazi Party. On February 12, 1938, eight days after assuming the supreme command, Hitler had summoned the Austrian Chancellor, Herr von Schuschnigg, to Berchtesgaden. He had obeyed, and was accompanied by his Foreign Minister, Guido Schmidt. We now have Schuschnigg's record, in which the following dialogue occurs.* Hitler had mentioned the defences of the Austrian frontier. These were no more than might be required to make a military operation necessary to overcome them, and thus raise major issues of peace and war.

Hitler: "I only need to give an order, and overnight all the ridiculous scarecrows on the frontier will vanish. You don't really believe that you could hold me up for half an hour? Who knows—perhaps I shall be suddenly overnight in Vienna: like a spring storm. Then you will really experience something. I would willingly spare the Austrians this; it will cost many victims. *After the troops will follow the S.A. and the Legion!* No one will be able to hinder their vengeance, not even myself. Do you want to turn Austria into another Spain? All this I would like if possible to avoid."

Schuschnigg: "I will obtain the necessary information and put a stop to the building of any defence works on the German frontier. Naturally I realise that you can march into Austria, but, Mr. Chancellor, whether we wish it or not, that would lead to the shedding of blood. We are not alone in the world. That probably means war."

Hitler: "That is very easy to say at this moment as we sit here in club armchairs, but behind it all there lies a sum of suffering and blood. Will you take the responsibility for that, Herr Schuschnigg? Don't believe that anyone in the world will hinder me in my decisions! Italy? I am quite clear with Mussolini: with Italy I am on the closest possible terms. England? England will not lift a finger for Austria. . . . And France? Well, two years ago when we marched into the Rhineland with a handful of

* Schuschnigg, *Ein Requiem in Rot-Weiss-Rot*, p. 37 ff.

battalions—at that moment I risked a great deal. If France had marched then we should have been forced to withdraw. . . . But for France it is now too late!"

This first interview took place at eleven in the morning. After a formal lunch the Austrians were summoned into a small room, and there confronted by Ribbentrop and Papen with a written ultimatum. The terms were not open to discussion. They included the appointment of the Austrian Nazi Seyss-Inquart as Minister of Security in the Austrian Cabinet, a general amnesty for all Austrian Nazis under detention, and the official incorporation of the Austrian Nazi Party in the Government-sponsored Fatherland Front.

Later Hitler received the Austrian Chancellor. "I repeat to you, this is the very last chance. Within three days I expect the execution of this agreement." In Jodl's diary the entry reads, "Von Schuschnigg together with Guido Schmidt are again being put under heaviest political and military pressure. At 11 p.m. Schuschnigg signs the 'protocol'."* As Papen drove back with Schuschnigg in the sledge which conveyed them over the snow-covered roads to Salzburg he commented, "Yes, that is how the Fuehrer can be; now you have experienced it for yourself. But when you next come you will have a much easier time. The Fuehrer can be really charming."†

On February 20 Hitler spoke to the Reichstag:

I am happy to be able to tell you, gentlemen, that during the past few days a further understanding has been reached with a country that is particularly close to us for many reasons. The Reich and German Austria are bound together not only because they are the same people, but also because they share a long history and a common culture. The difficulties which had been experienced in carrying out the Agreement of July 11, 1936, compelled us to make an attempt to clear out of the way misunderstandings and hindrances to a final conciliation. Had this not been done, it is clear that an intolerable situation might one day have developed, whether intentionally or otherwise, which might have brought about a very serious catastrophe. I am glad to be able to assure you that these considerations corresponded with the views of the Austrian Chancellor, whom I invited to come to visit me. The idea and the intention were to bring about a relaxation of the strain in our relations with one another by giving

* *Nuremberg Documents* (H.M. Stationery Office), Pt. I, p. 249.
† Schuschnigg, *op. cit.*, pp. 51-2.

under the existing legislation the same legal rights to citizens holding National-Socialist views as are enjoyed by the other citizens of German Austria. In conjunction with this there should be a practical contribution towards peace by granting a general amnesty, and by creating a better understanding between the two States through a still closer friendly co-operation in as many different fields as possible—political, personal, and economic—all complementary to and within the framework of the Agreement of July 11. I express in this connection before the German people my sincere thanks to the Austrian Chancellor for his great understanding and the warm-hearted willingness with which he accepted my invitation and worked with me, so that we might discover a way of serving the best interests of the two countries; for, after all, it is the interest of the whole German people, whose sons we all are, wherever we may have been born.*

One can hardly find a more perfect specimen of humbug and hypocrisy for British and American benefit. I only print it because of its unique quality in these respects. What is astounding is that it should have been regarded with anything but scorn by men and women of intelligence in any free country.

*   *   *   *   *

For a moment we must return to the serious British event which the previous chapter has described. On the next day, February 21, there was an imposing debate in the House of Commons on the resignation of the Foreign Secretary and his Under-Secretary, Lord Cranborne—a man in whom "still waters ran deep"—who acted with him in loyalty and conviction. Eden could of course make no open reference to President Roosevelt's overture and its discouragement. The differences about Italy were on a minor plane. Eden said:

I have spoken of the immediate difference which has divided me from my colleagues, and I should not be frank if I were to pretend that it is an isolated issue. It is not. *Within the last few weeks upon one most important decision of foreign policy which did not concern Italy at all the difference was fundamental.*

He concluded:

I do not believe that we can make progress in European appeasement if we allow the impression to gain currency abroad that we yield to constant pressure. . . . I am certain in my own mind that progress

* *Hitler's Speeches* (ed. N. H. Baynes), II, 1407–8.

depends above all on the temper of the nation, and that temper must find expression in a firm spirit. That spirit I am confident is there. Not to give voice to it is, I believe, fair neither to this country nor to the world.

Mr. Attlee made a searching point. The resignation of Mr. Eden was being proclaimed in Italy as "another great victory for the Duce". "All over the world we hear the story, 'You see how great is the power of our Leader; the British Foreign Secretary has gone.'"

I did not speak till the second day, when I paid my tribute to both the resigning Ministers. I also sustained Mr. Attlee's accusation:

This last week has been a good week for the Dictators—one of the best they have ever had. The German Dictator has laid his heavy hand upon a small but historic country, and the Italian Dictator has carried his vendetta against Mr. Eden to a victorious conclusion. The conflict between them has been long. There can be no doubt whatever that Signor Mussolini has won. All the majesty, power, and dominion of the British Empire have not been able to secure the success of the causes which were entrusted to the late Foreign Secretary by the general will of Parliament and of the country. . . . So that is the end of this part of the story, namely, the departure from power of the Englishman whom the British nation and the British Parliament entrusted with a certain task, and the complete triumph of the Italian Dictator, at a moment when he desperately needed success for domestic reasons. All over the world, in every land, under every sky and every system of government, wherever they may be, the friends of England are dismayed and the foes of England are exultant. . . .

The resignation of the late Foreign Secretary may well be a milestone in history. Great quarrels, it has been well said, arise from small occasions but seldom from small causes. The late Foreign Secretary adhered to the old policy which we have all forgotten for so long. The Prime Minister and his colleagues have entered upon another and a new policy. The old policy was an effort to establish the rule of law in Europe, and build up through the League of Nations effective deterrents against the aggressor. Is it the new policy to come to terms with the totalitarian Powers in the hope that by great and far-reaching acts of submission, not merely in sentiment and pride, but in material factors, peace may be preserved?

The other day Lord Halifax said that Europe was confused. The part of Europe which is confused is that part ruled by Parliamentary Governments. I know of no confusion on the side of the great

Dictators. They know what they want, and no one can deny that up to the present at every step they are getting what they want. The grave and largely irreparable injury to world security took place in the years 1932 to 1935. . . . The next opportunity when the Sibylline books were presented to us was the reoccupation of the Rhineland at the beginning of 1936. Now we know that a firm stand by France and Britain, under the authority of the League of Nations, would have been followed by the immediate evacuation of the Rhineland without the shedding of a drop of blood; and the effects of that *might have enabled the more prudent elements in the German Army to regain their proper position*, and would not have given to the political head of Germany that enormous ascendancy which has enabled him to move forward. Now we are at a moment when a third move is made, but when that opportunity does not present itself in the same favourable manner. Austria has been laid in thrall, *and we do not know whether Czechoslovakia will not suffer a similar attack.*

* * * * *

The Continental drama ran its course. Mussolini now sent a verbal message to Schuschnigg saying that he considered the Austrian attitude at Berchtesgaden to be both right and adroit. He assured him both of the unalterable attitude of Italy towards the Austrian question and of his personal friendship. On February 24 the Austrian Chancellor himself spoke to the Austrian Parliament, welcoming the settlement with Germany, but emphasising, with some sharpness, that beyond the specific terms of the agreement Austria would never go. On March 3 he sent a confidential message to Mussolini through the Austrian Military Attaché in Rome informing the Duce that he intended to strengthen the political position in Austria by holding a plebiscite. Twenty-four hours later he received a message from the Austrian Military Attaché in Rome describing his interview with Mussolini. In this the Duce expressed himself optimistically. The situation would improve. An imminent *détente* between Rome and London would ensure a lightening of the existing pressure. . . . As to the plebiscite Mussolini uttered a warning: "*E un errore*" (It's a mistake). "If the result is satisfactory, people will say that it is not genuine. If it is bad, the situation of the Government will be unbearable; and if it is indecisive, then it is worthless." But Schuschnigg was determined. On March 9 he announced officially that a plebiscite would be held throughout Austria on the following Sunday, March 13.

At first nothing happened. Seyss-Inquart seemed to accept the idea without demur. At 5.30 a.m. however on the morning of March 11 Schuschnigg was rung up on the telephone from Police Headquarters in Vienna. He was told: "The German frontier at Salzburg was closed an hour ago. The German customs officials have been withdrawn. Railway communications have been cut." The next message to reach the Austrian Chancellor was from his Consul-General in Munich saying that the German Army Corps there had been mobilised: supposed destination—Austria!

Later in the morning Seyss-Inquart came to announce that Goering had just telephoned to him that the plebiscite must be called off within an hour. If no reply was received within that time Goering would assume that Seyss-Inquart had been hindered from telephoning, and would act accordingly. After being informed by responsible officials that the police and Army were not entirely reliable, Schuschnigg informed Seyss-Inquart that the plebiscite would be postponed. A quarter of an hour later the latter returned with a reply from Goering scribbled on a message-pad:

The situation can only be saved if the Chancellor resigns immediately and if within two hours Dr. Seyss-Inquart is nominated Chancellor. If nothing is done within this period the German invasion of Austria will follow.*

Schuschnigg waited on President Miklas to tender his resignation. While in the President's room he received a deciphered message from the Italian Government that they could offer no counsel. The old President was obstinate: "So in the decisive hour I am left alone." He steadfastly refused to nominate a Nazi Chancellor. He was determined to force the Germans into a shameful and violent deed. But for this they were well prepared.

A vivid account of the German reaction is found again in Jodl's diary for March 10:

By surprise and without consulting his Ministers, von Schuschnigg ordered a plebiscite for Sunday, March 13, which should bring a strong majority for the legitimate party *in the absence of plan or preparation.* The Fuehrer is determined not to tolerate it. This very night, March 9-10, he calls for Goering. General von Reichenau is called back from the Cairo Olympic Committee; General von Schubert is

* Schuschnigg, *op. cit.,* pp. 66, 72.

ordered to come, as well as Minister Glaise-Horstenau, who is with the District Leader [Gauleiter Burckel] in the Palatinate. General Keitel communicates the facts at 1.45. He drives to the Reichskanzlei at ten o'clock. I follow at 10.15 to give him the old draft "Prepare Case Otto". 13.00 hours, General K. [Keitel] informs Chief of Operational Staff and Admiral Canaris; Ribbentrop is detained in London. Neurath takes over the Foreign Office. Fuehrer wants to transmit ultimatum to the Austrian Cabinet. A personal letter is dispatched to Mussolini, and the reasons are developed which forced the Fuehrer to take action.*

On the following day, March 11, orders were issued by Hitler to the German armed forces for the military occupation of Austria. Operation "Otto", so long studied, so carefully prepared, began. President Miklas confronted Seyss-Inquart and the Austrian Nazi leaders in Vienna with firmness throughout a hectic day. The telephone conversation between Hitler and Prince Philip of Hesse, his special envoy to the Duce, was quoted in evidence at Nuremberg, and is of interest:

Hesse: I have just come back from the Palazzo Venezia. The Duce accepted the whole thing in a very friendly manner. He sends you his regards. He had been informed from Austria; von Schuschnigg gave him the news. He had then said it [*i.e.*, Italian intervention] would be a complete impossibility; it would be a bluff; such a thing could not be done. So he [Schuschnigg] was told that it was unfortunately arranged thus, and it could not be changed any more. Then Mussolini said that Austria would be immaterial to him.

Hitler: Then please tell Mussolini I will never forget him for this.

Hesse: Yes.

Hitler: Never, never, never, whatever happens. I am still ready to make a quite different agreement with him.

Hesse: Yes, I told him that too.

Hitler: As soon as the Austrian affair has been settled I shall be ready to go with him through thick and thin; nothing matters.

Hesse: Yes, my Fuehrer.

Hitler: Listen. I will make any agreement—I am no longer in fear of the terrible position which would have existed militarily in case we had become involved in a conflict. You may tell him that I do thank him ever so much; never, never shall I forget that.

Hesse: Yes, my Fuehrer.

Hitler: I will never forget it, whatever may happen. If he should ever need any help or be in any danger he can be convinced that I shall

* *Nuremberg Documents*, I, p. 251.

stick to him whatever might happen, even if the whole world were against him.

Hesse: Yes, my Fuehrer.*

Certainly when he rescued Mussolini from the Italian Provisional Government in 1943 Hitler kept his word.

* * * * *

A triumphal entry into Vienna had been the Austrian Corporal's dream. On the night of Saturday, March 12, the Nazi Party in the capital had planned a torchlight procession to welcome the conquering hero. But nobody arrived. Three bewildered Bavarians of the supply services who had come by train to make billeting arrangements for the invading army had therefore to be carried shoulder-high through the streets. The cause of this hitch leaked out slowly. The German war machine had lumbered falteringly over the frontier and come to a standstill near Linz. In spite of perfect weather and road conditions the majority of the tanks broke down. Defects appeared in the motorised heavy artillery. The road from Linz to Vienna was blocked with heavy vehicles at a standstill. General von Reichenau, Hitler's special favourite, Commander of Army Group IV, was deemed responsible for a breakdown which exposed the unripe condition of the German Army at this stage in its reconstruction.

Hitler himself, motoring through Linz, saw the traffic jam, and was infuriated. The light tanks were disengaged from confusion and straggled into Vienna in the early hours of Sunday morning. The armoured vehicles and motorised heavy artillery were loaded on to the railway trucks, and only thus arrived in time for the ceremony. The pictures of Hitler driving through Vienna amid exultant or terrified crowds are well known. But this moment of mystic glory had an unquiet background. The Fuehrer was in fact convulsed with anger at the obvious shortcomings of his military machine. He rated his generals, and they answered back. They reminded him of his refusal to listen to Fritsch and his warnings that Germany was not in a position to undertake the risk of a major conflict. Appearances were preserved. The official celebrations and parades took place. On the Sunday, after large numbers of German troops and Austrian Nazis had taken possession of Vienna, Hitler declared the dissolu-

* Schuschnigg, *op. cit.*, pp. 102-3, and *Nuremberg Documents*, I, pp. 258-9.

tion of the Austrian Republic and the annexation of its territory to the German Reich.

*   *   *   *   *

Herr von Ribbentrop was at this time about to leave London to take up his duties as Foreign Secretary in Germany. Mr. Chamberlain gave a farewell luncheon in his honour at No. 10 Downing Street. My wife and I accepted the Prime Minister's invitation to attend. There were perhaps sixteen people present. My wife sat next to Sir Alexander Cadogan, near one end of the table. About half-way through the meal a Foreign Office messenger brought him an envelope. He opened it and was absorbed in the contents. Then he got up, walked round to where the Prime Minister was sitting, and gave him the message. Although Cadogan's demeanour would not have indicated that anything had happened, I could not help noticing the Prime Minister's evident preoccupation. Presently Cadogan came back with the paper and resumed his seat. Later I was told its contents. It said that Hitler had invaded Austria and that the German mechanised forces were advancing fast upon Vienna. The meal proceeded without the slightest interruption, but quite soon Mrs. Chamberlain, who had received some signal from her husband, got up, saying, "Let us *all* have coffee in the drawing-room." We trooped in there, and it was evident to me and perhaps to some others that Mr. and Mrs. Chamberlain wished to bring the proceedings to an end. A kind of general restlessness pervaded the company, and everyone stood about ready to say good-bye to the guests of honour.

However, Herr von Ribbentrop and his wife did not seem at all conscious of this atmosphere. On the contrary, they tarried for nearly half an hour engaging their host and hostess in voluble conversation. At one moment I came in contact with Frau von Ribbentrop, and in a valedictory vein I said, "I hope England and Germany will preserve their friendship." "Be careful you don't spoil it," was her graceful rejoinder. I am sure they both knew perfectly well what had happened, but thought it was a good manœuvre to keep the Prime Minister away from his work and the telephone. At length Mr. Chamberlain said to the Ambassador, "I am sorry I have to go now to attend to urgent business," and without more ado he left the room. The Ribben-trops lingered on, so that most of us made our excuses and our way home. Eventually I suppose they left. This was the last time I saw Herr von Ribbentrop before he was hanged.

* * * * *

The outrage against Austria and the subjugation of beautiful Vienna, with all its fame, culture, and contribution to the story of Europe, hit me hard. On the morrow of these events, March 14, I said in the House of Commons:

The gravity of the event of March 12 cannot be exaggerated. Europe is confronted with a programme of aggression, nicely calculated and timed, unfolding stage by stage, and there is only one choice open, not only to us but to other countries, either to submit like Austria, or else take effective measures while time remains to ward off the danger, and if it cannot be warded off to cope with it. . . . If we go on waiting upon events, how much shall we throw away of resources now available for our security and the maintenance of peace? How many friends will be alienated, how many potential allies shall we see go one by one down the grisly gulf? How many times will bluff succeed until behind bluff ever gathering forces have accumulated reality? . . . *Where are we going to be two years hence, for instance, when the German Army will certainly be much larger than the French Army*, and when all the small nations will have fled from Geneva to pay homage to the ever-waxing power of the Nazi system, and to make the best terms that they can for themselves?

And further:

Vienna is the centre of the communications of all the countries which formed the old Austro-Hungarian Empire, and of the countries lying to the south-east of Europe. A long stretch of the Danube is now in German hands. This mastery of Vienna gives to Nazi Germany military and economic control of the whole of the communications of South-eastern Europe, by road, by river, and by rail. What is the effect of this on the structure of Europe? What is the effect of it upon what is called the balance of power, such as it is—upon what is called the Little Entente? I must say a word about this group of Powers called the Little Entente. Taken singly, the three countries of the Little Entente may be called Powers of the second rank, but they are very powerful and vigorous States, and united they are a great Power. They have hitherto been, and are still, united by the closest military agreement. Together they make the complement of a great Power and of the military machinery of a great Power. Roumania has the oil, Yugoslavia has the minerals and raw materials. Both have large armies, both are mainly supplied with munitions from Czechoslovakia.

To English ears the name of Czechoslovakia sounds outlandish. No doubt they are only a small democratic State, no doubt they have an army only two or three times as large as ours, no doubt they have a munitions supply only three times as great as that of Italy, but still they are a virile people, they have their rights, they have their treaty rights, they have a line of fortresses, and they have a strongly-manifested will to live, a will to live freely.

Czechoslovakia is at this moment isolated, both in the economic and in the military sense. Her trade outlet through Hamburg, which is based upon the Peace Treaty, can of course be closed at any moment. Now her communications by rail and river to the south, and beyond the south to the south-east, are liable to be severed at any moment. Her trade may be subjected to tolls of a destructive character, of an absolutely strangling character. Here is a country which was once the greatest manufacturing area in the old Austro-Hungarian Empire. It is now cut off, or may be cut off at once, unless out of these discussions which must follow arrangements are made securing the communications of Czechoslovakia. She may be cut off at once from the sources of her raw materials in Yugoslavia and from the natural markets which she has established there. The economic life of this small State may be very largely strangled as a result of the act of violence which was perpetrated last Friday night. A wedge has been driven into the heart of what is called the Little Entente, this group of countries which have as much right to live in Europe unmolested as any of us have the right to live unmolested in our native land.

★　★　★　★　★

It was the Russians who now sounded the alarm, and on March 18 proposed a conference on the situation. They wished to discuss, if only in outline, ways and means of implementing the Franco-Soviet pact within the frame of League action in the event of a major threat to peace by Germany. This met with little warmth in Paris and London. The French Government was distracted by other preoccupations. There were serious strikes in the aircraft factories. Franco's armies were driving deep into the territory of Communist Spain. Chamberlain was both sceptical and depressed. He profoundly disagreed with my interpretation of the dangers ahead and the means of combating them. I had been urging the prospects of a Franco-British-Russian alliance as the only hope of checking the Nazi onrush.

Mr. Feiling tells us that the Prime Minister expressed his mood in a letter to his sister on March 20:

245

The plan of the "Grand Alliance", as Winston calls it, had occurred to me long before he mentioned it. . . . I talked about it to Halifax, and we submitted it to the Chiefs of Staff and F.O. experts. It is a very attractive idea; indeed, there is almost everything to be said for it until you come to examine its practicability. From that moment its attraction vanishes. You have only to look at the map to see that nothing that France or we could do could possibly save Czechoslovakia from being overrun by the Germans, if they wanted to do it. I have therefore abandoned any idea of giving guarantees to Czechoslovakia, or to the French in connection with her obligations to that country.*

Here was at any rate a decision. It was taken on wrong arguments. In modern wars of great nations or alliances particular areas are not defended only by local exertions. The whole vast balance of the war-front is involved. This is still more true of policy before war begins and while it may still be averted. It surely did not take much thought from the "Chiefs of Staff and F.O. experts" to tell the Prime Minister that the British Navy and the French Army could not be deployed on the Bohemian mountain front to stand between the Czechoslovak Republic and Hitler's invading army. This was indeed evident from the map. But the certainty that the crossing of the Bohemian frontier line would have involved a general European war might well even at that date have deterred or delayed Hitler's next assault. How erroneous Mr. Chamberlain's private and earnest reasoning appears when we cast our minds forward to the guarantee he was to give to Poland *within a year*, after all the strategic value of Czechoslovakia had been cast away, and Hitler's power and prestige had almost doubled!

* * * * *

On March 24, 1938, in the House of Commons, the Prime Minister gave us his view about the Russian move:

His Majesty's Government are of the opinion that the indirect but none the less inevitable consequence of such action as is proposed by the Soviet Government would be to aggravate the tendency towards the establishment of exclusive groups of nations, which must in the view of His Majesty's Government be inimical to the prospects of European peace.

Nevertheless the Prime Minister could not avoid facing the

* Feiling, *op. cit.*, pp. 347–8.

brutal fact that there existed a "profound disturbance of international confidence", and that the Government would have
sooner or later to decide upon a definition of Great Britain's
obligations in Europe. What would be our obligations in Central
Europe? "If war broke out it would be unlikely to be confined
to those who have assumed legal obligations. It would be quite
impossible to say where it would end and what Governments
might be involved." It must further be observed that the argument about the evils of "exclusive groups of nations" loses its
validity if the alternative is being mopped up one by one by the
aggressor. Moreover, it overlooks all questions of right and
wrong in international relationships. There was after all in
existence the League of Nations and its Charter.

The Prime Minister's course was now marked out: simultaneous diplomatic pressure on Berlin and Prague, appeasement
in regard to Italy, a strictly restrained definition of our obligations
to France. To carry out the first two moves, it was essential to
be careful and precise about the last.

*   *   *   *   *

The reader is now invited to move westward to the Emerald
Isle. "It's a long way to Tipperary," but a visit there is sometimes
irresistible. In the interval between Hitler's seizure of Austria and
his unfolding design upon Czechoslovakia we must turn to a
wholly different kind of misfortune which befell us.

Since the beginning of 1938 there had been negotiations between the British Government and that of Mr. de Valera in
Southern Ireland, and on April 25 an agreement was signed
whereby, among other matters, Great Britain renounced all
rights to occupy for naval purposes the two Southern Irish ports
of Queenstown and Berehaven, and the base in Lough Swilly.
The two southern ports were a vital feature in the naval defence
of our food supply. When in 1922 as Colonial and Dominions
Secretary I had dealt with the details of the Irish Settlement which
the Cabinet of those days had made, I brought Admiral Beatty to
the Colonial Office to explain to Michael Collins the importance
of these ports to our whole system of bringing supplies into
Britain. Collins was immediately convinced. "Of course you
must have the ports," he said; "they are necessary for your life."
Thus the matter was arranged, and everything had worked

smoothly in the sixteen years that had passed. The reason why Queenstown and Berehaven were necessary to our safety is easy to understand. They were the fuelling bases from which our destroyer flotillas ranged westward into the Atlantic to hunt U-boats and protect incoming convoys as they reached the throat of the narrow seas. Lough Swilly was similarly needed to protect the approaches to the Clyde and Mersey. To abandon these meant that our flotillas would have to start in the north from Lamlash and in the south from Pembroke Dock or Falmouth, thus decreasing their radius of action and the protection they could afford by more than 400 miles out and home.

It was incredible to me that the Chiefs of Staff should have agreed to throw away this major security, and to the last moment I thought that at least we had safeguarded our right to occupy these Irish ports in the event of war. However, Mr. de Valera announced in the Dail that no conditions of any kind were attached to the cession. I was later assured that Mr. de Valera was surprised at the readiness with which the British Government had deferred to his request. He had included it in his proposals as a bargaining counter which could be dispensed with when other points were satisfactorily settled.

Lord Chatfield has in his last book devoted a chapter to explaining the course he and the other Chiefs of Staff took.* This should certainly be read by those who wish to pursue the subject. Personally I remain convinced that the gratuitous surrender of our right to use the Irish ports in war was a major injury to British national life and safety. A more feckless act can hardly be imagined—and at such a time. It is true that in the end we survived without the ports. It is also true that if we had not been able to do without them we should have retaken them by force rather than perish by famine. But this is no excuse. Many a ship and many a life were soon to be lost as the result of this improvident example of appeasement.

The whole Conservative Party, except the handful of Ulster Members, supported the Prime Minister, and of course a step like this was meat and drink to the Labour and Liberal Opposition. I was therefore almost entirely alone when on May 5 I rose to make my protest. I was listened to with a patient air of scepticism. There was even a kind of sympathetic wonder that anyone

* Lord Chatfield, *It Might Happen Again*, Chapter XVIII.

of my standing should attempt to plead so hopeless a case. I never saw the House of Commons more completely misled. It was but fifteen months to the declaration of war. The Members were to feel very differently about it when our existence hung in the balance during the Battle of the Atlantic. As my speech has been fully published in *Into Battle* I do not quote it here save on one point. The issue of Southern Irish neutrality in time of war was not faced.

What guarantee [I asked] have you that Southern Ireland, or the Irish Republic as they claim to be, will not declare neutrality if we are engaged in war with some powerful nation? The first step certainly which such an enemy would take would be to offer complete immunity of every kind to Southern Ireland if she would remain neutral. . . . You cannot exclude this possibility of neutrality as being one which may come within the immediate sphere of our experience. The ports may be denied us in the hour of need, and we may be hampered in the gravest manner in protecting the British population from privation and even starvation. Who would wish to put his head in such a noose? Is there any other country in the world where such a step would even have been contemplated? It would be an easy step for a Dublin Government to deny the ports to us once we have gone. The cannon are there, the mines will be there. But more important for this purpose, the juridical right will be there. You had the rights; you have ceded them; you hope in their place to have goodwill strong enough to endure tribulation for your sake. Suppose you have it not. It will be no use saying, "Then we will retake the ports." You will have no right to do so. To violate Irish neutrality should it be declared at the moment of a Great War may put you out of court in the opinion of the world, and may vitiate the cause by which you may be involved in war. . . . You are casting away real and important means of security and survival for vain shadows and for ease.

The comment of the *Times* newspaper was illuminating.

The agreement on defence . . . releases the Government of the United Kingdom from the articles of the Anglo-Irish Treaty of 1921, by which they assumed the onerous and delicate task of defending the fortified harbours of Cork, Berehaven, and Lough Swilly in the event of war.

Further releases might have been obtained by handing over Gibraltar to Spain and Malta to Italy. Neither touched the actual existence of our population more directly.

With that I leave this lamentable and amazing episode.

CHAPTER XVI

CZECHOSLOVAKIA

An Unlikely Historical Controversy – Hitler's Next Objective – "No Evil Intentions towards Czechoslovakia" – M. Blum's Pledge – My Visit to Paris, March 1938 – M. Daladier succeeds M. Blum – The Anglo-Italian Pact – An Interview with the Sudeten Leader – Misgivings and Reluctance of the German Generals – The Relations of Soviet Russia with Czechoslovakia – Stalin and Beneš – Plot and Purge in Russia – M. Daladier's Declaration of June 12 – Hitler's Promise to Keitel – Captain Wiedemann's Mission to London – I Address My Constituents at Theydon Bois, August 27 – My Letter to Lord Halifax of August 31 – The Soviet Ambassador's Visit to Chartwell – My Report to the Foreign Office – The "Times" Leading Article of September 7 – M. Bonnet's Question and the British Answer – Hitler's Crisis Speech at Nuremberg.

FOR SOME YEARS it seemed that the question whether Britain and France were wise or foolish in the Munich episode would become a matter of long historical controversy. However, the revelations which have been made from German sources, and particularly at the Nuremberg Trials, have rendered this unlikely. The two main issues in dispute were: first, whether decisive action by Britain and France would have forced Hitler to recede or have led to his overthrow by a military conspiracy; secondly, whether the year that intervened between Munich and the outbreak of war placed the Western Powers relatively in a better or worse position, compared with Germany, than in September 1938.

Many volumes have been written, and will be written, upon the crisis that was ended at Munich by the sacrifice of Czechoslovakia, and it is only intended here to give a few of the cardinal facts and establish the main proportions of events. These follow inexorably from Hitler's resolve to reunite all Germans in a Greater Reich and to expand eastwards, and his conviction that the men at the head of France and Britain would not fight owing to their love of peace and failure to rearm. The usual technique was employed against Czechoslovakia. The grievances, which were not unreal, of the Sudeten Germans were magnified and exploited. The public case was opened against Czechoslovakia by Hitler in his speech to the Reichstag on February 20, 1938. "Over ten million Germans," he said, "live in two of the States adjoining our frontier." It was the duty of Germany to protect those fellow-Germans and secure to them "general freedom, personal, political, and ideological."

This public announcement of the intention of the German Government to interest themselves in the position of the German inhabitants of Austria and Czechoslovakia was intimately related to the secret planning of Germany's political offensive in Europe. The declared objectives of the Nazi German Government were twofold—the absorption by the Reich of all German minorities living beyond her frontiers, and thereby the extension of her living space in the East. The less publicised purpose of German policy was military in character—the liquidation of Czechoslovakia with its potentialities both as a Russian air-base and as an Anglo-French military make-weight in the event of war. As early as June 1937 the German General Staff had been, on Hitler's instructions, busy at work drafting plans for the invasion and destruction of the Czechoslovak State.

One draft reads:

The aim and object of this surprise attack by the German armed forces should be to eliminate from the very beginning and for the duration of the war the threat from Czechoslovakia to the rear of the operations in the West, and to take from the Russian Air Force the most substantial portion of its operational base in Czechoslovakia.*

The acceptance by the Western Democracies of the German subjugation of Austria encouraged Hitler to pursue his designs more sharply against Czechoslovakia. The military control of Austrian territory was in fact intended to be the indispensable preliminary to the assault on the Bohemian bastion. While the invasion of Austria was in full swing Hitler said in the motor-car

* *Nuremberg Documents*, II, p. 4.

to General von Halder, "This will be very inconvenient to the Czechs." Halder saw immediately the significance of this remark. To him it lighted up the future. It showed him Hitler's intentions, and at the same time, as he viewed it, Hitler's military ignorance. "It was practically impossible," he has explained, "for a German army to attack Czechoslovakia from the south. The single railway line through Linz was completely exposed, and surprise was out of the question." But Hitler's main political strategic conception was correct. The West Wall was growing, and, although far from complete, already confronted the French Army with horrible memories of the Somme and Passchendaele. He was convinced that neither France nor Britain would fight.

On the day of the march of the German armies into Austria the French Ambassador in Berlin reported that Goering had given a solemn assurance to the Czech Minister in Berlin that Germany had *"no evil intentions towards Czechoslovakia"*. On March 14 the French Premier, M. Blum, solemnly declared to the Czech Minister in Paris that France would unreservedly honour her engagement to Czechoslovakia. These diplomatic reassurances could not conceal the grim reality. The whole strategic position on the Continent had changed. The German arguments and armies could now concentrate directly upon the western frontiers of Czechoslovakia, whose border districts were German in racial character, with an aggressive and active German Nationalist Party eager to act as a fifth column in the event of trouble.

At the end of March I went to Paris and had searching conversations with the French leaders. The Government were agreeable to my going to vivify my French contacts. I stayed at our Embassy, and saw in a continued succession many of the principal French figures—Premier Léon Blum, Flandin, General Gamelin, Paul Reynaud, Pierre Cot, Herriot, Louis Marin, and others. To Blum I said at one moment, "The German field howitzer is believed to be superior in range and of course in striking power to the *soixante-quinze* even when relined." He replied, "Is it from you that I am to learn the state of the French artillery?" I said, "No, but ask your École Polytechnique, who are by no means convinced by the exposition lately given to them of the relative power of the modernised *soixante-quinze*." He was immediately genial and friendly. Reynaud said to me, "We quite understand that England will never have conscription. Why do you not therefore go in for a mechanical army? If you had six armoured divisions you would indeed be an effective Continental force," or words to that effect. It seemed that a Colonel de Gaulle had written a much-criticised book about the offensive power of modern armoured vehicles.

The Ambassador and I had a long luncheon alone with Flandin. He was quite a different man from the one I had known in 1936: then responsible and agitated; now out of office, cool, massive, and completely convinced that there was no hope for France except in an arrangement with Germany. We argued for two hours. Gamelin, who also visited me, was rightly confident in the strength of the French Army at the moment, but none too comfortable when I questioned him upon the artillery, about which he had precise knowledge. He was always trying his best within the limits of the French political system. But the attention of the French Government to the dangers of the European scene was distracted by the ceaseless whirlpool of internal politics at the moment and by the imminent fall of the Blum Government. It was all the more essential that our common and mutual obligations in the event of a general crisis should be established without any trace of misunderstanding. On April 10 the French Government was re-formed with M. Daladier as Premier and M. Bonnet as Minister for Foreign Affairs. These two men were to bear the responsibility for French policy in the critical months ahead.

In the hope of deterring Germany from a further aggression the British Government, in accordance with Mr. Chamberlain's resolve, sought a settlement with Italy in the Mediterranean. This would strengthen the position of France, and would enable both the French and British to concentrate upon events in Central Europe. Mussolini, to some extent placated by the fall of Eden, and feeling himself in a strong bargaining position, did not repulse the British repentance. On April 16, 1938, an Anglo-Italian agreement was signed, giving Italy in effect a free hand in Abyssinia and Spain in return for the imponderable value of Italian goodwill in Central Europe. The Foreign Office was sceptical of this transaction. Mr. Chamberlain's biographer tells us that he wrote in a personal and private letter, "You should have seen the draft put up to me by the F.O.; it would have frozen a Polar bear."*

* Feiling, *op. cit.*, p. 350.

I shared the misgivings of the Foreign Office at this move:

Mr. Churchill to Mr. Eden 18.IV.38

The Italian pact is of course a complete triumph for Mussolini, who gains our cordial acceptance for his fortification of the Mediterranean against us, for his conquest of Abyssinia, and for his violence in Spain. The fact that we are not to fortify Cyprus without "previous consultation" is highly detrimental. The rest of it is to my mind only padding.

Nevertheless I feel that considerable caution is necessary in opposing the agreement bluntly. It is a done thing. It is called a move towards peace. It undoubtedly makes it less likely that sparks from the Mediterranean should light a European conflagration. France will have to follow suit for her own protection, in order not to be divided from Britain. Finally, there is the possibility that Mussolini may be drawn by his interests to discourage German interference in the Danube basin.

Before making up my mind, I should like to know your views and intentions. I think the Anglo-Italian pact is only the first step, and that the second will be an attempt to patch up something even more specious with Germany, which will lull the British public while letting the German armed strength grow and German designs in the East of Europe develop.

Chamberlain last week told the Executive of the National Union [of Conservative Associations] in secret that he "had not abandoned hopes of similar arrangements with Germany". They took this rather coldly.

Meanwhile our progress in the air is increasingly disappointing. . . .

Mr. Eden to Mr. Churchill 28.IV.38

. . . With regard to the Italian pact, I agree with what you write. Mussolini gives us nothing more than the repetition of promises previously made and broken by him, except for the withdrawal of troops from Libya, troops which were probably originally sent there for their nuisance value. It has now become clear that, as I expected, Mussolini continued his intervention in Spain after the conversations in Rome had opened. He must be an optimist indeed who believes that Mussolini will cease increasing that intervention now, should it be required to secure Franco's victory.

As a diplomatic instrument the pact embodies a machinery which is likely to be found very troublesome to work. It is not to come into force until after the Italians leave Spain. It is almost certain however that many months will elapse before that occurs, and since what is important is not the presence of Italian infantry, but the assertions of their experts and the Germans, it will be difficult to establish with certainty that the withdrawal has taken place. But maybe some do not mind much about that.

Then there is the Italian position in Abyssinia, which, from what I hear, so far from improving grows steadily worse. I am afraid that the moment we are choosing for its recognition will not benefit our authority among the many millions of the King's coloured subjects.

None the less I equally agree as to the need for caution in any attitude taken up towards the agreement. After all it is not an Agreement yet, and it would be wrong certainly for me to say anything which could be considered as making its fruition more difficult. After all, this is precisely what I promised I would not do in my resignation speech and at Leamington.

The most anxious feature of the international situation, as I see it, is that temporary relaxation of tension may be taken as a pretext for the relaxation of national effort, which is already inadequate to the gravity of the times. . . .

Hitler was watching the scene with vigilance. To him also the ultimate alignment of Italy in a European crisis was important. In conference with his Chiefs of Staff at the end of April he was considering how to force the pace. Mussolini wanted a free hand in Abyssinia. In spite of the acquiescence of the British Government, he might ultimately need German support in this venture. If so, he should accept German action against Czechoslovakia. This issue must be brought to a head, and in the settling of the Czech question Italy would be involved on Germany's side. The declarations of British and French statesmen were of course studied in Berlin. The intention of these Western Powers to persuade the Czechs to be reasonable in the interests of European peace was noted with satisfaction. The Nazi Party of the Sudetenland, led by Henlein, now formulated their demands for autonomy in the German-border regions of that country. Their programme had been announced in Henlein's speech at Carlsbad on April 24. The British and French Ministers in Prague called on the Czech Foreign Minister shortly after this to "express the hope that the Czech Government will go to the furthest limit in order to settle the question".

During May the Germans in Czechoslovakia were ordered to increase their agitation. On May 12 Henlein visited London to acquaint the British Government with the wrongs inflicted upo his followers. He expressed a wish to see me. I therefore arranged a talk at Morpeth Mansions the next day, at which Sir Archibald

Sinclair was present, and Professor Lindemann was our interpreter.

Henlein's solution, as he described it, may be summed up as follows:

There should be a central Parliament in Prague, which should have control of foreign policy, defence, finance, and communications. All parties should be entitled to express their views there, and the Government would act on majority decisions. The frontier fortresses could be manned by Czech troops, who would of course have unhindered access thereto. The Sudeten German regions, and possibly the other minority districts, should enjoy local autonomy; that is to say, they should have their own town and county councils, and a Diet in which matters of common regional concern could be debated within definitely delimited frontiers. He would be prepared to submit questions of fact, *e.g.*, the tracing of the boundary, to an impartial tribunal, perhaps even appointed by the League of Nations. All parties would be free to organise and offer themselves for election, and impartial courts of justice would function in autonomous districts. The officials—*i.e.*, postal, railway, and police officers—in the German-speaking region would of course be German-speaking, and a reasonable proportion of the total taxes collected should be returned to these regions for their administration.

M. Masaryk, the Czech Minister in London, who was afterwards informed of this conversation, professed himself contented with a settlement on these lines. A peaceful solution of admitted racial and minority quarrels compatible with the independence of the Czech Republic was by no means impossible, if there were German good faith and goodwill. But on this condition I had no illusions.

On May 17 negotiations about the Sudeten question began between Henlein, who had visited Hitler on his return journey, and the Czech Government. Municipal elections were due in Czechoslovakia, and the German Government began a calculated war of nerves in preparation for them. Persistent rumours already circulated of German troop movements towards the Czech frontier. On May 20 Sir Nevile Henderson was requested to make inquiries in Berlin on this matter. German denials did not reassure the Czechs, who on the night of May 20–21 decreed a partial mobilisation of their Army.

★   ★   ★   ★   ★

It is important at this stage to consider the German intentions. Hitler had for some time been convinced that neither France nor Britain would fight for Czechoslovakia. On May 28 he called a meeting of his principal advisers and gave instructions for the preparations to attack Czechoslovakia. He declared this later in public in a speech to the Reichstag on January 30, 1939:

> In view of this intolerable provocation . . . I resolved to settle once and for all, and this time radically, the Sudeten-German question. On May 28 I ordered (1) that preparations should be made for military action against this State by October 2; and (2) the immense and accelerated expansion of our defensive front in the West.*

His Service advisers, however, did not share unanimously his overwhelming confidence. The German generals could not be persuaded, considering the still enormous preponderance of Allied strength except in the air, that France and Britain would submit to the Fuehrer's challenge. To break the Czech Army and pierce or turn the Bohemian fortress line would require practically the whole of thirty-five divisions. The German Chiefs of Staff informed Hitler that the Czech Army must be considered efficient and up-to-date in arms and equipment. The fortifications of the West Wall or Siegfried Line, though already in existence as field works, were far from completed. Thus at the moment of attacking the Czechs only five effective and eight Reserve divisions would be available to protect the whole of Germany's western frontier against the French Army, which could mobilise a hundred divisions. The generals were aghast at running such risks, when by waiting a few years the German Army would again be master. Although Hitler's political judgment had been proved correct by the pacifism and weakness of the Allies about conscription, the Rhineland, and Austria, the German High Command could not believe that Hitler's bluff would succeed a fourth time. It seemed so much beyond the bounds of reason that great victorious nations, possessing evident military superiority, would once again abandon the path of duty and honour, which was also for them the path of common sense and prudence. Besides all this there was Russia, with her Slav affinities with Czechoslovakia, and whose attitude towards Germany at this juncture was full of menace.

* *Hitler's Speeches*, II, 1571.

The relations of Soviet Russia with Czechoslovakia as a State, and personally with President Beneš, were those of intimate and solid friendship. The roots of this lay in a certain racial affinity, and also in comparatively recent events which require a brief digression. When President Beneš visited me at Marrakesh in January 1944 he told me this story. In 1935 he had received an offer from Hitler to respect in all circumstances the integrity of Czechoslovakia in return for a guarantee that she would remain neutral in the event of a Franco-German war. When Beneš pointed to his treaty obliging him to act with France in such a case, the German Ambassador replied that there was no need to denounce the treaty. It would be sufficient to break it, if and when the time came, by simply failing to mobilise or march. The small republic was not in a position to indulge in indignation at such a suggestion. Their fear of Germany was already very grave, more especially as the question of the Sudeten Germans might at any time be raised and fomented by Germany, to their extreme embarrassment and growing peril. They therefore let the matter drop without comment or commitment, and it did not stir for more than a year. In the autumn of 1936 a message from a high military source in Germany was conveyed to President Beneš to the effect that if he wanted to take advantage of the Fuehrer's offer he had better be quick, because events would shortly take place in Russia rendering any help he could give to Germany insignificant.

While Beneš was pondering over this disturbing hint he became aware that communications were passing through the Soviet Embassy in Prague between important personages in Russia and the German Government. This was a part of the so-called military and Old-Guard-Communist conspiracy to overthrow Stalin and introduce a new *régime* based on a pro-German policy. President Beneš lost no time in communicating all he could find out to Stalin.* Thereafter there followed the merciless, but perhaps not needless, military and political purge in Soviet Russia, and the series of trials in January 1937, in which Vyshinsky, the Public Prosecutor, played so masterful a part.

Although it is highly improbable that the Old-Guard Com-

---

* There is however some evidence that Beneš's information had previously been imparted to the Czech police by the Ogpu, who wished it to reach Stalin from a friendly foreign source. This did not detract from Beneš's service to Stalin, and is therefore irrelevant.

munists had made common cause with the military leaders, or *vice versa*, they were certainly filled with jealousy of Stalin, who had ousted them. It may therefore have been convenient to get rid of them at the same time, according to the standards maintained in a totalitarian State. Zinoviev, Bukharin, and others of the original leaders of the Revolution, Marshal Tukhachevsky, who had been invited to represent the Soviet Union at the Coronation of King George VI, and many other high officers of the Army, were shot. In all not less than five thousand officers and officials above the rank of Captain were "liquidated". The Russian Army was purged of its pro-German elements at a heavy cost to its military efficiency. The bias of the Soviet Government was turned in a marked manner against Germany. Stalin was conscious of a personal debt to President Beneš, and a very strong desire to help him and his threatened country against the Nazi peril animated the Soviet Government. The situation was of course thoroughly understood by Hitler; but I am not aware that the British and French Governments were equally enlightened. To Mr. Chamberlain and the British and French General Staffs the purge of 1937 presented itself mainly as a tearing to pieces internally of the Russian Army, and a picture of the Soviet Union as riven asunder by ferocious hatreds and vengeance. This was perhaps an excessive view; for a system of government founded on terror may well be strengthened by a ruthless and successful assertion of its power. The salient fact for the purposes of this account is the close association of Russia and Czechoslovakia, and of Stalin and Beneš.

But neither the internal stresses in Germany nor the ties between Beneš and Stalin were known to the outside world, or appreciated by the British and French Ministers. The Siegfried Line, albeit unperfected, seemed a fearful deterrent. The exact strength and fighting power of the German Army, new though it was, could not be accurately estimated and was certainly exaggerated. There were also the unmeasured dangers of air attack on undefended cities. Above all there was the hatred of war in the hearts of the Democracies.

Nevertheless on June 12 M. Daladier renewed his predecessor's pledge of March 14, and declared that France's engagements towards Czechoslovakia "are sacred, and cannot be evaded". This considerable statement sweeps away all chatter about the

Treaty of Locarno thirteen years before having by implication left everything in the East vague pending an Eastern Locarno. There can be no doubt before history that the treaty between France and Czechoslovakia of 1924 had complete validity not only in law but in fact, and that this was reaffirmed by successive heads of the French Government in all the circumstances of 1938.

But on this subject Hitler was convinced that his judgment alone was sound, and on June 18 he issued a final directive for the attack on Czechoslovakia, in the course of which he sought to reassure his anxious generals.

Hitler to Keitel

I will decide to take action against Czechoslovakia only if I am firmly convinced, as in the case of the demilitarised zone and the entry into Austria, that France will not march, and that therefore England will not intervene.*

With the object of confusing the issue Hitler at the beginning of July sent his personal aide, Captain Wiedemann, to London. This envoy was received by Lord Halifax on July 18, ostensibly without the knowledge of the German Embassy. The Fuehrer was, it was suggested, hurt at our lack of response to his overtures in the past. Perhaps the British Government would receive Goering in London for fuller discussions. The Germans might, in certain circumstances, be prepared to delay action against the Czechs for a year. A few days later Chamberlain took up this possibility with the German Ambassador. To clear the ground in Prague the British Prime Minister had already suggested to the Czechs the sending of an investigator to Czechoslovakia to promote a friendly compromise. The Royal visit to Paris on July 20 gave Halifax the opportunity of discussing this proposal with the French Government, and in a brief interchange of views both Governments agreed to make this effort at mediation.

On July 26, 1938, Chamberlain announced to Parliament the mission of Lord Runciman to Prague with the object of seeking a solution there by arrangements between the Czech Government and Herr Henlein. On the following day the Czechs issued a draft statute for national minorities to form a basis of negotiations. On the same day Lord Halifax stated in Parliament: "*I do not believe that those responsible for the Government of any country in*

---

* *Nuremberg Documents*, Pt. II, No. 10.

Europe to-day want war." On August 3 Lord Runciman reached Prague, and a series of interminable and complicated discussions took place with the various interested parties. Within a fortnight these negotiations broke down, and from this point events moved rapidly.

On August 27 Ribbentrop, now Foreign Minister, reported a visit which he had received from the Italian Ambassador in Berlin, who "had received another written instruction from Mussolini asking that Germany would communicate in time the probable date of action against Czechoslovakia". Mussolini asked for such notification in order "to be able to take in due time the necessary measures on the French frontier".

*   *   *   *   *

Anxiety grew steadily during August. To my constituents I said on the 27th:

> It is difficult for us in this ancient forest of Theyⁿ ⁿn Bois, the very name of which carries us back to Norman days—here, in the heart of peaceful, law-abiding England—to realise the ferocious passions which are rife in Europe. During this anxious month you have no doubt seen reports in the newspapers, one week good, another week bad; one week better, another week worse. But I must tell you that the whole state of Europe and of the world is moving steadily towards a climax which cannot be long delayed.
>
> War is certainly not inevitable. But the danger to peace will not be removed until the vast German armies which have been called from their homes into the ranks have been dispersed. For a country which is itself not menaced by anyone, in no fear of anyone, to place over fifteen hundred thousand soldiers upon a war footing is a very grave step. . . . It seems to me, and I must tell it to you plainly, that these great forces have not been placed upon a war footing without an intention to reach a conclusion within a very limited space of time. . . .
>
> We are all in full agreement with the course our Government have taken in sending Lord Runciman to Prague. We hope—indeed, we pray—that his mission of conciliation will be successful, and certainly it looks as if the Government of Czechoslovakia were doing their utmost to put their house in order, and to meet every demand which is not designed to compass their ruin as a State. . . . But larger and fiercer ambitions may prevent a settlement, and then Europe and the civilised world ¹ will have to face the demands of Nazi Germany, or perhaps be confronted with some sudden violent action on the part

of the German Nazi Party, carrying with it the invasion of a small country and its subjugation. Such an episode would not be simply an attack upon Czechoslovakia; it would be an outrage against the civilisation and freedom of the whole world. . . .

Whatever may happen, foreign countries should know—and the Government are right to let them know—that Great Britain and the British Empire must not be deemed incapable of playing their part and doing their duty as they have done on other great occasions which have not yet been forgotten by history.

I was in these days in some contact with Ministers. My relations with Lord Halifax were of course marked by the grave political differences which existed between me and His Majesty's Government, both in defence and foreign policy. In the main Eden and I meant the same thing. I could not feel the same about his successor. None the less, whenever there was any occasion we met as friends and former colleagues of many years' standing, and I wrote to him from time to time. Now and then he asked me to go to see him.

Mr. Churchill to Lord Halifax   31.VIII.38

If Beneš makes good, and Runciman thinks it a fair offer, yet nevertheless it is turned down, it seems to me there are two things which might have been done this week to increase the deterrents against violent action by Hitler, neither of which would commit you to the dread guarantee.

First, would it not be possible to frame a Joint Note between Britain, France, and Russia stating: (a) their desire for peace and friendly relations; (b) their deep anxiety at the military preparations of Germany; (c) their joint interest in a peaceful solution of the Czechoslovak controversy; and (d) that an invasion by Germany of Czechoslovakia would raise capital issues for all three Powers? This Note, when drafted, should be formally shown to Roosevelt by the Ambassadors of the three Powers, and we should use every effort to induce him to do his utmost upon it. It seems to me not impossible that he would then himself address Hitler, emphasising the gravity of the situation, and saying that it seemed to him that a world war would inevitably follow from an invasion of Czechoslovakia, and that he earnestly counselled a friendly settlement.

It seems to me that this process would give the best chance to the peaceful elements in German official circles to make a stand, and that Hitler might find a way out for himself by parleying with Roosevelt. However, none of these developments can be predicted; one only sees them as hopes. *The important thing is the Joint Note.*

The second step which might save the situation would be fleet movements, and the placing of the reserve flotillas and cruiser squadrons into full commission. I do not suggest calling out the Royal Fleet Reserve or mobilisation, but there are, I believe, five or six flotillas which could be raised to First Fleet scale, and also there are about two hundred trawlers which could be used for anti-submarine work. The taking of these and other measures would make a great stir in the naval ports, the effect of which could only be beneficial as a deterrent, and a timely precaution if the worst happened.

I venture to hope that you will not resent these suggestions from one who has lived through such days before. It is clear that speed is vital.

★　★　★　★　★

In the afternoon of September 2 I received a message from the Soviet Ambassador that he would like to come down to Chartwell and see me at once upon a matter of urgency. I had for some time had friendly personal relations with M. Maisky, who also saw a good deal of my son Randolph. I thereupon received the Ambassador, and after a few preliminaries he told me in precise and formal detail the story set out below. Before he had got very far I realised that he was making a declaration to me, a private person, because the Soviet Government preferred this channel to a direct offer to the Foreign Office, which might have encountered a rebuff. It was clearly intended that I should report what I was told to His Majesty's Government. This was not actually stated by the Ambassador, but it was implied by the fact that no request for secrecy was made. As the matter struck me at once as being of the first importance, I was careful not to prejudice its consideration by Halifax and Chamberlain by proceeding to commit myself in any way, or use language which would excite controversy between us.

Mr. Churchill to Lord Halifax　　　　　　　　　　　　　3.IX.38
I have received privately from an absolutely sure source the following information, which I feel it my duty to report to you, although I was not asked to do so.

Yesterday, September 2, the French Chargé d'Affaires in Moscow (the Ambassador being on leave) called upon M. Litvinov and, in the name of the French Government, asked him what aid Russia would give to Czechoslovakia against a German attack, having regard particularly to the difficulties which might be created by the neutrality of Poland or Roumania. Litvinov asked in reply what the French would do themselves, pointing out that the French had a direct obligation, whereas the Russian obligation was dependent on the action of France. The French Chargé d'Affaires did not reply to this question. Nevertheless, Litvinov stated to him, first, that the Russian Soviet Union had resolved to fulfil their obligations. He recognised the difficulties created by the attitude of Poland and Roumania, but he thought that in the case of Roumania these could be overcome.

In the last few months the policy of the Roumanian Government had been markedly friendly to Russia, and their relations had greatly improved. M. Litvinov thought that the best way to overcome the reluctance of Roumania would be through the agency of the League of Nations. If, for instance, the League decided that Czechoslovakia was the victim of aggression and that Germany was the aggressor, that would probably determine the action of Roumania in regard to allowing Russian troops and air forces to pass through her territory.

The French Chargé d'Affaires raised the point that the Council might not be unanimous, and was answered that M. Litvinov thought a majority decision would be sufficient, and that Roumania would probably associate herself with the majority in the vote of the Council. M. Litvinov therefore advised that the Council of the League should be invoked under Article 11, on the ground that there was danger of war, and that the League Powers should consult together. He thought the sooner this was done the better, as time might be very short. He next proceeded to tell the French Chargé d'Affaires that Staff conversations ought immediately to take place between Russia, France, and Czechoslovakia as to the means and measures of giving assistance. The Soviet Union was ready to join in such Staff conversations at once.

Fourthly, he recurred to his interview of March 17, of which you no doubt have a copy in the Foreign Office, advocating consultation among the peaceful Powers about the best method of preserving peace, with a view, perhaps, to a joint declaration including the three Great Powers concerned, France, Russia, and Great Britain. He believed that the United States would give moral support to such a declaration. All these statements were made on behalf of the Russian Government as what they think may be the best way of stopping a war.

I pointed out that the news to-day seemed to indicate a more peaceful attitude on the part of Herr Hitler, and that I thought it was unlikely that the British Government would consider any further steps until or unless there was a fresh breakdown in the Henlein-Beneš negotiations, in which the fault could not on any account be attributed to the Government of Czechoslovakia. We should not want to irritate Herr Hitler, if his mind was really turning towards a peaceful solution.

All this may of course have reached you through other channels,

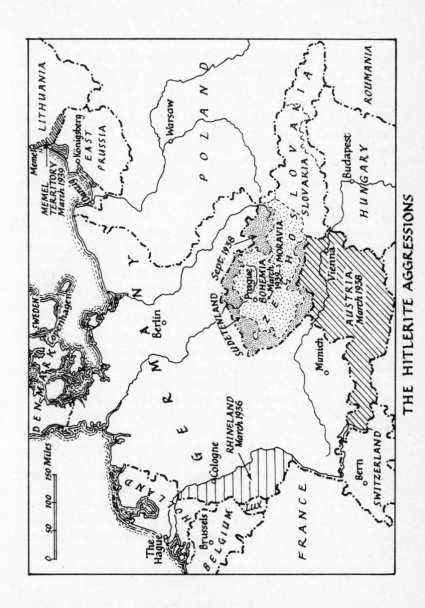

THE HITLERITE AGGRESSIONS

but I considered the declarations of M. Litvinov so important that I ought not to leave this to chance.

I sent the report to Lord Halifax as soon as I had dictated it, and he replied on September 5 in a guarded manner, that he did not at present feel that action of the kind proposed under Article 11 would be helpful, but that he would keep it in his mind. "For the present, I think, as you indicated, we must review the situation in the light of the report with which Henlein has returned from Berchtesgaden." He added that the situation remained very anxious.

★　★　★　★　★

In its leading article of September 7 the *Times* stated:

If the Sudetens now ask for more than the Czech Government are ready to give in their latest set of proposals, it can only be inferred that the Germans are going beyond the mere removal of disabilities for those who do not find themselves at ease within the Czechoslovak Republic. In that case it might be worth while for the Czechoslovak Government to consider whether they should exclude altogether the project, which has found favour in some quarters, of making Czechoslovakia a more homogeneous state by the cession of that fringe of alien populations who are contiguous to the nation to which they are united by race.

This of course involved the surrender of the whole of the Bohemian fortress line. Although the British Government stated at once that this *Times* article did not represent their views, public opinion abroad, particularly in France, was far from reassured. During the course of the same day—September 7—the French Ambassador in London called on Lord Halifax on behalf of his Government to ask for a clarification of the British position in the event of a German attack on Czechoslovakia.

M. Bonnet, then French Foreign Minister, declares that on September 10, 1938, he put the following question to our Ambassador in Paris, Sir Eric Phipps: "To-morrow Hitler may attack Czechoslovakia. If he does France will mobilise at once. She will turn to you, saying, 'We march: do you march with us?' What will be the answer of Great Britain?"

The following was the answer approved by the Cabinet, sent by Lord Halifax through Sir Eric Phipps on the 12th:

I naturally recognise of what importance it would be to the French

Government to have a plain answer to such a question. But, as you pointed out to Bonnet, the question itself, though plain in form, cannot be dissociated from the circumstances in which it might be posed, which are necessarily at this stage completely hypothetical.

Moreover, in this matter it is impossible for His Majesty's Government to have regard only to their own position, inasmuch as in any decision they may reach or action they may take they would, in fact, be committing the Dominions. Their Governments would quite certainly be unwilling to have their position in any way decided for them in advance of the actual circumstances, of which they would desire themselves to judge.

So far therefore as I am in a position to give any answer at this stage to M. Bonnet's question, it would have to be that while His Majesty's Government would never allow the security of France to be threatened, they are unable to make precise statements of the character of their future action, or the time at which it would be taken, in circumstances that they cannot at present foresee.*

Upon the statement that "His Majesty's Government would never allow the security of France to be threatened" the French asked what aid they could expect if it were. The reply from London was, according to Bonnet, two divisions, not motorised, and 150 aeroplanes during the first six months of the war. If M. Bonnet was seeking for an excuse for leaving the Czechs to their fate, it must be admitted that his search had met with some success.

On September 12 also Hitler delivered at a Nuremberg Party rally a violent attack on the Czechs, who replied on the following day by the establishment of martial law in certain districts of the republic. On September 14 negotiations with Henlein were definitely broken off, and on the 15th the Sudeten leader fled to Germany.

The summit of the crisis was now reached.

* Printed in Georges Bonnet, *De Washington au Quai d'Orsay*, pp. 360–1.

CHAPTER XVII

THE TRAGEDY OF MUNICH

Chamberlain in Control – He Visits Berchtesgaden – His Meeting with Hitler – The End of the Runciman Mission – Anglo-French Pressure upon Czechoslovakia – President Beneš's Submission – General Faucher Renounces French Citizenship – My Statement of September 21 – Litvinov's Formidable Declaration at the League Assembly – Soviet Power Ignored – The Vultures Gather Round the Doomed State – Chamberlain and Hitler at Godesberg – Hitler's Ultimatum – Rejection by the British and French Cabinets – Sir Horace Wilson's Mission to Berlin – My Visit to Downing Street on September 26 – Lord Halifax's Communiqué – Mobilisation of the British Navy – Behind the German Front – Dismissal of General von Beck – Hitler's Struggle with His Own Army Staff – General von Halder's Plot – Alleged Reason for its Collapse, September 14 – Memorial of the German General Staff to Hitler, September 26 – Admiral Raeder's Remonstrance – Hitler Wavers – Mr. Chamberlain's Broadcast of September 27 – His Third Offer to Visit Hitler – His Appeal to Mussolini – Drama in the House of Commons, September 28 – Conference at Munich – A Scrap of Paper – Chamberlain's Triumphant Return – "Peace with Honour!" – Marshal Keitel's Evidence at Nuremberg – Hitler's Judgment Again Vindicated – Some General Principles of Morals and Action – A Fatal Course for France and Britain.

MR. CHAMBERLAIN was now in complete control of British foreign policy, and Sir Horace Wilson was his principal confidant and agent. Lord Halifax, in spite of increasing doubts derived from the atmosphere of his department, followed the guidance of his chief. The Cabinet was deeply perturbed, but obeyed. The Government majority in the House of Commons was skilfully handled by the Whips. One man and one man only conducted our affairs. He did not shrink either from the responsibility which he incurred or from the personal exertions required.

During the night of September 13–14 M. Daladier got in touch with Mr. Chamberlain. The French Government were of the opinion that a joint approach to Hitler on a personal basis by the French and British leaders might be of value. Chamberlain however had been communing with himself. On his own initiative he telegraphed to Hitler proposing to come to see him. He informed the Cabinet of his action the next day, and in the afternoon received Hitler's reply inviting him to Berchtesgaden. Accordingly on the morning of September 15 the British Prime Minister flew to the Munich airfield. The moment was not in all respects well chosen. When the news reached Prague the Czech leaders could not believe it was true. They were astonished that at the very moment when for the first time they had the internal situation in the Sudeten areas in hand the British Prime Minister should himself pay a direct visit to Hitler. This, they felt, would weaken their position with Germany. Hitler's provocative speech of September 12, and the German-sponsored revolt of Henlein's adherents which had followed, had failed to gain local support. Henlein had fled to Germany, and the Sudeten German Party, bereft of his leadership, was clearly opposed to direct action. The Czech Government in the so-called "Fourth Plan" had officially proposed to the Sudeten German leaders administrative schemes for regional autonomy which not only exceeded Henlein's Carlsbad requests of April, but also fully met Chamberlain's view expressed in his speech of March 24, and Sir John Simon's statements in his speech of August 27. But even Lord Runciman realised that the last thing the Germans wanted was a satisfactory bargain between the Sudeten leaders and the Czech Government. Chamberlain's journeys gave them an opportunity to increase their demands; and on instructions from Berlin the extremists in the Sudeten Party now openly claimed union with the Reich.

★　　★　　★　　★　　★

The Prime Minister's plane arrived at Munich Airport in the afternoon of September 15; he travelled by train to Berchtesgaden. Meanwhile all the radio stations of Germany broadcast a proclamation by Henlein demanding the annexation of the

Sudeten areas to the Reich. This was the first news that reached Mr. Chamberlain when he landed. It was no doubt planned that he should know it before meeting Hitler. The question of ANNEXATION had never yet been raised either by the German Government or by Henlein; and a few days earlier the Foreign Office had stated that it was not the accepted policy of the British Government.

Mr. Feiling has already published such records as are extant of the conversations between Chamberlain and Hitler. The salient point we may derive from his account is this:

> In spite of the hardness and ruthlessness I thought I saw in his face, I got the impression that *here was a man who could be relied upon when he had given his word.**

In fact Hitler had for months past, as we have seen, resolved and prepared for the invasion of Czechoslovakia, which awaited only the final signal. When the Prime Minister reached London on Saturday, September 17, he summoned the Cabinet. Lord Runciman had now returned, and his report was assured of attention. He had all this time been failing in health, and the violent stress to which he had been exposed in his mission had reduced him to modest dimensions. He now recommended "a policy for immediate and drastic action", namely "the transfer of predominantly German districts to Germany." This at least had the merit of simplicity.

Both the Prime Minister and Lord Runciman were convinced that only the cession of the Sudeten areas to Germany would dissuade Hitler from ordering the invasion of Czechoslovakia. Mr. Chamberlain had been strongly impressed at his meeting with Hitler that he was "in a fighting mood". His Cabinet were also of the opinion that the French had no fight in them. There could therefore be no question of resisting Hitler's demands upon the Czech State. Some Ministers found consolation in such phrases as "the rights of self-determination", "the claims of a national minority to just treatment"; and even the mood appeared of "championing the small man against the Czech bully".

It was now necessary to keep in backward step with the French Government. On September 18 Daladier and Bonnet came to London. Chamberlain had already decided in principle to accept

* Feiling, *op. cit.*, p. 367.

Hitler's demands as explained to him at Berchtesgaden. There only remained the business of drafting the proposals to be presented to the Czech Government by the British and French representatives in Prague. The French Ministers brought with them a set of draft proposals which were certainly more skilfully conceived. They did not favour a plebiscite, because, they observed, there might be demands for further plebiscites in the Slovak and Ruthene areas. They favoured an outright cession of the Sudetenland to Germany. They added however that the British Government, with France *and with Russia*, whom they had not consulted, should guarantee the new frontiers of the mutilated Czechoslovakia.

Many of us, even outside Cabinet circles, had the sensation that Bonnet represented the quintessence of defeatism, and that all his clever verbal manœuvres had the aim of "peace at any price". In his book, written after the war, he labours naturally to thrust the whole burden upon Chamberlain and Halifax. There can be no doubt of what he had in his own mind. At all costs he wished to avoid having to fulfil the solemn, precise, and so recently renewed obligations of France to go to war in defence of Czechoslovakia. The British and French Cabinets at this time presented a front of two over-ripe melons crushed together; whereas what was needed was a gleam of steel. On one thing they were all agreed: there should be no consultation with the Czechs. These should be confronted with the decision of their guardians. The Babes in the Wood had no worse treatment.

In presenting their decision or ultimatum to the Czechs England and France said: "Both the French and British Governments recognise how great is the sacrifice thus required of Czechoslovakia. They have felt it their duty jointly to set forth frankly the conditions essential to security. . . . The Prime Minister must resume conversations with Herr Hitler not later than Wednesday, or sooner if possible. We therefore feel we must ask for your reply at the earliest possible moment." Proposals involving the immediate cession to Germany of all areas in Czechoslovakia containing over 50 per cent. of German inhabitants were therefore handed to the Czech Government in the afternoon of September 19.

Great Britain after all had no treaty obligation to defend Czechoslovakia, nor was she pledged in any informal way. But

France had definitely bound herself by treaty to make war upon Germany if she attacked Czechoslovakia. For twenty years President Beneš had been the faithful ally and almost vassal of France, always supporting French policies and French interests on the League of Nations and elsewhere. If ever there was a case of solemn obligation it was here and now. Fresh and vivid were the declarations of MM. Blum and Daladier. It was a portent of doom when a French Government failed to keep the word of France. I have always believed that Beneš was wrong to yield. He should have defended his fortress line. Once fighting had begun, in my opinion at that time, France would have moved to his aid in a surge of national passion, and Britain would have rallied to France almost immediately. At the height of this crisis (on September 20) I visited Paris for two days in order to see my friends in the French Government, Reynaud and Mandel. Both these Ministers were in lively distress and on the verge of resigning from the Daladier Cabinet. I was against this, as their sacrifice could not alter the course of events, and would only leave the French Government weakened by the loss of its two most capable and resolute men. I ventured even to speak to them in this sense. After this painful visit I returned to London.

*　*　*　*　*

At 2 a.m. on the night of September 20-21 the British and French Ministers in Prague called on President Beneš to inform him in effect that there was no hope of arbitration on the basis of the German Czechoslovak Treaty of 1925, and to urge upon him the acceptance of the Anglo-French proposals *"before producing a situation for which France and Britain could take no responsibility"*. The French Government at least was sufficiently ashamed of this communication to instruct its Minister to make it only verbally. Under this pressure on September 21 the Czech Government bowed to the Anglo-French proposals. There was in Prague at this moment a general of the French Army named Faucher. He had been in Czechoslovakia with the French Military Mission since 1919, and had been its chief since 1926. He now requested the French Government to relieve him of his duties, and placed himself at the disposal of the Czechoslovak Army. He also adopted Czech citizenship.

The following French defence has been made, and it cannot be lightly dismissed. If Czechoslovakia had refused to submit, and war had resulted, France would have fulfilled her obligations; but if the Czechs chose to give in under whatever pressures were administered French honour was saved. We must leave this to the judgment of history.

*   *   *   *   *

On the same day, September 21, I issued a statement on the crisis to the Press in London:

The partition of Czechoslovakia under pressure from England and France amounts to the complete surrender of the Western Democracies to the Nazi threat of force. Such a collapse will bring peace or security neither to England nor to France. On the contrary, it will place these two nations in an ever weaker and more dangerous situation. The mere neutralisation of Czechoslovakia means the liberation of twenty-five German divisions, which will threaten the Western front; in addition to which it will open up for the triumphant Nazis the road to the Black Sea. It is not Czechoslovakia alone which is menaced, but also the freedom and the democracy of all nations. The belief that security can be obtained by throwing a small State to the wolves is a fatal delusion. The war potential of Germany will increase in a short time more rapidly than it will be possible for France and Great Britain to complete the measures necessary for their defence.

*   *   *   *   *

At the Assembly of the League of Nations on September 21 an official warning was given by Litvinov:

... At the present time Czechoslovakia is suffering interference in its internal affairs at the hands of a neighbouring State, and is publicly and loudly menaced with attack. One of the oldest, most cultured, most hard-working of European peoples, who acquired their independence after centuries of oppression, to-day or to-morrow may decide to take up arms in defence of that independence. . . .

Such an event as the disappearance of Austria passed unnoticed by the League of Nations. Realising the significance of this event for the fate of the whole of Europe, and particularly of Czechoslovakia, the Soviet Government, immediately after the Anschluss, officially approached the other European Great Powers with a proposal for an immediate collective deliberation on the possible consequences of that event, in order to adopt collective preventive measures. To our regret, this proposal, which if carried out could have saved us from the alarm which all the world now feels for the fate of Czechoslovakia, did not receive its just appreciation. . . . When, a few days before I left for Geneva, the French Government for the first time inquired as to our attitude in the event of an attack on Czechoslovakia, I gave in the name of my Government the following perfectly clear and unambiguous reply:

"We intend to fulfil our obligations under the Pact, and together with France to afford assistance to Czechoslovakia by the ways open to us. Our War Department is ready immediately to participate in a conference with representatives of the French and Czechoslovak War Departments, in order to discuss the measures appropriate to the moment. . . ." It was only two days ago that the Czechoslovak Government addressed a formal inquiry to my Government as to whether the Soviet Union is prepared, in accordance with the Soviet-Czech Pact, to render Czechoslovakia immediate and effective aid if France, loyal to her obligations, will render similar assistance, to which my Government gave a clear answer in the affirmative.

It is indeed astonishing that this public, and unqualified, declaration by one of the greatest Powers concerned should not have played its part in Mr. Chamberlain's negotiations, or in the French conduct of the crisis. I have heard it suggested that it was geographically impossible for Russia to send troops into Czechoslovakia and that Russian aid in the event of war would have been limited to modest air support. The assent of Roumania, and also to a lesser extent of Hungary, to allow Russian forces to pass through their territory was of course necessary. This might well have been obtained from Roumania at least, as indicated to me by M. Maisky, through the pressures and guarantees of a Grand Alliance acting under the ægis of the League of Nations. There were two railways from Russia into Czechoslovakia through the Carpathian Mountains, the northerly from Czernowitz through the Bukovina, the southerly through Hungary by Debreczen. These two railways alone, which avoid both Bucharest and Budapest by good margins, might well have supported Russian armies of thirty divisions. As a counter for keeping the peace these possibilities would have been a substantial deterrent upon Hitler, and would almost certainly have led to far greater developments in the event of war. Stress has also been laid upon Soviet duplicity and bad faith. Anyhow, the Soviet offer was in effect ignored. They were not brought into the scale against Hitler, and were treated with an indifference—not to say disdain—which left a mark in Stalin's mind. Events took their course as if Soviet Russia did not exist. For this we afterwards paid dearly.

*　　*　　*　　*　　*

Mussolini, speaking at Treviso on September 21, said—not without some pith—"If Czechoslovakia finds herself to-day in what might be called a 'delicate situation', it is because she was— one may already say 'was', and I shall tell you why immediately— not just Czechoslovakia, but 'Czecho - Germano - Polono - Magyaro - Rutheno - Roumano-Slovakia', and I would em- phasise that now that this problem is being faced it is essential it should be solved in a general manner."*

Under the humiliation of the Anglo-French proposals the Czech Government resigned, and a non-party Administration was formed under General Syrovy, the commander of the Czecho- slovak legions in Siberia during the World War. On September 22 President Beneš broadcast to the Czech nation a dignified appeal for calm. While Beneš was preparing his broadcast Chamberlain had been flying to his second meeting with Hitler, this time at the Rhineland town of Godesberg. The British Prime Minister carried with him, as a basis for final discussion with the Fuehrer, the details of the Anglo-French proposals accepted by the Czech Government. The two men met in the hotel at Godesberg which Hitler had quitted in haste four years earlier for the Roehm purge. From the first Chamberlain realised that he was confronted with what he called in his own words "a totally unexpected situation". He described the scene in the House of Commons on his return:

I had been told at Berchtesgaden that if the principle of self-deter- mination were accepted Herr Hitler would discuss with me the ways and means of carrying it out. He told me afterwards that he never for one moment supposed that I should be able to come back and say that the principle was accepted. I do not want the House to think that he was deliberately deceiving me—I do not think so for one moment—but, for me, I expected that when I got back to Godesberg I had only to discuss quietly with him the proposals that I had brought with me; and it was a profound shock to me when I was told at the beginning of the conversation that these proposals were not acceptable, and that they were to be replaced by other proposals of a kind which I had not contemplated at all.

* Quoted in Ripka, *Munich and After*, p. 117.

I felt that I must have a little time to consider what I was to do. Consequently I withdrew, my mind full of foreboding as to the success of my mission. I first however obtained from Herr Hitler an extension of his previous assurance that he would not move his troops pending the results of the negotiations. I, on my side, undertook to appeal to the Czech Government to avoid any action which might provoke incidents.

Discussions were broken off until the next day. Throughout the morning of September 23 Chamberlain paced the balcony of his hotel. He sent a written message to Hitler after breakfast stating that he was ready to convey the new German proposals to the Czech Government, but pointing out grave difficulties. Hitler's reply in the afternoon showed little signs of yielding, and Chamberlain asked that a formal memorandum accompanied by maps should be handed to him at a final meeting that evening. The Czechs were now mobilising, and both the British and French Governments officially stated to their representatives in Prague that they could no longer take the responsibility of advising them not to. At 10.30 that night Chamberlain again met Hitler. The description of the meeting is best told in his own words:

The memorandum and the map were handed to me at my final interview with the Chancellor, which began at half-past ten that night and lasted into the small hours of the morning, an interview at which the German Foreign Secretary was present, as well as Sir Nevile Henderson and Sir Horace Wilson; and, for the first time, I found in the memorandum a time limit. Accordingly, on this occasion I spoke very frankly. I dwelt with all the emphasis at my command on the risks which would be incurred by insisting on such terms, and on the terrible consequences of a war, if war ensued. I declared that the language and the manner of the documents, which I described as an ultimatum rather than a memorandum, would profoundly shock public opinion in neutral countries, and I bitterly reproached the Chancellor for his failure to respond in any way to the efforts which I had made to secure peace.

I should add that Hitler repeated to me with great earnestness what he had said already at Berchtesgaden, namely, that this was the last of his territorial ambitions in Europe and that he had no wish to include in the Reich people of other races than Germans. In the second place he said, again very earnestly, that he wanted to be friends with England, and that if only this Sudeten question could be got out of the way in peace *he would gladly resume conversations*. It is true he said, "There is one awkward question, the Colonies; but that is not a matter for war."

On the afternoon of September 24 Mr. Chamberlain returned to London, and on the following day three meetings of the Cabinet were held. There was a noticeable stiffening of opinion both in London and in Paris. It was decided to reject the Godesberg terms, and this information was conveyed to the German Government. The French Cabinet agreed, and a partial French mobilisation was carried out promptly and with more efficiency than was expected. On the evening of September 25 the French Ministers came again to London and reluctantly accepted their obligations to the Czechs. During the course of the following afternoon Sir Horace Wilson was sent with a personal letter to Hitler in Berlin three hours before the latter was to speak in the Sports Palace. The only answer Sir Horace was able to obtain was that Hitler would not depart from the time limit set by the Godesberg ultimatum, namely, Saturday, October 1, on which day he would march into the territories concerned unless he received Czech acquiescence by 2 p.m. on Wednesday the 28th.

That evening Hitler spoke in Berlin. He referred to England and France in accommodating phrases, launching at the same time a coarse and brutal attack on Beneš and the Czechs. He said categorically that the Czechs must clear out of the Sudetenland by the 26th, but once this was settled he had no more interest in what happened to Czechoslovakia. "*This is the last territorial claim I have to make in Europe.*"

★   ★   ★   ★   ★

As on similar occasions, my contacts with His Majesty's Government became more frequent and intimate with the mounting of the crisis. On September 10 I had visited the Prime Minister at Downing Street for a long talk. Again on September 26 he either invited me or readily accorded me an interview. At 3.30 in the afternoon of this critical day I was received by him and Lord Halifax in the Cabinet Room. I pressed upon them the policy set forth in my letter to Lord Halifax of August 31, namely, a declaration showing the unity of sentiment and purpose between Britain, France, *and* Russia against Hitlerite aggression. We discussed at length and in detail a communique, and we seemed to be in complete agreement. Lord Halifax and I were at one, and

I certainly thought the Prime Minister was in full accord. There was present a high official of the Foreign Office, who built up the draft. When we separated I was satisfied and relieved.

About 8 o'clock that night Mr. Leeper, then Head of the Foreign Office Press Department, now Sir Reginald Leeper, presented to the Foreign Secretary a communiqué of which the following is the pith:

If, in spite of the efforts made by the British Prime Minister, a German attack is made upon Czechoslovakia, the immediate result must be that France will be bound to come to her assistance, and Great Britain *and Russia* will certainly stand by France.

This was approved by Lord Halifax and immediately issued.

When earlier I returned to my flat at Morpeth Mansions I found about fifteen gentlemen assembled. They were all Conservatives: Lord Cecil, Lord Lloyd, Sir Edward Grigg, Sir Robert Horne, Mr. Boothby, Mr. Bracken, and Mr. Law. The feeling was passionate. It all focused on the point, "We must get Russia in". I was impressed and indeed surprised by this intensity of view in Tory circles, showing how completely they had cast away all thoughts of class, party, or ideological interests, and to what a pitch their mood had come. I reported to them what had happened at Downing Street and described the character of the communiqué. They were all greatly reassured.

The French Right Press treated this communiqué with suspicion and disdain. The *Matin* called it "a clever lie". M. Bonnet, who is now very busy showing how forward in action he was, told several Deputies that he had no confirmation of it, leaving on them the impression that this was not the British pledge he was looking for. This was no doubt not difficult for him to convey.

I dined that night with Mr. Duff Cooper at the Admiralty. He told me that he was demanding from the Prime Minister the immediate mobilisation of the Fleet. I recalled my own experiences a quarter of a century before, when similar circumstances had presented themselves.

*　*　*　*　*

It seemed that the moment of clash had arrived and that the opposing forces were aligned. The Czechs had a million and a half men armed behind the strongest fortress line in Europe, and equipped by a highly organised and powerful industrial machine. The French Army was partly mobilised, and, albeit reluctantly, the French Ministers were prepared to honour their obligations to Czechoslovakia. Just before midnight on September 27 the warning telegram was sent out from the Admiralty ordering the mobilisation of the Fleet for the following day. This information was given to the British Press almost simultaneously (at 11.38 p.m.). At 11.20 a.m. on September 28 the actual orders to the British Fleet to mobilise were issued from the Admiralty.

★   ★   ★   ★   ★

We may now look behind the brazen front which Hitler presented to the British and French Governments. General Beck, the Chief of the Army General Staff, had become profoundly alarmed about Hitler's schemes. He entirely disapproved of them, and was prepared to resist. After the invasion of Austria in March he had sent a memorandum to Hitler arguing by detailed facts that the continuance of a programme of conquest must lead to world-wide catastrophe and the ruin of the now reviving Reich. To this Hitler did not reply. There was a pause. Beck refused to share the responsibility before history for the war plunge which the Fuehrer was resolved to make. In July a personal confrontation took place. When the imminence of an attack on Czechoslovakia became clear Beck demanded an assurance against further military adventures. Here was a crunch. Hitler rejoined that the Army was the instrument of the State, that he was the Head of the State, and that the Army and other forces owed unquestioning obedience to his will. On this Beck resigned. His request to be relieved of his post remained unanswered. But the General's decision was irrevocable. Henceforth he absented himself from the War Ministry. Hitler was therefore forced to dismiss him, and appointed Halder as his successor. For Beck there remained only a tragic but honourable fate.

All this was kept within a secret circle; but there now began an intense, unceasing struggle between the Fuehrer and his expert advisers. Beck was universally trusted and respected by the Army Staff, who were united not only in professional opinion but in resentment of civilian and party dictation. The September crisis seemed to provide all the circumstances which the German generals dreaded. Between thirty and forty Czech divisions were deploying upon Germany's eastern frontiers, and the weight of the French Army, at odds of nearly eight to one, began to lie heavy on the Western Wall. A hostile Russia might operate from Czech airfields, and Soviet armies might wend their way forward through Poland or Roumania. Finally, in the last stage the British Navy was said to be mobilising. As all this developed passions rose to fever-heat.

First we have the account, given by General Halder, of a definite plot to arrest Hitler and his principal associates. The evidence for this does not rest only on Halder's detailed statements. Plans were certainly made, but how far they were at the time backed by resolve cannot be judged precisely. The generals were repeatedly planning revolts, and as often drew back at the last moment for one reason or another. It was to the interest of the parties concerned after they were the prisoners of the Allies to dwell upon their efforts for peace. There can be no doubt however of the existence of the plot at this moment, and of serious measures taken to make it effective.

By the beginning of September [Halder says] we had taken the necessary steps to immunize Germany from this madman. At this time the prospect of war filled the great majority of the German people with horror. We did not intend to kill the Nazi leaders—merely to arrest them, establish a military Government, and issue a proclamation to the people that we had taken this action only because we were convinced they were being led to certain disaster.

The following were in the plot: Generals Halder, Beck, Stuelpnagel, Witzleben (Commander of the Berlin Garrison), Thomas (Controller of Armaments), Brockdorff (Commander of the Potsdam Garrison), and Graf von Helldorf, who was in charge of the Berlin police. The Commander-in-Chief, General von Brauchitsch, was informed, and approved.

It was easy, as a part of the troop movements against Czechoslovakia, and of ordinary military routine, to hold one Panzer division so near to Berlin that it could reach the capital by a night's march. The evidence is clear that the Third Panzer Division, commanded by General Hoeppner, was at the time of the Munich crisis stationed south of Berlin. General Hoeppner's secret mission was to occupy the capital, the Chancellery, and the important Nazi Ministries and offices at a given signal. For this purpose it was added to General Witzleben's command. According to Halder's account, Helldorf, Chief of the Berlin Police, then made meticulous arrangements to arrest Hitler, Goering, Goebbels, and Himmler. "There was no possibility of a hitch. All that was required for a completely successful *coup* was Hitler's presence in Berlin." He arrived there from Berchtesgaden on the morning of September 14. Halder heard of this at midday, and immediately went over to see Witzleben and complete the plans. It was decided to strike at 8 p.m. that same evening. At 4 p.m., according to Halder, a message was received in Witzleben's office that Mr. Chamberlain was going to fly to see the Fuehrer at Berchtesgaden. A meeting was at once held, at which he, Halder, told Witzleben that "if Hitler had succeeded in his bluff, he would not be justified, as Chief of Staff, in calling it." It was accordingly decided to defer action and await events.

Such is the tale, which historians should probe, of this internal crisis in Berlin as told by General Halder, at that time Chief of the Staff. It has since been confirmed by General Mueller-Hillebrandt, and has been accepted as genuine by various authorities who have examined it. If it should eventually be accepted as historical truth, it will be another example of the very small accidents upon which the fortunes of mankind turn.

Of other less violent but earnest efforts of the General Staff to restrain Hitler there can be no doubt. On September 26 a deputation, consisting of General von Hanneken, Ritter von Leeb, and Colonel Bodenschatz, called at the Chancellery of the Reich and requested to be received by Herr Hitler. They were sent away. At noon on the following day the principal generals held a meeting at the War Office. They agreed upon a memorial, which they left at the Chancellery. This document was published in France in November 1938.* It consisted of eighteen pages, divided into five chapters and three appendices. Chapter I stresses the divergences between the political and military leadership of the Third Reich, and declares that the low morale of the German population renders it incapable of sustaining a European war. It states that in the event of war breaking out exceptional powers must be given to the military authorities. Chapter II describes the bad

---

* Published by Professor Bernard Lavergne in *L'Année Politique Française et Étrangère* in November 1938. Quoted in Ripka, *op. cit.*, pp. 212 ff.

condition of the Reichswehr and mentions that the military authorities have felt obliged "to shut their eyes in many serious cases to the absence of discipline". Chapter III enumerates the various deficiencies in German armaments, dwelling upon the defects in the Siegfried Line, so hurriedly constructed, and the lack of fortifications in the Aix-la-Chapelle and Saarbruck areas. Fear is expressed of an incursion into Belgium by the French troops concentrated around Givet. Finally, emphasis is laid on the shortage of officers. No fewer than forty-eight thousand officers and a hundred thousand N.C.O.s were necessary to bring the Army up to war strength, and in the event of a general mobilisation no fewer than eighteen divisions would find themselves devoid of trained subordinate commanders.

The document presents the reasons why defeat must be expected in any but a strictly local war, and affirms that less than a fifth of the officers of the Reichswehr believed in the possibility of a victory for Germany. A military appreciation about Czechoslovakia in the Appendix states that the Czechoslovak Army, even if fighting without allies, could hold out for three months, and that Germany would need to retain covering forces on the Polish and French frontiers as well as on the Baltic and North Sea coasts, and to keep a force of at least a quarter of a million troops in Austria to guard against popular risings and a possible Czechoslovak offensive. Finally, the General Staff believed that it was highly improbable that hostilities would remain localised during the three-month period.

The warnings of the soldiers were finally reinforced by Admiral Raeder, Chief of the German Admiralty. At 10 p.m. on September 27 Raeder was received by the Fuehrer. He made a vehement appeal, which was emphasised a few hours later by the news that the British Fleet was being mobilised. Hitler now wavered. At 2 a.m. the German radio broadcast an official denial that Germany intended to mobilise on the 29th, and at 11.45 a.m. the same morning a statement of the German official news agency was given to the British Press again denying the reports of the intended German mobilisation. The strain upon this one man and upon his astounding will-power must at this moment have been most severe. Evidently he had brought himself to the brink of a general war. Could he take the plunge in the face of an unfavourable public opinion and of the solemn warning of the chiefs of his Army,

Navy, and Air Force? Could he, on the other hand, afford to retreat after living so long upon prestige?

* * * * *

While the Fuehrer was at grips with his generals Mr. Chamberlain himself was preparing to broadcast to the English nation. On the evening of September 27 he spoke as follows:

How horrible, fantastic, incredible, it is that we should be digging trenches and trying on gas-masks here because of a quarrel in a faraway country between people of whom we know nothing! . . . I would not hesitate to pay even a third visit to Germany if I thought it would do any good. . . . I am myself a man of peace to the depths of my soul. Armed conflict between nations is a nightmare to me; but if I were convinced that any nation had made up its mind to dominate the world by fear of its force I should feel that it must be resisted. Under such a domination life for people who believe in liberty would not be worth living; but war is a fearful thing, and we must be very clear, before we embark on it, that it is really the great issues that are at stake.

After delivering this balancing broadcast he received Hitler's reply to the letter he had sent through Sir Horace Wilson. This letter opened a chink of hope. Hitler offered to join in a guarantee of the new frontiers of Czechoslovakia, and was willing to give further assurances about the manner of carrying out the new plebiscite. There was little time to lose. The German ultimatum contained in the Godesberg memorandum was due to expire at 2 p.m. on the following day, Wednesday, September 28. Chamberlain therefore drafted a personal message to Hitler: "After reading your letter I feel certain that you can get all essentials without war, and without delay. I am ready to come to Berlin myself at once to discuss arrangements for transfer with you and representatives of the Czech Government, together with representatives of France and Italy, if you desire. I feel convinced that we could reach agreement in a week."* At the same time he telegraphed to Mussolini informing him of this last appeal to Hitler: "I trust your Excellency will inform the German Chancellor that you are willing to be represented, and urge him to agree to my proposal, which will keep our peoples out of war."

It is one of the remarkable features of this crisis that no close and

* Feiling, *op. cit.*, p. 372.

confidential consultation seems to have existed between London and Paris. There was a broad coincidence of view, but little or no personal contact. While Mr. Chamberlain, without consulting either the French Government or his own Cabinet colleagues, was drafting these two letters, the French Ministers were taking their own separate measures along parallel lines. We have seen the strength of the forces opposed to standing up to Germany in the French Press, and how the firm British communiqué, naming Russia, was suggested in Paris newspapers inspired by the French Foreign Office to be a forgery. The French Ambassador in Berlin was instructed on the night of the 27th to make yet further proposals extending the territory in the Sudetenland to be handed over for immediate German occupation. While M. François-Poncet was with Hitler a message arrived from Mussolini advising that Chamberlain's idea of a conference should be accepted and that Italy should take part. At 3 o'clock on the afternoon of September 28 Hitler sent messages to Chamberlain and Daladier proposing a meeting at Munich on the following day together with Mussolini. At that hour Mr. Chamberlain was addressing the House of Commons, giving them a general view of recent events. As he neared the end of his speech the message inviting him to Munich was passed down to him by Lord Halifax, who was sitting in the Peers' Gallery. Mr. Chamberlain was at that moment describing the letter which he had sent to Mussolini and the results of his move:

In reply to my message to Signor Mussolini, I was informed that instructions had been sent by the Duce . . . that while Italy would fulfil completely her pledges to stand by Germany, yet, in view of the great importance of the request made by His Majesty's Government to Signor Mussolini, the latter hoped Herr Hitler would see his way to postpone action which the Chancellor had told Sir Horace Wilson was to be taken at 2 p.m. to-day for at least twenty-four hours, so as to allow Signor Mussolini time to re-examine the situation and endeavour to find a peaceful settlement. In response, Herr Hitler has agreed to postpone mobilisation for twenty-four hours. . . . That is not all. I have something further to say to the House yet. I have now been informed by Herr Hitler that he invites me to meet him at Munich to-morrow morning. He has also invited Signor Mussolini and M. Daladier. Signor Mussolini has accepted, and I have no doubt M. Daladier will also accept. I need not say what my answer will be.

... I am sure that the House will be ready to release me now to go and see what I can make of this last effort.

Thus for the third time Mr. Chamberlain flew to Germany.

* * * * *

Many accounts have been written of this memorable meeting, and it is not possible here to do more than emphasise some special features. No invitation was extended to Russia. Nor were the Czechs themselves allowed to be present at the meetings. The Czech Government had been informed in bald terms on the evening of the 28th that a conference of the representatives of the four European Powers would take place the following day. Agreement was reached between "the Big Four" with speed. The conversations began at noon and lasted till two o'clock the next morning. A memorandum was drawn up and signed at 2 a.m. on September 30. It was in essentials the acceptance of the Godesberg ultimatum. The Sudetenland was to be evacuated in five stages beginning on October 1, and to be completed within ten days. An International Commission was to determine the final frontiers. The document was placed before the Czech delegates who had been allowed to come to Munich to receive the decisions.

While the four statesmen were waiting for the experts to draft the final document the Prime Minister asked Hitler whether he would care for a private talk. Hitler "jumped at the idea."* The two leaders met in Hitler's Munich flat on the morning of September 30, and were alone except for the interpreter. Chamberlain produced a draft declaration which he had prepared, as follows:

We, the German Fuehrer and Chancellor, and the British Prime Minister, have had a further meeting to-day, and are agreed in recognising that the question of Anglo-German relations is of the first importance for the two countries and for Europe.

We regard the Agreement signed last night, and the Anglo-German Naval Agreement, as symbolic of the desire of our two peoples never to go to war with one another again.

We are resolved that the method of consultation shall be the method adopted to deal with any other questions that may concern our two countries, and we are determined to continue our efforts to remove

---

* See Feiling, *op. cit.*, p. 376.

possible sources of difference, and thus to contribute to assure the peace of Europe.

Hitler read this Note and signed it without demur.

Chamberlain returned to England. At Heston, where he landed, he waved the joint declaration which he had got Hitler to sign, and read it to the crowd of notables and others who welcomed him. As his car drove through cheering crowds from the airport, he said to Halifax, sitting beside him, "All this will be over in three months"; but from the windows of Downing Street he waved his piece of paper again and used these words, "This is the second time in our history that there has come back from Germany to Downing Street peace with honour. I believe it is peace for our time."*

* * * * *

We have now also Marshal von Keitel's answer to the specific question put to him by the Czech representative at the Nuremberg trials:

Colonel Eger, representing Czechoslovakia, asked Marshal Keitel:
"Would the Reich have attacked Czechoslovakia in 1938 if the Western Powers had stood by Prague?"
Marshal Keitel answered:
"Certainly not. We were not strong enough militarily. The object of Munich [i.e., reaching an agreement at Munich] was to get Russia out of Europe, to gain time, and to complete the German armaments."†

* * * * *

Hitler's judgment had been once more decisively vindicated. The German General Staff was utterly abashed. Once again the Fuehrer had been right after all. He with his genius and intuition alone had truly measured all the circumstances, military and political. Once again, as in the Rhineland, the Fuehrer's leadership had triumphed over the obstruction of the German military chiefs. All these generals were patriotic men. They longed to see the Fatherland regain its position in the world. They were devoting themselves night and day to every process that could strengthen the German forces. They therefore felt smitten in their hearts at having been found so much below the level of the event,

---

* Feiling, *op. cit.*, p. 381.
† Quoted in Paul Reynaud's *La France a Sauvé l'Europe*, I, 561 ff.

and in many cases their dislike and their distrust of Hitler were overpowered by admiration for his commanding gifts and miraculous luck. Surely here was a star to follow, surely here was a guide to obey. Thus did Hitler finally become the undisputed master of Germany, and the path was clear for the great design. The conspirators lay low, and were not betrayed by their military comrades.

★ ★ ★ ★ ★

It may be well here to set down some principles of morals and action which may be a guide in the future. No case of this kind can be judged apart from its circumstances. The facts may be unknown at the time, and estimates of them must be largely guesswork, coloured by the general feeling and aims of whoever is trying to pronounce. Those who are prone by temperament and character to seek sharp and clear-cut solutions of difficult and obscure problems, who are ready to fight whenever some challenge comes from a foreign Power, have not always been right. On the other hand, those whose inclination is to bow their heads, to seek patiently and faithfully for peaceful compromise, are not always wrong. On the contrary, in the majority of instances they may be right, not only morally but from a practical standpoint. How many wars have been averted by patience and persisting goodwill! Religion and virtue alike lend their sanctions to meekness and humility, not only between men but between nations. How many wars have been precipitated by firebrands! How many misunderstandings which led to wars could have been removed by temporising! How often have countries fought cruel wars and then after a few years of peace found themselves not only friends but allies!

The Sermon on the Mount is the last word in Christian ethics. Everyone respects the Quakers. Still, it is not on these terms that Ministers assume their responsibilities of guiding States. Their duty is first so to deal with other nations as to avoid strife and war and to eschew aggression in all its forms, whether for nationalistic or ideological objects. But the safety of the State, the lives and freedom of their own fellow-countrymen, to whom they owe their position, make it right and imperative in the last resort, or when a final and definite conviction has been reached, that the use of force should not be excluded. If the circumstances are such as to warrant it, force may be used. And if this be so it should be used under the conditions which are most favourable. There is no merit in putting off a war for a year if, when it comes, it is a far worse war or one much harder to win. These are the tormenting dilemmas upon which mankind has throughout its history been so frequently impaled. Final judgment upon them can only be recorded by history in relation to the facts of the case as known to the parties at the time, and also as subsequently proved.

There is however one helpful guide, namely, for a nation to keep its word and to act in accordance with its treaty obligations to allies. This guide is called honour. It is baffling to reflect that what men call honour does not correspond always to Christian ethics. Honour is often influenced by that element of pride which plays so large a part in its inspiration. An exaggerated code of honour leading to the performance of utterly vain and unreasonable deeds could not be defended, however fine it might look. Here however the moment came when honour pointed the path of duty, and when also the right judgment of the facts at that time would have reinforced its dictates.

For the French Government to leave her faithful ally Czechoslovakia to her fate was a melancholy lapse from which flowed terrible consequences. Not only wise and fair policy, but chivalry, honour, and sympathy for a small threatened people made an overwhelming concentration. Great Britain, who would certainly have fought if bound by treaty obligations, was nevertheless now deeply involved, and it must be recorded with regret that the British Government not only acquiesced but encouraged the French Government in a fatal course.

CHAPTER XVIII

MUNICH WINTER

ON SEPTEMBER 30 Czechoslovakia bowed to the decisions of Munich. "They wished," they said, "to register their protest before the world against a decision in which they had no part." President Beneš resigned because "he might now prove a hindrance to the developments to which our new State must adapt itself". He departed from Czechoslovakia and found shelter in England. The dismemberment of the Czechoslovak State proceeded in accordance with the Agreement. But the Germans were not the only vultures upon the carcass. Immediately after the Munich Agreement on September 30 the Polish Government sent a twenty-four-hour ultimatum to the Czechs demanding the immediate handing over of the frontier district of Teschen. There was no means of resisting this harsh demand.

The heroic characteristics of the Polish race must not blind us to their errors, which over centuries have led them through measureless suffering. We see them in 1919, a people restored by the victory of the Western Allies after long generations of partition and servitude to be an independent republic and one of the main Powers in Europe. Now, in 1938, over a question so minor as Teschen, they sundered themselves from all those friends in France, Britain, and the United States who had lifted them once again to a national, coherent life, and whom they were soon to need so sorely. We see them hurrying, while the might of Germany glowered up against them, to grasp their share of the pillage and ruin of Czechoslovakia. During the crisis the door was shut in the face of the British and French Ambassadors, who were denied even access to the Foreign Secretary of the Polish State. It is a mystery and tragedy of European history that a people capable of every heroic virtue, gifted, valiant, charming, as individuals, should repeatedly show such inveterate faults in almost every aspect of their governmental life. All our hearts are with the Polish people in their new subjugation, and we are sure that we shall never seek in vain for their perennial impulse to strike against tyranny, and to suffer with invincible fortitude all the agonies which befall them. We all look forward to the dawn.

* * * * *

The Hungarians had also been on the fringe of the Munich discussions. Horthy had visited Germany at the end of August 1938, but Hitler had been very reserved in his attitude. Although he talked long with the Hungarian Regent on the afternoon of August 23, he did not reveal to him the date of his intended move against Czechoslovakia. "He himself did not know the time. Whoever wanted to join the meal would have to share in the cooking as well." But the hour of the meal had not been disclosed. Now however the Hungarians arrived with their claims.

* * * * *

It is not easy in these latter days, when we have all passed through years of intense moral and physical stress and exertion, to portray for another generation the passions which raged in Britain about the Munich Agreement. Among the Conservatives families and friends in intimate contact were divided to a degree the like of which I have never seen. Men and women, long bound together by party ties, social amenities, and family connections, glared upon one another in scorn and anger. The issue was not one to be settled by the cheering crowds which had welcomed Mr. Chamberlain back from the airport or blocked Downing

Street and its approaches, nor by the redoubtable exertions of the Ministerial Whips and partisans. We who were in a minority at the moment cared nothing for the jokes or scowls of the Government supporters. The Cabinet was shaken to its foundations, but the event had happened and they held together. One Minister alone stood forth. The First Lord of the Admiralty, Mr. Duff Cooper, resigned his great office, which he had dignified by the mobilisation of the Fleet. At the moment of Mr. Chamberlain's overwhelming mastery of public opinion he thrust his way through the exulting throng to declare his total disagreement with its leader.

At the opening of the three days' debate on Munich he made his resignation speech. This was a vivid incident in our Parliamentary life. Speaking with ease and without a note, for forty minutes he held the hostile majority of his party under his spell. It was easy for Labour men and Liberals in hot opposition to the Government of the day to applaud him. This was a rending quarrel within the Tory Party. Some of the truths he uttered must be recorded here:

I besought my colleagues not to see this problem always in terms of Czechoslovakia, not to review it always from the difficult strategic position of that small country, but rather to say to themselves, "A moment may come when, owing to the invasion of Czechoslovakia, a European war will begin, and when that moment comes we must take part in that war, we cannot keep out of it, and there is no doubt upon which side we shall fight." Let the world know that, and it will give those who are prepared to disturb the peace reason to hold their hand. . . .

Then came the last appeal from the Prime Minister on Wednesday morning. For the first time from the beginning to the end of the four weeks of negotiations Herr Hitler was prepared to yield an inch, an ell perhaps, but to yield some measure to the representations of Great Britain. But I would remind the House that the message from the Prime Minister was not the first news that he had received that morning. At dawn he had learned of the mobilisation of the British Fleet. It is impossible to know what are the motives of man, and we shall probably never be satisfied as to which of these two sources of inspiration moved him most when he agreed to go to Munich; but we do know that never before had he given in, and that then he did. I had been urging the mobilisation of the Fleet for many days. I had thought that this was the kind of language which would be easier for Herr

Hitler to understand than the guarded language of diplomacy or the conditional clauses of the Civil Service. I had urged that something in that direction might be done at the end of August and before the Prime Minister went to Berchtesgaden. I had suggested that it should accompany the mission of Sir Horace Wilson. I remember the Prime Minister stating it was the one thing that would ruin that mission, and I said it was the one thing that would lead it to success.

That is the deep difference between the Prime Minister and myself throughout these days. The Prime Minister has believed in addressing Herr Hitler through the language of sweet reasonableness. I have believed that he was more open to the language of the mailed fist. . . .

The Prime Minister has confidence in the goodwill and in the word of Herr Hitler, although when Herr Hitler broke the Treaty of Versailles he undertook to keep the Treaty of Locarno, and when he broke the Treaty of Locarno he undertook not to interfere further, or to have further territorial claims in Europe. When he entered Austria by force he authorised his henchmen to give an authoritative assurance that he would not interfere with Czechoslovakia. That was less than six months ago. Still the Prime Minister believes that he can rely upon the good faith of Hitler.

*　　*　　*　　*　　*

The long debate was not unworthy of the emotions aroused and the issues at stake. I well remember that when I said "We have sustained a total and unmitigated defeat" the storm which met me made it necessary to pause for a while before resuming. There was widespread and sincere admiration for Mr. Chamberlain's persevering and unflinching efforts to maintain peace, and for the personal exertions which he had made. It is impossible in this account to avoid marking the long series of miscalculations, and misjudgments of men and facts, on which he based himself; but the motives which inspired him have never been impugned, and the course he followed required the highest degree of moral courage. To this I paid tribute two years later in my speech after his death. The differences which arose between leading Conservatives, fierce though they were, carried with them no lack of mutual respect, nor in most cases did they sever, except temporarily, personal relations. It was common ground between us that the Labour and Liberal Oppositions, now so vehement for action, had never missed an opportunity of gaining popularity by resisting and denouncing even the half-measures for defence which the Government had taken.

There was also a serious and practical line of argument, albeit not to their credit, on which the Government could rest themselves. No one could deny that we were hideously unprepared for war. Who had been more forward in proving this than I and my friends? Great Britain had allowed herself to be far surpassed by the strength of the German Air Force. All our vulnerable points were unprotected. Barely a hundred anti-aircraft guns could be found for the defence of the largest city and centre of population in the world; and these were largely in the hands of untrained men. If Hitler was honest and lasting peace had in fact been achieved, Chamberlain was right. If, unhappily, he had been deceived, at least we should gain a breathing-space to repair the worst of our neglects. These considerations, and the general relief and rejoicing that the horrors of war had been temporarily averted, commanded the loyal assent of the mass of Government supporters. The House approved the policy of His Majesty's Government "by which war was averted in the recent crisis" by 366 to 144. The thirty or forty dissentient Conservatives could do no more than register their disapproval by abstention. This we did as a formal and united act.

In the course of my speech I said:

We really must not waste time after all this long debate upon the difference between the positions reached at Berchtesgaden, at Godesberg, and at Munich. They can be very simply epitomised, if the House will permit me to vary the metaphor. £1 was demanded at the pistol's point. When it was given, £2 were demanded at the pistol's point. Finally the Dictator consented to take £1 17s. 6d. and the rest in promises of goodwill for the future.

No one has been a more resolute and uncompromising struggler for peace than the Prime Minister. Everyone knows that. Never has there been such intense and undaunted determination to maintain and secure peace. Nevertheless, I am not quite clear why there was so much danger of Great Britain or France being involved in a war with Germany at this juncture if in fact they were ready all along to sacrifice Czechoslovakia. The terms which the Prime Minister brought back with him could easily have been agreed, I believe, through the ordinary diplomatic channels at any time during the summer. And I will say this, that I believe the Czechs, left to themselves, and told they were going to get no help from the Western Powers, would have been able to make better terms than they have got after all this tremendous perturbation. They could hardly have had worse.

All is over. Silent, mournful, abandoned, broken, Czechoslovakia recedes into the darkness. She has suffered in every respect by her associations with France, under whose guidance and policy she has been actuated for so long. . . .

I find unendurable the sense of our country falling into the power, into the orbit and influence of Nazi Germany, and of our existence becoming dependent upon their goodwill or pleasure. It is to prevent that that I have tried my best to urge the maintenance of every bulwark of defence—first, the timely creation of an Air Force superior to anything within striking distance of our shores; secondly, the gathering together of the collective strength of many nations; and, thirdly, the making of alliances and military conventions, all within the Covenant, in order to gather together forces at any rate to restrain the onward movement of this power. It has all been in vain. Every position has been successively undermined and abandoned on specious and plausible excuses.

I do not grudge our loyal, brave people, who were ready to do their duty no matter what the cost, who never flinched under the strain of last week, the natural, spontaneous outburst of joy and relief when they learned that the hard ordeal would no longer be required of them at the moment; but they should know the truth. They should know that there has been gross neglect and deficiency in our defences; they should know that we have sustained a defeat without a war, the consequences of which will travel far with us along our road; they should know that we have passed an awful milestone in our history, when the whole equilibrium of Europe has been deranged, and that the terrible words have for the time being been pronounced against the Western democracies: "Thou art weighed in the balance and found wanting." And do not suppose that this is the end. This is only the beginning of the reckoning. This is only the first sip, the first foretaste of a bitter cup which will be proffered to us year by year unless, by a supreme recovery of moral health and martial vigour, we arise again and take our stand for freedom as in the olden time.

*   *   *   *   *

Hitler's gratitude for British goodwill and for the sincere rejoicings that peace with Germany had been preserved at Munich found only frigid expression. On October 9, less than a fortnight after he had signed the declaration of mutual friendship which Mr. Chamberlain had pressed upon him, he said in a speech at Saarbrucken:

The statesmen who are opposed to us wish for peace . . . but they govern in countries whose domestic organisation makes it possible that at any time they may lose their position to make way for others who are not anxious for peace. And those others are there. It only needs that in England instead of Chamberlain Mr. Duff Cooper or Mr. Eden or Mr. Churchill should come to power, and then we know quite well that it would be the aim of these men immediately to begin a new World War. They make no secret of the fact: they admit it openly. We know further that now, as in the past, there lurks in the background the menacing figure of that Jewish-international foe who has found a basis and a form for himself in a State turned Bolshevist. And we know further the power of a certain international Press which lives only on lies and slander. That obliges us to be watchful and to remember the protection of the Reich. At any time ready for peace, but at every hour also ready to defend ourselves.

I have therefore decided, as I announced in my speech at Nuremberg, to continue the construction of our fortifications in the West with increased energy. I shall now also bring within the line of these fortifications the two large areas which up to the present lay in front of our fortifications—the district of Aachen [Aix-la-Chapelle] and the district of Saarbrucken.

He added:

It would be a good thing if in Great Britain people would gradually drop certain airs which they have inherited from the Versailles epoch. We cannot tolerate any longer *the tutelage of governesses*. Inquiries of British politicians concerning the fate of Germans within the frontiers of the Reich, or of others belonging to the Reich, are not in place. We for our part do not trouble ourselves about similar things in England. The outside world might often have reason enough to concern itself with its own national affairs, or, for instance, with affairs in Palestine.

After the sense of relief springing from the Munich agreement had worn off Mr. Chamberlain and his Government found themselves confronted by a sharp dilemma. The Prime Minister had said, "I believe it is peace for our time." But the majority of his colleagues wished to utilise "our time" to rearm as rapidly as possible. Here a division arose in the Cabinet. The sensations of alarm which the Munich crisis had aroused, the flagrant exposure of our deficiencies, especially in anti-aircraft guns, dictated vehement rearmament. Hitler, on the other hand, was shocked at such a mood. "Is this the trust and friendship," he might have pretended, "of our Munich pact? If we are friends and you trust us, why is it necessary for you to rearm? Let me have the arms, and you show the trust." But this view, though it would have been thoroughly justified on the data presented to Parliament, carried no conviction. There was a strong forward surge for invigorated rearmament. And this of course was criticised by the German Government and its inspired Press. However, there was no doubt of the opinion of the British nation. While rejoicing at being delivered from war by the Prime Minister and cheering peace slogans to the echo, they felt the need of weapons acutely. All the Service departments put in their claims and referred to the alarming shortages which the crisis had exposed. The Cabinet reached an agreeable compromise on the basis of all possible preparations without disturbing the trade of the country or irritating the Germans and Italians by large-scale measures.

★　★　★　★　★

It was to Mr. Chamberlain's credit that he did not yield to temptation and pressures to seek a General Election on the morrow of Munich. This could only have led to greater confusion. Nevertheless the winter months were anxious and depressing to those Conservatives who had criticised and refused to vote for the Munich settlement. Each of us was attacked in his constituency by the Conservative Party machine, and many there were who a year later were our ardent supporters who agitated against us. In my own constituency, the Epping Division, matters came to such a pass that I had to make it clear that if a resolution of censure were carried against me in my local association I should immediately resign my seat and fight a by-election. However, my ever-faithful and tireless champion and chairman, Sir James Hawkey, with a strong circle of determined men and women, fought the ground inch by inch and stood by me, and at the decisive meeting of the association I received in this murky hour a vote of confidence of three to two. But it was a gloomy winter.

In November we had another debate on national defence in which I spoke at length.

Mr. Duff Cooper to Mr. Churchill　　　　　　　　　　　19.XI.38
I am very distressed to hear that you resented the reference that I made to you in my speech in the House last Thursday. I cannot see why you should. I merely said that I thought that all the P.M. meant by his reference to 1914 was that any inquiry after mobilisation would

always show up gaps and deficiencies, and that therefore he had hardly merited the rebuke you delivered to him. I might of course have omitted all reference to you, but I think it is always a good thing in debate to hang one's arguments on to previous speeches. Nor was my position on Thursday quite simple. Your great philippic, which I enjoyed immensely and admired still more, was an onslaught on the Government's record over a period of three years, during the whole of which, except the last six weeks, I was a member of the Government. You could hardly expect me therefore to say that I entirely agreed with you and to vote accordingly. However, I am not the less sorry to have hurt you, whether your reasons for feeling hurt are good or bad, and I hope you will forgive me, because your friendship, your companionship, and your advice are very, very precious to me.

Mr. Churchill to Mr. Duff Cooper   22.XI.38

Thank you so much for your letter, which I was very glad to get. In the position in which our small band of friends now is it is a great mistake ever to take points off one another. The only rule is: Help each other when you can, but never harm.—Never help the Bear. With your facility of speech it ought to be quite easy to make your position clear without showing differences from me. I will always observe this rule. Although there was nothing in what you said to which I could possibly object, yet the fact that you went out of your way to answer me led several of my friends to wonder whether there was not some purpose behind it; for instance, the desire to isolate me as much as possible from the other Conservatives who disagree with the Government. I did not credit this myself, and I am entirely reassured by your charming letter. We are so few, enemies so many, the cause so great, that we cannot afford to weaken each other in any way.

I thought the parts of your speech which I heard very fine indeed, especially the catalogue of disasters which we have sustained in the last three years. I don't know how you remembered them all without a note.

I am of course sorry about the debate. Chamberlain has now got away with everything. Munich is dead, the unpreparedness is forgotten, and there is to be no real, earnest, new effort to arm the nation. Even the breathing-space, purchased at a hideous cost, is to be wasted. It was my distress at these public matters that made me grumpy when you suggested supper, for I did not then know what you had said in the early part of your speech.

But anyway count always upon your sincere friend.

*   *   *   *   *

On November 1 a nonentity, Dr. Hacha, was elected to the vacant Presidency of the remnants of Czechoslovakia. A new Government took office in Prague. "Conditions in Europe and the world in general," said the Foreign Minister of this forlorn Administration, "are not such that we should hope for a period of calm in the near future." Hitler thought so too. A formal division of the spoils was made by Germany at the beginning of November. Poland was not disturbed in her occupation of Teschen. The Slovaks, who had been used as a pawn by Germany, obtained a precarious autonomy. Hungary received a piece of flesh at the expense of Slovakia. When these consequences of Munich were raised in the House of Commons Mr. Chamberlain explained that the French and British offer of an international guarantee to Czechoslovakia which had been given after the Munich Pact did not affect the existing frontiers of that State, but referred only to the hypothetical question of unprovoked aggression. "What we are doing now," he said, with much detachment, "is witnessing the readjustment of frontiers laid down in the Treaty of Versailles. I do not know whether the people who were responsible for those frontiers thought they would remain permanently as they were laid down. I doubt very much whether they did. They probably expected that from time to time the frontiers would have to be adjusted. It is impossible to conceive that those people would be such supermen as to be able to see what would be the right frontiers for all time. The question is not whether those frontiers should be readjusted from time to time, but whether they should be readjusted by negotiation and discussion or be readjusted by war. Readjustment is going on, and in the case of the Hungarian frontier arbitration by Germany and Italy has been accepted by Czechoslovakia and Hungary for the final determination of the frontier between them. I think I have said enough about Czechoslovakia. . . ." There was, however, to be a later occasion.

* * * * *

I wrote on November 17, 1938:

Everyone must recognise that the Prime Minister is pursuing a policy of a most decided character and of capital importance. He has his own strong view about what to do, and about what is going to happen. He has his own standard of values; he has his own angle of vision. He believes that he can make a good settlement for Europe and for the British Empire by coming to terms with Herr Hitler and Signor Mussolini. No one impugns his motives. No one doubts his conviction or his courage. Besides all this, he has the power to do what he thinks best. Those who take a different view, both of the principles of our foreign policy and of the facts and probabilities with which our country has to deal, are bound to recognise that we have no power at all to prevent him, by the resources and methods which are at his disposal, from taking the course in which he sincerely believes. He is willing to take the responsibility; he has the right to take the responsibility; and we are going to learn, in a comparatively short time, what he proposes should happen to us.

The Prime Minister is persuaded that Herr Hitler seeks no further territorial expansion upon the Continent of Europe; that the mastering and absorption of the Republic of Czechoslovakia has satiated the appetite of the German Nazi *régime*. It may be that he wishes to induce the Conservative Party to return to Germany the mandated territories in British possession, or what is judged to be their full equivalent. He believes that this act of restoration will bring about prolonged good and secure relations between Great Britain and Germany. He believes further that these good relations can be achieved without weakening in any way the fundamental ties of self-preservation which bind us to the French Republic, which ties, it is common ground between us all, must be preserved. Mr. Chamberlain is convinced that all this will lead to general agreement, to the appeasement of the discontented Powers, and to a lasting peace.

But all lies in the regions of hope and speculation. A whole set of contrary possibilities must be held in mind. He may ask us to submit to things which we cannot endure; he may be forced to ask us to submit to things which we cannot endure. Or, again, the other side in this difficult negotiation may not act in the same spirit of goodwill and good faith as animates the Prime Minister. What we have to give, what we are made to give, may cost us dear, but it may not be enough. It may involve great injury and humbling to the British Empire, but it may not stay or even divert for more than a few months, if that, the march of events upon the Continent. *By this time next year we shall know whether the Prime Minister's view of Herr Hitler and the German Nazi Party is right or wrong. By this time next year we shall know whether the policy of appeasement has appeased, or whether it has only stimulated a more ferocious appetite.* All we can do in the meanwhile is to gather forces of resistance and defence, so that if the Prime Minister should unhappily be wrong, or misled, or deceived, we can at the worst keep body and soul together.

*   *   *   *   *

Whatever might be thought of "peace for our time", Mr. Chamberlain was more than ever alive to the need for dividing Italy from Germany. He hopefully believed that he had made friends with Hitler; to complete his work he must gain Mussolini's Italy as a counterpoise to the dear-bought reconciliation with Germany. In this renewed approach to the Italian Dictator he had to carry France with him. There must be love all round. We shall study the result of these overtures in the next chapter.

Late in November the Prime Minister and Lord Halifax visited Paris. The French Ministers agreed without enthusiasm to Mr. Chamberlain's suggestion of his visit to Rome; but he and Lord Halifax were glad to learn that the French were now planning to imitate the British declaration on the future of Anglo-German relations signed by Chamberlain and Hitler at Munich. On November 27, 1938, M. Bonnet sent a message to the French Ambassador in Washington describing this intention of the French Government: "Mr. Neville Chamberlain and Lord Halifax, in the course of discussions held in Paris yesterday, clearly expressed their satisfaction at a declaration which they regarded as being of a character, like that of the Anglo-German declaration, which would constitute an immediate contribution to the work of international appeasement."* For the purpose of these discussions Ribbentrop came to Paris, bringing with him Dr. Schacht. The Germans hoped not only for a general statement of good intentions, but for a concrete economic agreement. They obtained the former, which was signed in Paris on December 6, but even M. Bonnet was not prepared to accept the latter, in spite of considerable temptation to pose as the architect of Franco-German understanding.

The mission of Ribbentrop to Paris had also a deeper motive. Just as Mr. Chamberlain hoped to split Rome from Berlin, so Hitler believed that he could divide Paris from London. M. Bonnet's version of his talk with Ribbentrop on this subject is not without interest:

> In regard to Great Britain, I indicated to M. Ribbentrop the *rôle* which the improvement of Anglo-German relations must play in any developments of the policy of European appeasement, which was considered as the essential object of any Franco-German undertaking. The German Foreign Minister made efforts to throw upon the British

* *Livre Jaune Français*, pp. 35, 37.

Government the responsibility for the present state of affairs. The Government, and particularly the British Press, after having appeared to show, on the morrow of Munich, a certain comprehension, had adopted the most disappointing attitude towards the Government of Berlin. . . . The manifestations multiplied in Parliament by Messrs. Duff Cooper, Churchill, Eden, and Morrison, and certain newspaper articles, have been strongly resented in Germany, where one had not been able to restrain the reactions of the Press. I emphasised anew the fundamental and unshakable character of Anglo-French solidarity, indicating very clearly that a real Franco-German *détente* could not be conceivable in the long run without a parallel Anglo-German *détente*.*

The question has been debated whether Hitler or the Allies gained the more in strength in the year that followed Munich. Many persons in Britain who knew our nakedness felt a sense of relief as each month our Air Force developed and the Hurricane and Spitfire types approached issue. The number of formed squadrons grew and the ack-ack guns multiplied. Also the general pressure of industrial preparation for war continued to quicken. But these improvements, invaluable though they seemed, were petty compared with the mighty advance in German armaments. As has been explained, munitions production on a nation-wide plan is a four years' task. The first year yields nothing, the second very little, the third a lot, and the fourth a flood. Hitler's Germany in this period was already in the third or fourth year of intense preparation under conditions of grip and drive which were almost the same as those of war. Britain, on the other hand, had only been moving on a non-emergency basis, with a weaker impulse and on a far smaller scale. In 1938–39 British military expenditure of all kinds reached £304 millions,† and German was at least £1,500 millions. It is probable that in this last year before the outbreak Germany manufactured at least double, and possibly treble, the munitions of Britain and France put together, and also that her great plants for tank production reached full capacity. They were therefore getting weapons at a far higher rate than we.

The subjugation of Czechoslovakia robbed the Allies of the Czech Army of twenty-one regular divisions, fifteen or sixteen second-line divisions already mobilised, and also their mountain

---

* *Livre Jaune Français*, pp. 43–4.
† 1937–8, £234 millions. 1938–9, £304 millions. 1939–40, £367 millions.

fortress line, which in the days of Munich had required the deployment of thirty German divisions, or the main strength of the mobile and fully-trained German Army. According to Generals Halder and Jodl, there were but thirteen German divisions, of which only five were composed of first-line troops, left in the West at the time of the Munich arrangement. We certainly suffered a loss through the fall of Czechoslovakia equivalent to some thirty-five divisions. Besides this the Skoda works, the second most important arsenal in Central Europe, the production of which between August 1938 and September 1939 was in itself nearly equal to the actual output of British arms factories in that period, was made to change sides adversely. While all Germany was working under intense and almost war pressure, French labour had achieved as early as 1936 the long-desired forty-hours week.

Even more disastrous was the alteration in the relative strength of the French and German Armies. With every month that passed, from 1938 onwards, the German Army not only increased in numbers and formations and in the accumulation of reserves, but in quality and maturity. The advance in training and general proficiency kept pace with the ever-augmenting equipment. No similar improvement or expansion was open to the French Army. It was being overtaken along every path. In 1935 France, un-aided by her previous allies, could have invaded and reoccupied Germany almost without serious fighting. In 1936 there could still be no doubt of her overwhelmingly superior strength. We now know, from the German revelations, that this continued in 1938, and it was the knowledge of their weakness which led the German High Command to do their utmost to restrain Hitler from every one of the successful strokes by which his fame was enhanced. In the year after Munich, which we are now examin-ing, the German Army, though still weaker in trained reserves than the French, approached its full efficiency. As it was based upon a population double as large as that of France, it was only a question of time when it would become by every test the stronger In morale also the Germans had the advantage. The desertion of an ally, especially from fear of war, saps the spirit of any army. The sense of being forced to yield depresses both officers and men. While on the German side confidence, success, and the sense of growing power inflamed the martial instincts of the race, the admission of weakness discouraged the French soldiers of every rank.

★ ★ ★ ★ ★

There is however one vital sphere in which we began to overtake Germany and improve our own position. In 1938 the process of replacing British biplane fighters, like the Gladiators, by modern types of Hurricanes and later Spitfires had only just begun. In September of 1938 we had but five squadrons remounted on Hurricanes. Moreover, reserves and spares for the older aircraft had been allowed to drop, since they were going out of use. The Germans were well ahead of us in remounting with modern fighter types. They already had good numbers of the Me.109, against which our old aircraft would have fared very ill. Throughout 1939 our position improved as more squadrons were remounted. In July of that year we had twenty-six squadrons of modern eight-gun fighters, though there had been little time to build up a full scale of reserves and spares. By July 1940, at the time of the Battle of Britain, we had on the average forty-seven squadrons of modern fighters available.

On the German side the figures of strength increased as follows:

1938	Bombers	1,466
	Fighters	920
1939	Bombers	1,553
	Fighters	1,090
1940	Bombers	1,558
	Fighters	1,290

The Germans had in fact done most of their air expansion both in quantity and quality before the war began. Our effort was later than theirs by nearly two years. Between 1939 and 1940 they made a 20 per cent. increase only, whereas our increase in modern fighter aircraft was 80 per cent. The year 1938 in fact found us sadly deficient in quality, and although by 1939 we had gone some way towards meeting the disparity we were still relatively worse off than in 1940, when the test came.

We might in 1938 have had air raids on London, for which we were lamentably unprepared. There was however no possibility of a decisive Air Battle of Britain until the Germans had occupied France and the Low Countries, and thus obtained the necessary bases in close striking distance of our shores. Without these bases they could not have escorted their bombers with the fighter air-craft of those days. The German armies were not capable of defeating the French in 1938 or 1939.

The vast tank production with which they broke the French front did not come into existence till 1940, and, in the face of the French superiority in the West and an unconquered Poland in the East, they could certainly not have concentrated the whole of their air-power against England as they were able to do when France had been forced to surrender. This takes no account either of the attitude of Russia or of whatever resistance Czechoslovakia might have made. I have thought it right to set out the figures of relative air-power in the period concerned, but they do not in any way alter the conclusions which I have recorded.

For all the above reasons, the year's breathing-space said to be "gained" by Munich left Britain and France in a much worse position compared with Hitler's Germany than they had been at the Munich crisis.

★   ★   ★   ★   ★

Finally there is this staggering fact: that in the single year 1938 Hitler had annexed to the Reich and brought under his absolute rule 6,750,000 Austrians and 3,500,000 Sudetens, a total of over ten millions of subjects, toilers, and soldiers. Indeed the dread balance had turned in his favour.

CHAPTER XIX

PRAGUE, ALBANIA, AND THE POLISH
GUARANTEE

January – April 1939

Chamberlain's Visit to Rome – German Concentrations towards Czechoslovakia – Ministerial Optimism – Hitler Invades Czechoslovakia – Chamberlain's Speech at Birmingham – A Complete Change of Policy – My Letter to the Prime Minister of March 31 – The Soviet Government's Proposal for a Six-Power Conference – The British Guarantee to Poland – A Word with Colonel Beck – The Italian Landing in Albania, April 7, 1939 – Faulty Disposition of the British Mediterranean Fleet – My Speech in the House of Commons of April 13 – My Letter to Lord Halifax – Meeting of Goering, Mussolini, and Ciano on War Measures – German Strategic Advantages of the Annexation of Czechoslovakia – The Government Introduces Conscription – Weak Attitude of the Labour and Liberal Oppositions – Agitation for a National Government in Britain – Sir Stafford Cripps' Appeals – Mr. Stanley's Offer to Resign.

MR. CHAMBERLAIN continued to believe that he had only to form a personal contact with the Dictators to effect a marked improvement in the world situation. He little knew that their decisions were taken. In a hopeful spirit he proposed that he and Lord Halifax should visit Italy in January. After some delay an invitation was extended, and on January 11 the meeting took place. It makes one flush to read in Ciano's diary the comments which were made behind the Italian scene about our country and its representatives. "Essentially," writes Ciano, "the visit was kept in a minor key. . . . Effective contact has not been made. How far apart we are from these people! It is another world. We were talking about it after dinner to the

305

Duce. 'These men,' said Mussolini, 'are not made of the same stuff as Francis Drake and the other magnificent adventurers who created the Empire. They are after all the tired sons of a long line of rich men.'" "The British," noted Ciano, "do not want to fight. They try to draw back as slowly as possible, but they do not want to fight. . . . Our conversations with the British have ended. Nothing was accomplished. I have telephoned to Ribbentrop saying it was a fiasco, absolutely innocuous. . . . Chamberlain's eyes filled with tears as the train started moving and his countrymen started singing, 'For he's a jolly good fellow'. 'What is this little song?' asked Mussolini." And then a fortnight later: "Lord Perth has submitted for our approval the outlines of the speech that Chamberlain will make in the House of Commons in order that we may suggest changes if necessary." The Duce approved it, and commented: "I believe this is the first time that the head of the British Government has submitted to a foreign Government the outlines of one of his speeches. It's a bad sign for them."* However, in the end it was Ciano and Mussolini who went to their doom.

Meanwhile, on January 25 Ribbentrop was at Warsaw to continue the diplomatic offensive against Poland. The absorption of Czechoslovakia was to be followed by the encirclement of Poland. The first stage in this operation would be the cutting off of Poland from the sea by the assertion of German sovereignty in Danzig and by the prolongation of the German control of the Baltic to the vital Lithuanian port of Memel. The Polish Government displayed strong resistance to this pressure, and for a while Hitler watched and waited for the campaigning season.

During the second week of March rumours gathered of troop movements in Germany and Austria, particularly in the Vienna-Salzburg region. Forty German divisions were reported to be mobilised on a war footing. Confident of German support, the Slovaks were planning the separation of their territory from the Czechoslovak Republic. Colonel Beck, relieved to see the Teutonic wind blowing in another direction, declared publicly in Warsaw that his Government had full sympathy with the aspirations of the Slovaks. Father Tiso, the Slovak leader, was received by Hitler in Berlin with the honours due to a Prime Minister. On the 12th Mr. Chamberlain, questioned in Parlia-

* *Ciano's Diary*, 1939–43 (ed. Malcolm Muggeridge), pp. 9, 10.

ment about the guarantee of the Czechoslovak frontier, reminded the House that this proposal had been directed against unprovoked aggression. No such aggression had yet taken place. He did not have long to wait.

★   ★   ★   ★   ★

A wave of perverse optimism had swept across the British scene during these March days. In spite of the growing stresses in Czechoslovakia under intense German pressure from without and from within, the Ministers and newspapers identified with the Munich Agreement did not lose faith in the policy into which they had drawn the nation. For instance, on March 10 the Home Secretary addressed his constituents about his hopes of a Five Years' Peace Plan which would lead in time to the creation of "a Golden Age". A plan for a commercial treaty with Germany was still being hopefully discussed. The famous periodical *Punch* produced a cartoon showing John Bull waking with a gasp of relief from a nightmare, while all the evil rumours, fancies, and suspicions of the night were flying away out of the window. On the very day when this appeared Hitler launched his ultimatum to the tottering Czech Government, bereft of their fortified line by the Munich decisions. German troops, marching into Prague, assumed absolute control of the unresisting State. I remember sitting with Mr. Eden in the smoking-room of the House of Commons when the editions of the evening papers recording these events came in. Even those who like us had no illusions and had testified earnestly were surprised at the sudden violence of this outrage. One could hardly believe that with all their secret information His Majesty's Government could be so far adrift. March 14 witnessed the dissolution and subjugation of the Czechoslovak Republic. The Slovaks formally declared their independence. Hungarian troops, backed surreptitiously by Poland, crossed into the eastern province of Czechoslovakia, or the Carpatho-Ukraine, which they demanded. Hitler, having arrived in Prague, proclaimed a German protectorate over Czechoslovakia, which was thereby incorporated in the Reich.

On the 12th Mr. Chamberlain had to say to the House: "The occupation of Bohemia by German military forces began at six o'clock this morning. The Czech people have been ordered by their Government not to offer resistance." He then proceeded to state that the guarantee he had given Czechoslovakia no longer in his opinion had validity. After Munich, five months before, the Dominions Secretary, Sir Thomas Inskip, had said of this guarantee: "His Majesty's Government feel under a moral obligation to Czechoslovakia to keep the guarantee [as though it were technically in force]. . . . In the event therefore of an act of unprovoked aggression against Czechoslovakia His Majesty's Government would certainly be bound to take all steps in their power to see that the integrity of Czechoslovakia is preserved." "That," said the Prime Minister, "remained the position until yesterday. But the position has altered since the Slovak Diet declared the independence of Slovakia. The effect of this declaration put an end by internal disruption to the State whose frontiers we had proposed to guarantee, and His Majesty's Government cannot accordingly hold themselves bound by this obligation."

This seemed decisive. "It is natural," he said in conclusion, "that I should bitterly regret what has now occurred, but do not let us on that account be deflected from our course. Let us remember that the desire of all the peoples of the world still remains concentrated on the hopes of peace."

Mr. Chamberlain was due to speak at Birmingham two days later. I fully expected that he would accept what had happened with the best grace possible. This would have been in harmony with his statement to the House. I even imagined that he might claim credit for the Government for having, by its foresight at Munich, decisively detached Great Britain from the fate of Czechoslovakia, and indeed of Central Europe. "How fortunate," he might have said, "that we made up our minds in September last not to be drawn into the Continental struggle! We are now free to allow these broils between countries which mean nothing to us to settle themselves without expense in blood or treasure." This would, after all, have been a logical decision following upon the disruption of Czechoslovakia agreed to at Munich and endorsed by a majority of the British people, so far as they understood what was going on. This also was the view taken by some of the strongest supporters of the Munich Pact. I therefore awaited the Birmingham declaration with anticipatory contempt.

The Prime Minister's reaction surprised me. He had conceived himself as having a special insight into Hitler's character, and the power to measure with shrewdness the limits of German action. He believed, with hope, that there had been a true meeting of minds at Munich, and that he, Hitler, and Mussolini had together saved the world from the infinite horrors of war. Suddenly as by an explosion his faith and all that had followed from his actions and his arguments was shattered. Responsible as he was for grave misjudgments of facts, having deluded himself and imposed his errors on his subservient colleagues and upon the unhappy British public opinion, he none the less between night and morning turned his back abruptly upon his past. If Chamberlain failed to understand Hitler, Hitler completely underrated the nature of the British Prime Minister. He mistook his civilian aspect and passionate desire for peace for a complete explanation of his personality, and thought that his umbrella was his symbol. He did not realise that Neville Chamberlain had a very hard core, and that he did not like being cheated.

The Birmingham speech struck a new note. "His tone," says his biographer, "was very different. . . . Informed by fuller knowledge and by strong representations as to opinion in the House, the public, and the Dominions, he threw aside the speech long drafted on domestic questions and social service and grasped the nettle." He reproached Hitler with a flagrant personal breach of faith about the Munich Agreement. He quoted all the assurances Hitler had given. "This is the last territorial claim which I have to make in Europe." "I shall not be interested in the Czech State any more, and I can guarantee it. We don't want any more Czechs." "I am convinced," said the Prime Minister, "that after Munich the great majority of the British people shared my honest desire that that policy should be carried further, but to-day I share their disappointment, their indignation, that those hopes have been so wantonly shattered. How can these events this week be reconciled with those assurances which I have read out to you?

"Who can fail to feel his heart go out in sympathy to the proud, brave people who have so suddenly been subjected to this invasion, whose liberties are curtailed, whose national independence is gone? . . . Now we are told that this seizure of territory has been necessitated by disturbances in Czechoslovakia. . . . If there were disorders, were they not fomented from without? . . . Is this the last attack upon a small State, or is it to be followed by another? Is this in fact a step in the direction of an attempt to dominate the world by force?"

It is not easy to imagine a greater contradiction to the mood and policy of the Prime Minister's statement two days earlier in the House of Commons. He must have been through a period of intense stress. On the 15th he had said: "Do not let us be deflected from our course." But this was "Right-about-turn."

Moreover, Chamberlain's change of heart did not stop at words. The next "small State" on Hitler's list was Poland. When the gravity of the decision and all those who had to be consulted are borne in mind, the period must have been busy. Within a fortnight (March 31) the Prime Minister said in Parliament:

I now have to inform the House that . . . in the event of any action which clearly threatened Polish independence and which the Polish Government accordingly considered it vital to resist with their national forces, His Majesty's Government would feel themselves bound at once to lend the Polish Government all support in their power. They have given the Polish Government an assurance to this effect.

I may add that the French Government have authorised me to make it plain that they stand in the same position in this matter as do His Majesty's Government. . . . [And later] The Dominions have been kept fully informed.

This was no time for recriminations about the past. The guarantee to Poland was supported by the leaders of all parties and groups in the House. "God helping, we can do no other," was what I said. At the point we had reached it was a necessary action. But no one who understood the situation could doubt that it meant in all human probability a major war, in which we should be involved.

* * * * *

In this sad tale of wrong judgments formed by well-meaning and capable people we now reach our climax. That we should all have come to this pass makes those responsible, however honourable their motives, blameworthy before history. Look back and see what we had successively accepted or thrown away: a Germany disarmed by solemn treaty; a Germany rearmed in violation of a solemn treaty; air superiority or even air parity cast away; the Rhineland forcibly occupied and the Siegfried Line built or building; the Berlin-Rome Axis established; Austria devoured and digested by the Reich; Czechoslovakia deserted and ruined by the Munich Pact, its fortress line in German hands, its mighty arsenal of Skoda henceforward making munitions for the German armies; President Roosevelt's effort to stabilise or bring to a head the European situation by the intervention of the United States waved aside with one hand, and Soviet Russia's undoubted willingness to join the Western Powers and go all lengths to save Czechoslovakia ignored on the other; the services of thirty-five Czech divisions against the still unripened Germany Army cast away, when Great Britain could herself supply only two to strengthen the front in France; all gone with the wind.

And now, when every one of these aids and advantages has been squandered and thrown away, Great Britain advances, leading France by the hand, to guarantee the integrity of Poland— of that very Poland which with hyena appetite had only six months before joined in the pillage and destruction of the Czecho-slovak State. There was sense in fighting for Czechoslovakia in 1938, when the German Army could scarcely put half a dozen trained divisions on the Western Front, when the French with nearly sixty or seventy divisions could most certainly have rolled forward across the Rhine or into the Ruhr. But this had been judged unreasonable, rash, below the level of modern intellectual thought and morality. Yet now at last the two Western demo-cracies declared themselves ready to stake their lives upon the territorial integrity of Poland. History, which, we are told, is mainly the record of the crimes, follies, and miseries of mankind, may be scoured and ransacked to find a parallel to this sudden and complete reversal of five or six years' policy of easy-going placatory appeasement, and its transformation almost overnight into a readiness to accept an obviously imminent war on far worse conditions and on the greatest scale.

Moreover, how could we protect Poland and make good our guarantee? Only by declaring war upon Germany and attacking a stronger Western Wall and a more powerful German Army than those from which we had recoiled in September 1938. Here is a line of milestones to disaster. Here is a catalogue of surrenders, at first when all was easy and later when things were harder, to the ever-growing German power. But now at last was the end of British and French submission. Here was decision at last, taken at the worst possible moment and on the least satisfactory ground, which must surely lead to the slaughter of tens of millions of people. Here was the righteous cause deliberately and with a refinement of inverted artistry committed to mortal battle after its assets and advantages had been so improvidently squandered. Still, if you will not fight for the right when you can easily win without bloodshed, if you will not fight when your victory will be sure and not too costly, you may come to the moment when you will have to fight with all the odds against you and only a precarious chance of survival. There may even be a worse case. You may have to fight when there is no hope of victory, because it is better to perish than live as slaves.

★   ★   ★   ★   ★

The Birmingham speech brought me much closer to Mr. Chamberlain. I wrote to him:

I venture to reiterate the suggestion which I made to you in the Lobby yesterday afternoon, that the anti-aircraft defences should forthwith be placed in full preparedness. Such a step could not be deemed aggressive, yet it would emphasise the seriousness of the action H.M. Government are taking on the Continent. The bringing together of these officers and men would improve their efficiency with every day of their embodiment. The effect at home would be one of confidence rather than alarm. But it is of Hitler I am thinking mostly. He must be under intense strain at this moment. He knows we are endeavouring to form a coalition to restrain his further aggression. With such a man anything is possible. The temptation to make a surprise attack on London, or on the aircraft factories, about which I am even more anxious, would be removed if it was known that all was ready. There could, in fact, be no surprise, and therefore the incentive to the extremes of violence would be removed and more prudent counsels might prevail.

In August 1914 I persuaded Mr. Asquith to let me send the Fleet to the North so that it could pass the Straits of Dover and the Narrow Seas *before* the diplomatic situation had become hopeless. It seems to me that manning the anti-aircraft defences now stands in a very similar position, and I hope you will not mind my putting this before you.

★   ★   ★   ★   ★

The Poles had gained Teschen by their shameful attitude towards the liquidation of the Czechoslovak State. They were soon to pay their own forfeits. On March 21, when Ribbentrop saw M. Lipski, the Polish Ambassador in Berlin, he adopted a sharper tone than in previous discussions. The occupation of Bohemia and the creation of satellite Slovakia brought the German Army to the southern frontiers of Poland. Lipski told Ribbentrop that the Polish man-in-the-street could not understand why the Reich had assumed the protection of Slovakia, that protection being directed against Poland. He also inquired about the recent conversations between Ribbentrop and the Lithuanian Foreign Minister. Did they affect Memel? He received his answer two days later (March 23). German troops occupied Memel.

The means of organising any resistance to German aggression in Eastern Europe were now almost exhausted. Hungary was in the German camp. Poland had stood aside from the Czechs, and was unwilling to work closely with Roumania. Neither Poland nor Roumania would accept Russian intervention against Germany across their territories. The key to a Grand Alliance was an understanding with Russia. On March 19 the Russian Government, which was profoundly affected by all that was taking place, and in spite of having been left outside the door in the Munich crisis, proposed a Six-Power Conference. On this subject also Mr. Chamberlain had decided views. In a private letter he wrote on March 26:

I must confess to the most profound distrust of Russia. I have no belief whatever in her ability to maintain an effective offensive, even if she wanted to. And I distrust her motives, which seem to me to have little connection with our ideas of liberty, and to be concerned only with setting everyone else by the ears. Moreover, she is both hated and suspected by many of the smaller States, notably by Poland, Roumania, and Finland.*

The Soviet proposal for a Six-Power Conference was therefore coldly received and allowed to drop.

The possibilities of weaning Italy from the Axis, which had loomed so large in British official calculations, were also vanishing. On March 26 Mussolini made a violent speech asserting Italian claims against France in the Mediterranean. Secretly he was planning for the extension of Italian influence in the Balkans and the Adriatic, to balance the German advance in Central Europe. His plans for invading Albania were now ready.

On March 29 Mr. Chamberlain announced in Parliament the

---

* Feiling, *op. cit.*, p. 403.

planned doubling of the Territorial Army, including an increase
on paper of 210,000 men (unequipped). On April 3 Keitel,
Hitler's Chief of Staff, issued the secret "Directive for the Armed
Forces, 1939–40", in regard to Poland—"Case White" was the
code name. The Fuehrer added the following direction: "Preparations must be made in such a way that the operations can be
carried out at any time from September 1 onwards."

★   ★   ★   ★   ★

On April 4 the Government invited me to a luncheon at the
Savoy in honour of Colonel Beck, the Polish Foreign Minister,
who had come upon an official visit of significance. I had met
him the year before on the Riviera, when we had lunched alone
together. I now asked him: "Will you get back all right in your
special train through Germany to Poland?" He replied: "I think
we shall have time for that."

★   ★   ★   ★   ★

A new crisis now opened upon us.

At dawn on April 7, 1939, Italian forces landed in Albania, and
after a brief scuffle took over the country. As Czechoslovakia
was to be the base for aggression against Poland, so Albania would
be the springboard for Italian action against Greece and for the
neutralising of Yugoslavia. The British Government had already
undertaken a commitment in the interests of peace in North-
eastern Europe. What about the threat developing in the South-
east? The vessel of peace was springing a leak from every seam.

On April 9 I wrote to the Prime Minister:

I am hoping that Parliament will be recalled at the latest on Tuesday,
and I write to say how much I hope the statements which you will
be able to make will enable the same united front to be presented as
in the case of the Polish Agreement.

It seems to me however that hours now count. It is imperative for
us to recover the initiative in diplomacy. This can no longer be done
by declarations or by the denouncing of the Anglo-Italian Agreement
or by the withdrawal of our Ambassador.

It is freely stated in the Sunday papers that we are offering a guarantee to Greece and Turkey. At the same time I notice that several
newspapers speak of a British naval occupation of Corfu. Had this
step been already taken it would afford the best chance of maintaining
peace. If it is not taken by us, of course with Greek consent, it seems

to me that after the publicity given to the idea in the Press and the obvious needs of the situation Corfu will be speedily taken by Italy. Its recapture would then be impossible. On the other hand, if we are there first an attack even upon a few British ships would confront Mussolini with beginning a war of aggression upon England. This direct issue gives the best chance to all the forces in Italy which are opposed to a major war with England. So far from intensifying the grave risks now open, it diminishes them. But action ought to be taken to-night.

What is now at stake is nothing less than the whole of the Balkan Peninsula. If these States remain exposed to German and Italian pressure while we appear, as they may deem it, incapable of action, they will be forced to make the best terms possible with Berlin and Rome. How forlorn then will our position become! We shall be committed to Poland, and thus involved in the East of Europe, while at the same time cutting off from ourselves all hope of that large alliance which once effected might spell salvation.

I write the above without knowledge of the existing position of our Mediterranean Fleet, which should of course be concentrated and *at sea*, in a suitable but not too close supporting position.

The British Mediterranean Fleet was in fact scattered. Of our five great capital ships, one was at Gibraltar, another in the Eastern Mediterranean, and the remaining three were lolling about inside or outside widely-separated Italian ports, two of them not protected by their flotillas. The destroyer flotillas themselves were dispersed along the European and African shores, and a large number of cruisers were crowded in Malta harbour without the protection of the powerful anti-aircraft batteries of battleships. At the very time that the fleet was suffered to disperse in this manner it was known that the Italian Fleet was concentrated in the Straits of Otranto and that troops were being assembled and embarked for some serious enterprise.

I challenged these careless dispositions on April 13 in the House of Commons:

The British habit of the week-end, the great regard which the British pay to holidays which coincide with festivals of the Church, is studied abroad. Good Friday was also the first day after Parliament had dispersed. It was known too that on that day the British Fleet was carrying out in a routine manner a programme long announced. It would therefore be dispersed in all quarters. . . . I can well believe that if our Fleet had been concentrated and cruising in the southern parts of the

Ionian Sea the Albanian adventure would never have been under-taken. . . .

After twenty-five years' experience in peace and war, I believe the British Intelligence Service to be the finest of its kind in the world. Yet we have seen, both in the case of the subjugation of Bohemia and on the occasion of the invasion of Albania, that Ministers of the Crown had apparently no inkling, or at any rate no conviction, of what was coming. I cannot believe that this is the fault of the British Secret Service.

How was it that on the eve of the Bohemian outrage Ministers were indulging in what was called "sunshine talk" and predicting "the dawn of a Golden Age"? How was it that last week's holiday routine was observed at a time when clearly something of a quite exceptional character, the consequences of which could not be measured, was imminent? . . . It seems to me that Ministers run the most tremendous risks if they allow the information collected by the Intelligence Department, and sent to them, I am sure, in good time, to be sifted and coloured and reduced in consequence and importance, and if they ever get themselves into a mood of attaching weight only to those pieces of information which accord with their earnest and honourable desire that the peace of the world should remain unbroken.

All things are moving at the same moment. Year by year, month by month, they have all been moving forward together. While we have reached certain positions in thought, others have reached certain positions in fact. The danger is now very near, and a great part of Europe is to a very large extent mobilised. Millions of men are being prepared for war. Everywhere the frontier defences are being manned. Everywhere it is felt that some new stroke is impending. If it should fall, can there be any doubt that we should be involved? We are no longer where we were two or three months ago. We have committed ourselves in every direction, rightly in my opinion, having regard to all that has happened. It is not necessary to enumerate the countries to which directly or indirectly we have given or are giving guarantees. What we should not have dreamt of doing a year ago, when all was so much more powerful, what we should not have dreamt of doing even a month ago, we are doing now. Surely then when we aspire to lead all Europe back from the verge of the abyss on to the uplands of law and peace we must ourselves set the highest example. We must keep nothing back. How can we bear to continue to lead our comfortable, easy lives here at home, unwilling to pronounce even the word "compulsion", unwilling to take even the necessary measure by which the armies which we have promised can alone be recruited and equipped? The dark, bitter waters are rising fast on every side. How can we continue—let me say with particular frankness and sincerity—with less than the full force of the nation incorporated in the governing instrument?

I reiterated my complaints about the Fleet a few days later in a private letter to Lord Halifax:

The dispositions of our Fleet are inexplicable. First, on Tuesday night, April 4, the First Lord showed that the Home Fleet was in such a condition of preparedness that the men could not even leave the anti-aircraft guns to come below. This was the result of a scare telegram, and was, in my opinion, going beyond what vigilance requires. On the other hand, at the same time the Mediterranean Fleet was, as I described to the House, scattered in the most vulnerable disorder throughout the Mediterranean; and, as photographs published in the newspapers show, the *Barham* was actually moored alongside the Naples jetty. Now the Mediterranean Fleet has been concentrated and is at sea, where it should be. Therefore no doubt all is well in the Mediterranean. But the unpreparedness is transferred to home waters. The Atlantic Fleet, except for a few anti-aircraft guns, has been practically out of action for some days owing to very large numbers of men having been sent on leave. One would have thought at least the leave could be "staggered" in times like these. All the minesweepers are out of action refitting. How is it possible to reconcile this with the statement of tension declared to be existing on Tuesday week? It seems to be a grave departure from the procedure of continuous and reasonable vigilance. After all, the conditions prevailing now are not in principle different from those of last week. The First Sea Lord is seriously ill, so I expect a great deal falls upon Stanhope.

I write this to you for your own personal information, and in order that you can check the facts for yourself. Pray therefore treat my letter as strictly private, as I do not want to bother the Prime Minister with the matter, but I think you ought to know.

*   *   *   *   *

On April 15, 1939, after the declaration of the German protectorate of Bohemia and Moravia, Goering met Mussolini and Ciano in order to explain to the Italians the progress of German preparations for war. The minutes of this meeting have been found. One passage reads—it is Goering who is speaking: "The heavy armament of Czechoslovakia shows, in any case, how dangerous it could have been, even after Munich, in the event of a serious conflict. By German action the situation of both Axis countries was ameliorated because, among other reasons, of the economic possibilities which resulted from the transfer to Germany of the great productive capacity of Czechoslovakia. That contributes towards a considerable strengthening of the Axis against the Western Powers. Furthermore, Germany now need not keep ready a single division for protection against that country in case of a bigger conflict. This too is an advantage by which both Axis countries will, in the last analysis, benefit. . . . The action taken by Germany in Czechoslovakia is to be viewed as an advantage for the Axis Powers. Germany could now attack this country [Poland] from two flanks, and would be within only twenty-five minutes' flying distance from the new Polish industrial centre, which has been moved farther into the interior of the country, nearer to the other Polish industrial districts, because of its proximity to the border."*

"The bloodless solution of the Czech conflict in the autumn of 1938 and spring of 1939 and the annexation of Slovakia," said General von Jodl in a lecture some years after, "rounded off the territory of Greater Germany in such a way that it now became possible to consider the Polish problem on the basis of more or less favourable strategic premises."†

On the day of Goering's visit to Rome President Roosevelt sent a personal message to Hitler and Mussolini urging them to give a guarantee not to undertake any further aggression for ten "or even twenty-five years, if we are to look that far ahead." The Duce at first refused to read the document, and then remarked: "A result of infantile paralysis!" He little thought he was himself to suffer worse afflictions.

* * * * *

On April 27 the Prime Minister took the serious decision to introduce conscription, although repeated pledges had been given by him against such a step. To Mr. Hore-Belisha, the Secretary of State for War, belongs the credit of forcing this belated awakening. He certainly took his political life in his hands, and several of his interviews with his chief were of a formidable character. I saw something of him in this ordeal, and he was never sure that each day in office would not be his last.

Of course the introduction of conscription at this stage did not give us an army. It only applied to the men of twenty years of

* *Nuremberg Documents*, Pt. II, p. 106.
† *Ibid.*, p. 107.

age; they had still to be trained; and after they had been trained they had still to be armed. It was however a symbolic gesture of the utmost consequence to France and Poland, and to other nations on whom we had lavished our guarantees. In the debate the Opposition failed in their duty. Both Labour and Liberal Parties shrank from facing the ancient and deep-rooted prejudice which has always existed in England against compulsory military service. The Leader of the Labour Party moved that:

Whilst prepared to take all necessary steps to provide for the safety of the nation and the fulfilment of its international obligations, this House regrets that His Majesty's Government, in breach of their pledges, should abandon the voluntary principle, which has not failed to provide the man-power needed for defence, and is of opinion that the measure proposed is ill-conceived, and, so far from adding materially to the effective defence of the country, will promote division and discourage the national effort, and is further evidence that the Government's conduct of affairs throughout these critical times does not merit the confidence of the country or this House.

The Leader of the Liberal Party also found reasons for opposing this step. Both these men were distressed at the course they felt bound on party grounds to take. But they both took it, and adduced a wealth of reasons. The division was on party lines, and the Conservatives carried their policy by 380 to 143 votes. In my speech I tried my best to persuade the Opposition to support this indispensable measure; but my efforts were in vain. I understood fully their difficulties, especially when confronted with a Government to which they were opposed. I must record the event, because it deprives Liberal and Labour partisans of any right to censure the Government of the day. They showed their own measure in relation to events only too plainly. Presently they were to show a truer measure.

*　*　*　*　*

Though Mr. Chamberlain still hoped to avert war, it was plain that he would not shrink from it if it came. Mr. Feiling says that he noted in his diary, "Churchill's chances [of entering the Government] improve as war becomes more probable, and *vice versa*".* This was perhaps a somewhat disdainful epitome. There were many other thoughts in my mind besides those of becoming

* Feiling, *op. cit.*, p. 406.

once again a Minister. All the same, I understood the Prime Minister's outlook. He knew, if there was war, he would have to come to me, and he believed rightly that I would answer the call. On the other hand, he feared that Hitler would regard my entry into the Government as a hostile manifestation, and that it would thus wipe out all remaining chances of peace. This was a natural but a wrong view. None the less, no one can blame Mr. Chamberlain for not wishing to bring so tremendous and delicate a situation to a head for the sake of including any particular member of the House of Commons in his Government.

In March I had joined Mr. Eden and some thirty Conservative Members in tabling a resolution for a National Government. During the summer there arose a very considerable stir in the country in favour of this, or at least for my and Mr. Eden's inclusion in the Cabinet. Sir Stafford Cripps, in his independent position, became deeply distressed about the national danger. He visited me and various Ministers to urge the formation of what he called an "All-in Government". I could do nothing; but Mr. Stanley, President of the Board of Trade, was deeply moved. He wrote to the Prime Minister offering his own office if it would facilitate a reconstruction.

Mr. Stanley to the Prime Minister   *June* 30, 1939

I hesitate to write to you at a time like this, when you are overwhelmed with care and worry, and only the urgency of affairs is my excuse. I suppose we all feel that the only chance of averting war this autumn is to bring home to Hitler the certainty that we shall fulfil our obligations to Poland and that aggression on his part must inevitably mean a general conflagration. All of us, as well, must have been thinking whether there is any action we can take, which without being so menacing as to invite reprisal will be sufficiently dramatic to command attention. I myself can think of nothing which would be more effective, if it were found to be possible, than the formation now of the sort of Government which inevitably we should form at the outbreak of war It would be a dramatic confirmation of the national unity and determination, and would, I imagine, not only have a great effect in Germany, but also in the United States. It is also possible that if at the eleventh hour some possibility of a satisfactory settlement emerged it would be much easier for such a Government to be at all conciliatory. You of course must yourself have considered the possibility, and must be much more conscious of possible difficulties than I could be, but I thought I would write both to let you know my views and to assure you that if you did contemplate such a possibility, I—as I am sure all the rest of our colleagues—would gladly serve in any position, however small, either inside or outside the Government.

The Prime Minister contented himself with a formal acknowledgment.

As the weeks passed by almost all the newspapers, led by the *Daily Telegraph* (July 3), emphasised by the *Manchester Guardian*, reflected this surge of opinion. I was surprised to see its daily recurrent and repeated expression. Thousands of enormous posters were displayed for weeks on end on Metropolitan hoardings, "Churchill Must Come Back". Scores of young volunteer men and women carried sandwich-board placards with similar slogans up and down before the House of Commons. I had nothing to do with such methods of agitation, but I should certainly have joined the Government had I been invited. Here again my personal good fortune held, and all else flowed out in its logical, natural, and horrible sequence.

CHAPTER XX

THE SOVIET ENIGMA

Hitler Denounces the Anglo-German Naval Agreement – And the Polish Non-Aggression Pact – The Soviet Proposal of a Three-Power Alliance – Dilemma of the Border States – Soviet-German Contacts Grow – The Dismissal of Litvinov – Molotov – Anglo-Soviet Negotiations – Debate of May 19 – Mr. Lloyd George's Speech – My Statement on the European Situation – The Need of the Russian Alliance – Too Late – The "Pact of Steel" between Germany and Italy – Soviet Diplomatic Tactics.

WE have reached the period when all relations between Britain and Germany were at an end. We now know of course that there never had been any true relationship between our two countries since Hitler came into power. He had only hoped to persuade or frighten Britain into giving him a free hand in Eastern Europe, and Mr. Chamberlain had cherished the hope of appeasing and reforming him and leading him to grace. However, the time had come when the last illusions of the British Government had been dispelled. The Cabinet was at length convinced that Nazi Germany meant war, and the Prime Minister offered guarantees and contracted alliances in every direction still open, regardless of whether we could give any effective help to the countries concerned. To the Polish guarantee was added a Greek and Roumanian guarantee, and to these an alliance with Turkey.

We must now recall the sad piece of paper which Mr. Chamberlain had got Hitler to sign at Munich and which he waved triumphantly to the crowd when he quitted his aeroplane at Heston. In this he had invoked the two bonds which he assumed existed between him and Hitler and between Britain and Germany, namely, the Munich Agreement and the Anglo-German

Naval Treaty. The subjugation of Czechoslovakia had destroyed the first; Hitler now brushed away the second.

Addressing the Reichstag on April 28, he said:

Since England to-day, both through the Press and officially, upholds the view that Germany should be opposed in all circumstances, and confirms this by the policy of encirclement known to us, the basis of the Naval Treaty has been removed. I have therefore resolved to send to-day a communication to this effect to the British Government. This is to us not a matter of practical material importance—for I still hope that we shall be able to avoid an armaments race with England—but an action of self-respect. Should the British Government however wish to enter once more into negotiations with Germany on this problem, no one would be happier than I at the prospect of still being able to come to a clear and straightforward understanding.*

The Anglo-German Naval Agreement, which had been so marked a gain to Hitler at an important and critical moment in his policy, was now represented by him as a favour to Britain, the benefits of which would be withdrawn as a mark of German displeasure. The Fuehrer held out the hope to the British Government that he might be willing to discuss the naval problems further with His Majesty's Government, and he may even have expected that his former dupes would persist in their policy of appeasement. To him it now mattered nothing. He had Italy, and he had his air superiority; he had Austria and Czechoslovakia, with all that that implied. He had his Western Wall. In the purely naval sphere he had always been building U-boats as fast as possible irrespective of any agreement. He had already as a matter of form invoked his right to build 100 per cent. of the British numbers, but this had not limited in the slightest degree the German U-boat construction programme. As for the larger vessels, he could not nearly digest the generous allowance which had been accorded to him by the Naval Agreement. He therefore made fine impudent play with flinging it back in the face of the simpletons who made it.

In this same speech Hitler also denounced the German-Polish Non-Aggression Pact. He gave as his direct reason the Anglo-Polish Guarantee, "which would in certain circumstances compel Poland to take military action against Germany in the event of a conflict between Germany and any other Power, in which

* *Hitler's Speeches*, II, p. 1626.

England in her turn would be involved. This obligation is contrary to the Agreement which I made with Marshal Pilsudski some time ago. . . . I therefore look upon the Agreement as having been unilaterally infringed by Poland and thereby no longer in existence. I sent a communication to this effect to the Polish Government. . . ."

After studying this speech at the time, I wrote in one of my articles:

It seems only too probable that the glare of Nazi Germany is now to be turned on to Poland. Herr Hitler's speeches may or may not be a guide to his intentions, but the salient object of last Friday's performance was obviously to isolate Poland, to make the most plausible case against her, and to bring intensive pressure upon her. The German Dictator seemed to suppose that he could make the Anglo-Polish Agreement inoperative by focusing his demands on Danzig and the Corridor. He apparently expects that those elements in Great Britain which used to exclaim, "Who would fight for Czechoslovakia?" may now be induced to cry, "Who would fight for Danzig and the Corridor?" He does not seem to be conscious of the immense change which has been wrought in British public opinion by his treacherous breach of the Munich Agreement, and of the complete reversal of policy which this outrage brought about in the British Government, and especially in the Prime Minister.

The denunciation of the German-Polish Non-Aggression Pact of 1934 is an extremely serious and menacing step. That pact had been reaffirmed as recently as last January, when Ribbentrop visited Warsaw. Like the Anglo-German Naval Treaty, it was negotiated at the wish of Herr Hitler. Like the Naval Treaty, it gave marked advantages to Germany. Both Agreements eased Germany's position while she was weak. The Naval Agreement amounted in fact to a condonation by Great Britain of a breach of the military clauses of the Treaty of Versailles, and thus stultified both the decisions of the Stresa front and those which the Council of the League were induced to take. The German-Polish Agreement enabled Nazi attention to be concentrated first upon Austria and later upon Czechoslovakia, with ruinous results to those unhappy countries. It temporarily weakened the relations between France and Poland and prevented any solidarity of interests growing up among the States of Eastern Europe. Now that it has served its purpose for Germany, it is cast away by one-sided action. Poland is implicitly informed that she is now in the zone of potential aggression.

★ ★ ★ ★ ★

The British Government had to consider urgently the practical implications of the guarantees given to Poland and to Roumania. Neither set of assurances had any military value except within the framework of a general agreement with Russia. It was therefore with this object that talks at last began in Moscow on April 15 between the British Ambassador and M. Litvinov. Considering how the Soviet Government had hitherto been treated, there was not much to be expected from them now. However, on April 16 they made a formal offer, the text of which was not published, for the creation of a united front of mutual assistance between Great Britain, France, and the U.S.S.R. The three Powers, with Poland added if possible, were furthermore to guarantee those States in Central and Eastern Europe which lay under the menace of German aggression. The obstacle to such an agreement was the terror of these same border countries of receiving Soviet help in the shape of Soviet armies marching through their territories to defend them from the Germans, and incidentally incorporating them in the Soviet-Communist system, of which they were the most vehement opponents. Poland, Roumania, Finland, and the three Baltic States did not know whether it was German aggression or Russian rescue that they dreaded more. It was this hideous choice that paralysed British and French policy.

There can however be no doubt, even in the after-light, that Britain and France should have accepted the Russian offer, proclaimed the Triple Alliance, and left the method by which it could be made effective in case of war to be adjusted between allies engaged against a common foe. In such circumstances a different temper prevails. Allies in war are inclined to defer a great deal to each other's wishes; the flail of battle beats upon the front, and all kinds of expedients are welcomed which in peace would be abhorrent. It would not be easy in a Grand Alliance, such as might have been developed, for one ally to enter the territory of another unless invited.

But Mr. Chamberlain and the Foreign Office were baffled by this riddle of the Sphinx. When events are moving at such speed and in such tremendous mass as at this juncture it is wise to take one step at a time. The alliance of Britain, France, and Russia would have struck deep alarm into the heart of Germany in 1939, and no one can prove that war might not even then have been averted. The next step could have been taken with superior power on the side of the allies. The initiative would have been regained by their diplomacy. Hitler could afford neither to embark upon the war on two fronts, which he himself had so deeply condemned, nor to sustain a check. It was a pity not to have placed him in this awkward position, which might well have cost him his life. Statesmen are not called upon only to settle easy questions. These often settle themselves. It is where the balance quivers, and the proportions are veiled in mist, that the opportunity for world-saving decisions presents itself. Having got ourselves into this awful plight of 1939, it was vital to grasp the larger hope.

It is not even now possible to fix the moment when Stalin definitely abandoned all intention of working with the Western Democracies and considered coming to terms with Hitler. Indeed it seems probable that there never was such a moment. The publication in *Nazi-Soviet Relations, 1939-41*, by the American State Department of a mass of documents captured from the archives of the German Foreign Office gives us however a number of facts hitherto unknown. Apparently something happened as early as February 1939; but this was almost certainly concerned with trading and commercial questions affected by the status of Czechoslovakia after Munich which required discussion between the two countries. The incorporation of Czechoslovakia in the Reich in mid-March magnified these issues. Russia had some contracts with the Czechoslovak Government for munitions from the Skoda works. What was to happen to these contracts now that Skoda had become a German arsenal?

On April 17 the State Secretary in the German Foreign Office, Weizsaecker, records that the Russian Ambassador had visited him that day for the first time since he had presented his credentials nearly a year before. He asked about the Skoda contracts, and Weizsaecker pointed out that "a favourable atmosphere for the delivery of war materials to Soviet Russia was not exactly being created at present by reports of a Russian-British-French Air Pact and the like". On this the Soviet Ambassador turned at once from trade to politics and asked the State Secretary what he thought of German-Russian relations. Weizsaecker replied that it appeared to him that "the Russian Press lately was not fully participating in the anti-German tone of the American and some of the English papers". On this the Soviet Ambassador said,

"Ideological differences of opinion have hardly influenced the Russian-Italian relationship, and they need not prove a stumbling-block to Germany either. Soviet Russia has not exploited the present friction between Germany and the Western Democracies against her, nor does she desire to do so. There exists for Russia no reason why she should not live with Germany on a normal footing. And from normal relations might become better and better."

We must regard this conversation as significant, especially in view of the simultaneous discussions in Moscow between the British Ambassador and M. Litvinov and the formal offer of the Soviet on April 16 of a Three-Power Alliance with Great Britain and France. It is the first obvious move of Russia from one leg to the other. "Normalisation" of the relations between Russia and Germany was henceforward pursued, step by step, with the negotiations for a Triple Alliance against German aggression.

If, for instance, Mr. Chamberlain on receipt of the Russian offer had replied, "Yes. Let us three band together and break Hitler's neck," or words to that effect, Parliament would have approved, Stalin would have understood, and history might have taken a different course. At least it could not have taken a worse.

On May 4 I commented on the position in these terms:

Above all, time must not be lost. Ten or twelve days have already passed since the Russian offer was made. The British people, who have now, at the sacrifice of honoured, ingrained custom, accepted the principle of compulsory military service, have a right, in conjunction with the French Republic, to call upon Poland not to place obstacles in the way of a common cause. Not only must the full co-operation of Russia be accepted, but the three Baltic States, Lithuania, Latvia, and Esthonia, must also be brought into association. To these three countries of warlike peoples, possessing together armies totalling perhaps twenty divisions of virile troops, a friendly Russia supplying munitions and other aid is essential.

There is no means of maintaining an Eastern front against Nazi aggression without the active aid of Russia. Russian interests are deeply concerned in preventing Herr Hitler's designs on Eastern Europe. It should still be possible to range all the States and peoples from the Baltic to the Black Sea in one solid front against a new outrage or invasion. Such a front, if established in good heart, and with resolute and efficient military arrangements, combined with the strength of the Western Powers, may yet confront Hitler, Goering,

Himmler, Ribbentrop, Goebbels and Co. with forces the German people would be reluctant to challenge.

★　　★　　★　　★　　★

Instead, there was a long silence while half-measures and judicious compromises were being prepared. This delay was fatal to Litvinov. His last attempt to bring matters to a clear-cut decision with the Western Powers was deemed to have failed. Our credit stood very low. A wholly different foreign policy was required for the safety of Russia, and a new exponent must be found. On May 3 an official communiqué from Moscow announced that M. Litvinov had been released from the office of Foreign Commissar at his request and that his duties would be assumed by the Premier, M. Molotov. The German Chargé d'Affaires in Moscow reported on May 4 as follows: "Since Litvinov had received the English Ambassador as late as May 2 and had been named in the Press of yesterday as guest of honour at the parade, his dismissal appears to be the result of a spontaneous decision by Stalin. . . . At the last Party Congress Stalin urged caution lest the Soviet Union should be drawn into conflict. Molotov (no Jew) is held to be 'the most intimate friend and closest collaborator of Stalin'. His appointment is apparently the guarantee that the foreign policy will be continued strictly in accordance with Stalin's ideas."

Soviet diplomatic representatives abroad were instructed to inform the Government to which they were accredited that this change meant no alteration in Russian foreign policy. Moscow Radio announced on May 4 that Molotov would carry on the policy of Western security that for years had been Litvinov's aim. The eminent Jew, the target of German antagonism, was flung aside for the time being like a broken tool, and, without being allowed a word of explanation, was bundled off the world stage to obscurity, a pittance, and police supervision. Molotov, little known outside Russia, became Commissar for Foreign Affairs, in the closest confederacy with Stalin. He was free from all encumbrance of previous declarations, free from the League of Nations atmosphere, and able to move in any direction which the self-preservation of Russia might seem to require. There was in fact only one way in which he was now likely to move. He had always been favourable to an arrangement with Hitler. The Soviet Government were convinced by Munich and much else that neither Britain nor France would fight till they were attacked, and would not be much good then. The gathering storm was about to break. Russia must look after herself.

The dismissal of Litvinov marked the end of an epoch. It registered the abandonment by the Kremlin of all faith in a security pact with the Western Powers and in the possibility of organising an Eastern front against Germany. The German Press comments at the time, though not necessarily accurate, are interesting. A dispatch from Warsaw was published in the German newspapers on May 4 stating that Litvinov had resigned after a bitter quarrel with Marshal Voroshilov ("the Party boy", as cheeky and daring Russians called him in moments of relaxation). Voroshilov, no doubt on precise instructions, had declared that the Red Army was not prepared to fight for Poland, and, in the name of the Russian General Staff, condemned "excessively far-reaching military obligations". On May 7 the *Frankfurter Zeitung* was already sufficiently informed to state that Litvinov's resignation was extremely serious for the future of Anglo-French "encirclement", and its probable meaning was that those in Russia concerned with the military burden resulting from it had called a halt to Litvinov. All this was true; but for an interval it was necessary that a veil of deceit should cover the immense transaction, and that even up to the latest moment the Soviet attitude should remain in doubt. Russia must have a move both ways. How else could she drive her bargain with the hated and dreaded Hitler?

*   *   *   *   *

The Jew Litvinov was gone, and Hitler's dominant prejudice placated. From that moment the German Government ceased to define its foreign policy as anti-Bolshevism, and turned its abuse upon the "pluto-Democracies". Newspaper articles assured the Soviets that the German *Lebensraum* did not encroach on Russian territory; that indeed it stopped short of the Russian frontier at all points. Consequently there could be no cause of conflict between Russia and Germany unless the Soviets entered into "encirclement" engagements with England and France. The German Ambassador, Count Schulenburg, who had been summoned to Berlin for lengthy consultations, returned to Moscow with an offer of an advantageous goods-credit on a long-term basis. The movement on both sides was towards a compact.

This violent and unnatural reversal of Russian policy was a transmogrification of which only totalitarian States are capable. Barely two years since the leaders of the Russian Army, Tukhachevsky and several thousands of its most accomplished officers, had been slaughtered for the very inclinations which now became acceptable to the handful of anxious masters in the Kremlin. Then pro-Germanism had been heresy and treason. Now, overnight, it was the policy of the State, and woe was mechanically meted out to any who dared dispute it, and often to those not quick enough on the turn-about.

For the task in hand no one was better fitted or equipped than the new Foreign Commissar.

★   ★   ★   ★   ★

The figure whom Stalin had now moved to the pulpit of Soviet foreign policy deserves some description, not available to the British or French Governments at the time. Vyacheslav Molotov was a man of outstanding ability and cold-blooded ruthlessness. He had survived the fearful hazards and ordeals to which all the Bolshevik leaders had been subjected in the years of triumphant revolution. He had lived and thrived in a society where evervarying intrigue was accompanied by the constant menace of personal liquidation. His cannon-ball head, black moustache, and comprehending eyes, his slab face, his verbal adroitness and imperturbable demeanour, were appropriate manifestations of his qualities and skill. He was above all men fitted to be the agent and instrument of the policy of an incalculable machine. I have only met him on equal terms, in parleys where sometimes a strain of humour appeared, or at banquets where he genially proposed a long succession of conventional and meaningless toasts. I have never seen a human being who more perfectly represented the modern conception of a robot. And yet with all this there was an apparently reasonable and keenly-polished diplomatist. What he was to his inferiors I cannot tell. What he was to the Japanese Ambassador during the years when after the Teheran Conference Stalin had promised to attack Japan once the German Army was beaten can be deduced from his recorded conversations. One delicate, searching, awkward interview after another was conducted with perfect poise, impenetrable purpose, and bland, official correctitude. Never a chink was opened. Never a needless jar was made. His smile of Siberian winter, his carefully-measured and often wise words, his affable demeanour, combined to make him the perfect agent of Soviet policy in a deadly world.

Correspondence with him upon disputed matters was always useless, and, if pushed far, ended in lies and insults, of which this work will presently contain some examples. Only once did I seem to get a natural, human reaction. This was in the spring of 1942, when he alighted in Britain on his way back from the United States. We had signed the Anglo-Soviet Treaty, and he was about to make his dangerous flight home. At the garden gate of Downing Street, which we used for secrecy, I gripped his arm and we looked each other in the face. Suddenly he appeared deeply moved. Inside the image there appeared the man. He responded with an equal pressure. Silently we wrung each other's hands. But then we were all together, and it was life or death for the lot. Havoc and ruin had been around him all his days, either impending on himself or dealt by him to others. Certainly in Molotov the Soviet machine had found a capable and in many ways a characteristic representative—always the faithful Party man and Communist disciple. How glad I am at the end of my life not to have had to endure the stresses which he has suffered; better never be born. In the conduct of foreign affairs Mazarin, Talleyrand, Metternich, would welcome him to their company, if there be another world to which Bolsheviks allow themselves to go.

* * * * *

From the moment when Molotov became Foreign Commissar he pursued the policy of an arrangement with Germany at the expense of Poland. It was not very long before the French became aware of this. There is a remarkable dispatch by the French Ambassador in Berlin, dated May 7, published in the *French Yellow Book*, which states that on his secret information he was sure that a Fourth Partition of Poland was to be the basis of the German-Russian *rapprochement*. "Since the month of May," writes M. Daladier in April 1946, "the U.S.S.R. had conducted two negotiations, one with France, the other with Germany. She appeared to prefer to partition rather than to defend Poland. Such was the immediate cause of the Second World War."* But there were other causes too.

* * * * *

* Quoted by Reynaud, *op. cit.*, I, 585.

On May 8 the British Government at last replied to the Soviet Note of April 16. While the text of the British document was not published, the Tass Agency on May 9 issued a statement giving the main points of the British proposals. On May 10 the official organ *Isvestia* printed a communiqué to the effect that Reuter's statement of the British counter-proposals, namely, that "the Soviet Union must separately guarantee every neighbouring State, and that Great Britain must pledge herself to assist the U.S.S.R. if the latter becomes involved in war as a result of its guarantees", did not correspond to fact. The Soviet Government, said the communiqué, had received the British counter-proposals on May 8, but these did not mention the Soviet Union's obligation to give a separate guarantee to each of its neighbouring States, whereas they did state that the U.S.S.R. was obliged to render immediate assistance to Great Britain and France in the event of their being involved in war under their guarantees to Poland and Roumania. No mention however was made of any assistance on their part to the Soviet Union in the event of its being involved in war in consequence of its obligations towards any Eastern European State.

Later on the same day Mr. Chamberlain said that the Government had undertaken their new obligations in Eastern Europe without inviting the direct participation of the Soviet Government on account of various difficulties. His Majesty's Government had suggested that the Soviet Government should make, on their own behalf, a similar declaration, and express their readiness to lend assistance, if desired, to countries which might be victims of aggression and were prepared to defend their independence.

Almost simultaneously the Soviet Government presented a scheme at once more comprehensive and more rigid, which, whatever other advantages it might present, must in the view of His Majesty's Government inevitably raise the very difficulties which their own proposals had been designed to avoid. They accordingly pointed out to the Soviet Government the existence of these difficulties. At the same time they made certain modifications in their original proposals. In particular, they [H.M.G.] made it plain that *if the Soviet Government wished to make their own intervention contingent on that of Great Britain and France, His Majesty's Government for their part would have no objection.*

332

It was a pity that this had not been explicitly stated a fortnight earlier.

It should be mentioned here that on May 12 the Anglo-Turkish Agreement was formally ratified by the Turkish Parliament. By means of this addition to our commitments we hoped to strengthen our position in the Mediterranean in the event of a crisis. Here was our answer to the Italian occupation of Albania. Just as the period of talking with Germany was over, so now we reached in effect the same deadlock with Italy.

The Russian negotiations proceeded languidly, and on May 19 the whole issue was raised in the House of Commons. The debate, which was short and serious, was practically confined to the leaders of parties and to prominent ex-Ministers. Mr. Lloyd George, Mr. Eden, and I all pressed upon the Government the vital need of an immediate arrangement with Russia of the most far-reaching character and on equal terms. Mr. Lloyd George began, and painted a picture of gloom and peril in the darkest hues:

The situation reminds me very much of the feeling that prevailed in the early spring of 1918. We knew there was a great attack coming from Germany, but no one quite knew where the blow would fall. I remember that the French thought it would fall on their front, while our generals thought it would fall on ours. The French generals were not even agreed as to the part of their front on which the attack would fall, and our generals were equally divided. All that we knew was that there was a tremendous onslaught coming somewhere, and the whole atmosphere was filled with, I will not say fear, but with uneasiness. We could see the tremendous activities behind the German lines, and we knew that they were preparing something. That is more or less what seems to me to be the position to-day. . . . We are all very anxious; the whole world is under the impression that there is something preparing in the nature of another attack from the aggressors. Nobody quite knows where it will come. We can see that they are speeding up their armaments at a rate hitherto unprecedented, especially in weapons of the offensive—tanks, bombing aeroplanes, submarines. We know that they are occupying and fortifying fresh positions that will give them strategic advantages in a war with France and ourselves. . . . They are inspecting and surveying, from Libya to the North Sea, all sorts of situations that would be of vital importance in the event of war. There is a secrecy in the movements behind the lines which is very ominous.

There is the same kind of secrecy as in 1918, in order to baffle us as to their objects. They are not preparing for defence. . . . They are not preparing themselves against attack from either France, Britain, or Russia. That has never been threatened. I have never heard, either privately or publicly, any hint or suggestion that we were contemplating an attack upon Italy or Germany in any quarter, and they know it quite well. Therefore all these preparations are not for defence. They are for some contemplated offensive scheme against someone or other in whom we are interested.

*   *   *   *   *

Mr. Lloyd George then added some words of wisdom:

The main military purpose and scheme of the Dictators is to produce quick results, to avoid a prolonged war. A prolonged war never suits dictators. A prolonged war like the Peninsular War wears them down, and the great Russian defence, which produced no great military victory for the Russians, broke Napoleon. Germany's ideal is now, and always has been, a war which is brought to a speedy end. The war against Austria in 1866 did not last more than a few weeks, and the war in 1870 was waged in such a way that it was practically over in a month or two. In 1914 plans were made with exactly the same aim in view, and it was very nearly achieved; and they would have achieved it but for Russia. But from the moment they failed to achieve a speedy victory the game was up. You may depend upon it that the great military thinkers of Germany have been working out the problem, what was the mistake of 1914, what did they lack, how can they fill up the gaps and repair the blunders or avoid them in the next war?

Mr. Lloyd George, pressing on from fact to fancy, then suggested that the Germans had already got "twenty thousand tanks" and "thousands of bomber aeroplanes". This was far beyond the truth. Moreover, it was an undue appeal to the fear motive. And why had he not been busy all these years with my small group ingeminating rearmament? But his speech cast a chill over the assembly. Two years before, or better still three, such statements and all the pessimism of his speech would have been scorned and derided. But then there was time. Now, whatever the figures, it was all too late.

The Prime Minister replied, and for the first time revealed to us his views on the Soviet offer. His reception of it was certainly cool, and indeed disdainful:

If we can evolve a method by which we can enlist the co-operation and assistance of the Soviet Union in building up that peace front, we welcome it; we want it; we attach value to it. The suggestion that we despise the assistance of the Soviet Union is without foundation. Without accepting any view of an unauthorised character as to the precise value of the Russian military forces, or the way in which they would best be employed, no one would be so foolish as to suppose that that huge country, with its vast population and enormous resources, would be *a negligible factor* in such a situation as that with which we are confronted.

This seemed to show the same lack of proportion as we have seen in the rebuff to the Roosevelt proposals a year before.

I then took up the tale:

I have been quite unable to understand what is the objection to making the agreement with Russia which the Prime Minister professes himself desirous of doing, and making it in the broad and simple form proposed by the Russian Soviet Government.

Undoubtedly, the proposals put forward by the Russian Government contemplate a Triple Alliance against aggression between England, France, and Russia, which alliance may extend its benefits to other countries if and when those benefits are desired. The alliance is solely for the purpose of resisting further acts of aggression and of protecting the victims of aggression. I cannot see what is wrong with that. What is wrong with this simple proposal? It is said: "Can you trust the Russian Soviet Government?" I suppose in Moscow they say: "Can we trust Chamberlain?" I hope we may say that the answer to both questions is in the affirmative. I earnestly hope so. . . .

This Turkish proposal, which is universally accepted, is a great consolidating and stabilising force throughout the whole of the Black Sea area and the Eastern Mediterranean. Turkey, with whom we have made this agreement, is in the closest harmony with Russia. She is also in the closest harmony with Roumania. These Powers together are mutually protecting vital interests. . . .

There is a great identity of interests between Great Britain and the associated Powers in the South. Is there not a similar identity of interests in the North? Take the countries of the Baltic, Lithuania, Latvia, and Esthonia, which were once the occasion of the wars of Peter the Great. It is a major interest of Russia that these Powers should not fall into the hands of Nazi Germany. That is a vital interest in the North. I need not elaborate the arguments about [a German attack upon] the Ukraine, which means an invasion of Russian territory. All along the whole of this Eastern front you can see that the major interests of

Russia are definitely engaged, and therefore it seems you could fairly judge that they would pool their interests with other countries similarly affected. . . .

If you are ready to be an ally of Russia in time of war, which is the supreme test, the great occasion of all, if you are ready to join hands with Russia in the defence of Poland, which you have guaranteed, and of Roumania, why should you shrink from becoming the ally of Russia now, when you may by that very fact prevent the breaking out of war? I cannot understand all these refinements of diplomacy and delay. If the worst comes to the worst, you are in the midst of it with them, and you have to make the best of it with them. If the difficulties do not arise, well, you will have had the security in the preliminary stages. . . .

His Majesty's Government have given a guarantee to Poland. I was astounded when I heard them give this guarantee. I support it, but I was astounded by it, because nothing that had happened before led one to suppose that such a step would be taken. I want to draw the attention of the Committee to the fact that the question posed by Mr. Lloyd George ten days ago and repeated to-day has not been answered. The question was whether the General Staff was consulted before this guarantee was given as to whether it was safe and practical to give it, and whether there were any means of implementing it. The whole country knows that the question has been asked, and it has not been answered. That is disconcerting and disquieting. . . .

Clearly Russia is not going to enter into agreements unless she is treated as an equal, and not only is treated as an equal, but has confidence that the methods employed by the Allies—by the peace front —are such as would be likely to lead to success. No one wants to associate themselves with indeterminate leadership and uncertain policies. The Government must realise that none of these States in Eastern Europe can maintain themselves for, say, a year's war unless they have behind them the massive, solid backing of a friendly Russia, joined to the combination of the Western Powers. In the main, I agree with Mr. Lloyd George that if there is to be an effective Eastern front—an Eastern peace front, or a war front as it might become—it can be set up only with the effective support of a friendly Soviet Russia lying behind all those countries.

Unless there is an Eastern front set up, what is going to happen to the West? What is going to happen to those countries on the Western front to whom, if we have not given guarantees, it is admitted we are bound—countries like Belgium, Holland, Denmark, and Switzerland? Let us look back to the experiences we had in 1917. In 1917 the Russian front was broken and demoralised. Revolution and mutiny had sapped the courage of that great disciplined army, and the conditions at the front were indescribable; and yet, until the treaty was made closing the front down, more than 1,500,000 Germans were held upon that front, even in its most ineffectual and unhappy condition. Once that front was closed down, 1,000,000 Germans and 5,000 cannon were brought to the West, and at the last moment almost turned the course of the war and forced upon us a disastrous peace.

It is a tremendous thing, this question of the Eastern front. I am astonished that there is not more anxiety about it. Certainly, I do not ask favours of Soviet Russia. This is no time to ask favours of countries. But here is an offer, a fair offer, and a better offer, in my opinion, than the terms which the Government seek to get for themselves; a more simple, a more direct and a more effective offer. Let it not be put aside and come to nothing. I beg His Majesty's Government to get some of these brutal truths into their heads. Without an effective Eastern front there can be no satisfactory defence of our interests in the West, and without Russia there can be no effective Eastern front. If His Majesty's Government, having neglected our defences for a long time, having thrown away Czechoslovakia with all that Czechoslovakia meant in military power, having committed us, without examination of the technical aspects, to the defence of Poland and Roumania, now reject and cast away the indispensable aid of Russia, and so lead us in the worst of all ways into the worst of all wars, they will have ill deserved the confidence and, I will add, the generosity with which they have been treated by their fellow-countrymen.

There can be little doubt that all this was now too late. Attlee, Sinclair, and Eden spoke on the general line of the imminence of the danger and the need of the Russian alliance. The position of the leaders of the Labour and Liberal Parties was weakened by the vote against compulsory national service to which they had led their followers only a few weeks before. The plea, so often advanced, that this was because they did not like the foreign policy was feeble; for no foreign policy can have validity if there is no adequate force behind it, and no national readiness to make the necessary sacrifices to produce that force.

*　　*　　*　　*　　*

The efforts of the Western Powers to produce a defensive alignment against Germany were well matched by the other side. Conversations between Ribbentrop and Ciano at Como at the beginning of May came to formal and public fruition in the so-called "Pact of Steel", signed by the two Foreign Ministers in

Berlin on May 22. This was the challenging answer to the flimsy British network of guarantees in Eastern Europe. Ciano in his diary records a conversation with Hitler at the time of the signature of this alliance.

Hitler states that he is well satisfied with the Pact, and confirms the fact that Mediterranean policy will be directed by Italy. He takes an interest in Albania, and is enthusiastic about our programme for making of Albania a stronghold which will inexorably dominate the Balkans.*

Hitler's satisfaction was more clearly revealed when on the day following the signing of the Pact of Steel, May 23, he held a meeting with his Chiefs of Staff. The secret minutes of the conversations are on record:

We are at present in a state of patriotic fervour, which is shared by two other nations—Italy and Japan. The period which lies behind us has indeed been put to good use. All measures have been taken in the correct sequence and in harmony with our aims. The Pole is no "supplementary enemy". Poland will always be on the side of our adversaries. In spite of treaties of friendship, Poland has always had the secret intention of exploiting every opportunity to do us harm. Danzig is not the subject of the dispute at all. It is a question of expanding our living space in the East and of securing our food supplies. There is therefore no question of sparing Poland, and we are left with the decision to attack Poland at the first suitable opportunity. We cannot expect a repetition of the Czech affair. There will be war. Our task is to isolate Poland. The success of the isolation will be decisive.

If it is not certain that a German-Polish conflict will not lead to war in the West, then the fight must be primarily against England and France. If there were an alliance of France, England, and Russia against Germany, Italy, and Japan, I should be constrained to attack England and France with a few annihilating blows. I doubt the possibility of a peaceful settlement with England. We must prepare ourselves for the conflict. England sees in our development the foundation of a hegemony which would weaken her. England is therefore our enemy, and the conflict with England will be a life-and-death struggle. The Dutch and Belgian air bases must be occupied by armed force. Declarations of neutrality must be ignored.

If England intends to intervene in the Polish war we must occupy Holland with lightning speed. We must aim at securing a new defence line on Dutch soil up to the Zuyder Zee. The idea that we can get off

* *Ciano's Diary*, p. 90.

cheaply is dangerous; there is no such possibility. We must burn our boats, and it is no longer a question of justice or injustice, but of life or death for eighty million human beings. Every country's armed forces or Government must aim at a short war. The Government however must also be prepared for a war of ten or fifteen years' duration.

England knows that to lose a war will mean the end of her world-power. England is the driving force against Germany.

The British themselves are proud, courageous, tenacious, firm in resistance, and gifted as organisers. They know how to exploit every new development. They have the love of adventure and the bravery of the Nordic race. The German average is higher. But if in the first World War we had had two battleships and two cruisers more, and if the Battle of Jutland had begun in the morning, the British Fleet would have been defeated* and England brought to her knees. In addition to the surprise attack, preparations for a long war must be made, while opportunities on the Continent for England are eliminated. The Army will have to hold positions essential to the Navy and Air Force. If Holland and Belgium are successfully occupied and held, and if France is also defeated, the fundamental conditions for a successful war against England will have been secured.†

On May 30 the German Foreign Office sent the following instructions to their Ambassador in Moscow: "Contrary to the policy previously planned, we have now decided to undertake definite negotiations with the Soviet Union."† While the ranks of the Axis closed for military preparation, the vital link of the Western Powers with Russia had perished. The underlying discordance of view can be read into Foreign Commissar Molotov's speech of May 31 in reply to Mr. Chamberlain's speech in the Commons of May 19.

As far back [he said] as the middle of April the Soviet Government entered into negotiations with the British and French Governments about the necessary measures to be taken. The negotiations started then are not yet concluded. It became clear some time ago that if there was any real desire to create an efficient front of peaceable countries against the advance of aggression the following minimum conditions were imperative:

---

* *Nuremberg Documents*, I, pp. 167–8. Hitler was evidently quite ignorant of the facts of Jutland, which was from beginning to end an unsuccessful effort by the British Fleet to bring the Germans to a general action, in which the overwhelming gunfire of the British line of battle would soon have been decisive.

† *Nazi-Soviet Relations*, p. 15.

The conclusion between Great Britain, France, and the U.S.S.R. of an effective pact of mutual assistance against aggression, of an exclusively defensive character.

A guarantee on the part of Great Britain, France, and the U.S.S.R. of the States of Central and Eastern Europe, including, without exception, all the European countries bordering on the U.S.S.R., against an attack by aggressors.

The conclusion between Great Britain, France, and the U.S.S.R. of a definite agreement on the forms and extent of the immediate and effective assistance to be rendered to one another and to the guaranteed States in the event of an attack by aggressors.

The negotiations had come to a seemingly unbreakable deadlock. The Polish and Roumanian Governments, while accepting the British guarantee, were not prepared to accept a similar undertaking in the same form from the Russian Government. A similar attitude prevailed in another vital strategic quarter—the Baltic States. The Soviet Government made it clear that they would only adhere to a pact of mutual assistance if Finland and the Baltic States were included in a general guarantee. All four countries now refused, and perhaps in their terror would for a long time have refused such a condition. Finland and Esthonia even asserted that they would consider a guarantee extended to them without their assent as an act of aggression. On June 7 Esthonia and Latvia signed non-aggression pacts with Germany. Thus Hitler penetrated with ease into the final defences of the tardy, irresolute coalition against him.

CHAPTER XXI

ON THE VERGE

*The Threat to Danzig – General Gamelin Invites Me to Visit the
Rhine Front – A Tour with General Georges – Some Impressions –
French Acceptance of the Defensive – The Position of Atomic Research
– My Note on Air Defence – Renewed Efforts to Agree with Soviet
Russia – Polish Obstruction – The Military Conversations in Moscow
– Stalin's Account to Me in 1942 – A Record in Deceit – Ribbentrop
Invited to Moscow – The Russo-German Non-Aggression Treaty
– The News Breaks upon the World – Hitler's Army Orders –
"Honesty is the Best Policy" – British Precautionary Measures – The
Prime Minister's Letter to Hitler – An Insolent Reply – Hitler Post-
pones D-Day – Hitler's Letter to Mussolini – The Duce's Reply –
The Last Few Days.*

SUMMER advanced, preparations for war continued
throughout Europe, and the attitudes of diplomatists, the
speeches of politicians, and the wishes of mankind counted
each day for less. German military movements seemed to portend
the forcible settlement of the dispute with Poland over Danzig as
a preliminary to the assault on Poland itself. Mr. Chamberlain
expressed his anxieties to Parliament on June 10, and repeated his
intention to stand by Poland if her independence were threatened.
In a spirit of detachment from the facts the Belgian Government,
largely under the influence of their King, announced on June 23
that they were opposed to Staff talks with England and France
and that Belgium intended to maintain a strict neutrality. The
tide of events brought with it a closing of the ranks between
England and France, and also at home. There was much coming
and going between Paris and London during the month of July.
The celebrations of the Fourteenth of July were an occasion for a

display of Anglo-French union. I was invited by the French Government to attend this brilliant spectacle.

As I was leaving Le Bourget after the parade General Gamelin suggested that I should visit the French front. "You have never seen the Rhine sector," he said. "Come then in August; we will show you everything." Accordingly a plan was made, and on August 15 General Spears and I were welcomed by his close friend, General Georges, Commander-in-Chief of the armies on the North-eastern front and *Successeur Eventuel* to the Supreme Command. I was delighted to meet this most agreeable and competent officer, and we passed the next ten days in his company, revolving military problems and making contacts with Gamelin, who was also inspecting certain points on this part of the front.

Beginning at the angle of the Rhine near Lauterbourg, we traversed the whole section to the Swiss frontier. In England, as in 1914, the carefree people were enjoying their holidays and playing with their children on the sands. But here along the Rhine a different light glared. All the temporary bridges across the river had been removed to one side or the other. The permanent bridges were heavily guarded and mined. Trusty officers were stationed night and day to press at a signal the buttons which would blow them up. The great river, swollen by the melting Alpine snows, streamed along in sullen, turgid flow. The French outposts crouched in their rifle-pits amid the brushwood. Two or three of us could stroll together to the water's edge, but nothing like a target, we were told, must be presented. Three hundred yards away on the farther side, here and there among the bushes, German figures could be seen working rather leisurely with pick and shovel at their defences. All the riverside quarter of Strasbourg had already been cleared of civilians. I stood on its bridge for some time and watched one or two motor-cars pass over it. Prolonged examination of passports and character took place at either end. Here the German post was little more than a hundred yards away from the French. There was no intercourse between them. Yet Europe was at peace. There was no dispute between Germany and France. The Rhine flowed on, swirling and eddying, at six or seven miles an hour. One or two canoes with boys in them sped past on the current. I did not see the Rhine again until more than five years later, in March 1945, when

I crossed it in a small boat with Field-Marshal Montgomery. But that was near Wesel, far to the north.

On my return I sent a few notes of what I had gathered to the Secretary of State for War, and perhaps to some other Ministers with whom I was in touch:

The French Front cannot be surprised. It cannot be broken at any point except by an effort which would be enormously costly in life, and would take so much time that the general situation would be transformed while it was in progress. The same is true, though to a lesser extent, of the German side.

The flanks of this front however rest upon two small neutral States. The attitude of Belgium is thought to be profoundly unsatisfactory. At present there are no military relations of any kind between the French and the Belgians.

At the other end of the line, about which I was able to learn a good deal, the French have done everything in their power to prepare against an invasion through Switzerland. This operation would take the form of a German advance up the Aar, protected on its right by a movement into or towards the Belfort Gap. I personally think it extremely unlikely that any heavy German attempt will be made either against the French Front or against the two small countries on its flanks in the opening phase.

It is not necessary for Germany to mobilise before attacking Poland. They have enough divisions already on a war footing to act upon their eastern front, and would have time to reinforce the Siegfried Line by mobilising simultaneously with the beginning of a heavy attack on Poland. Thus a German mobilisation is a warning signal which may not be forthcoming in advance of war. The French, on the other hand, may have to take extra measures in the period of extreme tension now upon us.

As to date, it is thought Hitler would be wise to wait until the snow falls in the Alps and gives the protection of winter to Mussolini. During the first fortnight of September, or even earlier, these conditions would be established. There would still be time for Hitler to strike heavily at Poland before the mud period of late October or early November would hamper a German offensive there. Thus this first fortnight in September seems to be particularly critical, and the present German arrangements for the Nuremberg demonstration—propaganda, etc.— seem to harmonise with such a conclusion.

*   *   *   *   *

What was remarkable about all I learned on my visit was the complete acceptance of the defensive which dominated my most responsible French hosts, and imposed itself irresistibly upon me. In talking to all these highly competent French officers one had the sense that the Germans were the stronger, and that France had no longer the life-thrust to mount a great offensive. She would fight for her existence—*Voilà tout!* There was the fortified Siegfried Line, with all the increased fire-power of modern weapons. In my own bones, too, was the horror of the Somme and Passchendaele offensives. The Germans were of course far stronger than in the days of Munich. We did not know the deep anxieties which rent their High Command. We had allowed ourselves to get into such a condition, physically and psychologically, that no responsible person—and up to this point I had no responsibilities—could act on the assumption—which was true—that only forty-two half-equipped and half-trained German divisions guarded their long front from the North Sea to Switzerland. This compared with thirteen at the time of Munich.

* * * * *

In these final weeks my fear was that His Majesty's Government, in spite of our guarantee, would recoil from waging war upon Germany if she attacked Poland. There is no doubt that at this time Mr. Chamberlain had resolved to take the plunge, bitter though it was to him. But I did not know him so well as I did a year later. I feared that Hitler might try a bluff about some novel agency or secret weapon which would baffle or puzzle the overburdened Cabinet. From time to time Professor Lindemann had talked to me about Atomic Energy. I therefore asked him to let me know how things stood in this sphere, and after a conversation I wrote the following letter to Kingsley Wood, with whom my fairly intimate relations have been mentioned:

Mr. Churchill to Secretary of State for Air   *August 5, 1939*
Some weeks ago one of the Sunday papers splashed the story of the immense amount of energy which might be released from uranium by the recently discovered chain of processes which take place when this particular type of atom is split by neutrons. At first sight this might seem to portend the appearance of new explosives of devastating power. *In view of this it is essential to realise that there is no danger that this discovery, however great its scientific interest, and perhaps ultimately its practical importance, will lead to results capable of being put into operation on a large scale for several years.*

There are indications that tales will be deliberately circulated when international tension becomes acute about the adaptation of this process to produce some terrible new secret explosive, capable of wiping out London. Attempts will no doubt be made by the Fifth Column to induce us by means of this threat to accept another surrender. For this reason it is imperative to state the true position.

First, the best authorities hold that only a minor constituent of uranium is effective in these processes, and that it will be necessary to extract this before large-scale results are possible. This will be a matter of many years. Secondly, the chain process can take place only if the uranium is concentrated in a large mass. As soon as the energy develops it will explode with a mild detonation before any really violent effects can be produced.* It might be as good as our present-day explosives, but it is unlikely to produce anything very much more dangerous. Thirdly, these experiments cannot be carried out on a small scale. If they had been successfully done on a big scale (*i.e.*, with the results with which we shall be threatened unless we submit to blackmail) it would be impossible to keep them secret. Fourthly, only a comparatively small amount of uranium in the territories of what used to be Czechoslovakia is under the control of Berlin.

For all these reasons the fear that this new discovery has provided the Nazis with some sinister, new, secret explosive with which to destroy their enemies is clearly without foundation. Dark hints will no doubt be dropped and terrifying whispers will be assiduously circulated, but it is hoped that nobody will be taken in by them.

It is remarkable how accurate this forecast was. Nor was it the Germans who found the path. Indeed they followed the wrong trail, and had actually abandoned the search for the Atomic Bomb in favour of rockets or pilotless aeroplanes at the moment when President Roosevelt and I were taking the decisions and reaching the memorable agreements, which will be described in their proper place, for the large-scale manufacture of atomic bombs.

I also wrote in my final paper for the Air Defence Research Committee:

August 10, 1939

The main defence of England against air raids is the toll which can be extracted from the raiders. One-fifth knocked out each go will soon bring the raids to an end. . . . We must imagine the opening attack as a large affair crossing the sea in relays for many hours. But it is not the first results of the air attack which will govern the future of the air war. It is not child's-play to come and attack England. A

---

* This difficulty was of course overcome later by only very elaborate methods **after** several years of research.

heavy proportion of casualties will lead the enemy to make severe calculations of profit and loss. As daylight raiding will soon become too expensive, we have chiefly to deal with random night-bombing of the built-up areas.

*   *   *   *   *

"Tell Chamberlain," said Mussolini to the British Ambassador on July 7, "that if England is ready to fight in defence of Poland, Italy will take up arms with her ally, Germany." But behind the scenes his attitude was the opposite. He sought at this time no more than to consolidate his interests in the Mediterranean and North Africa, to cull the fruits of his intervention in Spain, and to digest his Albanian conquest. He did not like being dragged into a European war for Germany to conquer Poland. For all his public boastings, he knew the military and political fragility of Italy better than anyone. He was willing to talk about a war in 1942, if Germany would give him the munitions; but in 1939—no!

As the pressure upon Poland sharpened during the summer, Mussolini turned his thoughts upon repeating his Munich *rôle* of mediator, and he suggested a World Peace Conference. Hitler curtly dispelled such ideas. On August 11 Ciano met Ribbentrop at Salzburg. According to Ciano's diary:

The Duce is anxious for me to prove by documentary evidence that an outbreak of war at this time would be folly. . . . It would be impossible to localise it in Poland, and a general war would be disastrous for everyone. Never has the Duce spoken of the need for peace so unreservedly and with so much warmth. . . . Ribbentrop is evasive. Whenever I ask him for particulars about German policy his conscience troubles him. He has lied too many times about German intentions towards Poland not to feel uneasy now about what he must tell me, and what they are really planning to do. . . . The German decision to fight is implacable. Even if they were given more than they ask they would attack just the same, because they are possessed by the demon of destruction. . . . At times our conversation becomes very tense. I do not hesitate to express my thoughts with brutal frankness. But this does not move him. I am becoming aware how little we are worth in the opinion of the Germans.*

Ciano went on to see Hitler the next day. We have the German minutes of this meeting. Hitler made it clear that he intended to

* *Ciano's Diary*, p. 123.

settle with Poland, that he would be forced to fight England and France as well, and that he wanted Italy to come in. He said, "If England keeps the necessary troops in her own country, she can send to France at the most two infantry divisions and one armoured division. For the rest she could supply a few bomber squadrons, but hardly any fighters, because the German Air Force would at once attack England, and the English fighters would be urgently needed for its defence." About France he said that after the destruction of Poland—which would not take long—Germany would be able to assemble hundreds of divisions along the West Wall, and France would thus be compelled to concentrate all her available forces from the colonies and from the Italian frontier and elsewhere on her Maginot Line for the life-and-death struggle. Ciano in reply expressed his surprise at the gravity of what he had been told. There had, he complained, never been any previous sign from the German side that the Polish quarrel was so serious and imminent. On the contrary, Ribbentrop had said that the Danzig question would be settled in the course of time. The Duce, convinced that a conflict with the Western Powers was unavoidable, had assumed that he should make plans for this event during a period of two or three years.

After these interchanges Ciano returned gloomily to report to his master, whom he found more deeply convinced that the Democracies would fight, and even more resolved to keep out of the struggle himself.

*   *   *   *   *

A renewed effort to come to an arrangement with Soviet Russia was made by the British and French Governments. It was decided to send a special envoy to Moscow. Mr. Eden, who had made useful contacts with Stalin some years before, volunteered to go. This generous offer was declined by the Prime Minister. Instead on June 12 Mr. Strang, an able official but without any special standing outside the Foreign Office, was entrusted with this momentous mission. This was another mistake. The sending of so subordinate a figure gave actual offence. It is doubtful whether he was able to pierce the outer crust of the Soviet organism. In any case all was now too late. Much had happened since M. Maisky had been sent to see me at Chartwell in August 1938. Munich had happened. Hitler's armies had had a year more to mature. His munitions factories, reinforced by the Skoda works, were all in full blast. The Soviet Government cared much for Czechoslovakia; but Czechoslovakia was gone. Beneš was in exile. A German Gauleiter ruled in Prague.

On the other hand, Poland presented to Russia an entirely different set of age-long political and strategic problems. Their last major contact had been the Battle of Warsaw in 1920, when the Bolshevik armies under Kamieniev had been hurled back from their invasion by Pilsudski, aided by the advice of General Weygand and the British Mission under Lord D'Abernon, and thereafter pursued with bloody vengeance. All these years Poland had been a spear-point of anti-Bolshevism. With her left hand she joined and sustained the anti-Soviet Baltic States. But with her right hand, at Munich-time, she had helped to despoil Czechoslovakia. The Soviet Government were sure that Poland hated them, and also that Poland had no power to withstand a German onslaught. They were however very conscious of their own perils and of their need for time to repair the havoc in the High Commands of their armies. In these circumstances the prospects of Mr. Strang's mission were not exuberant.

The negotiations wandered around the question of the reluctance of Poland and the Baltic States to be rescued from Germany by the Soviets; and here they made no progress. In its leading article of June 13 *Pravda* had already declared that an effective neutrality of Finland, Esthonia, and Latvia was vital to the safety of the U.S.S.R. The security of such States, it said, was of prime importance for Britain and France, as "even such a politician as Mr. Churchill" had recognised. The issue was discussed in Moscow on June 15. On the following day the Russian Press declared that "in the circles of the Soviet Foreign Ministry results of the first talks are regarded as not entirely favourable". All through July the discussions continued fitfully, and eventually the Soviet Government proposed that conversations should be continued on a military basis with both French and British representatives. The British Government therefore dispatched Admiral Drax with a mission to Moscow on August 10. These officers possessed no written authority to negotiate. The French Mission was headed by General Doumenc. On the Russian side Marshal Voroshilov officiated. We now know that at this same time the Soviet Government agreed to the journey of a German negotiator to Moscow. The military conference soon foundered upon the refusal of Poland and Roumania to allow the transit of Russian troops. The Polish atttitude was, "With the Germans we risk losing our liberty; with the Russians our soul".*

*   *   *   *   *

At the Kremlin in August 1942 Stalin, in the early hours of the morning, gave me one aspect of the Soviet position. "We formed the impression," said Stalin, "that the British and French Governments were not resolved to go to war if Poland were attacked, but that they hoped the diplomatic line-up of Britain, France, and Russia would deter Hitler. We were sure it would not." "How many divisions," Stalin had asked, "will France send against Germany on mobilisation?" The answer was, "About a hundred." He then asked, "How many will England send?" The answer was, "Two, and two more later." "Ah, two, and two more later," Stalin had repeated. "Do you know," he asked, "how many divisions we shall have to put on the Russian front if we go to war with Germany?" There was a pause. "More than three hundred." I was not told with whom this conversation took place or its date. It must be recognised that this was solid ground, but not favourable for Mr. Strang of the Foreign Office.

It was judged necessary by Stalin and Molotov for bargaining purposes to conceal their true intentions till the last possible moment. Remarkable skill in duplicity was shown by Molotov and his subordinates in all their contacts with both sides. As late as August 4 the German Ambassador Schulenburg could only telegraph from Moscow: "From Molotov's whole attitude it was evident that the Soviet Government was in fact more prepared for improvement in German-Soviet relations, but that the old mistrust of Germany persists. My overall impression is that the Soviet Government is at present determined to sign with England and France if they fulfil all Soviet wishes. Negotiations, to be sure, might still last a long time, especially since the mistrust of England is also great. . . . It will take a considerable effort on our part to cause the Soviet Government to swing about."† He need not have worried. The die was cast.

*   *   *   *   *

---

* Quoted in Reynaud, *op. cit.*, I, p. 587.
† *Nazi-Soviet Relations*, p. 41.

On the evening of August 19 Stalin announced to the Politburo his intention to sign a pact with Germany. On August 22 Marshal Voroshilov was not to be found by the Allied missions until evening. He then said to the head of the French Mission, "The question of military collaboration with France has been in the air for several years, but has never been settled. Last year, when Czechoslovakia was perishing, we waited for a signal from France, but none was given. Our troops were ready. . . . The French and English Governments have now dragged out the political and military discussions too long. For that reason the possibility is not to be excluded that certain political events may take place. . . ." The next day Ribbentrop arrived in Moscow.[*]

*   *   *   *   *

We now possess in the Nuremberg documents, and in those captured and recently published by the United States, the details of this never-to-be-forgotten transaction. According to Ribbentrop's chief assistant, Gauss, who flew with him to Moscow, "On the evening of August 23 the first conversation between Ribbentrop and Stalin took place. . . . The Reich Foreign Minister returned very satisfied from this long conference. . . ." Later in the day an agreement on the text of the Soviet-German Non-Aggression Pact was reached quickly and without difficulties. "Ribbentrop himself," says Gauss, "had inserted in the preamble a rather far-reaching phrase concerning the formation of friendly German-Soviet relations. To this Stalin objected, remarking that the Soviet Government could not suddenly present to their public a German-Soviet declaration of friendship after they had been covered with *pails of manure* by the Nazi Government for six years. Thereupon this phrase in the preamble was deleted." In a secret agreement Germany declared herself politically disinterested in Latvia, Esthonia, and Finland, but considered Lithuania to be in her sphere of influence. A demarcation line was drawn for the Polish partition. In the Baltic countries Germany claimed only economic interests. The Non-Aggression Pact and the secret agreement were signed rather late on the night of August 23.[†]

*   *   *   *

[*] Reynaud, *op. cit.*, I, p. 588.
[†] *Nuremberg Documents*, Pt. X, pp. 210 ff.

Despite all that has been dispassionately recorded in this and the foregoing chapter, only totalitarian despotism in both countries could have faced the odium of such an unnatural act. It is a question whether Hitler or Stalin loathed it most. Both were aware that it could only be a temporary expedient The antagonisms between the two empires and systems were mortal. Stalin no doubt felt that Hitler would be a less deadly foe to Russia after a year of war with the Western Powers. Hitler followed his method of "One at a time". The fact that such an agreement could be made marks the culminating failure of British and French foreign policy and diplomacy over several years.

On the Soviet side it must be said that their vital need was to hold the deployment positions of the German armies as far to the west as possible, so as to give the Russians more time for assembling their forces from all parts of their immense empire. They had burnt in their minds the disasters which had come upon their armies in 1914, when they had hurled themselves forward to attack the Germans while still themselves only partly mobilised. But now their frontiers lay far to the east of those of the previous war. They must be in occupation of the Baltic States and a large part of Poland by force or fraud before they were attacked. If their policy was cold-blooded, it was also at the moment realistic in a high degree.

The sinister news broke upon the world like an explosion. On August 21-22 the Soviet Tass Agency stated that Ribbentrop was flying to Moscow to sign a Non-Aggression Pact with the Soviet Union. Whatever emotions the British Government may have experienced, fear was not among them. They lost no time in declaring that "such an event would in no way affect their obligations, which they were determined to fulfil". Nothing could now avert or delay the conflict.

* * * * *

It is still worth while to record the terms of the Pact.

Both High Contracting Parties obligate themselves to desist from any act of violence, any aggressive action, and any attack on each other, either individually or jointly with other Powers.

This treaty was to last ten years, and if not denounced by either side one year before the expiration of that period would be auto-

matically extended for another five years. There was much
jubilation and many toasts around the conference table. Stalin
spontaneously proposed the toast of the Fuehrer, as follows: "I
know how much the German nation loves its Fuehrer; I should
therefore like to drink his health." A moral may be drawn from
all this, which is of homely simplicity. "Honesty is the best
policy." Several examples of this will be shown in these pages.
Crafty men and statesmen will be shown misled by all their
elaborate calculations. But this is the signal instance. Only
twenty-two months were to pass before Stalin and the Russian
nation in its scores of millions were to pay a frightful forfeit. If a
Government has no moral scruples it often seems to gain great
advantages and liberties of action, but "All comes out even at the
end of the day, and all will come out yet more even when all the
days are ended."

*　　*　　*　　*　　*

Hitler was sure from secret interchanges that the Russian Pact
would be signed on August 22. Even before Ribbentrop returned
from Moscow or the public announcement was made he addressed
his Commanders-in-Chief as follows:

We must be determined from the beginning to fight the Western
Powers. . . . The conflict with Poland was bound to come sooner or
later. I had already made this decision in the spring, but I thought I
would first turn against the West and only afterwards against the
East. . . . We need not be afraid of a blockade. The East will supply
us with grain, cattle, coal. . . . I am only afraid that at the last minute
some *Schweinhund* will make a proposal for mediation. . . . The political
aim is set further. A beginning has been made for the destruction of
England's hegemony. The same is open for the soldier, after I have
made the political preparations.

*　　*　　*　　*　　*

On the news of the German-Soviet Pact the British Government at once took precautionary measures. Orders were issued
for key parties of the coast and anti-aircraft defences to assemble,
and for the protection of vulnerable points. Telegrams were sent
to Dominion Governments and to the Colonies warning them
that it might be necessary in the very near future to institute the
precautionary stage. The Lord Privy Seal was authorised to
bring the Regional Organisations to a war footing. On August 23
the Admiralty received Cabinet authority to requisition twenty-

five merchantmen for conversion to armed merchant cruisers, and thirty-five trawlers to be fitted with Asdics. Six thousand reservists for the overseas garrisons were called up. The anti-aircraft defence of the Radar stations and the full deployment of the anti-aircraft forces were approved. Twenty-four thousand reservists of the Air Force and all the Air Auxiliary Force, including the balloon squadrons, were called up. All leave was stopped throughout the fighting services. The Admiralty issued warnings to merchant shipping. Many other steps were taken.

* * * * *

The Prime Minister decided to write to Hitler about these preparatory measures. This letter does not appear in Mr. Feiling's biography, but has been printed elsewhere. In justice to Mr. Chamberlain it should certainly be widely read.

Your Excellency will have already heard of certain measures taken by His Majesty's Government and announced in the Press and on the wireless this evening.

These steps have, in the opinion of His Majesty's Government, been rendered necessary by the military movements which have been reported from Germany, and by the fact that apparently the announcement of a German-Soviet Agreement is taken in some quarters in Berlin to indicate that intervention by Great Britain on behalf of Poland is no longer a contingency that need be reckoned with. No greater mistake could be made. Whatever may prove to be the nature of the German-Soviet Agreement, it cannot alter Great Britain's obligation to Poland, which His Majesty's Government have stated in public repeatedly and plainly, and which they are determined to fulfil.

It has been alleged that if His Majesty's Government had made their position more clear in 1914 the great catastrophe would have been avoided. Whether or not there is any force in that allegation, His Majesty's Government are resolved that on this occasion there shall be no such tragic misunderstanding. If the need should arise, they are resolved and prepared to employ without delay all the forces at their command, and it is impossible to foresee the end of hostilities once engaged. It would be a dangerous delusion to think that, if war once starts, it will come to an early end, even if a success on any one of the several fronts on which it will be engaged should have been secured.

At this time I confess I can see no other way to avoid a catastrophe that will involve Europe in war. In view of the grave consequences to humanity which may follow from the action of their rulers, I trust that Your Excellency will weigh with the utmost deliberation the considerations which I have put before you.*

Hitler's reply, after dwelling on the "unparalleled magnanimity" with which Germany was prepared to settle the question of Danzig and the Corridor, contained the following piece of lying effrontery:

The unconditional assurance given by England to Poland that she would render assistance to that country in all circumstances, regardless of the causes from which a conflict might spring, could only be interpreted in that country as an encouragement henceforward to unloose, under cover of such a charter, a wave of appalling terrorism against the million and a half German inhabitants living in Poland.†

On August 25 the British Government proclaimed a formal treaty with Poland, confirming the guarantees already given. It was hoped by this step to give the best chance of a settlement by direct negotiation between Germany and Poland in the face of the fact that if this failed Britain would stand by Poland. Said Goering at Nuremberg:

On the day when England gave her official guarantee to Poland the Fuehrer called me on the telephone and told me that he had stopped the planned invasion of Poland. I asked him then whether this was just temporary or for good. He said, "No, I shall have to see whether we can eliminate British intervention."‡

In fact Hitler postponed D-Day from August 25 to September 1, and entered into direct negotiation with Poland, as Chamberlain desired. His object was not however to reach an agreement with Poland, but to give His Majesty's Government every opportunity to escape from their guarantee. Their thoughts, like those of Parliament and the nation, were upon a different plane. It is a curious fact about the British Islanders, who hate drill and have not been invaded for nearly a thousand years, that as danger comes nearer and grows they become progressively less nervous; when it is imminent they are fierce; when it is mortal they are fearless. These habits have led them into some very narrow escapes.

*　　*　　*　　*　　*

* *Nuremberg Documents*, Pt. II, pp. 157-8.
† *Ibid.*, p. 158.
‡ *Ibid.*, p. 166.

A letter from Hitler to Mussolini at this time has recently been published in Italy:

Duce,

For some time Germany and Russia have been meditating upon the possibility of placing their mutual political relations upon a new basis. The need to arrive at concrete results in this sense has been strengthened by:

1. The condition of the world political situation in general.

2. The continued procrastination of the Japanese Cabinet in taking up a clear stand. Japan was ready for an alliance against Russia in which Germany—and in my view Italy—could only be interested in the present circumstances as a secondary consideration. She was not agreeable however to assuming any clear obligations regarding England—a decisive question from the German side, and I think also from Italy's. . . .

3. The relations between Germany and Poland have been unsatisfactory since the spring, and in recent weeks have become simply intolerable, not through the fault of the Reich, but principally because of British action. . . . These reasons have induced me to hasten on a conclusion of the Russian-German talks. I have not yet informed you, Duce, in detail on this question. But now in recent weeks the disposition of the Kremlin to engage in an exchange of relations with Germany—a disposition produced from the moment of the dismissal of Litvinov—has been increasingly marked, and has now made it possible for me, after having reached a preliminary clarification, to send my Foreign Minister to Moscow to draw up a treaty which is far and away the most extensive non-aggression pact in existence to-day, and the text of which will be made public. The pact is unconditional, and establishes in addition the commitment to consult on all questions which interest Germany and Russia. I can also inform you, Duce, that, given these undertakings, the benevolent attitude of Russia is assured, and *that above all there now exists no longer the possibility of any attack whatsoever on the part of Roumania in the event of a conflict.**

To this Mussolini sent an immediate answer:

I am replying to your letter, which has just been delivered to me by Ambassador Mackensen.

1. As far as the agreement with Russia is concerned, I completely approve.

2. I feel it would be useful to avoid a rupture or coolness with Japan and her consequent drawing together with the group of democratic States. . . .

* *Hitler e Mussolini, Lettere e Documenti*, p. 7.

3. The Moscow Pact blocks Roumania, and may change the position of Turkey, who has accepted an English loan but who has not yet signed the alliance. A new attitude on the part of Turkey would upset the strategic disposition of the French and English in the Eastern Mediterranean.

4. About Poland, I understand completely the German position and the fact that such a tense situation cannot continue indefinitely.

5. Regarding the practical attitude of Italy in the event of military action my point of view is the following:

If Germany attacks Poland and the conflict is localised Italy will give Germany every form of political and economic aid which may be required.

If Germany attacks Poland and the allies of the latter counter-attack Germany I must emphasise to you that I cannot assume the initiative of warlike operations, given the actual conditions of Italian military preparations, which have been repeatedly and in timely fashion pointed out to you, Fuehrer, and to von Ribbentrop.

Our intervention could however be immediate if Germany were to give us at once the munitions and raw materials to sustain the shock which the French and British would probably inflict upon us. In our previous meetings war was envisaged after 1942, and on this date I should have been ready on land, by sea, and in the air, according to our agreed plans.*

From this point Hitler knew, if he had not divined it already, that he could not count upon the armed intervention of Italy if war came. Any last-minute attempts by Mussolini to repeat his performance of Munich were brushed aside. It seems to have been from English rather than from German sources that the Duce learnt of the final moves. Ciano records in his diary on August 27, "The English communicate to us the text of the German proposals to London, about which we are kept entirely in the dark".† Mussolini's only need now was Hitler's acquiescence in Italy's neutrality. This was accorded to him.

* * * * *

On August 31 Hitler issued his "Directive Number 1 for the Conduct of the War".

1. Now that all the political possibilities of disposing by peaceful means of a situation on the Eastern frontier which is intolerable for Germany are exhausted, I have determined on a solution by force.

* *Hitler e Mussolini, Lettere e Documenti,* p. 10.
† *Ciano's Diary,* p. 136.

2. The attack on Poland is to be carried out in accordance with the preparation made for "Fall Weiss" [Case White], with the alterations which result, where the Army is concerned, from the fact that it has in the meantime almost completed its dispositions. Allotment of tasks and the operational targets remain unchanged.

The date of attack—September 1, 1939. Time of attack—04.45 [inserted in red pencil].

3. In the West it is important that the responsibility for the opening of hostilities should rest unequivocally with England and France. At first purely local action should be taken against insignificant frontier violations.*

* * * * *

On my return from the Rhine front I passed some sunshine days at Madame Balsan's place, with a pleasant but deeply anxious company, in the old château where King Henry of Navarre had slept the night before the Battle of Ivry. Mrs. Euan Wallace and her sons were with us. Her husband was a Cabinet Minister. She was expecting him to join her. Presently he telegraphed he could not come, and would explain later why. Other signs of danger drifted in upon us. One could feel the deep apprehension brooding over all, and even the light of this lovely valley of the Eure seemed robbed of its genial ray. I found painting hard work in this uncertainty. On August 26 I decided to go home, where at least I could find out what was going on. I told my wife I would send her word in good time. On my way through Paris I gave General Georges luncheon. He produced all the figures of the French and German Armies, and classified the divisions in quality. The result impressed me so much that for the first time I said, "But you are the masters." He replied, "The Germans have a very strong Army, and we shall never be allowed to strike first. If they attack, both our countries will rally to their duty."

That night I slept at Chartwell, where I had asked General Ironside to stay with me next day. He had just returned from Poland, and the reports he gave of the Polish Army were most favourable. He had seen a divisional attack exercise under a live barrage, not without casualties. Polish morale was high. He stayed three days with me, and we tried hard to measure the un-knowable. Also at this time I completed bricklaying the kitchen

---

* _Nuremberg Documents_, Pt. II, p. 172.

of the cottage which during the year past I had prepared for our family home in the years which were to come. My wife, on my signal, came over *via* Dunkirk on August 30.

* * * * *

There were known to be twenty thousand organised German Nazis in England at this time, and it would only have been in accord with their procedure in other friendly countries that the outbreak of war should be preceded by a sharp prelude of sabotage and murder. I had at that time no official protection, and I did not wish to ask for any; but I thought myself sufficiently prominent to take precautions. I had enough information to convince me that Hitler recognised me as a foe. My former Scotland Yard detective, Inspector Thompson, was in retirement. I told him to come along and bring his pistol with him. I got out my own weapons, which were good. While one slept the other watched. Thus nobody would have had a walk-over. In these hours I knew that if war came—and who could doubt its coming?—a major burden would fall upon me.

BOOK II

THE TWILIGHT WAR

September 3, 1939 – May 10, 1940

CHAPTER XXII

WAR

Mr. Chamberlain's Invitation – The Pause of September 2 – War Declared, September 3 – The First Air Alarm – At the Admiralty Once More – Admiral Sir Dudley Pound – My Knowledge of Naval Matters – Contrast between 1914 and 1939 – The Naval Strategic Situation – The Baltic – The Kiel Canal – The Attitude of Italy– Our Mediterranean Strategy – The Submarine Menace – The Air Menace – The Attitude of Japan – Singapore – The Security of Australia and New Zealand – Composition of the War Cabinet – Mr. Chamberlain's First Selections – An Antediluvian – The Virtues of Siesta.

POLAND was attacked by Germany at dawn on September 1. The mobilisation of all our forces was ordered during the morning. The Prime Minister asked me to visit him in the afternoon at Downing Street. He told me that he saw no hope of averting a war with Germany, and that he proposed to form a small War Cabinet of Ministers without departments to conduct it. He mentioned that the Labour Party were not, he understood, willing to share in a national coalition. He still had hopes that the Liberals would join him. He invited me to become a member of the War Cabinet. I agreed to his proposal without comment, and on this basis we had a long talk on men and measures.

After some reflection, I felt that the average age of the Ministers who were to form the supreme executive of war direction would be thought too high, and I wrote to Mr. Chamberlain after midnight accordingly:

2.9.39

Aren't we a very old team? I make out that the six you mentioned to me yesterday aggregate 386 years, or an average of over sixty-four! Only one year short of the Old Age Pension! If however you added Sinclair (49) and Eden (42) the average comes down to 57½.

If the *Daily Herald* is right that Labour will not come in, we shall certainly have to face a constant stream of criticism, as well as the many disappointments and surprises of which war largely consists. Therefore it seems to me all the more important to have the Liberal Opposition firmly incorporated in our ranks. Eden's influence with the section of Conservatives who are associated with him, as well as with moderate Liberal elements, also seems to me to be a very necessary reinforcement.

The Poles have now been under heavy attack for thirty hours, and I am much concerned to hear that there is talk in Paris of a further note. I trust you will be able to announce our Joint Declaration of War at *latest* when Parliament meets this afternoon.

The *Bremen* will soon be out of the interception zone unless the Admiralty take special measures and the signal is given to-day. This is only a minor point, but it may well be vexatious.

I remain here at your disposal.*

I was surprised to hear nothing from Mr. Chamberlain during the whole of September 2, which was a day of intense crisis. I thought it probable that a last-minute effort was being made to preserve peace; and this proved true. However, when Parliament met in the evening a short but very fierce debate occurred, in which the Prime Minister's temporising statement was ill-received by the House. When Mr. Greenwood rose to speak on behalf of the Labour Opposition Mr. Amery from the Conservative benches cried out to him, "Speak for England." This was received with loud cheers. There was no doubt that the temper of the House was for war. I even deemed it more resolute and united than in the similar scene on August 3, 1914, in which I had also taken part. In the evening a number of gentlemen of importance in all parties called upon me at my flat opposite the Westminster Cathedral, and all expressed deep anxiety lest we should fail in our obligations to Poland. The House was to meet again at noon the next day. I wrote that night as follows to the Prime Minister:

2.9.39

I have not heard anything from you since our talks on Friday, when I understood that I was to serve as your colleague, and when you told me that this would be announced speedily. I really do not know what has happened during the course of this agitated day; though it seems to me that entirely different ideas have ruled from those which you

* Printed in Feiling, *op. cit.*, p. 420.

expressed to me when you said "the die was cast": I quite realise that in contact with this tremendous European situation changes of method may become necessary, but I feel entitled to ask you to let me know how we stand, both publicly and privately, before the debate opens at noon.

It seems to me that if the Labour Party, and as I gather the Liberal Party, are estranged, it will be difficult to form an effective War Government on the limited basis you mentioned. I consider that a further effort should be made to bring in the Liberals, and in addition that the composition and scope of the War Cabinet you discussed with me requires review. There was a feeling to-night in the House that injury had been done to the spirit of national unity by the apparent weakening of our resolve. I do not underrate the difficulties you have with the French; but I trust that we shall now take our decision independently, and thus give our French friends any lead that may be necessary. In order to do this we shall need the strongest and most integral combination that can be formed. I therefore ask that there should be no announcement of the composition of the War Cabinet until we have had a further talk.

As I wrote to you yesterday morning, I hold myself entirely at your disposal, with every desire to aid you in your task.

I learnt later that a British ultimatum had been given to Germany at 9.30 p.m. on September 1, and that this had been followed by a second and final ultimatum at 9 a.m. on September 3. The early broadcast of the 3rd announced that the Prime Minister would speak on the radio at 11.15 a.m. As it now seemed certain that war would be immediately declared by Great Britain and also by France, I prepared a short speech which I thought would be becoming to the solemn and awful moment in our lives and history.

The Prime Minister's broadcast informed us that we were already at war, and he had scarcely ceased speaking when a strange, prolonged, wailing noise, afterwards to become familiar, broke upon the ear. My wife came into the room braced by the crisis and commented favourably upon the German promptitude and precision, and we went up to the flat top of the house to see what was going on. Around us on every side, in the clear, cool September light, rose the roofs and spires of London. Above them were already slowly rising thirty or forty cylindrical balloons. We gave the Government a good mark for this evident sign of preparation, and as the quarter of an hour's notice which we had been led to expect we should receive was now running out we made our way to the shelter assigned to us, armed with a bottle of brandy and other appropriate medical comforts.

Our shelter was a hundred yards down the street, and consisted merely of an open basement, not even sand-bagged, in which the tenants of half a dozen flats were already assembled. Everyone was cheerful and jocular, as is the English manner when about to encounter the unknown. As I gazed from the doorway along the empty street and at the crowded room below my imagination drew pictures of ruin and carnage and vast explosions shaking the ground; of buildings clattering down in dust and rubble, of fire-brigades and ambulances scurrying through the smoke, beneath the drone of hostile aeroplanes. For had we not all been taught how terrible air raids would be? The Air Ministry had, in natural self-importance, greatly exaggerated their power. The pacifists had sought to play on public fears, and those of us who had so long pressed for preparation and a superior Air Force, while not accepting the most lurid forecasts, had been content that they should act as a spur. I knew that the Government were prepared, in the first few days of the war, with over 250,000 beds for air-raid casualties. Here at least there had been no under-estimation. Now we should see what were the facts.

After about ten minutes had passed the wailing broke out again. I was myself not sure that this was not a reiteration of the previous warning, but a man came running along the street shouting "All clear", and we dispersed to our dwellings and went about our business. Mine was to go to the House of Commons, which duly met at noon with its unhurried procedure and brief, stately prayers. There I received a note from the Prime Minister asking me to come to his room as soon as the debate died down. As I sat in my place, listening to the speeches, a very strong sense of calm came over me, after the intense passions and excitements of the last few days. I felt a serenity of mind and was conscious of a kind of uplifted detachment from human and personal affairs. The glory of Old England, peace-loving and ill-prepared as she was, but instant and fearless at the call of honour, thrilled my being and seemed to lift our fate to those spheres far removed from earthly facts and physical sensation. I tried to convey some of this mood to the House when I spoke, not without acceptance.

Mr. Chamberlain told me that he had considered my letters,

that the Liberals would not join the Government, that he was able to meet my views about the average age to some extent by bringing the three Service Ministers into the War Cabinet in spite of their executive functions, and that this would reduce the average age to less than sixty. This, he said, made it possible for him to offer me the Admiralty as well as a seat in the War Cabinet. I was very glad of this, because, though I had not raised the point, I naturally preferred a definite task to that exalted brooding over the work done by others which may well be the lot of a Minister, however influential, who has no department. It is easier to give directions than advice, and more agreeable to have the right to act, even in a limited sphere, than the privilege to talk at large. Had the Prime Minister in the first instance given me the choice between the War Cabinet and the Admiralty, I should of course have chosen the Admiralty. Now I was to have both.

Nothing had been said about when I should formally receive my office from the King, and in fact I did not kiss hands till the 5th. But the opening hours of war may be vital with navies. I therefore sent word to the Admiralty that I would take charge forthwith and arrive at 6 o'clock. On this the Board were kind enough to signal to the Fleet, "Winston is back." So it was that I came again to the room I had quitted in pain and sorrow almost exactly a quarter of a century before, when Lord Fisher's resignation had led to my removal from my post as First Lord and ruined irretrievably, as it proved, the important conception of forcing the Dardanelles. A few feet behind me, as I sat in my old chair, was the wooden map-case I had had fixed in 1911, and inside it still remained the chart of the North Sea on which each day, in order to focus attention on the supreme objective, I had made the Naval Intelligence Branch record the movements and dispositions of the German High Seas Fleet. Since 1911 much more than a quarter of a century had passed, and still mortal peril threatened us at the hands of the same nation. Once again defence of the rights of a weak State, outraged and invaded by unprovoked aggression, forced us to draw the sword. Once again we must fight for life and honour against all the might and fury of the valiant, disciplined, and ruthless German race. Once again! So be it.

* * * * *

Presently the First Sea Lord came to see me. I had known Dudley Pound slightly in my previous tenure of the Admiralty as one of Lord Fisher's trusted Staff officers. I had strongly condemned in Parliament the dispositions of the Mediterranean Fleet when he commanded it, at the moment of the Italian descent upon Albania. Now we met as colleagues upon whose intimate relations and fundamental agreement the smooth working of the vast Admiralty machine would depend. We eyed each other amicably if doubtfully. But from the earliest days our friendship and mutual confidence grew and ripened. I measured and respected the great professional and personal qualities of Admiral Pound. As the war, with all its shifts and fortunes, beat upon us with clanging blows we became ever truer comrades and friends. And when, four years later, he died at the moment of the general victory over Italy I mourned with a personal pang for all the Navy and the nation had lost.

I spent a good part of the night of the 3rd meeting the Sea Lords and heads of the various departments, and from the morning of the 4th I laid my hands upon the naval affair. As in 1914, precautionary measures against surprise had been taken in advance of general mobilisation. As early as June 15 large numbers of officers and men of the Reserves had been called up. The Reserve fleet, fully manned for exercises, had been inspected by the King on August 9, and on the 22nd various additional classes of reservists had been summoned. On the 24th an Emergency Powers Defence Bill was passed through Parliament, and at the same time the Fleet was ordered to its war stations; in fact, our main forces had been at Scapa Flow for some weeks. After the general mobilisation of the Fleet had been authorised the Admiralty war plan had unfolded smoothly, and in spite of certain serious deficiencies, notably in cruisers and anti-submarine vessels, the challenge, as in 1914, found the Fleet equal to the immense tasks before it.

* * * * *

I had, as the reader may be aware, a considerable knowledge of the Admiralty and of the Royal Navy. The four years from 1911 to 1915, when I had the duty of preparing the Fleet for war and the task of directing the Admiralty during the first ten critical months, had been the most vivid of my life. I had amassed an immense amount of detailed information and had learned many lessons about the Fleet and war at sea. In the interval I had studied and written much about naval affairs. I had spoken repeatedly upon them in the House of Commons. I had always preserved a close contact with the Admiralty, and, although their foremost critic in these years, I had been made privy to many of their secrets. My four years' work on the Air Defence Research Committee had given me access to all the most modern developments in Radar, which now vitally affected the naval service. I have mentioned how in June 1938 Lord Chatfield, the First Sea Lord, had himself shown me over the Anti-Submarine School at Portland, and how we had gone to sea in destroyers on an exercise in submarine detection by the use of the Asdic apparatus. My intimacy with the late Admiral Henderson, Controller of the Navy till 1938, and the discussions which the First Lord of those days had encouraged me to have with Lord Chatfield upon the design of new battleships and cruisers, gave me a full view over the sphere of new construction. I was of course familiar from the published records with the strength, composition, and structure of the Fleet, actual and prospective, and with those of the German, Italian, and Japanese Navies.

As a critic and a spur, my public speeches had naturally dwelt upon weaknesses and shortcomings, and, taken by themselves, had by no means portrayed either the vast strength of the Royal Navy or my own confidence in it. It would be unjust to the Chamberlain Administration and their Service advisers to suggest that the Navy had not been adequately prepared for a war with Germany, or with Germany and Italy. The effective defence of Australasia and India in the face of a simultaneous attack by Japan raised more serious difficulties; but such an assault—which was at the moment unlikely—might well have involved the United States. I therefore felt, when I entered upon my duties, that I had at my disposal what was undoubtedly the finest-tempered instrument of naval war in the world, and I was sure that time would be granted to make good the oversights of peace and to cope with the equally certain unpleasant surprises of war.

*   *   *   *   *

The tremendous naval situation of 1914 in no way repeated itself. Then we had entered the war with a ratio of sixteen to ten in capital ships and two to one in cruisers. In those days we had mobilised eight battle squadrons of eight battleships, with a cruiser squadron and a flotilla assigned to each, together with important detached cruiser forces, and I looked forward to a general action with a weaker but still formidable fleet. Now the German Navy had only begun their rebuilding and had no power even to form a line of battle. Their two great battleships, *Bismarck* and *Tirpitz*, both of which, it must be assumed, had transgressed the agreed Treaty limits in tonnage, were at least a year from completion. The light battle-cruisers, *Scharnhorst* and *Gneisenau*, which had been fraudulently increased by the Germans from 10,000 tons to 26,000 tons, had been completed in 1938. Besides this Germany had available the three "pocket-battleships" of 10,000 tons, *Admiral Graf Spee*, *Admiral Scheer*, and *Deutschland*, together with two fast 8-inch-gun cruisers of 10,000 tons, six light cruisers, and sixty destroyers and smaller vessels. Thus there was no challenge in surface craft to our command of the seas. There was no doubt that the British Navy was overwhelmingly superior to the German in strength and in numbers, and no reason to assume that its science training or skill was in any way defective. Apart from the shortage of cruisers and destroyers, the Fleet had been maintained at its customary high standard. It had to face enormous and innumerable duties, rather than an antagonist.

*   *   *   *   *

My views on the naval strategic situation were already largely formed when I went to the Admiralty. The command of the Baltic was vital to the enemy. Scandinavian supplies, Swedish ore, and above all protection against Russian descents on the long, undefended northern coastline of Germany—in one place little more than a hundred miles from Berlin—made it imperative for the German Navy to dominate the Baltic. I was therefore sure that in this opening phase Germany would not compromise her command of that sea. Thus, while submarines and raiding cruisers, or perhaps one pocket-battleship, might be sent out to disturb our traffic, no ships would be risked which were necessary to the Baltic control. The German Fleet, as at this moment developed, must aim at this as its prime and almost its sole objective. For the main purposes of sea-power and for the enforcement of our principal naval offensive measure, the blockade, we must of course maintain a superior fleet in our northern waters; but no very large British naval forces were, it seemed, needed to watch the debouches from the Baltic or from the Heligoland Bight.

British security would be markedly increased if an air attack upon the Kiel Canal rendered that side-door from the Baltic useless, even if only at intervals.

A year before I had sent a note upon this special operation to Sir Thomas Inskip.

October 29, 1938

In a war with Germany the severance of the Kiel Canal would be an achievement of the first importance. I do not elaborate this, as I assume it to be admitted. Plans should be made to do this, and, if need be, all the details should be worked out in their variants by a special technical committee. Owing to there being few locks, and no marked difference of sea-level at the two ends of the canal, its interruption by H.E. bombs, even of the heaviest type, could swiftly be repaired. If however many bombs of medium size fitted with time fuzes, some set for a day, others for a week, and others for a month, etc., could be dropped in the canal, their explosions at uncertain intervals and in uncertain places would close the canal to the movement of warships or valuable vessels until the whole bottom had been deeply dredged. Alternatively, *special fuzes with magnetic actuation* should be considered.

The phrase about magnetic mines is interesting in view of what was soon to come upon us. No special action had however been taken.

*   *   *   *   *

The British merchant fleet on the outbreak of war was about the same size as in 1914. It was over twenty-one million tons. The average size of the ships had increased, and thus there were fewer. This tonnage was not however all available for trade. The Navy required auxiliary warships of various types which must be drawn chiefly from the highest class of liners. All the defence Services needed ships for special purposes: the Army and R.A.F. for the movement of troops and equipment overseas, and the Navy for all the work at fleet bases and elsewhere, and particularly for providing oil fuel at strategic points all over the world. Demands for tonnage for all these objects amounted to nearly three million tons, and to these must be added the shipping requirements of the Empire overseas. At the end of 1939, after balancing gains and losses, the total British tonnage available for commercial use was about fifteen and a half million tons.

* * * * *

Italy had not declared war, and it was already clear that Mussolini was waiting upon events. In this uncertainty and as a measure of precaution till all our arrangements were complete we thought it best to divert our shipping round the Cape. We had however already on our side, in addition to our own preponderance over Germany and Italy combined, the powerful fleet of France, which by the remarkable capacity and long administration of Admiral Darlan had been brought to the highest strength and degree of efficiency ever attained by the French Navy since the days of the Monarchy. Should Italy become hostile our first battlefield must be the Mediterranean. I was entirely opposed, except as a temporary convenience, to all plans for quitting the centre and merely sealing up the ends of the great inland sea. Our forces alone, even without the aid of the French Navy and its fortified harbours, were sufficient to drive the Italian ships from the sea, and should secure complete naval command of the Mediterranean within two months, and possibly sooner.

The British domination of the Mediterranean would inflict injuries upon an enemy Italy which might be fatal to her power of continuing the war. All her troops in Libya and in Abyssinia would be cut flowers in a vase. The French and our own people in Egypt could be reinforced to any extent desired, while theirs would be overweighted, if not starved. Not to hold the Central Mediterranean would be to expose Egypt and the Canal, as well as the French possessions, to invasion by Italian troops with German leadership. Moreover, a series of swift and striking victories in this theatre, which might be obtainable in the early weeks of a war, would have a most healthy and helpful bearing upon the main struggle with Germany. Nothing should stand between us and these results, both naval and military.

* * * * *

I had accepted too readily when out of office the Admiralty view of the extent to which the submarine had been mastered. Whilst the technical efficiency of the Asdic apparatus was proved in many early encounters with U-boats, our anti-U-boat resources were far too limited to prevent our suffering serious losses. My opinion recorded at the time, "The submarine should be quite controllable in the outer seas, and certainly in the Mediterranean. There will be losses, but nothing to affect the scale of events," was not incorrect. Nothing of major importance occurred in the first year of the U-boat warfare. The Battle of the Atlantic was reserved for 1941 and 1942.

In common with prevailing Admiralty belief before the war, I did not sufficiently measure the danger to, or the consequent deterrent upon, British warships from air attack. "In my opinion," I had written a few months before the war, "given with great humility (because these things are very difficult to judge), an air attack upon British warships, armed and protected as they now are, will not prevent full exercise of their superior sea-power." However, the deterrents, albeit exaggerated, upon our mobility soon become grave. The air almost immediately proved itself a formidable menace, especially in the Mediterranean. Malta, with its almost negligible air defences, presented a problem for which there was no immediate solution. On the other hand, in the first year no British capital ship was sunk by air attack.

★　　★　　★　　★　　★

There was no sign at this moment of any hostile action or intent upon the part of Japan. The main preoccupation of Japan was naturally America. It did not seem possible to me that the United States could sit passive and watch a general assault by Japan upon all European establishments in the Far East, even if they themselves were not for the moment involved. In this case we should gain far more from the entry of the United States, perhaps only against Japan, if that were possible, than we should suffer from the hostility of Japan, vexatious though that would be. On no account must anything which threatened in the Far East divert us from our prime objectives in Europe. We could not protect our interests and possessions in the Yellow Sea from Japanese attack. The farthest point we could defend if Japan came in would be the fortress of Singapore. Singapore must hold out until the Mediterranean was safe and the Italian Fleet liquidated.

I did not fear at the moment of the outbreak that Japan would send a fleet and army to conquer Singapore, provided that fortress were adequately garrisoned and supplied with food and ammunition for at least six months. Singapore was as far from Japan as

Southampton from New York. Over these three thousand miles of salt water Japan would have to send the bulk of her Fleet, escort at least sixty thousand men in transports in order to effect a landing, and begin a siege which would end only in disaster if the Japanese sea-communications were cut at any stage. These views of course ceased to apply once the Japanese had occupied Indo-China and Siam and had built up a powerful army and very heavy air forces only three hundred miles away across the Gulf of Siam. This however did not occur for more than a year and a half.

As long as the British Navy was undefeated, and as long as we held Singapore, no invasion of Australia or New Zealand by Japan was deemed possible. We could give Australasia a good guarantee to protect them from this danger, but we must do it in our own way, and in the proper sequence of operations. It seemed unlikely that a hostile Japan, exulting in the mastery of the Yellow Sea, would send afloat a conquering and colonising expedition to Australia. A large and well-equipped army would be needed for a long time to make any impression upon Australian manhood. Such an undertaking would require the improvident diversion of the Japanese Fleet, and its engagement in a long, desultory struggle in Australia. At any moment a decision in the Mediterranean would liberate very powerful naval forces to cut invaders from their base. It would be easy for the United States to tell Japan that they would regard the sending of Japanese fleets and transports south of the equator as an act of war. They might well be disposed to make such a declaration, and there would be no harm in sounding them upon this very remote contingency.

The actual strength of the British and German Fleets, built and building, on the night of September 3, 1939, and that of the American, French, Italian, and Japanese Fleets on the same basis, is set forth in Appendix F. It was my recorded conviction that *in the first year of a world war* Australia and New Zealand would be in no danger whatever in their homeland, and by the end of the first year we might hope to have cleaned up the seas and oceans. As a forecast of *the first year of the naval war* these thoughts proved true. I shall in their proper place recount the tremendous events which occurred in 1941 and 1942 in the Far East.

* * * * *

Newspaper opinion, headed by the *Times*, favoured the principle of a War Cabinet of not more than five or six Ministers, all of whom should be free from departmental duties. Thus alone, it was argued, could a broad and concerted view be taken upon war policy, especially in its larger aspects. Put shortly, "Five men with nothing to do but to run the war" was deemed the ideal. There are however many practical objections to such a course. A group of detached statesmen, however high their nominal authority, are at a serious disadvantage in dealing with the Ministers at the head of the great departments vitally concerned. This is especially true of the Service departments. The War Cabinet personages can have no direct responsibility for day-to-day events. They may take major decisions, they may advise in general terms beforehand or criticise afterwards, but they are no match, for instance, for a First Lord of the Admiralty or a Secretary of State for War or Air, who, knowing every detail of the subject and supported by his professional colleagues, bears the burden of action. United, there is little they cannot settle, but usually there are several opinions among them. Words and arguments are interminable, and meanwhile the torrent of war takes its headlong course. The War Cabinet Ministers themselves would naturally be diffident of challenging the responsible Minister, armed with all his facts and figures. They feel a compunction in adding to the strain upon those actually in executive control. They tend therefore to become more and more theoretical supervisors and commentators, reading an immense amount of material every day, but doubtful how to use their knowledge without doing more harm than good. Often they can do little more than arbitrate or find a compromise in inter-departmental disputes. It is therefore necessary that the Ministers in charge of the Foreign Office and the fighting departments should be integral members of the supreme body. Usually some at least of the "Big Five" are chosen for their political influence, rather than for their knowledge of and aptitude for warlike operations. The numbers therefore begin to grow far beyond the limited circle originally conceived. Of course, where the Prime Minister himself becomes Minister of Defence a strong compression is obtained. Personally, when I was placed in charge I did not like having unharnessed Ministers around me. I preferred to deal with chiefs of organisations rather than counsellors. Everyone should do a good day's work and be accountable for some definite task, and then they do not make trouble for trouble's sake or to cut a figure.

Mr. Chamberlain's original War Cabinet plan was almost immediately expanded, by the force of circumstances, to include Lord Halifax, Foreign Secretary; Sir Samuel Hoare, Lord Privy Seal; Sir John Simon, Chancellor of the Exchequer; Lord Chatfield, Minister for the Co-ordination of Defence; and Lord Hankey, Minister without Portfolio. To these were added the Service Ministers, of whom I was now one, with Mr. Hore Belisha, Secretary of State for War, and Sir Kingsley Wood, Secretary of State for Air. In addition it was necessary that the Dominions Secretary, Mr. Eden, and Sir John Anderson as Home Secretary and Minister of Home Security, though not actual members of the War Cabinet, should be present on all occasions. Thus our total was eleven. The decision to bring in the three Service Ministers profoundly affected Lord Chatfield's authority as Minister for the Co-ordination of Defence. He accepted the position with his customary good-nature.

Apart from myself all the other Ministers had directed our affairs for a good many recent years or were involved in the situation we now had to face both in diplomacy and war. Mr. Eden had resigned on foreign policy in February 1938. I had not held public office for nearly eleven years. I had therefore no responsibility for the past or for any want of preparation now apparent. On the contrary, I had for the last six or seven years been a continual prophet of evils which had now in large measure come to pass. Thus, armed as I now was with the mighty machine of the Navy, on which fell in this phase the sole burden of active fighting, I did not feel myself at any disadvantage, and had I done so it would have been removed by the courtesy and loyalty of the Prime Minister and his colleagues. All these men I knew very well. Most of us had served together for five years in Mr. Baldwin's Cabinet, and we had of course been constantly in contact, friendly or controversial, through the changing scenes of Parliamentary life. Sir John Simon and I however represented an older political generation. I had served, off and on, in British Governments for fifteen years, and he for almost as long, before any of the others had gained public office. I had been at the head of the Admiralty or Ministry of Munitions through the stresses of the First World War. Although the Prime Minister was my senior by some years in age, I was almost the only antediluvian. This might well have been a matter of reproach in a time of crisis, when it was natural and popular to demand the force of young men and new ideas. I saw therefore that I should have to strive my utmost to keep pace with the generation now in power and with fresh young giants who might at any time appear. In this I relied upon knowledge as well as upon all possible zeal and mental energy.

For this purpose I had recourse to a method of life which had been forced upon me at the Admiralty in 1914 and 1915, and which I found greatly extended my daily capacity for work. I always went to bed at least for one hour as early as possible in the afternoon, and exploited to the full my happy gift of falling almost immediately into deep sleep. By this means I was able to press a day and a half's work into one. Nature had not intended mankind to work from eight in the morning until midnight without that refreshment of blessed oblivion which, even if it only lasts twenty minutes, is sufficient to renew all the vital forces. I regretted having to send myself to bed like a child every after-noon, but I was rewarded by being able to work through the night until two or even later—sometimes much later—in the morning, and begin the new day between eight and nine o'clock. This routine I observed throughout the war, and I commend it to others if and when they find it necessary for a long spell to get the last scrap out of the human structure. The First Sea Lord, Ad-miral Pound, as soon as he had realised my technique, adopted it himself, except that he did not actually go to bed, but dozed off in his arm-chair. He even carried the policy so far as often to go to sleep during the Cabinet meetings. One word about the Navy was however sufficient to awaken him to the fullest activity. Nothing slipped past his vigilant ear, or his comprehending mind.

CHAPTER XXIII

THE ADMIRALTY TASK

Sea War Alone – The Admiralty War Plan – The U-boat Attack –
The Asdic Trawlers – Control of Merchant Shipping – The Convoy
System – Blockade – Record of My First Conference – Need of the
Southern Irish Ports – The Main Fleet Base – Inadequate Precautions
– "Hide-and-Seek" – My Visit to Scapa Flow – Reflections at Loch
Ewe – Loss of the "Courageous" – Cruiser Policy – The First Month
of the U-Boat War – A Fruitful September – Wider Naval Operations
– Ardour of the Polish Navy – President Roosevelt's Letter.

ASTONISHMENT was world-wide when Hitler's crashing onslaught upon Poland and the declarations of war upon Germany by Britain and France were followed only by a prolonged and oppressive pause. Mr. Chamberlain in a private letter published by his biographer described this phase as "Twilight War";* and I find the expression so just and expressive that I have adopted it as the title for this Book. The French armies made no attack upon Germany. Their mobilisation completed, they remained in contact motionless along the whole front. No air action, except reconnaissance, was taken against Britain; nor was any air attack made upon France by the Germans. The French Government requested us to abstain from air attack on Germany, stating that it would provoke retaliation upon their war factories, which were unprotected. We contented ourselves with dropping pamphlets to rouse the Germans to a higher morality. This strange phase of the war on land and in the air astounded everyone. France and Britain remained impassive while Poland was in a few weeks destroyed or subjugated by the whole might of the German war machine. Hitler had no reason to complain of this.

The war at sea, on the contrary, began from the first hour with

* Feiling, *op. cit.*, p. 424.

full intensity, and the Admiralty therefore became the active centre of events. On September 3 all our ships were sailing about the world on their normal business. Suddenly they were set upon by U-boats carefully posted beforehand, especially in the Western Approaches. At 9 p.m. that very night the outward-bound passenger liner *Athenia*, of 13,500 tons, was torpedoed, and foundered with a loss of 112 lives, including twenty-eight American citizens. This outrage broke upon the world within a few hours. The German Government, to prevent any misunderstanding in the United States, immediately issued a statement that I personally had ordered a bomb to be placed on board this vessel in order by its destruction to prejudice German-American relations. This falsehood received some credence in unfriendly quarters.* On the 5th and 6th the *Bosnia*, *Royal Sceptre*, and *Rio Claro* were sunk off the coast of Spain. All these were important vessels.

My first Admiralty minute was concerned with the probable scale of the U-boat menace in the immediate future:

Director of Naval Intelligence   4.IX.39

Let me have a statement of the German U-boat forces, actual and prospective, for the next few months. Please distinguish between ocean-going and small-size U-boats. Give the estimated radius of action in days and miles in each case.

I was at once informed that the enemy had sixty U-boats and that a hundred would be ready early in 1940. A detailed answer was returned on the 5th, which should be studied.† The numbers

---

* See also *Nuremberg Documents*, Pt. IV, p. 267 ff.—the confession of the U-boat captain.

† GERMAN SUBMARINES

Type	Tonnage	Numbers in Service August 1939	Numbers expected to be in Service December 1939	Numbers expected to be in Service by early 1940	Estimated Radius of Action	
					Miles	Days
Coastal ..	250	30	32	32	4,000	33 at 5 knots
Ocean ..	500	10	10	23	} 7,200	30 at 10 knots
Ocean ..	517	9	15	17		
Ocean ..	712	2	2	..	} 8,400	35 at 10 knots
Ocean ..	740	8	13	16		
Ocean ..	1,060	..	2	11	10,000	42 at 10 knots
Ocean ..	1,028	1 (Built for Turkey, not delivered)			8,000	33 at 10 knots
Grand totals ..	..	60	74	99	..	..

of long-range endurance vessels were formidable, and revealed the intentions of the enemy to work far out in the oceans as soon as possible.

* * * * *

Comprehensive plans existed at the Admiralty for multiplying our anti-submarine craft. In particular preparations had been made to take up eighty-six of the largest and fastest trawlers and to equip them with Asdics; the conversion of many of these was already well advanced. A war-time building programme of destroyers, both large and small, and of cruisers, with many ancillary vessels, was also ready in every detail, and this came into operation automatically with the declaration of war. The previous war had proved the sovereign merits of convoy. The Admiralty had for some days assumed control of the movements of all merchant shipping, and shipmasters were required to obey orders about their routes or about joining convoy. Our weakness in escort vessels had however forced the Admiralty to devise a policy of evasive routing on the oceans, unless and until the enemy adopted unrestricted U-boat warfare, and to confine convoys in the first instance to the east coast of Britain. But the sinking of the *Athenia* upset these plans, and we adopted convoy in the North Atlantic forthwith.

The organisation of convoy had been fully prepared, and shipowners had already been brought into regular consultation on matters of defence which affected them. Furthermore instructions had been issued for the guidance of shipmasters in the many unfamiliar tasks which would inevitably fall upon them in war, and special signalling as well as other equipment had been provided to enable them to take their place in convoy. The men of the Merchant Navy faced the unknown future with determination. Not content with a passive *rôle*, they demanded weapons. The use of guns in self-defence by merchant ships has always been recognised as justifiable by international law, and the defensive arming of all seagoing merchant ships, together with the training of the crews, formed an integral part of the Admiralty plans, which were at once put into effect. To force the U-boat to attack submerged and not merely by gun-fire on the surface not only gave a greater chance for a ship to escape, but caused the attacker to expend his precious torpedoes more lavishly and often fruitlessly. Foresight had preserved the guns of the previous war for use against U-boats, but there was a grave shortage of anti-aircraft weapons. It was very many months before adequate self-protection against air attack could be provided for merchant ships, which suffered severe losses meanwhile. We planned from these first days to equip during the first three months of war a thousand ships with at least an anti-submarine gun each. This was in fact achieved.

Besides protecting our own shipping, we had to drive German commerce off the seas and stop all imports into Germany. Blockade was enforced with full rigour. A Ministry of Economic Warfare was formed to guide the policy, whilst the Admiralty controlled its execution. Enemy shipping, as in 1914, virtually vanished almost at once from the high seas. The German ships mostly took refuge in neutral ports, or, when intercepted, scuttled themselves. None the less, fifteen ships, totalling 75,000 tons, were captured and put into service by the Allies before the end of 1939. The great German liner *Bremen*, after sheltering in the Soviet port of Murmansk, reached Germany only because she was spared by the British submarine *Salmon*, which observed rightly and punctiliously the conventions of international law.*

★ ★ ★ ★ ★

I held my first Admiralty conference on the night of September 4. On account of the importance of the issues, before going to bed in the small hours I recorded its conclusions for circulation and action in my own words:

5.IX.39

1. In this first phase, with Japan placid, and Italy neutral though indeterminate, the prime attack appears to fall on the approaches to Great Britain from the Atlantic.

2. The convoy system is being set up. By convoy system is meant only anti-submarine convoy. All question of dealing with raiding cruisers or heavy ships is excluded from this particular paper.

3. The First Sea Lord is considering a movement to the Western Approaches of Great Britain of whatever destroyers and escort vessels can be scraped from the Eastern and Mediterranean theatres, with the object of adding, if possible, twelve to the escorts for convoys. These

---

* This submarine was commanded by Lieutenant-Commander Bickford, who was specially promoted for his numerous exploits, but was soon afterwards lost with his vessel.

should be available during the period of, say, a month, until the flow of Asdic trawlers begins. A statement should be prepared showing the prospective deliveries during October of these vessels. It would seem well, at any rate in the earliest deliveries, not to wait for the arming of them with guns, but to rely upon depth-charges. Gun-arming can be reconsidered when the pressure eases.

4. The Director of the Trade Division (D.T.D.) should be able to report daily the inward movement of all British merchant ships approaching the Island. For this purpose, if necessary, a room and additional staff should be provided. A chart of large size should show at each morning all vessels within two, or better still three, days' distance from our shores. The guidance or control of each of these vessels must be foreseen and prescribed so that there is not one whose case has not been individually dealt with, as far as our resources allow. Pray let me have proposals to implement this, which should come into being within twenty-four hours, and work up later. The necessary contacts with the Board of Trade or other departments concerned should be effected and reported upon.

5. The D.T.D. should also prepare to-morrow a scheme under which every captain or master of a merchant ship from the Atlantic (including the Bay) is visited on arrival by a competent naval authority, who in the name of the D.T D. will examine the record of the course he has steered, including zigzags. All infractions or divergences from Admiralty instructions should be pointed out, and all serious departures should be punished, examples being made by dismissal. The Admiralty assume responsibility, and the merchant skippers must be made to obey. Details of this scheme should be worked out in personnel and regulations, together with appropriate penalties.

6. For the present it would seem wise to maintain the diversion of merchant traffic from the Mediterranean to the Cape route. This would not exclude the passage of convoys for troops, to which of course merchant vessels which were handy might add themselves. But these convoys can only be occasional, i.e., not more than once a month or three weeks, and they must be regarded not as part of the trade protection, but as naval operations.

7. It follows from the above that in this period, i.e., the first six weeks or two months of the war, the Red Sea will also be closed to everything except naval operations, or perhaps coastal traffic to Egypt.

8. This unpleasant situation would be eased by the deliveries of the Asdic trawlers and other reliefs. Secondly, by the determination of the attitude of Italy. We cannot be sure that the Italian uncertainty will be cleared up in the next six weeks, though we should press His

Majesty's Government to bring it to a head in a favourable sense as soon as possible. Meanwhile the heavy ships in the Mediterranean will be on the defensive, and can therefore spare some of the destroyer protection they would need if they were required to approach Italian waters.

9. The question of a breaking out of any of the five (or seven) German ships of weight would be a major naval crisis requiring a special plan. It is impossible for the Admiralty to provide escorts for convoys of merchant ships against serious surface attack. These raids, if they occur, could only be dealt with as a naval operation by the main Fleet, which would organise the necessary hunting parties to attack the enemy, the trade being cleared out of the way so far as possible till results were obtained.

The First Lord submits these notes to his naval colleagues for consideration, *for criticism and correction*, and hopes to receive proposals for action in the sense desired.

The organisation of outward-bound convoys was brought into force almost at once. By September 8 three main routes had begun to work, namely, from Liverpool and from the Thames to the western ocean, and a coastal convoy between the Thames and the Forth. Staffs for the control of convoys at these ports and many others at home and abroad were included in the War Plan, and had already been dispatched. Meanwhile all ships outward bound in the Channel and Irish Sea and not in convoy were ordered to Plymouth and Milford Haven, and all independent outward sailings were cancelled. Overseas arrangements for forming homeward-bound convoys were pressed forward. The first of them sailed from Freetown on September 14 and from Halifax, Nova Scotia, on the 16th. Before the end of the month regular ocean convoys were in operation, outward from the Thames and Liverpool and homeward from Halifax, Gibraltar, and Freetown.

Upon all the vital need of feeding the Island and developing our power to wage war there now at once fell the numbing loss of the Southern Irish ports. This imposed a grievous restriction on the radius of action of our already scarce destroyers:

First Sea Lord and others   5.IX.39
A special report should be drawn up by the heads of departments concerned and sent to the First Lord through the First Sea Lord and the Naval Staff upon the questions arising from the so-called neutrality

of the so-called Eire. Various considerations arise: (1) What does Intelligence say about possible succouring of U-boats by Irish malcontents in West of Ireland inlets? If they throw bombs in London,* why should they not supply fuel to U-boats? Extreme vigilance should be practised.

Secondly, a study is required of the addition to the radius of our destroyers through not having the use of Berehaven or other South Irish anti-submarine bases; showing also the advantage to be gained by our having these facilities.

The Board must realise that we may not be able to obtain satisfaction, as the question of Eirish neutrality raises political issues which have not yet been faced, and which the First Lord is not certain he can solve. But the full case must be made for consideration.

★　★　★　★　★

After the institution of the convoy system the next vital naval need was a safe base for the Fleet. At 10 p.m. on September 5 I held a lengthy conference on this. It recalled many old memories. In a war with Germany Scapa Flow is the true strategic point from which the British Navy can control the exits from the North Sea and enforce blockade. It was only in the last two years of the previous war that the Grand Fleet was judged to have sufficient superiority to move south to Rosyth, where it had the advantage of lying at a first-class dockyard. But Scapa, on account of its greater distance from German air bases, was now plainly the best position, and had been definitely chosen in the Admiralty War Plan.

In the autumn of 1914 a wave of uneasiness had swept the Grand Fleet. The idea had got round that *"the German submarines were coming after them into the harbours"*. Nobody at the Admiralty then believed that it was possible to take a submarine, submerged, through the intricate and swirling channels by which the great lake of Scapa can alone be entered. The violent tides and currents of the Pentland Firth, often running eight or ten knots, had seemed in those days to be an effective deterrent. But a mood of doubt spread through the mighty array of perhaps a hundred large vessels which in those days composed the Grand Fleet. On two or three occasions, notably on October 17, 1914, the alarm was given that there was a U-boat inside the anchorage. Guns were fired, destroyers thrashed the waters, and the whole

---

* This referred to a criminal act unconnected with the war.

gigantic armada put to sea in haste and dudgeon. In the final result the Admiralty were proved right. No German submarine in that war ever overcame the terrors of the passage. It was only in 1918, at the very end of the war, that a U-boat made the attempt, and perished in this final desperate effort. Nevertheless, I retained a most vivid and unpleasant memory of those days and of the extreme exertions we made to block all the entrances and reassure the Fleet.

There were now in 1939 two dangers to be considered: the first, the old one of submarine incursion; the second, the new one of the air. I was surprised to learn at my conference that more precautions had not been taken in both cases to prepare the defences against modern forms of attack. Anti-submarine booms of new design were in position at each of the three main entrances, but these consisted merely of single lines of net. The narrow and tortuous approaches on the east side of the Flow, defended only by remnants of the blockships placed in the former war, and reinforced now by two or three recent additions, remained a source of anxiety. On account of the increased size, speed, and power of modern submarines, the old belief that the strong tidal streams made these passages impassable to a submarine no longer carried conviction in responsible quarters. As a result of the conference on my second evening at the Admiralty many orders were given for additional nets and blockings.

The new danger from the air had been almost entirely ignored. Except for two batteries of anti-aircraft guns to defend the naval oil tanks at Hoy and the destroyer anchorage, there were no air defences at Scapa. One airfield near Kirkwall was available for the use of naval aircraft when the Fleet was present, but no provision had been made for immediate R.A.F. participation in the defence, and the shore Radar station, although operative, was not wholly effective. Plans for basing two R.A.F. fighter squadrons at Wick had been approved, but this measure could not become effective before 1940. I called for an immediate plan of action. Our air defence was so strained, our resources so limited, and our vulnerable points—including all vast London—so numerous that it was no use asking for much. On the other hand, protection from air attack was now needed only for five or six great ships, each carrying a powerful anti-aircraft armament of its own. To keep things going the Admiralty undertook to provide two squadrons of naval fighter aircraft whilst the Fleet was in Scapa.

It seemed most important to have the artillery in position at the shortest interval, and meanwhile there was nothing for it but to adopt the same policy of "hide-and-seek" to which we had been forced in the autumn days of 1914. The west coast of Scotland had many landlocked anchorages easy to protect from U-boat by indicator nets and ceaseless patrolling. We had found concealment in the previous war a good security; but even in those days the curiosity of a wandering aeroplane, perhaps fuelled by traitor hands, had filled our hearts with fear. Now that the range of aircraft exposed the whole British Islands at any time to photographic reconnaissance machines, there was no sure concealment against large-scale attack either by U-boats or from the air. However, there were so few ships to cover, and they could be moved so often from one place to another, that, having no alternative, we accepted the hazard with as good a grace as possible.

*   *   *   *   *

I felt it my duty to visit Scapa at the earliest moment. I had not met the Commander-in-Chief, Sir Charles Forbes, since Lord Chatfield had taken me to the Anti-Submarine School at Portland in June 1938. I therefore obtained leave from our daily Cabinets, and started for Wick with a small personal staff on the night of September 14. I spent most of the next two days inspecting the harbour and the entrances, with their booms and nets. I was assured that they were as good as in the last war, and that important additions and improvements were being made or were on the way. I stayed with the Commander-in-Chief in his flagship, *Nelson*, and discussed not only Scapa but the whole naval problem with him and his principal officers. The rest of the Fleet was hiding in Loch Ewe, and on the 17th the Admiral took me to them in the *Nelson*. As we came out through the gateway into the open sea I was surprised to see no escort of destroyers for this great ship. "I thought," I remarked, "you never went to sea without at least two, even for a single battleship." But the Admiral replied, "Of course, that is what we should like; but we haven't got the destroyers to carry out any such rule. There are a lot of patrolling craft about, and we shall be into the Minches in a few hours."

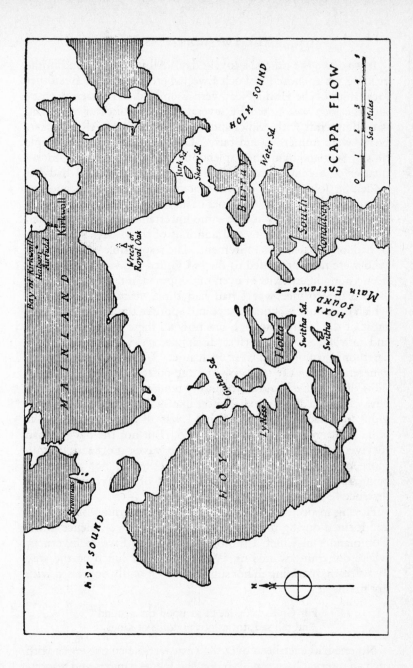

SCAPA FLOW

0 1 2 3 4 5
Sea Miles

HOLM SOUND

Water Sd.

Kirk Sd.
Sherry Sd.

Burra

South
Ronaldsay

Bay of Kirkwall
Hatston
Airfield
Kirkwall

Wreck of
Royal Oak

MAINLAND

Main Entrance
HOXA SOUND

Switha Sd.
Switha

Flotta

Gutter Sd.

HOY

Lyness

Stromness

HOY SOUND

N

It was, like the others, a lovely day. All went well, and in the evening we anchored in Loch Ewe, where the four or five other great ships of the Home Fleet were assembled. The narrow entry into the loch was closed by several lines of indicator nets, and patrolling craft with Asdics and depth-charges, as well as picket-boats, were numerous and busy. On every side rose the purple hills of Scotland in all their splendour. My thoughts went back a quarter of a century to that other September when I had last visited Sir John Jellicoe and his captains in this very bay, and had found them with their long lines of battleships and cruisers drawn out at anchor, a prey to the same uncertainties as now afflicted us. Most of the captains and admirals of those days were dead, or had long passed into retirement. The responsible senior officers who were now presented to me as I visited the various ships had been young lieutenants or even midshipmen in those far-off days. Before the former war I had had three years' preparation in which to make the acquaintance and approve the appointments of most of the high personnel, but now all these were new figures and new faces. The perfect discipline, style, and bearing, the ceremonial routine—all were unchanged. But an entirely different generation filled the uniforms and the posts. Only the ships had most of them been laid down in my tenure. None of them was new. It was a strange experience, like suddenly resuming a previous incarnation. It seemed that I was all that survived in the same position I had held so long ago. But no; the dangers had survived too. Danger from beneath the waves, more serious with more powerful U-boats; danger from the air, not merely of being spotted in your hiding-place, but of heavy and perhaps destructive attack!

Having inspected two more ships on the morning of the 18th, and formed during my visit a strong feeling of confidence in the Commander-in-Chief, I motored from Loch Ewe to Inverness, where our train awaited us. We had a picnic lunch on the way by a stream, sparkling in hot sunshine. I felt oddly oppressed with my memories.

> For God's sake, let us sit upon the ground
> And tell sad stories of the death of kings.

No one had ever been over the same terrible course twice with such an interval between. No one had felt its dangers and responsibilities from the summit as I had, or, to descend to a small point, understood how First Lords of the Admiralty are treated when great ships are sunk and things go wrong. If we were in fact going over the same cycle a second time, should I have once again to endure the pangs of dismissal? Fisher, Wilson, Battenberg, Jellicoe, Beatty, Pakenham, Sturdee, all gone!

> I feel like one
> Who treads alone
> Some banquet-hall deserted,
> Whose lights are fled,
> Whose garlands dead,
> And all but he departed.

And what of the supreme, measureless ordeal in which we were again irrevocably plunged? Poland in its agony; France but a pale reflection of her former warlike ardour; the Russian Colossus no longer an ally, not even neutral, possibly to become a foe. Italy no friend. Japan no ally. Would America ever come in again? The British Empire remained intact and gloriously united, but ill-prepared, unready. We still had command of the sea. We were woefully outmatched in numbers in this new mortal weapon of the air. Somehow the light faded out of the landscape.

We joined our train at Inverness and travelled through the afternoon and night to London. As we got out at Euston the next morning I was surprised to see the First Sea Lord on the platform. Admiral Pound's look was grave. "I have bad news for you, First Lord. The *Courageous* was sunk yesterday evening in the Bristol Channel." The *Courageous* was one of our oldest aircraft-carriers, but a very necessary ship at this time. I thanked him for coming to break it to me himself, and said, "We can't expect to carry on a war like this without that sort of thing happening from time to time. I have seen lots of it before." And so to bath and the toil of another day.

In order to bridge the gap of two or three weeks between the outbreak of war and the completion of our auxiliary anti-U-boat flotillas, we had decided to use the aircraft-carriers with some freedom in helping to bring in the unarmed, unorganised, and unconvoyed traffic which was then approaching our shores in large numbers. This was a risk which it was right to run. The *Courageous*, attended by four destroyers, had been thus employed.

Towards evening on the 17th two of these had to go to hunt a U-boat which was attacking a merchant ship. When the *Courageous* turned into the wind at dusk, in order to enable her own aircraft to alight upon her landing-deck, she happened, in her unpredictable course, by what may have been a hundred-to-one chance, to meet a U-boat. Out of the crew of 1,260 over 500 were drowned, including Captain Makeig-Jones, who went down with his ship. Three days before another of our aircraft-carriers, later to become famous, H.M.S. *Ark Royal*, had also been attacked by a submarine while similarly engaged. Mercifully the torpedoes missed, and her assailant was promptly sunk by her escorting destroyers.

* * * * *

Outstanding among our naval problems was that of dealing effectively with surface raiders, which would inevitably make their appearance in the near future as they had done in 1914.

On September 12 I issued the following minute:

First Lord to First Sea Lord 12.IX.39

CRUISER POLICY

In the past we have sought to protect our trade against sudden attack by [means of] cruisers; having regard to the vast ocean spaces to be controlled, the principle was "the more the better". In the search for enemy raiders or cruisers even small cruisers could play their part, and in the case of the *Emden* we were forced to gather over twenty ships before she was rounded up. However, a long view of cruiser policy would seem to suggest that a new Unit of Search is required. Whereas a cruiser squadron of four ships could search on a front of, say, 80 miles, a single cruiser accompanied by an aircraft-carrier could cover at least 300 miles, or, if the movement of the ship is taken into account, 400 miles. On the other hand, we must apprehend that the raiders of the future will be powerful vessels, eager to fight a single-ship action if a chance is presented. The mere multiplication of small, weak cruisers is no means of ridding the seas of powerful raiders. Indeed they are only an easy prey. The raider, cornered at length, will overwhelm one weak vessel and escape from the cordon.

Every Unit of Search must be able to find, to catch, and to kill. For this purpose we require a number of cruisers superior to the 10,000-ton type, or else pairs of our own 10,000-ton type. These must be accompanied by small aircraft-carriers carrying perhaps a dozen or two dozen machines, and of the smallest possible displacement. The ideal Unit of Search would be one killer or two three-quarter killers, plus one aircraft-carrier, plus four ocean-going destroyers, plus two or three specially-constructed tankers of good speed. Such a formation cruising would be protected against submarines, and could search an enormous area and destroy any single raider when detected.

The policy of forming hunting groups as discussed in this minute, comprising balanced forces capable of scouring wide areas and overwhelming any raider within the field of search, was developed so far as our limited resources allowed, and I shall refer to this subject again in a later chapter. The same idea was afterawards more fully expanded by the United States in their Task Force system, which made an important contribution to the art of sea warfare.

*   *   *   *   *

Towards the end of the month I thought it would be well for me to give the House some coherent story of what was happening and why.

First Lord to Prime Minister   24.IX.39

Would it not be well for me to make a statement to the House on the anti-submarine warfare and general naval position, more at length than what you could give in your own speech? I think I could speak for twenty-five or thirty minutes on the subject, and that this would do good. At any rate, when I saw in confidence sixty Press representatives the other day they appeared vastly relieved by the account I was able to give. If this idea commended itself to you, you would perhaps say in your speech that I would give a fuller account later on in the discussion, which I suppose will take place on Thursday, as the Budget is on Wednesday.

Mr. Chamberlain readily assented, and accordingly in his speech on the 26th he told the House that I would make a statement on the sea war as soon as he sat down. This was the first time. apart from answering questions, that I had spoken in Parliament since I had entered the Government. I had a good tale to tell. In the first seven days our losses in tonnage had been half the weekly losses of the month of April 1917, which was the peak year of the U-boat attack in the first war. We had already made progress by setting in motion the convoy system; secondly, by pressing on with the arming of all our merchant ships; and, thirdly, by our counter-attack upon the U-boats. "In the first week our losses by U-boat sinkings amounted to 65,000 tons; in the second week they were 46,000 tons; and in the third week they were 21,000 tons. In the last six days we have lost only 9,000 tons."* I observed throughout that habit of understatement and of avoiding all optimistic forecasts which had been inculcated upon me by the hard experiences of the past. "One must not dwell," I said, "upon these reassuring figures too much, for war is full of unpleasant surprises. But certainly I am entitled to say that so far as they go these figures need not cause any undue despondency or alarm."

Meanwhile [I continued] the whole vast business of our world-wide trade continues without interruption or appreciable diminution. Great convoys of troops are escorted to their various destinations. The enemy's ships and commerce have been swept from the seas. Over 2,000,000 tons of German shipping is now sheltering in German, or interned in neutral, harbours. . . . In the first fortnight of the war we have actually arrested, seized, and converted to our own use 67,000 tons more German merchandise than has been sunk in ships of our own. . . . Again I reiterate my caution against over-sanguine conclusions. We have in fact however got more supplies in this country this afternoon than we should have if no war had been declared and no U-boat had come into action. It is not going beyond the limits of prudent statement if I say that at that rate it will take a long time to starve us out.

* The following are the corrected figures:

BRITISH MERCHANT SHIPPING LOSSES BY ENEMY ACTION, SEPTEMBER 1939
(Numbers of ships shown in brackets)

	Submarine (Gross tons)	Other Causes (Gross tons)
1st WEEK (September 3–9) ..	64,595 (11)	—
2nd WEEK (September 10–16)	53,561 (11)	11,437 (2) (mine)
3rd WEEK (September 17–23)	12,750 (3)	—
4th WEEK (September 24–30)	4,646 (1)	5,051 (1) (surface raider)
Total	135,552 (26)	16,488 (3)
	152,040 (29)	

In addition there were losses in neutral and Allied shipping amounting to 15 ships, of 33,527 tons.

From time to time the German U-boat commanders have tried their best to behave with humanity. We have seen them give good warning and also endeavour to help the crews to find their way to port. One German captain signalled to me personally the position of a British ship which he had just sunk, and urged that rescue should be sent. He signed his message "German Submarine". I was in some doubt at the time to what address I should direct a reply. However, he is now in our hands, and is treated with all consideration.

Even taking six or seven U-boats sunk as a safe figure,* that is one-tenth of the total enemy submarine fleet as it existed at the declaration of war, destroyed during the first fortnight of the war, and it is probably one-quarter, or perhaps even one-third, of all the U-boats which are being employed actively. But the British attack upon the U-boats is only just beginning. Our hunting force is getting stronger every day. By the end of October we expect to have three times the hunting force which was operating at the beginning of the war.

This speech, which lasted only twenty-five minutes, was extremely well received by the House, and in fact it recorded the failure of the first German U-boat attack upon our trade. My fears were for the future, but our preparations for 1941 were now proceeding with all possible speed and on the largest scale which our vast resources would allow.

*　　*　　*　　*　　*

By the end of September we had little cause for dissatisfaction with the results of the first impact of the war at sea. I could feel that I had effectively taken over the great department which I knew so well and loved with a discriminating eye. I now knew what there was in hand and on the way. I knew where everything was. I had visited all the principal naval ports and met all the Commanders-in-Chief. By the Letters Patent constituting the Board, the First Lord is "responsible to Crown and Parliament for all the business of the Admiralty", and I certainly felt prepared to discharge that duty in fact as well as in form.

On the whole the month of September had been prosperous and fruitful for the Navy. We had made the immense, delicate, and hazardous transition from peace to war. Forfeits had to be paid in the first few weeks by a world-wide commerce suddenly attacked contrary to formal international agreement by indis-

---

* We now know that only two U-boats were sunk in September 1939.

criminate U-boat warfare; but the convoy system was now in full flow, and merchant ships were leaving our ports every day by scores with a gun, sometimes high-angle, mounted aft, and a nucleus of trained gunners. The Asdic-equipped trawlers and other small craft armed with depth-charges, all well prepared by the Admiralty before the outbreak, were now coming daily into commission in a growing stream, with trained crews. We all felt sure that the first attack of the U-boat on British trade had been broken and that the menace was in thorough and hardening control. It was obvious that the Germans would build submarines by hundreds, and no doubt numerous shoals were upon the slips in various stages of completion. In twelve months, certainly in eighteen, we must expect the main U-boat war to begin. But by that time we hoped that our mass of new flotillas and anti-U-boat craft, which was our First Priority, would be ready to meet it with a proportionate and effective predominance. The painful dearth of anti-aircraft guns, especially 3.7-inch and Bofors, could, alas, only be relieved after many months; but measures had been taken within the limits of our resources to provide for the defence of our naval harbours; and meanwhile the Fleet, while ruling the oceans, would have to go on playing hide-and-seek.

* * * * *

In the wider sphere of naval operations no definite challenge had yet been made to our position. After the temporary suspension of traffic in the Mediterranean our shipping soon moved again through this invaluable corridor. Meanwhile the transport of the Expeditionary Force to France was proceeding smoothly. The Home Fleet itself, "somewhere in the North", was ready to intercept any sortie by the few heavy ships of the enemy. The blockade of Germany was being enforced by similar methods to those employed in the previous war. The Northern Patrol had been established between Scotland and Iceland, and by the end of the first month a total of nearly three thundred thousand tons of goods destined for Germany had been seized in prize, against a loss to ourselves of a hundred and forty thousand tons by enemy action at sea. Overseas our cruisers were hunting down German ships, while at the same time providing cover against attack on our shipping by raiders. German shipping had thus come to a standstill. By the end of September 325 German ships, totalling nearly 750,000 tons, were immobilised in foreign ports. Few therefore fell into our hands.

Our Allies also played their part. The French took an important share in the control of the Mediterranean. In home waters and the Bay of Biscay they also helped in the battle against the U-boats, and in the central Atlantic a powerful force based on Dakar formed part of the Allied plans against surface raiders.

The young Polish Navy distinguished itself. Early in the war three modern destroyers and two submarines, *Wilk* and *Orzel*, escaped from Poland, and, defying the German forces in the Baltic, succeeded in reaching England. The escape of the submarine *Orzel* is an epic. Sailing from Gdynia when the Germans invaded Poland, she first cruised in the Baltic, putting into the neutral port of Tallinn on September 15 to land her sick captain. The Esthonian authorities decided to intern the vessel, placed a guard on board, and removed her charts and the breech-blocks of her guns. Undismayed, her commanding officer put to sea, after overpowering the guard. In the ensuing weeks the submarine was continually hunted by sea and air patrols, but eventually, without even charts, made her escape from the Baltic into the North Sea. Here she was able to transmit a faint wireless signal to a British station giving her supposed position, and on October 14 was met and escorted into safety by a British destroyer.

* * * * *

In September I was delighted to receive a personal letter from President Roosevelt. I had only met him once in the previous war. It was at a dinner at Gray's Inn, and I had been struck by his magnificent presence in all his youth and strength. There had been no opportunity for anything but salutations.

President Roosevelt to Mr. Churchill 11.IX.39

It is because you and I occupied similar positions in the World War that I want you to know how glad I am that you are back again in the Admiralty. Your problems are, I realise, complicated by new factors, but the essential is not very different. What I want you and the Prime Minister to know is that I shall at all times welcome it if you will keep me in touch personally with anything you want me to know about. You can always send sealed letters through your pouch or my pouch.

I am glad you did the Marlborough volumes before this thing started —and I much enjoyed reading them.

I responded with alacrity, using the signature of "Naval Person", and thus began that long and memorable correspondence —covering perhaps a thousand communications on each side, and lasting till his death more than five years later.

CHAPTER XXIV

THE RUIN OF POLAND

The German Plan of Invasion – Unsound Polish Dispositions – Inferiority in Artillery and Tanks – Destruction of the Polish Air Force – The First Week – The Second Week – The Heroic Polish Counter-Attack – Extermination – The Turn of the Soviets – The Warsaw Radio Silent – The Modern Blitzkrieg – My Memorandum of September 21 – Our Immediate Dangers – My Broadcast of October 1.

MEANWHILE around the Cabinet table we were witnessing the swift and almost mechanical destruction of a weaker State according to Hitler's method and long design. Poland was open to German invasion on three sides. In all fifty-six divisions, including all his nine armoured and motorised divisions, composed the invading armies. From East Prussia the Third Army (eight divisions) advanced southwards on Warsaw and Bialystok. From Pomerania the Fourth Army (twelve divisions) was ordered to destroy the Polish troops in the Danzig Corridor, and then move south-eastward to Warsaw along both banks of the Vistula. The frontier opposite the Posen bulge was held defensively by German reserve troops, but on their right to the southward lay the Eighth Army (seven divisions), whose task was to cover the left flank of the main thrust. This thrust was assigned to the Tenth Army (seventeen divisions), directed straight upon Warsaw. Further south again the Fourteenth Army (fourteen divisions) had a dual task, first to capture the important industrial area west of Cracow, and then, if the main front prospered, to make direct for Lemberg (Lwow), in Southeast Poland.

Thus the Polish forces on the frontiers were first to be penetrated, and then overwhelmed and surrounded by two pincer movements, the first from the north and south-west on Warsaw, the second and more far-reaching, "outer" pincers, formed by the Third Army advancing by Brest-Litovsk, to be joined by the Fourteenth Army after Lemberg was gained. Those who escaped the closing of the Warsaw pincers would thus be cut off from retreat into Roumania. Over fifteen hundred modern aircraft were hurled on Poland. Their first duty was to overwhelm the Polish Air Force, and thereafter to support the Army on the battle-field, and beyond it to attack military installations and all com-munications by road and rail. They were also to spread terror far and wide.

In numbers and equipment the Polish Army was no match for their assailants, nor were their dispositions wise. They spread all their forces along the frontiers of their native land. They had no central reserve. While taking a proud and haughty line against German ambitions, they had nevertheless feared to be accused of provocation by mobilising in good time against the masses gathering around them. Thirty divisions, representing only two-thirds of their active army, were ready or nearly ready to meet the first shock. The speed of events and the violent intervention of the German Air Force prevented the rest from reaching the forward positions till all was broken, and they were only involved in the final disasters. Thus the thirty Polish divisions faced nearly double their numbers around a long perimeter with nothing behind them. Nor was it in numbers alone that they were inferior. They were heavily out-classed in artillery, and had but a single armoured brigade to meet the nine German Panzers, as they were already called. Their horse cavalry, of which they had twelve brigades, charged valiantly against the swarming tanks and armoured cars, but could not harm them with their swords and lances. Their nine hundred first-line aircraft, of which per-haps half were modern types, were taken by surprise, and many were destroyed before they even got into the air.

According to Hitler's plan the German armies were unleashed on September 1, and ahead of them his Air Force struck the Polish squadrons on their airfields. In two days the Polish air power was virtually annihilated. Within a week the German armies had bitten deep into Poland. Resistance everywhere was brave but vain. All the Polish armies on the frontiers, except the Posen group, whose flanks were deeply turned, were driven backwards. The Lodz group was split in twain by the main thrust of the German Tenth Army; one half withdrew eastwards to Radom, the other was forced north-westward; and through this gap darted two Panzer divisions, making straight for Warsaw. Farther north the German Fourth Army reached and crossed the Vistula, and turned along it in their march on Warsaw. Only the Polish northern group was able to inflict a check upon the German Third Army. They were soon outflanked and fell back to the river Narew, where alone a fairly strong defensive system had been prepared in advance. Such were the results of the first week of the Blitzkrieg.

The second week was marked by bitter fighting, and by its end the Polish Army, nominally of about two million men, ceased to exist as an organised force. In the south the Fourteenth German Army drove on to reach the river San. North of them the four Polish divisions which had retreated to Radom were there en-

THE GERMAN AND POLISH CONCENTRATIONS   Sept. 1ˢᵗ 1939

circled and destroyed. The two armoured divisions of the Tenth Army reached the outskirts of Warsaw, but having no infantry with them could not make headway against the desperate resistance organised by the townsfolk. North-east of Warsaw the Third Army encircled the capital from the east, and its left column reached Brest-Litovsk, a hundred miles behind the battle-front.

It was within the claws of the Warsaw pincers that the Polish Army fought and died. Their Posen group had been joined by divisions from the Thorn and Lodz groups, forced towards them by the German onslaught. It now numbered twelve divisions, and across its southern flank the German Tenth Army was streaming towards Warsaw, protected only by the relatively weak Eighth Army. Although already virtually surrounded, the Polish commander of the Posen group, General Kutrzea, resolved to strike south against the flank of the main German drive. This audacious Polish counter-attack, called the Battle of the River Bzura, created a crisis which drew in not only the German Eighth Army but a part of the Tenth, deflected from their Warsaw objective, and even a corps of the Fourth Army from the north. Under the assault of all these powerful bodies, and overwhelmed by unresisted air bombardment, the Posen group maintained its ever-glorious struggle for ten days. It was blotted out on September 19.

In the meantime the outer pincers had met and closed. The Fourteenth Army reached the outskirts of Lemberg on September 12, and, striking north, joined hands on the 17th with the troops of the Third Army, which had passed through Brest-Litovsk. There was now no loophole of escape save for straggling and daring individuals. On the 20th the Germans announced that the Battle of the Vistula was "one of the greatest battles of extermination of all time".

It was now the turn of the Soviets. What they now call "Democracy" came into action. On September 17 the Russian armies swarmed across the almost undefended Polish eastern frontier and rolled westward on a broad front. On the 18th they occupied Vilna, and met their German collaborators at Brest-Litovsk. Here in the previous war the Bolsheviks, in breach of their solemn agreements with the Western Allies, had made their separate peace with the Kaiser's Germany and had bowed to its harsh terms. Now in Brest-Litovsk it was with Hitler's Germany that the Russian Communists grinned and shook hands. The ruin of Poland and its entire subjugation proceeded apace. Warsaw and Modlin still remained unconquered. The resistance of Warsaw, largely arising from the surge of its citizens, was magnificent and forlorn. After many days of violent bombardment from the air and by heavy artillery, much of which was rapidly transported across the great lateral highways from the idle Western Front, the Warsaw radio ceased to play the Polish National Anthem, and Hitler entered the ruins of the city. Modlin, a fortress twenty miles down the Vistula, had taken in the remnants of the Thorn group, and fought on until the 28th. Thus in one month all was over, and a nation of thirty-five millions fell into the merciless grip of those who sought not only conquest, but enslavement and indeed extinction for vast numbers.

We had seen a perfect specimen of the modern Blitzkrieg; the close interaction on the battlefield of Army and Air Force; the violent bombardment of all communications and of any town that seemed an attractive target; the arming of an active Fifth Column; the free use of spies and parachutists; and above all the irresistible forward thrusts of great masses of armour. The Poles were not to be the last to endure this ordeal.

★　　★　　★　　★　　★

The Soviet armies continued to advance up to the line they had settled with Hitler, and on the 29th the Russo-German treaty partitioning Poland was formally signed. I was still convinced of the profound, and, as I believed, quenchless antagonism between Russia and Germany, and I clung to the hope that the Soviets would be drawn to our side by the force of events. I did not therefore give way to the indignation which I felt and which surged around me in our Cabinet at their callous, brutal policy. I had never had any illusions about them. I knew that they accepted no moral code, and studied their own interests alone. But at least they owed us nothing. Besides, in mortal war anger must be subordinated to defeating the main immediate enemy. I was determined to put the best construction on their odious conduct. Therefore in a paper which I wrote for the War Cabinet on September 25 I struck a cool note.

THE INNER PINCERS CLOSE Sept. 13th

Although the Russians were guilty of the grossest bad faith in the recent negotiations, their demand made by Marshal Voroshilov that Russian armies should occupy Vilna and Lemberg if they were to be allies of Poland was a perfectly valid military request. It was rejected by Poland on grounds which, though natural, can now be seen to have been insufficient. In the result Russia has occupied the same line and positions as the enemy of Poland which possibly she might have occupied as a very doubtful and suspected friend. The difference in fact is not so great as it might seem. The Russians have mobilised very large forces and have shown themselves able to advance fast and far from their pre-war positions. They are now limitrophe with Germany, and it is quite impossible for Germany to denude the Eastern front. A large German army must be left to watch it. I see General Gamelin puts it at at least twenty divisions. It may well be twenty-five or more. An Eastern front is therefore potentially in existence.

But it is possible that a South-eastern front may also be built up in which Russia, Britain, and France will have a common interest. The left paw of the Bear has already closed the pathway from Poland to Roumania. Russian interest in the Slavonic peoples of the Balkans is traditional. The arrival of the Germans on the Black Sea would be a deadly threat to Russia. And also to Turkey. That these two countries should make common cause to prevent this is a direct fulfilment of our wishes. It in no way conflicts with our policy towards Turkey. It may well be that Russia will deprive Roumania of Bessarabia; but this does not necessarily conflict with our major interest, which is to arrest the German movement towards the east and south-east of Europe. Roumania, which gained enormously from the late war, in which she was rescued from utter defeat by the Allied victory, will be lucky if she gets out of this war with no greater losses than Bessarabia and the southern part of the Dobrudja, which latter she ought willingly to cede to Bulgaria in the interests of a Balkan *bloc*. The reactions of the Russian movement, so far as it can at present be

THE OUTER PINCERS CLOSE · THE RUSSIANS ADVANCE Sept. 17th

judged, should be favourable throughout the Balkans, and particularly in Yugoslavia. Thus, besides the potential Eastern front a potential South-eastern front may be coming into existence, reaching in a crescent from the Gulf of Riga to the head of the Adriatic (and thence perhaps on across the Brenner to the Alps).

Of course we should much prefer that all these countries should fall at once upon the sole and common foe, Nazi Germany; and this possibility should not be excluded as time goes on. It would come very near if Germany struck through Hungary at Roumania, and to a lesser degree if she struck at Yugoslavia. The policy we are pursuing of fostering this front, of strengthening it and endeavouring to throw it into simultaneous action should any part of it be attacked, seems absolutely right. This policy implies a renewal of relations with Russia, as the Foreign Secretary has swiftly foreseen. It also compels our adherence to the policy declared by the Prime Minister of not committing ourselves to particular territorial solutions and concentrating the whole effort of Britain and France upon smashing Hitlerism, and also of making sure that the German Terror is not renewed upon the Western democracies for a long time to come. This last point, which appeals so much to the French, is exactly expressed by the Prime Minister's words: "Our general purpose . . . is to redeem Europe from the perpetual and recurring fear of German aggression, and enable the peoples of Europe to preserve their liberties and their independence." This cannot be repeated too often or too widely.

Upon this general appreciation our handling of the Turkish negotiations can more easily be considered. I cannot feel that there is the same urgency about them as there was when Hitler was reported to be about to invade Roumania with twenty-eight divisions, etc. It now seems possible that that man may be warned off his Eastern career; but of course he may renew his threat at any time, and we besides have a main interest in bringing all the Balkans and Eastern front into hostile action against Germany. It therefore seems most important to make the Turkish treaty.*

If it should turn out that Hitler is barred in the East, which, of course, is not yet certain, three courses are open to him:

(1) A major attack on the Western front, probably through Belgium, collecting Holland on the way.
(2) An intensive attack by air upon British factories, naval ports, etc., or perhaps on the French air factories.
(3) What the Prime Minister calls "the peace offensive".

Personally, I shall believe that (1) is imminent only when at least thirty divisions have been concentrated opposite Belgium and Luxem-

* See Appendix K.

burg. As to (2), it seems a very likely thing for that man to do; but he may not do it, or he may not be allowed to do it by his generals, who now are presumably more powerful, for fear of making a mortal blood-feud with Great Britain, and perhaps drawing in the United States by the air massacres which would be inevitable. As for (3), if he has not tried (2) it would seem our duty and policy to agree to nothing that will help him out of his troubles, and to leave him to stew in his own juice during the winter while speeding forward our armaments and weaving up our alliances. The general outlook, therefore, seems far more favourable than it did in the autumn of 1914, when a large part of France was occupied and Russia had been shattered at Tannenberg.

But there always remains No. 2. That is the immediate pinch.

In a broadcast on October 1 I said:

Poland has again been overrun by two of the great Powers which held her in bondage for a hundred and fifty years but were unable to quench the spirit of the Polish nation. The heroic defence of Warsaw shows that the soul of Poland is indestructible, and that she will rise again like a rock, which may for a time be submerged by a tidal wave, but which remains a rock.

Russia has pursued a cold policy of self-interest. We could have wished that the Russian armies should be standing on their present line as the friends and allies of Poland instead of as invaders. But that the Russian armies should stand on this line was clearly necessary for the safety of Russia against the Nazi menace. At any rate, the line is there, and an Eastern front has been created which Nazi Germany does not dare assail. . . .

I cannot forecast to you the action of Russia. It is a riddle wrapped in a mystery inside an enigma. But perhaps there is a key. That key is Russian national interest. It cannot be in accordance with the interest or the safety of Russia that Germany should plant herself upon the shores of the Black Sea, or that she should overrun the Balkan States and subjugate the Slavonic peoples of South-eastern Europe. That would be contrary to the historic life-interests of Russia.

The Prime Minister was in full agreement. "I take the same view as Winston," he said, in a letter to his sister, "to whose excellent broadcast we have just been listening. I believe Russia will always act as she thinks her own interests demand, and I cannot believe she would think her interests served by a German victory followed by a German domination of Europe."*

* Feiling, *op. cit.*, p. 425.

CHAPTER XXV

WAR CABINET PROBLEMS

Our Daily Meetings – A Fifty-Five Division Army for Britain – Our Heavy Artillery – My Letter to the Prime Minister, September 10 – To the Minister of Supply, September 10, and His Answer – Need for a Ministry of Shipping – My Letter to the Prime Minister, September 15 – His Reply, September 16 – Further Correspondence about Munitions and Man-power – My Letter to the Chancellor of the Exchequer, September 24 – An Economy Campaign – The Search for a Naval Offensive – The Baltic – "Catherine the Great" – Plans for Forcing Entry – Technical and Tactical Aspects – The Prize – Views of the First Sea Lord – Lord Cork's Appointment – Progress of the Plan – The Veto of the Air – The New Construction Programme – Cruisers – Destroyers – Numbers Versus Size – Long- and Short-Term Policies – Speeding the Programme – Need of an Air-Proof Battle Squadron – The Waste of the "Royal Sovereigns" – I Establish My Own Statistical Department.

THE *WAR CABINET* and its additional members, with the Chiefs of Staff for the three Services and a number of secretaries, had met together for the first time on September 4. Thereafter we met daily, and often twice a day. I do not recall any period when the weather was so hot—I had a black alpaca jacket made to wear over only a linen shirt. It was indeed just the weather that Hitler wanted for his invasion of Poland. The great rivers on which the Poles had counted in their defensive plan were nearly everywhere fordable, and the ground was hard and firm for the movement of tanks and vehicles of all kinds. Each morning the C.I.G.S., General Ironside, standing before the map, gave long reports and appreciations which very soon left no doubt in our minds that the resistance of Poland would speedily be crushed. Each day I reported to the Cabinet the Admiralty tale, which usually consisted of a list of British merchant ships sunk by the U-boats. The British Expeditionary Force of four divisions began its movement to France, and the Air Ministry deplored the fact that they were not allowed to bombard military objectives in Germany. For the rest a great deal of business was transacted on the Home Front, and there were of course lengthy discussions about foreign affairs, particularly concerning the attitude of Soviet Russia and Italy and the policy to be pursued in the Balkans.

The most important step was the setting up of the Land Forces Committee, under Sir Samuel Hoare, at this time Lord Privy Seal, in order to advise the War Cabinet upon the scale and organisation of the Army we should form. I was a member of this small body, which met at the Home Office, and in one single sweltering afternoon agreed, after hearing the generals, that we should forthwith begin the creation of a fifty-five-division Army, together with all the munition factories, plants, and supply services of every kind necessary to sustain it in action. It was hoped that by the eighteenth month two-thirds of this, a considerable force, would either already have been sent to France or be fit to take the field. Sir Samuel Hoare was clear-sighted and active in all this, and I gave him my constant support. The Air Ministry, on the other hand, feared that so large an Army and its supplies would be an undue drain upon our skilled labour and man-power, and would hamper them in the vast plans they had formed on paper for the creation of an all-powerful overwhelming Air Force in two or three years. The Prime Minister was impressed by Sir Kingsley Wood's arguments, and hesitated to commit himself to an Army of this size and all that it entailed. The War Cabinet was divided upon the issue, and it was a week or more before a decision was reached to adopt the advice of the Land Forces Committee for a fifty-five-division Army, or rather target.

I felt that as a member of the War Cabinet I was bound to take a general view, and I did not fail to subordinate my own departmental requirements for the Admiralty to the main design. I was anxious to establish a broad basis of common ground with the Prime Minister, and also to place him in possession of my knowledge in this field, which I had trodden before; and, being encouraged by his courtesy, I wrote him a series of letters on the various problems as they arose. I did not wish to be drawn into arguments with him at Cabinets, and always preferred putting things down on paper. In nearly all cases we found ourselves in agreement, and although at first he gave me the impression of being very much on his guard, yet I am glad to say that month by month his confidence and goodwill seemed to grow. His biographer has borne testimony to this. I also wrote to other members of the War Cabinet and to various Ministers with whom I had departmental or other business. The War Cabinet was hampered somewhat by the fact that they seldom sat together alone without secretaries or military experts. It was an earnest and workmanlike body, and the advantages of free discussion among men bound so closely together in a common task, without any formality and without any record being kept, are very great. Such meetings are an essential counterpart to the formal meetings where business is transacted and decisions are recorded for guidance and action. Both processes are indispensable to the handling of the most difficult affairs.

I was deeply interested in the fate of the great mass of heavy artillery which as Minister of Munitions I had made in the previous war. Such weapons take a year and a half to manufacture, but it is of great value to an army, whether in defence or offence, to have at its disposal a mass of heavy batteries. I remembered the struggles which Mr. Lloyd George had had with the War Office in 1915 and all the political disturbance which had arisen on this subject of the creation of a dominating Very Heavy Artillery, and how he had been vindicated by events. The character of the war on land, when it eventually manifested itself eight months later in 1940, proved utterly different from that of 1914–18. As will be seen, however, a vital need in Home Defence was met by these great cannon. At this time I conceived we had a buried treasure which it would be folly to neglect.

I wrote to the Prime Minister on this and other matters:

First Lord to Prime Minister   10.IX.39

I hope you will not mind my sending you a few points privately.

1. I am still inclined to think that we should not take the initiative in bombing, except in the immediate zone in which the French armies are operating, where we must of course help. It is to our interest that the war should be conducted in accordance with the more humane conceptions of war, and that we should follow and not precede the Germans in the process, no doubt inevitable, of deepening severity

406

and violence. Every day that passes gives more shelter to the population of London and the big cities, and in a fortnight or so there will be far more comparatively safe refuges than now.

2. You ought to know what we were told about the condition of our small Expeditionary Force and their deficiencies in tanks, in trained trench-mortar detachments, and above all in heavy artillery. There will be a just criticism if it is found that the heavy batteries are lacking. . . . In 1919, after the war, when I was S. of S. for War, I ordered a mass of heavy cannon to be stored, oiled, and carefully kept; and I also remember making in 1918 two 12-inch hows. at the request of G.H.Q. to support their advance into Germany in 1919. These were never used, but they were the last word at the time. They are not easy things to lose. . . . It seems to me most vitally urgent, first, to see what there is in the cupboard, secondly, to recondition it at once and make the ammunition of a modern character. Where this heavy stuff is concerned I may be able to help at the Admiralty, because of course we are very comfortable in respect of everything big. . . .

3. You may like to know the principles I am following in recasting the Naval programme of new construction. I propose to suspend work upon all except the first three or perhaps four of the new battleships, and not to worry at the present time about vessels that cannot come into action until 1942. This decision must be reviewed in six months. It is by this change that I get the spare capacity to help the Army. On the other hand, I must make a great effort to bring forward the smaller anti-U-boat fleet. Numbers are vital in this sphere. A good many are coming forward in 1940, but not nearly enough considering that we may have to face an attack by 200 or 300 U-boats in the summer of 1940. . . .

4. With regard to the supply of the Army and its relation to the Air Force, pardon me if I put my experience and knowledge, which were bought, not taught, at your disposal. The making by the Minister of Supply of a lay-out on the basis of fifty-five divisions at the present time would not prejudice Air or Admiralty, because (a) the preliminary work of securing the sites and building the factories will not for many months require skilled labour; here are months of digging foundations, laying concrete, bricks and mortar, drainage, etc., for which the ordinary building-trade labourers suffice; and (b) even if you could not realise a fifty-five-division front by the twenty-fourth month because of other claims, you could alter the time to the thirty-sixth month or even later without affecting the scale. On the other hand, if he does not make a big lay-out at the beginning there will be vexatious delays when existing factories have to be enlarged. Let him make his lay-out on the large scale, and protect the needs of the Air

Force and Army by varying the time factor. A factory once set up need not be used until it is necessary, but if it is not in existence you may be helpless if you need a further effort. It is only when these big plants get into work that you can achieve adequate results.

5. Up to the present (noon) no further losses by U-boats are reported—*i.e.*, thirty-six hours blank. Perhaps they have all gone away for the week-end! But I pass my time waiting to be hit. Nevertheless I am sure all will be well.

I also wrote to Dr. Burgin:

First Lord to Minister of Supply   10.IX.39
In 1919, when I was at the War Office, I gave careful instructions to store and oil a mass of heavy artillery. Now it appears that this has been discovered. It seems to me the first thing you should do would be to get hold of this store and recondition them with the highest priority, as well as make the heavy ammunition. The Admiralty might be able to help with the heavy shells. Do not hesitate to ask.

The reply was most satisfactory:

Minister of Supply to First Lord   11.IX.39
The preparation for use of the super-heavy artillery of which you write has been the lively concern of the War Office since the September crisis of 1938, and work actually started on the reconditioning of guns and mountings, both of the 9.2-inch guns and the 12-inch howitzers, last January.

These equipments were put away in 1919 with considerable care, and as a result they are proving to be, on the whole, not in bad condition. Certain parts of them have however deteriorated and require renewal, and this work has been going on steadily throughout this year. We shall undoubtedly have some equipments ready during this month, and of course I am giving the work a high priority. . . .

I am most grateful for your letter. You will be glad to see how much has already been done on the lines you recommend.

*   *   *   *   *

First Lord to Prime Minister   11.IX.39
Everyone says there ought to be a Ministry of Shipping. The President of the Chamber of Shipping to-day pressed me strongly for it at our meeting with the shipowners. The President of the Board of Trade asked me to associate him with this request, which of course entails a curtailment of his own functions. I am sure there will be a strong Parliamentary demand. Moreover, the measure seems to me good on the merits. The functions are threefold:

(a) To secure the maximum fertility and economy of freights in accordance with the war policy of the Cabinet and the pressure of events.

(b) To provide and organise the very large shipbuilding programme necessary as a safeguard against the heavy losses of tonnage we may expect from a U-boat attack apprehended in the summer of 1940. This should certainly include the study of concrete ships, thus relieving the strain on our steel during a period of steel stringency.

(c) The care, comfort, and encouragement of the merchant seamen who will have to go to sea repeatedly after having been torpedoed and saved. These merchant seamen are a most important and potentially formidable factor in this kind of war.

The President of the Board of Trade has already told you that two or three weeks would be required to disentangle the branches of his department which would go to make up the Ministry of Shipping from the parent office. It seems to me very wise to allow this period of transition. If a Minister were appointed and announced, he would gather to himself the necessary personal staff and take over gradually the branches of the Board of Trade which are concerned. It also seems important that the step of creating a Ministry of Shipping should be taken by the Government before pressure is applied in Parliament and from shipping circles, and before we are told that there is valid complaint against the existing system.

* * * * *

This Ministry was formed after a month's discussion and announced on October 13. Mr. Chamberlain selected Sir John Gilmour as its first head. The choice was criticised as being inadequate. Gilmour was a most agreeable Scotsman and a well-known Member of Parliament. He had held Cabinet office under Mr. Baldwin and Mr. Chamberlain. His health was declining, and he died within a few months of his appointment, and was succeeded by Mr. Ronald Cross.

First Lord to Prime Minister   15.IX.39

As I shall be away till Monday I give you my present thought on the main situation.

It seems to me most unlikely that the Germans will attempt an offensive in the West at this late season. . . . Surely his obvious plan should be to press on through Poland, Hungary, and Roumania to the Black Sea, and it may be that he has some understanding with Russia by which she will take part of Poland and recover Bessarabia. . . .

It would seem wise for Hitler to make good his Eastern connections and feeding-grounds during these winter months, and thus give his people the spectacle of repeated successes, and the assurance of weakening our blockade. I do not therefore apprehend that he will attack in the West until he has collected the easy spoils which await him in the East. None the less, I am strongly of opinion that we should make every preparation to defend ourselves in the West. Every effort should be made to make Belgium take the necessary precautions in conjunction with the French and British Armies. Meanwhile the French frontier behind Belgium should be fortified night and day by every conceivable resource. In particular the obstacles to tank attack, planting railway rails upright, digging deep ditches, erecting concrete dolls, land-mines in some parts and inundations all ready to let out in others, etc., should be combined in a deep system of defence. The attack of three or four German armoured divisions, which has been so effective in Poland, can only be stopped by physical obstacles defended by resolute troops and a powerful artillery. . . . Without physical obstacles the attack of armoured vehicles cannot be effectively resisted.

I am very glad to find that the mass of war-time artillery which I stored in 1919 is all available. It comprises 32 12-inch, 145 9-inch, a large number of 8-inch, nearly 200 6-inch, howitzers, together with very large quantities of ammunition; in fact, it is the heavy artillery, not of our small Expeditionary Force, but of a great army. No time should be lost in bringing some of these guns into the field, so that whatever else our troops will lack they will not suffer from want of heavy artillery.

I hope you will consider carefully what I write to you. I do so only in my desire to aid you in your responsibilities and discharge my own.

The Prime Minister wrote back on the 16th, saying:

All your letters are carefully read and considered by me, and if I have not replied to them it is only because I am seeing you every day, and moreover because, as far as I have been able to observe, your views and mine have very closely coincided. . . . To my mind the lesson of the Polish campaign is the power of the Air Force, when it has obtained complete mastery in the air, to paralyse the operations of land forces. . . . Accordingly, as it seems to me, although I shall of course await the report of the Land Forces Committee before making up my mind, absolute priority ought to be given to our plans for rapidly accelerating the strength of our Air Force, and the extent of our effort on land should be determined by our resources *after* we have provided for Air Force extension.

I am entirely with you in believing that Air Power stands foremost in our requirements, and indeed I sometimes think that it may be the ultimate path by which victory will be gained. On the other hand, the Air Ministry paper, which I have just been studying, seems to peg out vast and vague claims which are not at present substantiated, and which, if accorded absolute priority, would overlay other indispensable forms of war effort. I am preparing a note upon this paper, and will only quote one figure which struck me in it.

If the aircraft industry with its present 360,000 men can produce nearly 1,000 machines a month, it seems extraordinary that 1,050,000 men should be required for a monthly output of 2,000. One would expect a very large "reduction on taking a quantity", especially if mass-production is used. I cannot believe the Germans will be using anything like 1,000.000 men to produce 2,000 machines a month. While, broadly speaking, I should accept an output of 2,000 machines a month as the objective, I am not at present convinced that it would make anything like so large a demand upon our war-making capacity as is implied in this paper.

The reason why I am anxious that the Army should be planned upon a fifty- or fifty-five-division scale is that I doubt whether the French would acquiesce in a division of effort which gave us the sea and air and left them to pay almost the whole blood-tax on land. Such an arrangement would certainly be agreeable to us; but I do not like the idea of our having to continue the war single-handed.

There are great dangers in giving absolute priority to any department. In the late war the Admiralty used their priority arbitrarily and selfishly, especially in the last year, when they were overwhelmingly strong, and had the American Navy added to them. I am every day restraining such tendencies in the common interest.

As I mentioned in my first letter to you, the lay-out of the shell, gun, and filling factories, and the provision for explosives and steel, does not compete directly while the plants are being made with the quite different class of labour required for aeroplane production. It is a question of clever dovetailing. The provision of mechanical vehicles, on the other hand, is directly competitive, and must be carefully adjusted. It would be wise to bring the Army munitions plants into existence on a large scale, and then to let them begin to eat only as our resources allow and the character of the war requires. The time factor is the regulator which you would apply according to circumstances. If however the plants are not begun now, you will no longer have the option.

I thought it would be a wise thing to state to the French our intention to work up to an army of fifty or fifty-five divisions. But whether this could be reached at the twenty-fourth month or at the thirtieth or fortieth month should certainly be kept fluid.

At the end of the late war we had about ninety divisions in all theatres, and we were producing aircraft at the rate of 2,000 a month, as well as maintaining a Navy very much larger than was needed, and far larger than our present plans contemplate. I do not therefore feel that fifty or fifty-five divisions and 2,000 aircraft per month are incompatible aims, although of course the modern divisions and modern aircraft represent a much higher industrial effort—everything having become so much more complicated.

First Lord to Prime Minister   21.IX.39

I wonder if you would consider having an occasional meeting of the War Cabinet Ministers to talk among themselves without either secretaries or military experts. I am not satisfied that the large issues are being effectively discussed in our formal sessions. We have been constituted the responsible Ministers for the conduct of the war, and I am sure it would be in the public interest if we met as a body from time to time. Much is being thrown upon the Chiefs of Staff which falls outside the professional sphere. We have had the advantage of many valuable and illuminating reports from them. But I venture to represent to you that we ought sometimes to discuss the general position alone. I do not feel that we are getting to the root of the matter on many points.

I have not spoken to any colleague about this, and have no idea what their opinions are. I give you my own, as in duty bound.

*   *   *   *   *

On September 24 I wrote to the Chancellor of the Exchequer:

I am thinking a great deal about you and your problem, as one who has been through the Exchequer mill. I look forward to a severe Budget based upon the broad masses of the well-to-do. But I think you ought to couple with this a strong anti-waste campaign. Judging by the small results achieved for our present gigantic expenditure, I think there never was so little *"value for money"* as what is going on now. In 1918 we had a lot of unpleasant regulations in force for the prevention of waste, which after all was part of the winning of victory. Surely you ought to make a strong feature of this in your Wednesday's statement. An effort should be made to tell people the things they ought to try to avoid doing. This is by no means a doctrine of abstention from expenditure. Everything should be eaten up prudently, even luxuries, *so long as no more are created.* Take stationery, for example

—this should be regulated at once in all departments. Envelopes should be pasted up and re-directed again and again. Although this seems a small thing, it teaches every official—and we now have millions of them—to think of saving.

An active "saving campaign" was inculcated at the Front in 1918, and people began to take a pride in it, and look upon it as part of the show. Why not inculcate these ideas in the B.E.F. from the outset in all zones not actually under fire?

I am trying to prune the Admiralty of large schemes of naval improvement which cannot operate till after 1941, or even in some cases (when they cannot operate) till after the end of 1940. Beware lest these fortifications people and other departmentals do not consume our strength upon long-scale developments which cannot mature till after the climax which settles our fate.

I see the departments full of loose fat, following on undue starvation. It would be much better from your point of view to come along with your alguazils *as critics* upon wasteful exhibitions, rather than delaying action. Don't hamper departments acting in a time of crisis; give them the responsibility; but call them swiftly to account for any failure in thrift.

I hope you will not mind me writing to you upon this subject, because I feel just as strongly about the husbanding of the money-power as I do about the war effort, of which it is indeed an integral part. In all these matters you can count on my support, and also, as the head of a spending department, upon my submission to searching superintendence.

★　　★　　★　　★　　★

In every war in which the Royal Navy has claimed the command of the seas it has had to pay the price of exposing immense targets to the enemy. The privateer, the raiding cruiser, and above all the U-boat, have in all the varying forms of war exacted a heavy toll upon the life-lines of our commerce and food-supply. A prime function of defence has therefore always been imposed upon us. From this fact the danger arises of our being driven or subsiding into a defensive naval strategy and habit of mind. Modern developments have aggravated this tendency. In the two Great Wars, during parts of which I was responsible for the control of the Admiralty, I always sought to rupture this defensive obsession by searching for forms of counter-offensive. To make the enemy wonder where he is going to be hit next may bring immeasurable relief to the process of shepherding hundreds of convoys and thousands of merchantmen safely into port. In the First World War I hoped to find in the Dardanelles, and later in an attack upon Borkum and other Frisian islands, the means of regaining the initiative and forcing the weaker naval Power to study his own problems rather than ours. Called to the Admiralty again in 1939, and as soon as immediate needs were dealt with and perils warded off, I could not rest content with the policy of "Convoy and blockade". I sought earnestly for a way of attacking Germany by naval means.

First and foremost gleamed the Baltic. The command of the Baltic by a British fleet carried with it possibly decisive gains. Scandinavia, freed from the menace of German invasion, would thereby naturally be drawn into our system of war trade, if not indeed into actual co-belligerency. A British fleet in mastery of the Baltic would hold out a hand to Russia in a manner likely to be decisive upon the whole Soviet policy and strategy. These facts were not disputed among responsible and well-informed men. The command of the Baltic was the obvious supreme prize, not only for the Royal Navy but for Britain. Could it be won? In this new war the German Navy was no obstacle. Our superiority in heavy ships made us eager to engage them wherever and whenever there was opportunity. Minefields could be swept by the stronger naval Power. The U-boats imposed no veto upon a fleet guarded by efficient flotillas. But now, instead of the powerful German Navy of 1914 and 1915, there was the Air Arm, formidable, unmeasured, and certainly increasing in importance with every month that passed.

If two or three years earlier it had been possible to make an alliance with Soviet Russia, this might have been implemented by a British battle squadron joined to the Russian Fleet and based on Cronstadt. I commended this to my circle of friends at the time. Whether such an arrangement was ever within the bounds of action cannot be known. It was certainly one way of restraining Germany; but there were also easier methods which were not taken. Now in the autumn of 1939 Russia was an adverse neutral, balancing between antagonism and actual war. Sweden had several suitable harbours on which a British fleet could be based. But Sweden could not be expected to expose herself to invasion by Germany. Without the command of the Baltic we could not ask for a Swedish harbour. Without a Swedish harbour we could not have the command of the Baltic. Here was a deadlock in strategic thought. Was it possible to break it? It is always right to probe. During the war, as will be seen, I forced long Staff studies of various operations, as the result of which I was usually convinced that they were better left alone, or else that they could not be fitted in with the general conduct of the struggle. Of these the first was the Baltic domination.

*   *   *   *   *

On the fourth day after I reached the Admiralty I asked that a plan for forcing a passage into the Baltic should be prepared by the Naval Staff. The Plans Division replied quickly that Italy and Japan must be neutral; that the threat of air attack appeared prohibitive; but that apart from this the operation justified detailed planning, and should, if judged practicable, be carried out in March 1940, or earlier. Meanwhile I had long talks with the Director of Naval Construction, Sir Stanley Goodall, one of my friends from 1911-12, who was immediately captivated by the idea. I named the plan "Catherine", after Catherine the Great, because Russia lay in the background of my thought. On September 12 I was able to write a detailed minute to the authorities concerned.*

Admiral Pound replied on the 20th that success would depend on Russia not joining Germany and on the assurance of co-operation by Norway and Sweden; and that we must be able to win the war against any probable combination of Powers without counting upon whatever force was sent into the Baltic. He was all for the exploration. On September 21 he agreed that Admiral of the Fleet the Earl of Cork and Orrery, an officer of the highest attainments and distinction, should come to work at the Admiralty, with quarters and a nucleus staff, and all information necessary for exploring and planning the Baltic offensive project. There was an apt precedent for this in the previous war, when I had brought back the famous Admiral "Tug" Wilson to the Admiralty for special duties of this kind with the full agreement of Lord Fisher; and there are several instances in this war where, in an easy and friendly manner, large issues of this kind were tested without any resentment being felt by the Chiefs of Staff concerned.

* See Appendix G.

Both Lord Cork's ideas and mine rested upon the construction of capital ships specially adapted to withstand air and torpedo attack. As is seen from the minute in the Appendix, I wished to convert two or three ships of the *Royal Sovereign* class for action inshore or in narrow waters by giving them super-bulges against torpedoes and strong armour-plated decks against air-bombs. For this I was prepared to sacrifice one or even two turrets and seven or eight knots speed. Quite apart from the Baltic, this would give us facilities for offensive action both off the enemy's North Sea coast and even more in the Mediterranean. Nothing could be ready before the late spring of 1940, even if the earliest estimates of the naval construction and the dockyards were realised. On this basis therefore we proceeded.

On the 26th Lord Cork presented his preliminary appreciation, based of course on a purely military study of the problem. He considered the operation, which he would of course have commanded, perfectly feasible but hazardous. He asked for a margin of at least 30 per cent. over the German Fleet on account of expected losses in the passage. If we were to act in 1940 the assembly of the fleet and all necessary training must be complete by the middle of February. Time did not therefore permit the deck-armouring and side-blistering of the *Royal Sovereigns* on which I counted. Here was another deadlock. Still, if this kind of thing goes working on one may get into position—maybe a year later—to act. But in war, as in life, all other things are moving too. If one can plan calmly with a year or two in hand better solutions are open.

I had strong support in all this from the Deputy Chief of Staff, Admiral Tom Phillips (who perished in the *Prince of Wales* at the end of 1941 near Singapore), and from Admiral Fraser, the Controller and Third Sea Lord. He advised the addition to the assault fleet of the four fast merchant ships of the Glen Line, which were to play their part in other events.

* * * * *

One of my first duties at the Admiralty was to examine the existing programmes of new construction and war expansion which had come into force on the outbreak.

At any given moment there are at least four successive annual programmes running at the Admiralty. In 1936 and 1937 five new battleships had been laid down which would come into service in 1940 and 1941. Four more battleships had been authorised by Parliament in 1938 and 1939, which could not be finished for five or six years from the date of order. Nineteen cruisers were in various stages of construction. The constructive genius and commanding reputation of the Royal Navy in design had been distorted and hampered by the treaty restrictions for twenty years. All our cruisers were the result of trying to conform to treaty limitations and "gentleman's agreements". In peace-time vessels had thus been built to keep up the strength of the Navy from year to year amid political difficulties. In war-time a definite tactical object must inspire all construction. I greatly desired to build a few 14,000-ton cruisers carrying 9.2-inch guns, with good armour against 8-inch projectiles, wide radius of action, and superior speed to any existing *Deutschland* or other German cruiser. Hitherto the treaty restrictions had prevented such a policy. Now that we were free from them, the hard priorities of war interposed an equally decisive veto on such long-term plans.

Destroyers were our most urgent need, and also our worst feature. None had been included in the 1938 programme, but sixteen had been ordered in 1939. In all, thirty-two of these indispensable craft were in the yards, and only nine could be delivered before the end of 1940. The irresistible tendency to make each successive flotilla an improvement upon the last had lengthened the time of building to nearer three than two years. Naturally the Navy liked to have vessels capable of riding out the Atlantic swell and large enough to carry all the modern improvements in gunnery, and especially anti-aircraft defence. It is evident that along this line of solid argument a point is soon reached where one is no longer building a destroyer but a small cruiser. The displacement approaches or even exceeds 2,000 tons, and a crew of more than two hundred sail the seas in these unarmoured ships, themselves an easy prey to any regular cruiser. The destroyer is the chief weapon against the U-boat, but as it grows ever larger it becomes itself a worth-while target. The line is passed where the hunter becomes the hunted. We could not have too many destroyers, but their perpetual improvements and growth imposed severe limitation on the numbers the yards could build, and deadly delay in completion.

On the other hand, there are seldom less than two thousand

British merchant ships at sea, and the sailings in and out of our home ports amounted each week to several hundreds of ocean-going vessels and several thousands of coastwise traders. To bring the convoy system into play, to patrol the Narrow Seas, to guard the hundreds of ports of the British Isles, to serve our bases all over the world, to protect the minesweepers in their ceaseless task, all required an immense multiplication of small armed vessels. Numbers and speed of construction were the dominating conditions.

It was my duty to readjust our programmes to the need of the hour and to enforce the largest possible expansion of anti-U-boat vessels. For this purpose two principles were laid down. Firstly, the long-term programme should be either stopped or severely delayed, thus concentrating labour and materials upon what we could get in the first year or year and a half. Secondly, new types of anti-submarine craft must be devised which were good enough for work on the approaches to the Island, thus setting free our larger destroyers for more distant duties.

On all these questions I addressed a series of minutes to my naval colleagues*:

Having regard to the U-boat menace, which must be expected to renew itself on a much larger scale towards the end of 1940, the type of destroyer to be constructed must aim at numbers and celerity of construction rather than size and power. It ought to be possible to design destroyers which can be completed in under a year, in which case fifty at least should be begun forthwith. I am well aware of the need of a proportion of flotilla leaders and large destroyers capable of ocean service, but the arrival in our fleets of fifty destroyers of the medium emergency type I am contemplating would liberate all larger vessels for ocean work and for combat.

The usual conflict between long-term and short-term policy rises to intensity in war. I prescribed that all work likely to compete with essential construction should be stopped on large vessels which could not come into service before the end of 1940, and that the multiplication of our anti-submarine fleets must be effected by types capable of being built within twelve months, or, if possible, eight. For the first type we revived the name corvette. Orders for fifty-eight of these had been placed shortly before the outbreak of war, but none were yet laid down. Later and im-

* See Appendix H.

proved vessels of a similar type, ordered in 1940, were called frigates. Besides this, a great number of small craft of many kinds, particularly trawlers, had to be converted with the utmost dispatch and fitted with guns, depth-charges, and Asdics; motor launches of new Admiralty design were also required in large numbers for coastal work. Orders were placed to the limit of our shipbuilding resources, including those of Canada. Even so we did not achieve all that we hoped, and delays arose which were inevitable under the prevailing conditions and which caused the deliveries from the shipyards to fall considerably short of our expectations.*

* * * * *

Eventually my view about Baltic strategy and battleship reconstruction prevailed in the protracted discussions. The designs were made and the orders were given. However, one reason after another was advanced, some of them well-founded, for not putting the work in hand. The *Royal Sovereigns*, it was said, might be needed for convoy in case the German pocket-battleships or 8-inch-gun cruisers broke loose. It was represented that the scheme involved unacceptable interference with other vital work, and a plausible case could be shown for alternative priorities for our labour and armour. I deeply regretted that I was never able to achieve my conception of a squadron of very heavily deck-armoured ships of no more than fifteen knots, bristling with anti-aircraft guns and capable of withstanding to a degree not enjoyed by any other vessel afloat both air and under-water attack. When in 1941 and 1942 the defence and succouring of Malta became so vital, when we had every need to bombard Italian ports, and above all Tripoli, others felt the need as much as I. It was then too late.

Throughout the war the *Royal Sovereigns* remained an expense and an anxiety. They had none of them been rebuilt like their sisters the *Queen Elizabeths*, and when, as will be seen in due course, the possibility of bringing them into action against the Japanese fleet which entered the Indian Ocean in April 1942 presented itself the only thought of the Admiral on the spot, of Admiral Pound, and the Minister of Defence, was to put as many thousands of miles as possible between them and the enemy in the shortest possible time.

* * * * *

One of the first steps I took on taking charge of the Admiralty and becoming a member of the War Cabinet was to form a statistical department of my own. For this purpose I relied on Professor Lindemann, my friend and confidant of so many years. Together we had formed our views and estimates about the whole story. I now installed him at the Admiralty with half a dozen statisticians and economists whom we could trust to pay no attention to anything but realities. This group of capable men, with access to all official information, was able, under Lindemann's guidance, to present me continually with tables and diagrams, illustrating the whole war so far as it came within our knowledge. They examined and analysed with relentless pertinacity all the departmental papers which were circulated to the War Cabinet, and also pursued all the inquiries which I wished to make myself.

At this time there was no general Government statistical organisation. Each department presented its tale on its own figures and data. The Air Ministry counted one way, the War Office another. The Ministry of Supply and the Board of Trade, though meaning the same thing, talked different dialects. This led sometimes to misunderstandings and waste of time when some point or other came to a crunch in the Cabinet. I had however from the beginning my own sure, steady source of information, every part of which was integrally related to all the rest. Although at first this covered only a portion of the field, it was most helpful to me in forming a just and comprehensible view of the innumerable facts and figures which flowed out upon us.

CHAPTER XXVI

THE FRONT IN FRANCE

Movement of the B.E.F. to France – Fortification of the Belgian Frontier – Advantages of Aggression – Belgian Neutrality – France and the Offensive – The Maginot Line – Accepted Power of the Defensive – Unattractive French Alternatives – Estimates of the British Chiefs of Staff – Hitler's Error – Relative Strength in the West – Possible German Lines of Attack – Opinion of the British Chiefs of Staff; their Paper of September 18, 1939 – Gamelin Develops Plan D – Instruction No. 8 – Meeting of Allied Supreme Council in Paris on November 17 – Plan D Adopted – Extension of Plan D to Holland.

*I*MMEDIATELY upon the outbreak our Expeditionary Army began to move to France. Whereas before the previous war at least three years had been spent in making the preparations, it was not till the spring of 1938 that the War Office set up a special section for this purpose. Two serious new factors were now present. First, the equipment and organisation of a modern army was far less simple than in 1914. Every division had mechanical transport, was more numerous, and had a much higher proportion of non-fighting elements. Secondly, the extravagant fear of air attack on the troopships and landing-ports led the War Office to use only the southern French harbours and St. Nazaire, which became the principal base. This lengthened the communications of the Army, and in consequence retarded the arrival, deployment, and maintenance of the British troops, and consumed profuse additional numbers along the route.*

Oddly enough, it had not been decided before war on which

---

* Advance parties of the British Expeditionary Force began to land in France on September 4. The 1st Corps were ashore by September 19, and the 2nd Corps by October 3. General Headquarters (G.H.Q.) was set up initially at Le Mans on September 15. The principal movement of troops was made through Cherbourg, with vehicles and stores, through Brest and Nantes, and assembly-points at Le Mans and Laval.

sector of the front our troops should be deployed, but the strong presumption was that it would be south of Lille; and this was confirmed on September 22. By mid-October four British divisions, formed into two Army Corps of professional quality, were in their stations along the Franco-Belgian frontier. This involved a road-and-rail movement of 250 miles from the remote ports which had been chosen for landing. Three infantry brigades which arrived separately during October and November were formed into the 5th Division in December 1939. The 48th Division went out in January 1940, followed by the 50th and 51st Divisions in February, and the 42nd and 44th in March, making a total of ten. As our numbers grew we took over more line. We were not of course at any point in contact with the enemy.

When the B.E.F. reached their prescribed positions they found ready-prepared a fairly complete artificial anti-tank ditch along the front line, and every thousand yards or so was a large and very visible pillbox giving enfilade fire along the ditch for machine and anti-tank guns. There was also a continuous belt of wire. Much of the work of our troops during this strange autumn and winter was directed to improving the French-made defences and organising a kind of Siegfried Line. In spite of frost progress was rapid. Air photographs showed the rate at which the Germans were extending their own Siegfried Line northwards from the Moselle. Despite the many advantages they enjoyed in home resources and forced labour, we seemed to be keeping pace with them. By the time of the May offensive, 1940, our troops had completed 400 new pillboxes. Forty miles of revetted anti-tank ditch had been dug and great quantities of wire spread. Immense demands were made by the long line of communications stretching back to Nantes. Large base installations were created, roads improved, a hundred miles of broad-gauge railway-line laid, an extensive system of buried cable dug in, and several tunnelled headquarters for the corps and Army commands almost completed. Nearly fifty new airfields and satellites were developed or improved with runways, involving over 50,000 tons of concrete.

On all these tasks the Army laboured industriously, and to vary their experiences moved brigades by rotation to a sector of the French front in contact with the enemy near Metz, where there was at least some patrol activity. All the rest of the time was spent by our troops in training. This was indeed necessary. A far lower scale of preparation had been reached when war broke out than that attained by Sir John French's Army a quarter of a century before. For several years no considerable exercise with troops had been held at home. The Regular Army was 20,000 short of establishment, including 5,000 officers, and under the Cardwell system, which had to provide for the defence of India, the greater part of this fell upon the home units, which in consequence became hardly more than cadres. The little-considered, though well-meant, doubling of the Territorial Army in March 1939, and the creation of the Militia in May of that year, both involved drawing heavily upon the Regular Army for instructors. The winter months in France were turned to good account, and every kind of training programme was woven into the prime work of fortification. It is certain that our Army advanced markedly in efficiency during the breathing-space which was granted it, and in spite of exacting toils and the absence of any kind of action its morale and spirit grew.

Behind our front immense masses of stores and ammunition were accumulated in the depots all along the communications. Ten days' supply was gathered between the Seine and the Somme, *and seven days additional north of the Somme.* This latter provision saved the Army after the German break-through. Gradually, in view of the prevailing tranquillity, other ports north of Havre were brought into use in succession. Dieppe became a hospital base, Fécamp was concerned with ammunition; and in the end we were making use in all of thirteen French harbours.

★　　★　　★　　★　　★

The advantage which a Government bound by no law or treaty has over countries which derive their war-impulse only after the criminal has struck, and have to plan accordingly, cannot be measured. It is enormous. On the other hand, unless the victory of the aggressors is absolute and final there may be some day a reckoning. Hitler, unhampered by any restraint except that of superior force, could strike when and where he chose; but the two Western democracies could not violate Belgium's neutrality. The most they could do was to be ready to come to the rescue when called upon by the Belgians, and it was probable that this would never happen until it was too late. Of course, if British and French policy during the five years preceding the war had been of a manly and resolute character, within the sanctity of treaties and the approval of the League of Nations, Belgium might have adhered to her old allies and allowed a common front to be formed. This would have brought immense security, and might perhaps have averted the disasters which were to come.

Such an alliance properly organised would have erected a shield along the Belgian frontier to the sea against that terrible turning movement which had nearly compassed our destruction in 1914 and was to play its part in the ruin of France in 1940. It would also have opened the possibility of a rapid advance from Belgium into the heart-centre of German industry in the Ruhr, and thus added a powerful deterrent upon German aggression. At the worst Belgium could have suffered no harder fate than actually befell her. When we recall the aloofness of the United States; Mr. Ramsay MacDonald's campaign for the disarmament of France; the repeated rebuffs and humiliations which we had accepted in the various German breaches of the Disarmament Clauses of the Treaty; our submission to the German violation of the Rhineland; our acquiescence in the absorption of Austria; our pact at Munich and acceptance of the German occupation of Prague—when we recall all this, no man in Britain or France who in those years was responsible for public action has a right to blame Belgium. In a period of vacillation and appeasement the Belgians clung to neutrality, and vainly comforted themselves with the belief that they could hold the German invader on their fortified frontiers until the British and French Armies could come to their aid.

* * * * *

In 1914 the spirit of the French Army and nation, burning from sire to son since 1870, was vehemently offensive. Their doctrine was that the numerically weaker Power could only meet invasion by the counter-offensive, not only strategic but tactical at every point. At the beginning the French, with their blue tunics and red trousers, marched forward while their bands played the Marseillaise. Wherever this happened the Germans, although invading, sat down and fired upon them with devastating effect. The apostle of the offensive creed, Colonel Grandmaison, had perished in the forefront of the battle for his country and his theme. I have explained in *The World Crisis* why the power of the defensive was predominant from 1914 to 1916 or 1917. The magazine

rifle, which we ourselves had seen used with great effect by hand-
fuls of Boers in the South African War, could take a heavy if
not decisive toll from troops advancing across the open. Besides
this there were the ever-multiplying machine-guns.

Then had come the great battles of the artillery. An area was
pulverised by hundreds and presently by thousands of guns. But
if after heroic sacrifices the French and British advanced together
against the strongly-entrenched Germans, successive lines of
fortifications confronted them; and the crater-fields which their
bombardment had created to quell the first lines of the enemy
became a decisive obstacle to their further progress, even when
they were successful. The only conclusion to be drawn from these
hard experiences was that the defensive was master. Moreover,
in the quarter of a century that had passed the fire-power of
weapons had enormously increased. But this cut both ways; as
will later be apparent.

It was now a very different France from that which had hurled
itself upon its ancient foe in August 1914. The spirit of *revanche*
had exhausted its mission and itself in victory. The chiefs who had
nursed it were long dead. The French people had undergone the
frightful slaughter of a million and a half of their manhood.
Offensive action was associated in the great majority of French
minds with the initial failures of the French onslaught of 1914,
with General Nivelle's repulse in 1917, with the long agonies of
the Somme and Passchendaele, and above all with the sense that
the fire-power of modern weapons was devastating to the
attacker. Neither in France nor in Britain had there been any
effective comprehension of the consequences of the new fact that
armoured vehicles could be made capable of withstanding artillery fire, and could advance a hundred miles a day. An illuminating book on this subject, published some years before by a Commander de Gaulle, had met with no response. The authority of
the aged Marshal Pétain in the Conseil Supérieur de la Guerre had
weighed heavily upon French military thought in closing the
door to new ideas, and especially in discouraging what had been
quaintly called "offensive weapons".

In the after-light the policy of the Maginot Line has often been
condemned. It certainly engendered a defensive mentality. Yet
it is always a wise precaution in defending a frontier of hundreds
of miles to bar off as much as possible by fortifications, and thus

economise in the use of troops in sedentary *rôles* and "canalise" potential invasion. Properly used in the French scheme of war, the Maginot Line would have been of immense service to France. It could have been viewed as presenting a long succession of invaluable sally-ports, and above all as blocking off large sections of the front as a means of accumulating the general reserves or "mass of manœuvre". Having regard to the disparity of the population of France to that of Germany, the Maginot Line must be regarded as a wise and prudent measure. Indeed, it was extraordinary that it should not have been carried forward at least along the river Meuse. It could then have served as a trusty shield, freeing a heavy, sharp, offensive French sword. But Marshal Pétain had opposed this extension. He held strongly that the Ardennes could be ruled out as a channel of invasion on account of the nature of the ground. Ruled out accordingly it was. The offensive conceptions of the Maginot Line were explained to me by General Giraud when I visited Metz in 1937. They were however not carried into effect, and the Line not only absorbed very large numbers of highly-trained regular soldiers and technicians, but exercised an enervating effect both upon military strategy and national vigilance.

The new air-power was justly esteemed a revolutionary factor in all operations. Considering the comparatively small numbers of aircraft available on either side at this time, its effects were even exaggerated, and were held in the main to favour the defensive by hampering the concentrations and communications of great armies once launched in attack. Even the period of the French mobilisation was regarded by the French High Command as most critical on account of the possible destruction of railway centres, although the numbers of German aircraft, like those of the Allies, were far too few for such a task. These thoughts expressed by Air Chiefs followed correct lines, and were justified in the later years of the war, when the air strength had grown ten- or twenty-fold. At the outbreak they were premature.

*   *   *   *   *

It is a joke in Britain to say that the War Office is always preparing for the last war. But this is probably true of other departments and of other countries, and it was certainly true of the French Army. I also rested under the impression of the superior power of the defensive provided it were actively conducted. I had neither the responsibility nor the continuous information to make a new measurement. I knew that the carnage of the previous war had bitten deeply into the soul of the French people. The Germans had been given the time to build the Siegfried Line. How frightful to hurl the remaining manhood of France against this wall of fire and concrete! I print in Appendix O one kind of long-term method (called Cultivator No. 6) by which I then thought the fire-power of the defensive could be overcome. But in my mind's outlook in the opening months of this Second World War I did not dissent from the general view about the defensive, and I believed that anti-tank obstacles and field guns, cleverly posted and with suitable ammunition, could frustrate or break up tanks except in darkness or fog, real or artificial.

In the problems which the Almighty sets his humble servants things hardly ever happen the same way twice over, or if they seem to do so there is some variant which stultifies undue generalisation. The human mind, except when guided by extraordinary genius, cannot surmount the established conclusions amid which it has been reared. Yet we are to see, after eight months of inactivity on both sides, the Hitler inrush of a vast offensive, led by spear-point masses of cannon-proof or heavily-armoured vehicles, breaking up all defensive opposition, and for the first time for centuries, and even perhaps since the invention of gunpowder, making artillery for a while almost impotent on the battlefield. We are also to see that the increase of fire-power made the actual battles less bloody by enabling the necessary ground to be held with very small numbers of men, thus offering a far smaller human target.

* * * * *

No frontier has ever received the same strategic attention and experiment as that which stretches through the Low Countries between France and Germany. Every aspect of the ground, its heights and its waterways, has been studied for centuries in the light of the latest campaign by all the generals and military colleges in Western Europe. At this period there were two lines to which the Allies could advance if Belgium were invaded by Germany and they chose to come to her succour, or which they could occupy by a well-planned secret and sudden scheme, if invited by Belgium. The first of these lines was what may be called the line of the Scheldt. This was no great march from the French frontier and involved little serious risk. At the worst it would do no harm to hold it as a "false front". At the best it might be built up according to events. The second line was far more ambitious. It followed the Meuse through Givet, Dinant, and Namur by Louvain to Antwerp. If this adventurous line was seized by the Allies and held in hard battles the German right-handed swing of invasion would be heavily checked; and if their armies were proved inferior it would be an admirable prelude to the entry and control of the vital centre of Germany's munitions production in the Ruhr.

Since the case of an advance through Belgium without Belgian consent was excluded on grounds of international morality, there only remained an advance from the common Franco-German frontier. An attack due eastwards across the Rhine, north and south of Strasbourg, opened mainly into the Black Forest, which, like the Ardennes, was at that time regarded as bad ground for offensive operations. There was however the question of an advance from the front Strasbourg-Metz north-eastward into the Palatinate. Such an advance, with its right on the Rhine, might gain the control of that river as far north as Coblenz or Cologne. This led into good fighting country; and these possibilities, with many variants, had been a part of the war-games in the Staff Colleges of Western Europe for a good many years. In this sector however the Siegfried Line, with its well-built concrete pillboxes mutually supporting one another and organised in depth with masses of wire, was in September 1939 already formidable. The earliest date at which the French could have mounted a big attack was perhaps at the end of the third week of September. But by that time the Polish campaign had ended. By mid-October the Germans had seventy divisions on the Western Front. The fleeting French numerical superiority in the West was passing. A French offensive from their eastern frontier would have denuded their far more vital northern front. Even if an initial success had been gained by the French armies at the outset, within a month they would have had extreme difficulty in maintaining their conquests in the east, and would have been exposed to the whole force of the German counter-stroke to the north.

This is the answer to the question "Why remain passive till

Poland was destroyed?" But this battle had been lost some years before. In 1938 there was a good chance of victory while Czechoslovakia still existed. In 1936 there could have been no effective opposition. In 1933 a rescript from Geneva would have procured bloodless compliance. General Gamelin cannot be the only one to blame because in 1939 he did not run the risks which had so enormously increased since the previous crises, from which both the French and British Governments had recoiled.

The British Chiefs of Staff Committee estimated that the Germans had by September 18 mobilised at least 116 divisions of all classes, distributed as follows: Western front, 42 divisions; Central Germany, 16 divisions; Eastern front, 58 divisions. We now know from enemy records that this estimate was almost exactly correct. Germany had in all from 108 to 117 divisions. Poland was attacked by 58 of the most matured. There remained 50 or 60 divisions of varying quality. Of these, along the Western front from Aix-la-Chapelle to the Swiss frontier there stood 42 German divisions (14 active, 25 reserve, and 3 Landwehr). The German armour was either engaged in Poland or had not yet come into being, and the great flow of tanks from the factories had hardly begun. The British Expeditionary Force was no more than a symbolic contribution. It was able to deploy two divisions by the first and two more by the second week in October. In spite of the enormous improvement since Munich in their relative strength, the German High Command regarded their situation in the West while Poland was unconquered with profound anxiety, and only Hitler's despotic authority, will-power, and five-times-vindicated political judgment about the unwillingness of France and Great Britain to fight induced or compelled them to run what they deemed an unjustified risk.

Hitler was sure that the French political system was rotten to the core, and that it had infected the French Army. He knew the power of the Communists in France, and that it would be used to weaken or paralyse action once Ribbentrop and Molotov had come to terms and Moscow had denounced the French and British Governments for entering upon a capitalist and imperialist war. He was convinced that Britain was pacifist and degenerate. In his view, though Mr. Chamberlain and M. Daladier had been brought to the point of declaring war by a bellicose minority in England, they would both wage as little of it as they could, and once Poland had been crushed would accept the accomplished fact, as they had done a year before in the case of Czechoslovakia. On the repeated occasions which have been set forth Hitler's instinct had been proved right and the arguments and fears of his generals wrong. He did not understand the profound change which takes place in Great Britain and throughout the British Empire once the signal for war has been given; nor how those who have been the most strenuous for peace turn overnight into untiring toilers for victory. He could not comprehend the mental or spiritual force of our Island people, who, however much opposed to war or military preparation, had through the centuries come to regard victory as their birthright. In any case, the British Army could be no factor at the outset, and he was certain that the French nation had not thrown its heart into the war. This was indeed true. He had his way, and his orders were obeyed.

★   ★   ★   ★   ★

It was thought by our officers that when Germany had completely defeated the Polish Army she would have to keep in Poland some 15 divisions, of which a large proportion might be of low category. If she had any doubts about the Russian pact this total might have to be increased to upwards of 30 divisions in the East. On the least favourable assumption Germany would therefore be able to draw over 40 divisions from the Eastern front, making 100 divisions available for the West. By that time the French would have mobilised 72 divisions in France, in addition to fortress troops equivalent to 12 or 14 divisions, and there would be 4 divisions of the British Expeditionary Force. Twelve French divisions would be required to watch the Italian frontier, making 76 against Germany. The enemy would thus have a superiority of four to three over the Allies, and might also be expected to form additional reserve divisions, bringing this total up to 130 in the near future. Against this the French had 14 additional divisions in North Africa, some of which could be drawn upon, and whatever further forces Great Britain could gradually supply.

In air-power our Chiefs of Staff estimated that Germany could concentrate, after the destruction of Poland, over 2,000 bombers in the West as against a combined Franco-British total of 950.★

★ Actually the German bomber strength at that date was 1,546.

It was therefore clear that once Hitler had disposed of Poland he would be far more powerful on the ground and in the air than the British and French combined. There could therefore be no question of a French offensive against Germany. What then were the probabilities of a German offensive against France?

There were of course three methods open. First, invasion through Switzerland. This might turn the southern flank of the Maginot Line, but had many geographical and strategic difficulties. Secondly, invasion of France across the common frontier. This appeared unlikely, as the German Army was not believed to be fully equipped or armed for a heavy attack on the Maginot Line. And, thirdly, invasion of France through Holland and Belgium. This would turn the Maginot Line, and would not entail the losses likely to be sustained in a frontal attack against permanent fortifications. The Chiefs of Staff estimated that for this attack Germany would require to bring from the Eastern front twenty-nine divisions for the initial phase, with fourteen echelonned behind, as reinforcements to her troops already in the West. Such a movement could not be completed and the attack mounted with full artillery support under three weeks, and its preparation should be discernible by us a fortnight before the blow fell. It would be late in the year for the Germans to undertake so great an operation; but the possibility could not be excluded.

We should of course try to retard the German movement from east to west by air attack upon the communications and concentration areas. Thus a preliminary air battle to reduce or eliminate the Allied air forces by attacks on airfields and aircraft factories might be expected, and so far as England was concerned would not be unwelcome. Our next task would be to deal with the German advance through the Low Countries. We could not meet their attack so far forward as Holland, but it would be in the Allied interest to stem it, if possible, in Belgium. "We understand," wrote the Chiefs of Staff, "that the French idea is that, provided the Belgians are still holding out on the Meuse, the French and British Armies should occupy the line Givet-Namur, the British Expeditionary Force operating on the left. *We consider it would be unsound to adopt this plan unless plans are concerted with the Belgians for the occupation of this line in sufficient time before the Germans advance.... Unless the present Belgian attitude alters and plans can be prepared for early occupation of the Givet-Namur* [also called Meuse-Antwerp] *line, we are strongly of opinion that the German advance should be met in prepared positions on the French frontier."* In this case it would of course be necessary to bomb Belgian and Dutch towns and railway centres used or occupied by German troops.

The subsequent history of this important issue must be recorded. It was brought before the War Cabinet on September 20, and after a brief discussion was remitted to the Supreme War Council. In due course the Supreme War Council invited General Gamelin's comments. In his reply General Gamelin said merely that the question of Plan D (*i.e.*, the advance to the Meuse-Antwerp line) had been dealt with in a report by the French Delegation. In this report the operative passage was, "If the call is made in time the Anglo-French troops will enter Belgium, but not to engage in an encounter battle. Among the recognised lines of defence are the line of the Scheldt and the line Meuse-Namur-Antwerp". After considering the French reply the British Chiefs of Staff submitted another paper to the Cabinet, which discussed the alternative of an advance to the Scheldt, but made no mention at all of the far larger commitments of an advance to the Meuse-

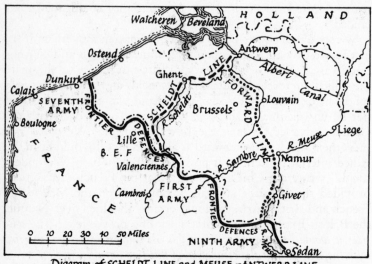

Diagram of SCHELDT LINE and MEUSE–ANTWERP LINE

Antwerp line. When the second report was presented to the Cabinet on October 4 by the Chiefs of Staff no reference was made by them to the all-important alternative of Plan D. It was therefore taken for granted by the War Cabinet that the views of the British Chiefs of Staff had been met and that no further action or decision was required. I was present at both these Cabinets, and was not aware that any significant issue was still pending. During October, there being no effective arrangements with the Belgians, it was assumed that the advance was limited to the Scheldt.

Meanwhile General Gamelin, negotiating secretly with the Belgians, stipulated, first, that the Belgian Army should be maintained at full strength, and, secondly, that Belgian defences should be prepared on the more advanced line from Namur to Louvain. By early November agreement was reached with the Belgians on these points, and from November 5 to 14 a series of conferences was held at Vincennes and La Ferté, at which, or at some of which, Ironside, Newall, and Gort were present. On November 15 General Gamelin issued his Instruction No. 8, confirming the agreements of the 14th, whereby support would be given to the Belgians "if circumstances permitted" by an advance to the line Meuse-Antwerp. The Allied Supreme Council met in Paris on November 17. Mr. Chamberlain took with him Lord Halifax, Lord Chatfield, and Sir Kingsley Wood. I had not at that time reached the position where I should be invited to accompany the Prime Minister to these meetings. The decision was taken: "Given the importance of holding the German forces as far east as possible, it is essential to make every endeavour to hold the line Meuse-Antwerp in the event of a German invasion of Belgium." At this meeting Mr. Chamberlain and M. Daladier insisted on the importance which they attached to this resolution, and thereafter it governed action. This was, in fact, a decision in favour of Plan D, and it superseded the arrangements hitherto accepted of the modest forward move to the Scheldt.

As a new addition to Plan D there presently appeared the task of a Seventh French Army. The idea of an advance of this army on the seaward flank of the Allied armies first came to light early in November 1939. General Giraud, who was restless with a reserve army around Rheims, was put in command. The object of this excursion of Plan D was to move into Holland *via* Antwerp so as to help the Dutch, and secondly to occupy some parts of the Dutch islands Walcheren and Beveland. All this would have been good if the Germans had already been stopped on the Albert Canal. General Gamelin wanted it. General Georges thought it beyond our scope, and preferred that the troops involved should be brought into reserve behind the centre of the line. Of these differences we knew nothing.

In this posture therefore we passed the winter and awaited the spring. No new decisions of strategic principle were taken by the French and British Staffs or by their Governments in the six months which lay between us and the German onslaught.

CHAPTER XXVII

THE COMBAT DEEPENS

Peace Suggestions – The Anglo-French Rejection – Soviet Absorption of the Baltic States – My Views on British Military Preparations – Possible Détente with Italy in the Mediterranean – The Home Front – The Sinking of the "Royal Oak" – My Second Visit to Scapa Flow, October 31 – Decision about the Main Fleet Base – Mr. and Mrs. Chamberlain Dine at Admiralty House – The Loss of the "Rawalpindi" – A False Alarm.

HITLER took advantage of his successes to propose his Peace Plan to the Allies. One of the unhappy consequences of our appeasement policy and generally of our attitude in the face of his rise to power had been to convince him that neither we nor France were capable of fighting a war. He had been unpleasantly surprised by the declaration of Great Britain and France on September 3, but he firmly believed that the spectacle of the swift and crashing destruction of Poland would make the decadent democracies realise that the day when they could exercise influence over the fate of Eastern and Central Europe was gone for ever. He felt very sure at this time of the Russians, gorged as they were with Polish territory and the Baltic States. Indeed, during this month of October he was able to send the captured American merchantman *City of Flint* into the Soviet port of Murmansk under a German prize crew. He had no wish at this stage to continue a war with France and Britain. He felt sure His Majesty's Government would be very glad to accept the decision reached by him in Poland, and that a peace offer would enable Mr. Chamberlain and his old colleagues, having vindicated their honour by a declaration of war, to get out of the scrape into which they had been forced by the war-mongering elements in Parliament. It never occurred to him for a moment that Mr.

Chamberlain and the rest of the British Empire and Common-wealth of Nations now meant to have his blood or perish in the attempt.

The next step taken by Russia after partitioning Poland with Germany was to make three "Mutual Assistance Pacts" with Esthonia, Latvia, and Lithuania. These Baltic States had broken themselves free from the Soviet Government in the War of Liberation of 1918 and 1920. Carrying through drastic land reform largely at the expense of the former German landowners, these small countries evolved a nationalist and peasant way of life strongly anti-Communist in outlook. Ever fearful of their power-ful Soviet neighbour, and desperately anxious to maintain their neutrality, these States attempted to avoid provocations in any direction. Their geographical situation made their task unen-viable. Riga, for example, became a listening-post for news from Russia and an international anti-Bolshevik meeting-place. But the Germans had been content to throw them into their Russian deal, and the Soviet Government now advanced with pent-up hate and eager appetite upon their prey. These three States had formed a part of the Tsarist Empire, and were the old conquests of Peter the Great. They were immediately occupied by strong Russian forces, against which they had no means of effectual resistance. A ferocious liquidation of all anti-Communist and anti-Russian elements was carried through by the usual methods. Great numbers of people who for twenty years had lived in freedom in their native land and had represented the dominant majority of its people disappeared. A large proportion of these were transported to Siberia. The rest went farther. Such was the process described as "Mutual Assistance".

* * * * *

At home we busied ourselves with the expansion of the Army and the Air Force and with all the necessary measures to strengthen our naval power. I continued to submit my ideas to the Prime Minister, and pressed them upon other colleagues as might be acceptable.

First Lord to Prime Minister   1.X.39
  This week-end I venture to write to you about several large issues.
  1. When the peace offensive opens upon us it will be necessary to sustain the French. Although we have nearly a million men under

arms, our contribution is, and must for many months remain, petty. We should tell the French that we are making as great a war effort, though in a different form, as in 1918; that we are constructing an Army of fifty-five divisions, which will be brought into action wherever needed, as fast as it can be trained and supplied, having regard to our great contribution in the air.

At present we have our Regular Army, which produces four or five divisions probably superior to anything in the field. But do not imagine that Territorial divisions will be able after six months' training or so to take their part without needless losses and bad results against German regular troops with at least two years' service and better equipment; or stand at the side of French troops many of whom have had three years' service. The only way in which our forces in France can be rapidly expanded is by bringing the professional troops from India, and using them as the cadre upon which the Territorials and conscripts will form. I do not attempt to go into details now, but in principle 60,000 Territorials should be sent to India to maintain internal security and complete their training, and 40,000 or 45,000 Regular troops should *pari passu* be brought back to Europe. These troops should go into camps in the South of France, where the winter weather is more favourable to training than here, and where there are many military facilities, and become the nucleus and framework of eight or ten good field divisions. The texture of these troops would, by the late spring, be equal to those they will have to meet or stand beside. The fact of this force developing in France during the winter months would be a great encouragement and satisfaction to the French.

2. I was much concerned at the figures put forward by the Air Ministry of their fighting strength. They had 120 squadrons at the outbreak of war, but this actually boiled down to 96 able to go into action. One usually expects that on mobilisation there will be a large expansion. In this case there has been a severe contraction. What has happened is that a large number of squadrons have had to be gutted of trained air personnel, of mechanics, or spare parts, etc., in order to produce a fighting force, and that the débris of these squadrons has been thrown into a big pool called the Reserve. Into this pool will also flow, if the winter months pass without heavy attack, a great mass of new machines and large numbers of trained pilots. Even after making every deduction which is reasonable, we ought to be able to form at least six squadrons a month. It is much better to form squadrons which are held back in reserve than merely to have a large pool of spare pilots, spare machines, and spare parts. The disparity at the present time with Germany is shocking. I am sure this expansion could be achieved if you gave the word.

3. The A.R.P. [Air Raid Precautions] defences and expense are founded upon a wholly fallacious view of the degree of danger to each part of the country which they cover. Schedules should be made of the target areas and of the paths of flight by which they may be approached. In these areas there must be a large proportion of whole-time employees. London is of course the chief [target], and others will readily occur. In these target areas the street-lighting should be made so that it can be controlled by the Air Wardens on the alarm signal being given; and while shelters should be hurried on with and strengthened, night and day, the people's spirits should be kept up by theatres and cinemas until the actual attack begins. Over a great part of the countryside modified lighting should be at once allowed, and places of entertainment opened. No paid A.R.P. personnel should be allowed in these [areas]. All should be on a voluntary basis, the Government contenting itself with giving advice and leaving the rest to local effort. In these areas, which comprise at least seven-eighths of the United Kingdom, gas-masks should be kept at home, and only carried in the target areas as scheduled. There is really no reason why orders to this effect should not be given during the coming week.

*　　*　　*　　*　　*

The disasters which had occurred in Poland and the Baltic States made me all the more anxious to keep Italy out of the war and to build up by every possible means some common interest between us. In the meantime the war went on, and I was busy over a number of administrative matters.

First Lord to Home Secretary   7.X.39

In spite of having a full day's work usually here, I cannot help feeling anxious about the Home Front. You know my views about the needless, and in most parts of the country senseless, severities of these black-outs, entertainment restrictions, and the rest.* But what about petrol? Have the Navy failed to bring in the supplies? Are there not more supplies on the water approaching and probably arriving than would have been ordered had peace remained unbroken? I am told that very large numbers of people and a large part of the business of the country is hampered by the stinting. Surely the proper way to deal with this is to have a ration at the standard price, and allow free purchasing, subject to a heavy tax, beyond it. People will pay for locomotion, the Revenue will benefit by the tax, more cars will come out with registration fees, and the business of the country can go forward.

Then look at these rations, all devised by the Ministry of Food to

* See Appendix L.

win the war. By all means have rations, but I am told that the meat ration for instance is very little better than that of Germany. Is there any need of this when the seas are open?

If we have a heavy set-back from air attack or surface attack, it might be necessary to inflict these severities. Up to the present there is no reason to suppose that the Navy has failed in bringing in the supplies, or that it will fail.

Then what about all these people of middle age, many of whom served in the last war, who are full of vigour and experience, and who are being told by tens of thousands that they are not wanted, and that there is nothing for them except to register at the local Labour Exchange? Surely this is very foolish. Why do we not form a Home Guard of half a million men over forty (if they like to volunteer), and put all our elderly stars at the head and in the structure of these new formations? Let these five hundred thousand men come along and push the young and active out of all the home billets. If uniforms are lacking a brassard would suffice, and I am assured there are plenty of rifles at any rate. I thought from what you said to me the other day that you liked this idea. If so, let us make it work.

I hear continual complaints from every quarter of the lack of organisation on the Home Front. Can't we get at it?

★ ★ ★ ★ ★

Amidst all these preoccupations there burst upon us suddenly an event which touched the Admiralty in a most sensitive spot.

I have mentioned the alarm caused by the report that a U-boat was *inside Scapa Flow*, which had driven the Grand Fleet to sea on the night of October 17, 1914. That alarm was premature. Now, after exactly a quarter of a century almost to a day, it came true. At 1.30 a.m. on October 14, 1939, a German U-boat braved the tides and currents, penetrated our defences, and sank the battleship *Royal Oak* as she lay at anchor. At first, out of a salvo of torpedoes, only one hit the bow, and caused a muffled explosion. So incredible was it to the Admiral and captain on board that a torpedo could have struck them, safe in Scapa Flow, that they attributed the explosion to some internal cause. Twenty minutes passed before the U-boat, for such she was, had reloaded her tubes and fired a second salvo. Then three or four torpedoes, striking in quick succession, ripped the bottom out of the ship. In ten minutes she capsized and sank. Most of the men were at action stations, but the rate at which the ship turned over made it almost impossible for anyone below to escape.

An account based on a German report written at the time may be recorded.

At 01.30 on October 14, 1939, H.M.S. *Royal Oak*, lying at anchor in Scapa Flow, was torpedoed by U.47 (Lieutenant Prien). The operation had been carefully planned by Admiral Doenitz himself, the Flag Officer (Submarines). Prien left Kiel on October 8, a clear, bright autumn day, and passed through Kiel Canal—course N.N.W., Scapa Flow. On October 13, at 4 a.m., the boat was lying off the Orkneys. At 7 p.m.—surface; a fresh breeze blowing, nothing in sight; looming in the half darkness the line of the distant coast; long streamers of Northern Lights flashing blue wisps across the sky. Course west. The boat crept steadily closer to Holm Sound, the eastern approach to Scapa Flow. Unfortunate it was that these channels had not been completely blocked. A narrow passage lay open between two sunken ships. With great skill Prien steered through the swirling waters. The shore was close. A man on a bicycle could be seen going home along the coast road. Then suddenly the whole bay opened out. Kirk Sound was passed. They were in. There under the land to the north could be seen the great shadow of a battleship lying on the water, with the great mast rising above it like a piece of filigree on a black cloth. Near, nearer—all tubes clear—no alarm, no sound but the lap of the water, the low hiss of air pressure and the sharp click of a tube lever. *Los!* [Fire!]—five seconds—ten seconds—twenty seconds. Then came a shattering explosion, and a great pillar of water rose in the darkness. Prien waited some minutes to fire another salvo. Tubes ready. Fire! The torpedoes hit amidships, and there followed a series of crashing explosions. H.M.S. *Royal Oak* sank, with the loss of 786 officers and men, including Rear-Admiral H. E. C. Blagrove (Rear-Admiral Second Battle Squadron). U.47 crept quietly away back through the gap. A blockship arrived twenty-four hours later.

This episode, which must be regarded as a feat of arms on the part of the German U-boat commander, gave a shock to public opinion. It might well have been politically fatal to any Minister who had been responsible for the pre-war precautions. Being a newcomer I was immune from such reproaches in these early months, and, moreover, the Opposition did not attempt to make capital out of the misfortune. On the contrary, Mr. A. V. Alexander was restrained and sympathetic. I promised the strictest inquiry.

On this occasion the Prime Minister also gave the House an account of the German air raids which had been made on October

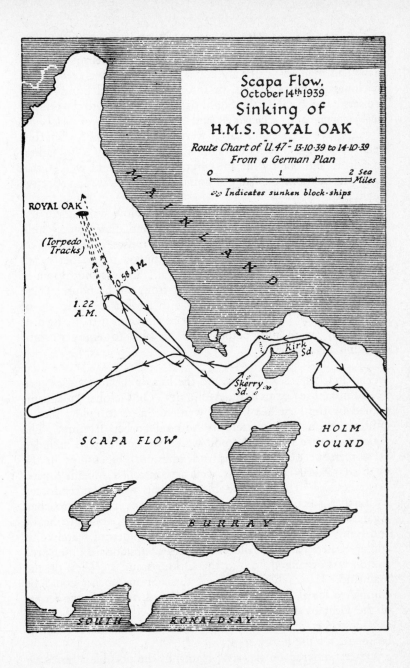

Scapa Flow,
October 14th 1939
Sinking of
H.M.S. ROYAL OAK
Route Chart of "U.47" 13·10·39 to 14·10·39
From a German Plan

0   1   2 Sea Miles

Indicates sunken block-ships

ROYAL OAK

(Torpedo Tracks)

0·58 A.M.

1.22 A.M.

MAINLAND

Kirk Sd.

Skerry Sd.

SCAPA FLOW

HOLM SOUND

BURRAY

SOUTH RONALDSAY

16 upon the Firth of Forth. This was the first attempt the Germans had made to strike by air at our Fleet. Twelve or more machines in flights of two or three at a time had bombed our cruisers lying in the Firth. Slight damage was done to the cruisers *Southampton* and *Edinburgh* and to the destroyer *Mohawk*. Twenty-five officers and sailors were killed or wounded; but four enemy bombers were brought down, three by our fighter squadrons and one by the anti-aircraft fire. It might well be that only half the bombers had got home safely. This was an effective deterrent.

The following morning, the 17th, Scapa Flow was raided, and the old *Iron Duke*, now a demilitarised and disarmoured hulk used as a depot ship, was injured by near misses. She settled on the bottom in shallow water and continued to do her work throughout the war. Another enemy aircraft was shot down in flames. The Fleet was happily absent from the harbour. These events showed how necessary it was to perfect the defences of Scapa against all forms of attack before allowing it to be used. It was nearly six months before we were able to enjoy its commanding advantages.

★   ★   ★   ★   ★

The attack on Scapa Flow and the loss of the *Royal Oak* provoked instant reactions in the Admiralty. On October 31, accompanied by the First Sea Lord, I went to Scapa to hold a second conference on these matters in Admiral Forbes' flagship. The scale of defence for Scapa upon which we now agreed included reinforcement of the booms and additional blockships in the exposed eastern channels, as well as controlled minefields and other devices. These formidable deterrents would be reinforced by further patrol craft and guns sited to cover all approaches. Against air attack it was planned to mount eighty-eight heavy and forty light A.A. guns, together with numerous searchlights and increased barrage-balloon defences. Substantial fighter protection was organised both in the Orkneys and at Wick on the mainland. It was hoped that all these arrangements could be completed, or at least sufficiently advanced, to justify the return of the Fleet by March 1940. Meanwhile Scapa could be used as a destroyer refuelling base; but other accommodation had to be found for the heavy ships.

Experts differed on the rival claims of the possible alternative bases. Admiralty opinion favoured the Clyde, but Admiral Forbes demurred on the ground that this would involve an extra day's steaming each way to his main operational area. This in turn would require an increase in his destroyer forces and would necessitate the heavy ships working in two divisions. The other alternative was Rosyth, which had been our main base in the latter part of the previous war. It was more suitably placed geographically, but was more vulnerable to air attack. The decisions eventually reached at this conference were summed up in a minute which I prepared on my return to London.*

On Friday, November 13, my relations with Mr. Chamberlain had so far ripened that he and Mrs. Chamberlain came to dine with us at Admiralty House, where we had a comfortable flat in the attics. We were a party of four. Although we had been colleagues under Mr. Baldwin for five years, my wife and I had never met the Chamberlains in such circumstances before. By happy chance I turned the conversation on to his life in the Bahamas, and I was delighted to find my guest expand in personal reminiscence to a degree I had not noticed before. He told us the whole story, of which I knew only the barest outline, of his six years' struggle to grow sisal on a barren West Indian islet near Nassau. His father, the great "Joe", was firmly convinced that here was an opportunity at once to develop an Empire industry and fortify the family fortunes. His father and Austen had summoned him in 1890 from Birmingham to Canada, where they had long examined the project. About forty miles from Nassau in the Caribbean Gulf there was a small desert island, almost uninhabited, where the soil was reported to be suitable for growing sisal. After careful reconnaissance by his two sons, Mr. Joseph Chamberlain had acquired a tract on the island of Andros, and assigned the capital required to develop it. All that remained was to grow the sisal. Austen was dedicated to the House of Commons. The task therefore fell to Neville.

Not only in filial duty but with conviction and alacrity he obeyed, and the next five years of his life were spent in trying to grow sisal in this lonely spot, swept by hurricanes from time to time, living nearly naked, struggling with labour difficulties and every other kind of obstacle, and with the town of Nassau as the only gleam of civilisation. He had insisted, he told us, on three

* See Appendix J.

months' leave in England each year. He built a small harbour and landing-stage and a short railroad or tramway. He used all the processes of fertilisation which were judged suitable to the soil and generally led a completely primitive, open-air existence. But no sisal! Or at any rate no sisal that would face the market. At the end of five years he was convinced that the plan could not succeed. He came home and faced his formidable parent, who was by no means contented with the result. I gathered that in the family the feeling was that though they loved him dearly they were sorry to have lost £50,000.

I was fascinated by the way Mr. Chamberlain warmed as he talked, and by the tale itself, which was one of gallant endeavour. I thought to myself, "What a pity Hitler did not know when he met this sober English politician with his umbrella at Berchtesgaden, Godesberg, and Munich that he was actually talking to a hard-bitten pioneer from the outer marches of the British Empire!" This was really the only intimate social conversation that I can remember with Neville Chamberlain amid all the business we did together over nearly twenty years.

During dinner the war went on and things happened. With the soup an officer came up from the War Room below to report that a U-boat had been sunk. With the sweet he came again and reported that a second U-boat had been sunk; and just before the ladies left the dining-room he came a third time reporting that a third U-boat had been sunk. Nothing like this had ever happened before in a single day, and it was more than a year before such a record was repeated. As the ladies left us, Mrs. Chamberlain, with a naïve and charming glance, said to me, "Did you arrange all this on purpose?" I assured her that if she would come again we would produce a similar result.*

* * * * *

Our long, tenuous blockade-line north of the Orkneys, largely composed of armed merchant-cruisers with supporting warships at intervals, was of course always liable to a sudden attack by German capital ships, and particularly by their two fast and most powerful battle-cruisers, the *Scharnhorst* and the *Gneisenau*. We could not prevent such a stroke being made. Our hope was to bring the intruders to decisive action.

* Alas, these hopeful reports are not confirmed by the post-war analysis.

Late in the afternoon of November 23 the armed merchant-cruiser *Rawalpindi*, on patrol between Iceland and the Faroes, sighted an enemy warship which closed her rapidly. She believed the stranger to be the pocket-battleship *Deutschland*, and reported accordingly. Her commanding officer, Captain Kennedy, could have had no illusions about the outcome of such an encounter. His ship was but a converted passenger liner with a broadside of four old 6-inch guns, and his presumed antagonist mounted six 11-inch guns, besides a powerful secondary armament. Nevertheless he accepted the odds, determined to fight his ship to the last. The enemy opened fire at 10,000 yards, and the *Rawalpindi* struck back. Such a one-sided action could not last long, but the fight continued until, with all her guns out of action, the *Rawalpindi* was reduced to a blazing wreck. She sank some time after dark, with the loss of her captain and 270 of her gallant crew. Only 38 survived, 27 of whom were made prisoners by the Germans, the remaining 11 being picked up alive after thirty-six hours in icy water by another British ship.

In fact it was not the *Deutschland* but the two battle-cruisers *Scharnhorst* and *Gneisenau* which were engaged. These ships had left Germany two days before to attack our Atlantic convoys, but having encountered and sunk the *Rawalpindi*, and fearing the consequences of the exposure, they abandoned the rest of their mission and returned at once to Germany. The *Rawalpindi's* heroic fight was not therefore in vain. The cruiser *Newcastle*, near by on patrol, saw the gun-flashes, and responded at once to the *Rawalpindi's* first report, arriving on the scene with the cruiser *Delhi* to find the burning ship still afloat. She pursued the enemy, and at 6.15 p.m. sighted two ships in gathering darkness and heavy rain. One of these she recognised as a battle-cruiser, but lost contact in the gloom, and the enemy made good his escape.

The hope of bringing these two vital German ships to battle dominated all concerned, and the Commander-in-Chief put to sea at once with his whole fleet. When last seen the enemy was retiring to the eastward, and strong forces, including submarines, were promptly organised to intercept him in the North Sea. However, we could not ignore the possibility that having shaken off the pursuit the enemy might renew his advance to the west-ward and enter the Atlantic. We feared for our convoys, and the situation called for the use of all available forces. Sea and air patrols were established to watch all the exits from the North Sea, and a powerful force of cruisers extended this watch to the coast of Norway. In the Atlantic the battleship *Warspite* left her convoy to search the Denmark Strait, and, finding nothing, continued round the north of Iceland to link up with the watchers in the North Sea. The *Hood*, the French battle-cruiser *Dunkerque*, and two French cruisers were dispatched to Icelandic waters, and the *Repulse* and *Furious* sailed from Halifax for the same destination. By the 25th fourteen British cruisers were combing the North Sea, with destroyers and submarines co-operating and with the battle-fleet in support. But fortune was adverse; nothing was found, nor was there any indication of an enemy move to the west. Despite very severe weather the arduous search was maintained for seven days.

On the fifth day, while we were waiting anxiously in the Admiralty and still cherishing the hope that this splendid prize would not be denied us, a German U-boat was heard by our D.F. stations making a report. We judged from this that an attack had been made on one of our warships in the North Sea. Soon the German broadcast claimed that Captain Prien, the sinker of the *Royal Oak*, had sunk an 8-inch-gun cruiser to the eastward of the Shetlands. Admiral Pound and I were together when this news came in. British public opinion is extremely sensitive when British ships are sunk, and the loss of the *Rawalpindi*, after a gallant fight and with a heavy toll in life, would tell seriously against the Admiralty if it remained unavenged. "Why," it would be demanded, "was so weak a ship exposed without effective support? Could the German cruisers range at will even in the blockade zone in which our main forces were employed? Were the raiders to escape unscathed?"

We made a signal at once to clear up the mystery. When we met again an hour later without any reply, we passed through a very bad moment. I recall it because it marked the strong comradeship that had grown up between us and with Admiral Tom Phillips, who was also there. "I take full responsibility," I said, as was my duty. "No, it is mine," said Pound. We wrung each other's hands in lively distress. Hardened as we both were in war, it is not possible to sustain such blows without the most bitter pangs.

But it proved to be nobody's fault. Eight hours later it appeared that the *Norfolk* was the ship involved and that she was undamaged. She had not encountered any U-boat, but said that an air bomb had fallen close astern. However, Captain Prien was no braggart.* What the *Norfolk* thought to be an air bomb from a clouded sky was in fact a German torpedo, which had narrowly missed its target and exploded in the ship's wake. Peering through the periscope, Prien had seen the great upheaval of water, blotting out the ship from his gaze. He dived to avoid an expected salvo. When, after half an hour, he rose for another peep the visibility was poor and no cruiser was to be seen. Hence his report. Our relief after the pain we had suffered took some of the sting out of the news that the *Scharnhorst* and the *Gneisenau* had safely re-entered the Baltic. It is now known that the *Scharnhorst* and *Gneisenau* passed through our cruiser line patrolling near the Norwegian coast on the morning of November 26. The weather was thick and neither saw the other. Modern Radar would have ensured contact, but then it was not available. Public impressions were unfavourable to the Admiralty. We could not bring home to the outside world the vastness of the seas or the intense exertions which the Navy was making in so many areas. After more than two months of war and several serious losses we had nothing to show on the other side. Nor could we yet answer the question, "What is the Navy doing?"

* See Appendix N.

CHAPTER XXVIII

THE MAGNETIC MINE

November and December 1939

Conference with Admiral Darlan – The Anglo-French Naval Position – M. Campinchi – The Northern Barrage – The Magnetic Mine – A Devoted Deed – Technical Aspects – Mine-sweeping Methods – "Degaussing" – The Magnetic Mining Attack Mastered and under Control – Retaliation – Fluvial Mines in the Rhine – Operation "Royal Marine".

IN THE FIRST DAYS of November I paid a visit to France for a conference on our joint operations with the French naval authorities. Admiral Pound and I drove out about forty miles from Paris to the French Marine Headquarters, which were established in the park around the ancient château of the Duc de Noailles. Before we went into the conference Admiral Darlan explained to me how Admiralty matters were managed in France. The Minister of Marine, M. Campinchi, was not allowed by him to be present when operational matters were under discussion. These fell into the purely professional sphere. I said that the First Sea Lord and I were one. Darlan said he recognised this, but in France it was different. "However," he said, "Monsieur le Ministre will arrive for luncheon." We then ranged over naval business for two hours with a great measure of agreement. At luncheon M. Campinchi turned up. He knew his place, and now presided affably over the meal. My son-in-law, Duncan Sandys, who was acting as my *aide*, sat next to Darlan. The Admiral spent most of luncheon explaining to him the limits to which the civilian Minister was restricted by the French system. Before leaving I called on the Duke in his château. He and his family seemed plunged in melancholy, but showed us their very beautiful house and its art treasures.

In the evening I gave a small dinner in a private room at the Ritz to M. Campinchi. I formed a high opinion of this man. His patriotism, his ardour, his acute intelligence, and above all his resolve to conquer or die, hit home. I could not help mentally comparing him with the Admiral, who, jealous of his position, was fighting on quite a different front from ours. Pound's valuation was the same as mine, although we both realised all that Darlan had done for the French Navy. One must not underrate Darlan, nor fail to understand the impulse that moved him. He deemed himself the French Navy, and the French Navy acclaimed him their chief and their reviver. For seven years he had held his office while shifting Ministerial phantoms had filled the office of Minister of Marine. It was his obsession to keep the politicians in their place as chatterboxes in the Chamber. Pound and I got on very well with Campinchi. This tough Corsican never flinched or failed. When he died, broken and under the scowl of Vichy, towards the beginning of 1941, his last words were of hope in me. I shall always deem them an honour.

Here is the statement summing up our naval position at this moment, which I made at the conference:

STATEMENT TO THE FRENCH ADMIRALTY BY THE FIRST LORD

The naval war alone has opened at full intensity. The U-boat attack on commerce, so nearly fatal in 1917, has been controlled by the Anglo-French anti-submarine craft. We must expect a large increase in German U-boats (and possibly some will be lent to them by Russia). This need cause no anxiety provided that all our counter-measures are taken at full speed and on the largest scale. The Admiralty representatives will explain in detail our large programmes. But the full development of these will not come till late in 1940. In the meanwhile it is indispensable that every anti-submarine craft available should be finished and put in commission.

2. There is no doubt that our Asdic method is effective, and far better than anything known in the last war. It enables two torpedo-boats to do what required ten in 1917-18. But this applies only to hunting. For convoys numbers are still essential. One is only safe when escorted by vessels fitted with Asdics. This applies to warships equally with merchant convoys. The defeat of the U-boat will be achieved when it is certain that any attack on French or British vessels will be followed by an Asdic counter-attack.

The British Admiralty is prepared to supply and fit every French anti-submarine craft with Asdics. The cost is small, and accounts can be regulated later on. But any French vessels sent to England for fitting will be immediately taken in hand; and also we will arrange for the imparting of the method and for training to be given in each case. It would be most convenient to do this at Portland, the home of the Asdics, where all facilities are available. We contemplate making provision for equipping fifty French vessels if desired.

3. But we earnestly hope that the French Marine will multiply their Asdic vessels, and will complete with the utmost rapidity all that can enter into action during 1940. After this is arranged for it will be possible six months hence to consider 1941. For the present let us aim at 1940, and especially at the spring and summer. The six large destroyers laid down in 1936 and 1937 will be urgently needed for ocean convoys before the climax of the U-boat warfare is reached in 1940. There are also fourteen small destroyers laid down in 1939, or now projected, which will play an invaluable part without making any great drain on labour and materials. Total, twenty vessels, which could be completed during 1940, and which, fitted with Asdics by us, would be weapons of high consequence in the destruction of the U-boat offensive of 1940. We also venture to mention as most desirable vessels the six sloop-minesweepers laid down in 1936, and twelve laid down in 1937, and also the sixteen submarine-chasers of the programme of 1938. For all these we offer Asdics and every facility. We will fit them as they are ready, as if it were a field operation. We cannot however consider these smaller vessels in the same order of importance as the large and small destroyers mentioned above.

4. It must not be forgotten that defeat of the U-boats carries with it the sovereignty of all the oceans of the world for the Allied Fleets, and the possibility of powerful neutrals coming to our aid, as well as the drawing of resources from every part of the French and British Empires, and the maintenance of trade, gathering with it the necessary wealth to continue the war.

5. At the British Admiralty we have drawn a sharp line between large vessels which can be finished in 1940 and those of later periods. In particular, we are straining every nerve to finish the *King George V* and the *Prince of Wales* battleships within that year, if possible by the autumn. This is necessary because the arrival of the *Bismarck* on the oceans before these two ships were completed would be disastrous in the highest degree, as it can neither be caught nor killed, and would therefore range freely throughout the oceans, rupturing all communications. But France has also a vessel of the highest importance in the *Richelieu*, which might be ready in the autumn of 1940 or even earlier, and will certainly be needed if the two new Italian ships should be

finished by the dates in 1940 at which they profess to aim. Not to have these three capital ships in action before the end of 1940 would be an error in naval strategy of the gravest character, and might entail not only naval but diplomatic consequences extremely disagreeable. It is hoped therefore that every effort will be made to complete the *Richelieu* at the earliest possible date.

With regard to later capital ships of the British and French Navies, it would be well to discuss these in April or May next year, when we shall see much more clearly the course and character of the war.

6. The British Admiralty now express their gratitude to their French colleagues and comrades for the very remarkable assistance which they have given to the common cause since the beginning of this war. This assistance has gone far beyond any promises or engagements made before the war. In escorting home the convoys from Sierra Leone the French cruisers and destroyers have played a part which could not otherwise have been supplied, and which, if not forthcoming, would simply have meant more slaughter of Allied merchantmen. The cruisers and contre-torpilleurs which, with the *Dunkerque*, have covered the arrival of convoys in the Western Approaches were at the time the only means by which the German raiders could be warded off. The maintenance of the French submarines in the neighbourhood of Trinidad has been a most acceptable service. Above all, the two destroyers which constantly escort the homeward- and outward-bound convoys between Gibraltar and Brest are an important relief to our resources, which, though large and ever-growing, are at full strain.

Finally, we are extremely obliged by the facilities given to the *Argus* aircraft-carrier to carry out her training of British naval aircraft pilots under the favourable conditions of Mediterranean weather.

7. Surveying the more general aspects of the war: the fact that the enemy have no line of battle has enabled us to disperse our naval forces widely over the oceans, and we have seven or eight British hunting units, joined by two French hunting units, each capable of catching and killing a *Deutschland*. We are now cruising in the North Atlantic, the South Atlantic, and the Indian Oceans. The result has been that the raiders have not chosen to inflict the losses upon the convoys which before the war it had been supposed they could certainly do. The fact that certainly one, and perhaps two, *Deutschlands* have been upon the main Atlantic trade routes for several weeks without achieving anything makes us feel easier about this form of attack, which had formerly been rated extremely dangerous. We cannot possibly exclude its renewal in a more energetic form. The British Admiralty think it is not at all objectionable to keep large vessels in suitable units ranging widely over the oceans, where they are safe from air attack, and

make effective and apparent the control of the broad waters for the Allies.

8. We shall shortly be engaged in bringing the leading elements of the Canadian and Australian armies to France, and for this purpose a widespread disposition of all our hunting groups is convenient. It will also be necessary to give battleship escorts to many of the largest convoys crossing the Atlantic Ocean. We intend to maintain continually the Northern Blockade from Greenland to Scotland, in spite of the severities of the winter. Upon this blockade twenty-five armed merchant-cruisers will be employed in reliefs, supported by four 8-inch-gun 10,000-ton cruisers, and behind these we always maintain the main fighting forces of the British Navy, to wit, the latest battleships, and either the *Hood* or another great vessel, the whole sufficient to engage or pursue the *Scharnhorst* and the *Gneisenau* should they attempt to break out. We do not think it likely, in view of the situation in the Baltic, that these two vessels will be so employed. Nevertheless, we maintain continually the forces necessary to cope with them.

It is hoped that by a continuance of this strategy by the two Allied Navies no temptation will be offered to Italy to enter the war against us, and that the German power of resistance will certainly be brought to an end.

The French Admiralty in their reply explained that they were in fact proceeding with the completion of the vessels specified, and that they gladly accepted our Asdic offer. Not only would the *Richelieu* be finished in the summer of 1940, but also in the autumn the *Jean Bart*.

* * * * *

In mid-November Admiral Pound presented me with proposals for re-creating the minefield barrage between Scotland and Norway which had been established by the British and American Admiralties in 1917–18. I did not like this kind of warfare, which is essentially defensive, and seeks to substitute material on a vast scale for dominating action. However, I was gradually worn down and reconciled. I submitted the project to the War Cabinet on November 19.

THE NORTHERN BARRAGE
MEMORANDUM BY THE FIRST LORD OF THE ADMIRALTY

After much consideration I commend this project to my colleagues. There is no doubt that, as it is completed, it will impose a very great deterrent upon the exit and return of U-boats and surface raiders. It appears to be a prudent provision against an intensification of the U-boat warfare, and an insurance against the danger of Russia joining our enemy. By this we coop the lot in, and have complete control of all approaches alike to the Baltic and the North Sea. The essence of this offensive minefield is that the enemy will be prevented by the constant vigilance of superior naval force from sweeping channels through it. When it is in existence we shall feel much freer in the outer seas than at present. Its gradual but remorseless growth, which will be known to the enemy, will exercise a depressive effect upon his morale. The cost is deplorably heavy, but a large provision has already been made by the Treasury, and the Northern Barrage is far the best method of employing this means of war [*i.e.*, mining].

This represented the highest professional advice, and of course is just the kind of thing that passes easily through a grave, wise Cabinet. Events swept it away; but not until a great deal of money had been spent. The barrage mines came in handy later on for other tasks.

*　　*　　*　　*　　*

Presently a new and formidable danger threatened our life. During September and October nearly a dozen merchant ships were sunk at the entrance of our harbours, although these had been properly swept for mines. The Admiralty at once suspected that a magnetic mine had been used. This was no novelty to us; we had even begun to use it on a small scale at the end of the previous war. In 1936 an Admiralty committee had studied countermeasures against magnetic-firing devices, but their work had dealt chiefly with countering magnetic torpedoes or buoyant mines, and the terrible damage that could be done by large ground-mines laid in considerable depth by ships or aircraft had not been fully realised. Without a specimen of the mine it was impossible to devise the remedy. Losses by mines, largely Allied and neutral, in September and October had amounted to 56,000 tons, and in November Hitler was encouraged to hint darkly at his new "secret weapon" to which there was no counter. One night when I was at Chartwell Admiral Pound came down to see me in serious anxiety. Six ships had been sunk in the approaches to the Thames. Every day hundreds of ships went in and out of British harbours, and our survival depended on their movement. Hitler's experts may well have told him that this form of attack would compass our ruin. Luckily he began on a small scale, and with limited stocks and manufacturing capacity.

Fortune also favoured us more directly. On November 22 between 9 and 10 p.m. a German aircraft was observed to drop a large object attached to a parachute into the sea near Shoeburyness. The coast here is girdled with great areas of mud which uncover with the tide, and it was immediately obvious that whatever the object was it could be examined and possibly recovered at low water. Here was our golden opportunity. Before midnight that same night two highly-skilled officers, Lt.-Commanders Ouvry and Lewis, from H.M.S. *Vernon*, the naval establishment responsible for developing underwater weapons, were called to the Admiralty, where the First Sea Lord and I interviewed them and heard their plans. By 1.30 in the morning they were on their way by car to Southend to undertake the hazardous task of recovery. Before daylight on the 23rd, in pitch-darkness, aided only by a signal lamp, they found the mine some 500 yards below high-water mark, but as the tide was then rising they could only inspect it and make their preparations for attacking it after the next high water.

The critical operation began early in the afternoon, by which time it had been discovered that a second mine was also on the mud near the first. Ouvry with Chief Petty Officer Baldwin tackled the first, whilst their colleagues, Lewis and Able Seaman Vearncombe, waited at a safe distance in case of accidents. After each prearranged operation Ouvry would signal to Lewis, so that the knowledge gained would be available when the second mine came to be dismantled. Eventually the combined efforts of all four men were required on the first, and their skill and devotion were amply rewarded. That evening some of the party came to the Admiralty to report that the mine had been recovered intact and was on its way to Portsmouth for detailed examination. I received them with enthusiasm. I gathered together eighty or a hundred officers and officials in our largest room, and a thrilled audience listened to the tale, deeply conscious of all that was at stake. From this moment the whole position was transformed. Immediately the knowledge derived from past research could be applied to devising practical measures for combating the particular characteristics of the mine.

The whole power and science of the Navy were now applied;

and it was not long before trial and experiment began to yield practical results. Rear-Admiral Wake-Walker was appointed to co-ordinate all technical measures which the occasion demanded. We worked all ways at once, devising first active means of attacking the mine by new methods of minesweeping and fuze-provocation, and, secondly, passive means of defence for all ships against possible mines in unswept, or ineffectually swept, channels. For this second purpose a most effective system of demagnetising ships by girdling them with an electric cable was developed. This was called "degaussing", and was at once applied to ships of all types. Merchant ships were thus equipped in all our major ports without appreciably delaying their turn-round. In the Fleet progress was simplified by the presence of the highly-trained technical staffs of the Royal Navy. The reader who does not shrink from technical details will find an account of these developments in Appendix M.

*　*　*　*　*

Serious casualties continued. The new cruiser *Belfast* was mined in the Firth of Forth on November 21, and on December 4 the battleship *Nelson* was mined whilst entering Loch Ewe. Both ships were however able to reach a dockyard port. Two destroyers were lost, and two others, besides the minelayer *Adventure*, were damaged on the East Coast during this period. It is remarkable that German Intelligence failed to pierce our security measures covering the injury to the *Nelson* until the ship had been repaired and was again in service. Yet from the first many thousands in England had to know the true facts.

Experience soon gave us new and simpler methods of degaussing. The moral effect of its success was tremendous, but it was on the faithful, courageous, and persistent work of the mine-sweepers and the patient skill of the technical experts, who devised and provided the equipment they used, that we relied chiefly to defeat the enemy's efforts. From this time onward, despite many anxious periods, the mine menace was always under control, and eventually the danger began to recede. By Christmas Day I was able to write to the Prime Minister:

December 25, 1939
Everything is very quiet here, but I thought you would like to know that we have had a marked success against the magnetic mines. The

first two devices for setting them off which we have got into action have both proved effective. Two mines were blown up by the magnetic sweep and two by lighters carrying heavy coils. This occurred at Port A [Loch Ewe], where our interesting invalid [the *Nelson*] is still waiting for a clear passage to be swept for her to the convalescent home at Portsmouth. It also looks as if the demagnetisation of warships and merchant ships can be accomplished by a simple, speedy, and inexpensive process. All our best devices are now approaching [completion]. The aeroplanes and the magnetic ship—the *Borde*—will be at work within the next ten days, and we all feel pretty sure that the danger from magnetic mines will soon be out of the way.

We are also studying the possible varying of this form of attack, viz., acoustic mines and supersonic mines. Thirty ardent experts are pursuing these possibilities, but I am not yet able to say that they have found a cure. . . .

It is well to ponder this side of the naval war. In the event a significant proportion of our whole war effort had to be devoted to combating the mine. A vast output of material and money was diverted from other tasks, and many thousands of men risked their lives night and day in the minesweepers alone. The peak figure was reached in June 1944, when nearly sixty thousand were thus employed. Nothing daunted the ardour of the Merchant Navy, and their spirits rose with the deadly complications of the mining attack and our effective measures for countering it. Their toils and tireless courage were our salvation. The sea traffic on which we depended for our existence proceeded without interruption.

*   *   *   *   *

The first impact of the magnetic mine had stirred me deeply, and, apart from all the protective measures which had been enforced upon us, I sought for a means of retaliation. My visit to the Rhine on the eve of the war had focused my mental vision upon this supreme and vital German artery. Even in September I had raised discussion in the Admiralty about the launching or dropping of fluvial mines in the Rhine. Considering that this river was used by the traffic of many neutral nations, we could not of course take action unless and until the Germans had taken the initiative in this form of indiscriminate warfare against us. Now that they had done so, it seemed to me that the proper retort for indiscriminate sinkings by mines at the mouths of the British harbours was a similar and if possible more effective mining attack upon the Rhine.

Accordingly on November 19 I issued several minutes, of which the following gives the most precise account of the plan:

Controller [and others]

1. As a measure of retaliation it may become necessary to feed large numbers of floating mines into the Rhine. This can easily be done at any point between Strasbourg and the Lauter, where the left bank is French territory. General Gamelin was much interested in this idea, and asked me to work it out for him.

2. Let us clearly see the object in view. The Rhine is traversed by an enormous number of very large barges, and is the main artery of German trade and life. These barges, built only for river work, have not got double keels or any large subdivision by bulkheads. It is easy to check these details. In addition there are at least twelve bridges of boats recently thrown across the Rhine upon which the German armies concentrated in the Saarbruck-Luxemburg area depend.

3. The type of mine required is therefore a small one, perhaps no bigger than a football. The current of the river is at most about seven miles an hour, and three or four at ordinary times, but it is quite easy to verify this. There must therefore be a clockwork apparatus in the mine which makes it dangerous only after it has gone a certain distance, so as to be clear of French territory and also so as to spread the terror farther down the Rhine to its confluence with the Moselle and beyond. The mine should automatically sink, or preferably explode, by this apparatus before reaching Dutch territory. After the mine has proceeded the required distance, which can be varied, it should explode on a light contact. It would be a convenience if, in addition to the above, the mine could go off if stranded after a certain amount of time, as it might easily spread alarm on either of the German banks.

4. It would be necessary in addition that the mine should float a convenient distance beneath the surface so as to be invisible in the turgid waters. A hydrostatic valve actuated by a small cylinder of compressed air should be devised. I have not made the calculations, but I should suppose 48 hours would be the maximum for which it would have to work. An alternative would be to throw very large numbers of camouflage globes—tin shells—into the river, which would spread confusion and exhaust remedial activities.

5. What can they do against this? Obviously nets would be put across; but wreckage passing down the river would break these nets, and except at the frontier they would be a great inconvenience to the traffic. Anyhow, when our mine fetched up against them it would

explode, breaking a large hole in the nets, and after a dozen or more of these explosions the channel would become free again, and other mines would jog along. Specially large mines might be used to break the nets. I cannot think of any other method of defence, but perhaps some may occur to the officers entrusted with this study.

6. Finally, as very large numbers of these mines would be used and the process kept up night after night for months on end, so as to deny the use of the waterway, it is necessary to bear in mind the simplification required for mass production.

The War Cabinet liked this plan. It seemed to them only right and proper that when the Germans were using the magnetic mine to waylay and destroy all traffic, Allied or neutral, entering British ports we should strike back by paralysing, as we might well do, the whole of their vast traffic on the Rhine. The necessary permissions and priorities were obtained, and work started at full speed. In conjunction with the Air Ministry we developed a plan for mining the Ruhr section of the Rhine by discharge from aeroplanes. I entrusted all this work to Rear-Admiral FitzGerald, serving under the First Sea Lord. This brilliant officer, who perished later in command of an Atlantic convoy, made an immense personal contribution. The technical problems were solved. A good supply of mines was assured; and several hundred ardent British sailors and marines were organised to handle them when the time should come. All this was in November, and we could not be ready before March. It is always agreeable in peace or war to have something positive coming along on your side.

CHAPTER XXIX

THE ACTION OFF THE RIVER PLATE

Surface Raiders – The German Pocket-Battleship – Orders of the German Admiralty – British Hunting Groups – The American Three-Hundred-Mile Limit – Offer of Our Asdics to the United States – Anxieties at Home – Caution of the "Deutschland" – Daring of the "Graf Spee" – Captain Langsdorff's Manœuvres – Commodore Harwood's Squadron off the Plate – His Foresight and Fortune – Collision on December 13 – Langsdorff's Mistake – The "Exeter" Disabled – Retreat of the German Pocket-Battleship – Pursuit by "Ajax" and "Achilles" – The "Spee" Takes Refuge in Montevideo – My Letter of December 17 to the Prime Minister – British Concentration on Montevideo – Langsdorff's Orders from the Fuehrer – Scuttling of the "Spee" – Langsdorff's Suicide – End of the First Surface Challenge to British Commerce – The "Altmark" – The "Exeter" – Effects of the Action off the Plate – My Telegram to President Roosevelt.

LTHOUGH it was the U-boat menace from which we suffered most and ran the greatest risks, the attack on our ocean commerce by surface raiders would have been even more formidable could it have been sustained. The three German pocket-battleships permitted by the Treaty of Versailles had been designed with profound thought as commerce-destroyers. Their six 11-inch guns, their 26-knot speed, and the armour they carried had been compressed with masterly skill into the limits of a 10,000-ton displacement. No single British cruiser could match them. The German 8-inch-gun cruisers were more modern than ours, and if employed as commerce-raiders would also be a formidable threat. Besides this the enemy might use disguised heavily-armed merchantmen. We had vivid memories of the depredations of the *Emden* and *Koenigsberg* in 1914, and of the thirty or more warships and armed merchantmen they had forced us to combine for their destruction.

There were rumours and reports before the outbreak of the new war that one or more pocket-battleships had already sailed from Germany. The Home Fleet searched but found nothing. We now know that both the *Deutschland* and the *Admiral Graf Spee* sailed from Germany between August 21 and 24, and were already through the danger zone and loose in the oceans before our blockade and northern patrols were organised. On September 3 the *Deutschland*, having passed through the Denmark Strait, was lurking near Greenland. The *Graf Spee* had crossed the North Atlantic trade route unseen and was already far south of the Azores. Each was accompanied by an auxiliary vessel to replenish fuel and stores. Both at first remained inactive and lost in the ocean spaces. Unless they struck they won no prizes. Until they struck they were in no danger.

The orders of the German Admiralty issued on August 4 were well conceived:

TASK IN THE EVENT OF WAR

Disruption and destruction of enemy merchant shipping by all possible means. . . . Enemy naval forces, even if inferior, are only to be engaged if it should further the principal task. . . .

Frequent changes of position in the operational areas will create uncertainty and will restrict enemy merchant shipping, even without tangible results. A temporary departure into distant areas will also add to the uncertainty of the enemy.

If the enemy should protect his shipping with superior forces so that direct successes cannot be obtained, then the mere fact that his shipping is so restricted means that we have greatly impaired his supply situation. Valuable results will also be obtained if the pocket-battleships continue to remain in the convoy area.

With all this wisdom the British Admiralty would have been in rueful agreement.

★   ★   ★   ★   ★

On September 30 the British liner *Clement*, of 5,000 tons, sailing independently, was sunk by the *Graf Spee* off Pernambuco. The news electrified the Admiralty. It was the signal for which we had been waiting. A number of hunting groups were immediately formed, comprising all our available aircraft-carriers, supported by battleships, battle-cruisers, and cruisers.

Each group of two or more ships was judged to be capable of catching and destroying a pocket-battleship.

In all, during the ensuing months the search for two raiders entailed the formation of nine hunting groups, comprising twenty-three powerful ships. We were also compelled to provide three battleships and two cruisers as additional escorts with the important North Atlantic convoys. These requirements represented a very severe drain on the resources of the Home and Mediterranean Fleets, from which it was necessary to withdraw twelve ships of the most powerful types, including three aircraft-carriers. Working from widely-dispersed bases in the Atlantic and Indian Oceans, the hunting groups could cover the main focal areas traversed by our shipping. To attack our trade the enemy must place himself within reach of at least one of them. To give an idea of the scale of these operations, I set out overleaf the full list of the hunting groups at their highest point.

* * * * *

At this time it was the prime objective of the American Government to keep the war as far from their shores as possible. On October 3 delegates of twenty-one American republics, assembled at Panama, decided to declare an American Security Zone, proposing to fix a belt of from three hundred to six hundred miles from their coasts within which no warlike act should be committed. We were anxious to help in keeping the war out of American waters—to some extent, indeed, this was to our advantage. I therefore hastened to inform President Roosevelt that if America asked all belligerents to respect such a zone we should immediately declare our readiness to fall in with their wishes— subject of course to our rights under international law. We should not mind how far south the Security Zone went, provided that it was effectively maintained. We should have found great difficulty in accepting a Security Zone which was to be policed only by some weak neutral; but if the United States Navy was to take care of it we should feel no anxiety. The more United States warships there were cruising along the South American coast the better we should be pleased; for the German raider which we were hunting might then prefer to leave American waters for the South African trade route, where we were ready to deal with him. But if a surface raider operated from the American Security

Zone or took refuge in it we should expect either to be protected or to be allowed to protect ourselves from the mischief which he might do.

At this date we had no definite knowledge of the sinking of three ships on the Cape of Good Hope route which occurred between October 5 and 10. All three were sailing homeward independently. No distress messages were received, and suspicion was only aroused when they became overdue. It was some time before it could be assumed that they had fallen victims to a raider.

The necessary dispersion of our forces caused me and others anxiety, especially as our main Fleet was sheltering on our west coast.

First Sea Lord and Deputy Chief of the Naval Staff 21.X.39

The appearance of *Scheer* off Pernambuco and subsequent mystery of her movements, and why she does not attack trade, make one ask, did the Germans want to provoke a widespread dispersion of our surplus vessels, and if so why? As the First Sea Lord has observed, it would be more natural they should wish to concentrate them in home waters in order to have targets for air attack. Moreover, how could they have foreseen the extent to which we should react on the rumour of *Scheer* in South Atlantic? It all seems quite purposeless; yet the Germans are not the people to do things without reason. Are you sure it was *Scheer* and not a plant, or a fake?

I see the German wireless boast they are driving the Fleet out of the North Sea. At present this is less mendacious than most of their stuff. There may therefore be danger on the East Coast from surface ships. Could not submarine flotillas of our own be disposed well out at sea across a probable line of hostile advance? They would want a parent destroyer perhaps to scout for them. They should be well out of our line of watching trawlers. It may well be there is something going to happen, now that we have retired to a distance to gain time.

I should be the last to raise those "invasion scares", which I combated so constantly during the early days of 1914–15. Still, it might be well for the Chiefs of Staff to consider what would happen if, for instance, 20,000 men were run across and landed, say, at Harwich, or at Webburn Hook, where there is deep water close inshore. These 20,000 men might make the training of Mr. Hore-Belisha's masses very much more realistic than is at present expected. The long dark nights would help such designs. Have any arrangements been made by the War Office to provide against this contingency? Remember how we stand in the North Sea at the present time. I do not think it likely, but it is physically possible.

ORGANISATION OF HUNTING GROUPS—OCTOBER 31ST, 1939

Force	Composition			Area
	Battleships and Battle-cruisers	Cruisers	Aircraft-carriers	
F		*Berwick* *York*		North America and West Indies
G		*Cumberland* *Exeter* *Ajax* *Achilles*		East coast of South America
H		*Sussex* *Shropshire*		Cape of Good Hope
I		*Cornwall* *Dorsetshire*	*Eagle*	Ceylon
	Malaya		*Glorious*	Gulf of Aden
K	*Renown*		*Ark Royal*	Pernambuco-Freetown
L	*Repulse*		*Furious*	Atlantic convoys
X		Two French 8-inch cruisers	*Hermes*	Pernambuco-Dakar
Y	*Strasbourg*	*Neptune* One French 8-inch cruiser		Pernambuco-Dakar

Additional escorts with North Atlantic convoys:

Battleships: *Revenge*
Resolution
Warspite
Cruisers: *Emerald*
Enterprise

The *Deutschland*, which was to have harassed our lifeline across the North-west Atlantic, interpreted her orders with comprehending caution. At no time during her two and a half months' cruise did she approach a convoy. Her determined efforts to avoid British forces prevented her from making more than two kills, one being a small Norwegian ship. A third ship, the United States *City of Flint*, carrying a cargo for Britain, was captured, but was eventually released by the Germans from a Norwegian port. Early in November the *Deutschland* slunk back to Germany, passing again through Arctic waters. The mere presence of this powerful ship upon our main trade route had however imposed, as was intended, a serious strain upon our escorts and hunting groups in the North Atlantic. We should in fact have preferred her activity to the vague menace she embodied.

The *Graf Spee* was more daring and imaginative, and soon became the centre of attention in the South Atlantic. In this vast area powerful Allied forces came into play by the middle of October. One group consisted of the aircraft-carrier *Ark Royal* and the battle-cruiser *Renown*, working from Freetown, in conjunction with a French group of two heavy cruisers and the British aircraft-carrier *Hermes*, based on Dakar. At the Cape of Good Hope were the two heavy cruisers *Sussex* and *Shropshire*, while on the east coast of South America, covering the vital traffic with the River Plate and Rio de Janeiro, ranged Commodore Harwood's group, comprising the *Cumberland*, *Exeter*, *Ajax*, and *Achilles*. The *Achilles* was a New Zealand ship manned mainly by New Zealanders.

The *Spee's* practice was to make a brief appearance at some point, claim a victim, and vanish again into the trackless ocean wastes. After a second appearance farther south on the Cape route, in which she sank only one ship, there was no further sign of her for nearly a month, during which our hunting groups were searching far and wide in all areas, and special vigilance was enjoined in the Indian Ocean. This was in fact her destination, and on November 15 she sank a small British tanker in the Mozambique Channel, between Madagascar and the mainland. Having thus registered her appearance as a feint in the Indian Ocean, in order to draw the hunt in that direction, her captain—Langsdorff, a high-class person—promptly doubled back and, keeping well south of the Cape, re-entered the Atlantic. This move had not been unforeseen; but our plans to intercept him were foiled by the quickness of his withdrawal. It was by no means clear to the Admiralty whether in fact one raider was on the prowl or two, and exertions were made both in the Indian and Atlantic Oceans. We also thought that the *Spee* was her sister ship, the *Scheer*. The disproportion between the strength of the enemy and the counter-measures forced upon us was vexatious. It recalled to me the anxious weeks before the action at Coronel and later at the Falkland Islands in December 1914, when we had to be prepared at seven or eight different points, in the Pacific and South Atlantic, for the arrival of Admiral von Spee with the earlier edition of the *Scharnhorst* and *Gneisenau*. A quarter of a century had passed, but the puzzle was the same. It was with a definite sense of relief that we learnt that the *Spee* had appeared once more on the Cape-Freetown route, sinking two more ships on December 2 and one on the 7th.

★ ★ ★ ★ ★

From the beginning of the war Commodore Harwood's special care and duty had been to cover British shipping off the River Plate and Rio de Janeiro. He was convinced that sooner or later the *Spee* would come towards the Plate, where the richest prizes were offered to her. He had carefully thought out the tactics which he would adopt in an encounter. Together, his 8-inch cruisers *Cumberland* and *Exeter*, and his 6-inch cruisers *Ajax* and *Achilles*, could not only catch but kill. However, the needs of fuel and refit made it unlikely that all four would be present "on the day". If they were not the issue was disputable. On hearing that the *Doric Star* had been sunk on December 2, Harwood guessed right. Although she was over 3,000 miles away he assumed that the *Spee* would come towards the Plate. He estimated with luck and wisdom that she might arrive by the 13th. He ordered all his available forces to concentrate there by December 12. Alas, the *Cumberland* was refitting at the Falklands; but on the morning of the 13th *Exeter*, *Ajax*, and *Achilles* were in company at the centre of the shipping routes off the mouth of the river. Sure enough, at 6.14 a.m. smoke was sighted to the east. The longed-for collision had come.

Harwood, in the *Ajax*, disposing his forces so as to attack the pocket-battleship from widely-divergent quarters and thus confuse her fire, advanced at the utmost speed of his small squadron. Captain Langsdorff thought at the first glance that he had only to deal with one light cruiser and two destroyers, and he too went full speed ahead; but a few moments later he recognised the quality of his opponents, and knew that a mortal action impended. The two forces were now closing at nearly fifty miles an hour. Langsdorff had but a minute to make up his mind. His right course would have been to turn away immediately so as to keep his assailants as long as possible under the superior range and weight of his 11-inch guns, to which the British could not at first have replied. He would thus have gained for his undisturbed firing the difference between adding speeds and subtracting them. He might well have crippled one of his foes before any could fire at him. He decided, on the contrary, to hold on his course and make for the *Exeter*. The action therefore began almost simultaneously on both sides.

Commodore Harwood's tactics proved advantageous. The 8-inch salvoes from the *Exeter* struck the *Spee* from the earliest stages of the fight. Meanwhile the 6-inch cruisers were also hitting hard and effectively. Soon the *Exeter* received a hit which, besides knocking out B turret, destroyed all the communications on the bridge, killed or wounded nearly all upon it, and put the ship temporarily out of control. By this time however the 6-inch cruisers could no longer be neglected by the enemy, and the *Spee* shifted her main armament to them, thus giving respite to the *Exeter* at a critical moment. The German battleship, plastered from three directions, found the British attack too hot, and soon afterwards turned away under a smoke-screen with the apparent intention of making for the River Plate. Langsdorff had better have done this earlier.

After this turn the *Spee* once more engaged the *Exeter*, hard hit by the 11-inch shells. All her forward guns were out of action. She was burning fiercely amidships and had a heavy list. Captain Bell, unscathed by the explosion on the bridge, gathered two or three officers round him in the after control-station, and kept his ship in action with her sole remaining turret, until at 7.30 failure of pressure put this too out of action. He could do no more. At 7.40 the *Exeter* turned away to effect repairs and took no further part in the fight.

The *Ajax* and *Achilles*, already in pursuit, continued the action

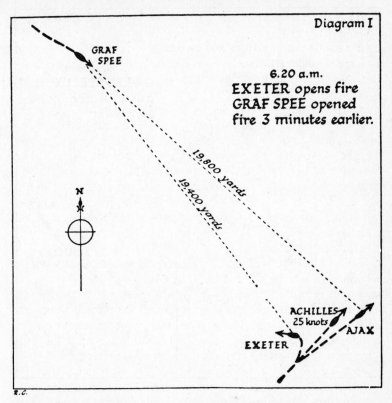

Diagram I

GRAF SPEE

6.20 a.m.
EXETER opens fire
GRAF SPEE opened
fire 3 minutes earlier.

19,800 yards

19,400 yards

N

ACHILLES
25 knots

AJAX

EXETER

in the most spirited manner. The *Spee* turned all her heavy guns
upon them. By 7.25 the two after-turrets in the *Ajax* had been
knocked out, and the *Achilles* had also suffered damage. These
two light cruisers were no match for the enemy in gun-power,
and, finding that his ammunition was running low, Harwood in
the *Ajax* decided to break off the fight till dark, when he would
have better chances of using his lighter armament effectively,
and perhaps his torpedoes. He therefore turned away under
cover of smoke, and the enemy did not follow. This fierce action
had lasted an hour and twenty minutes. During all the rest of
the day the *Spee* made for Montevideo, the British cruisers hanging grimly on her heels, with only occasional interchanges of fire.
Shortly after midnight the *Spee* entered Montevideo, and lay
there repairing damage, taking in stores, landing wounded,

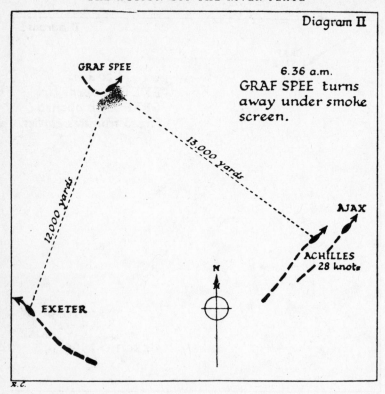

Diagram II

GRAF SPEE

6.36 a.m.
GRAF SPEE turns
away under smoke
screen.

13,000 yards

12,000 yards

AJAX

ACHILLES
28 knots

EXETER

N

transhipping personnel to a German merchant ship, and reporting
to the Fuehrer. *Ajax* and *Achilles* lay outside, determined to dog
her to her doom should she venture forth. Meanwhile on the
night of the 14th the *Cumberland*, which had been steaming at full
speed from the Falklands, took the place of the utterly crippled
Exeter. The arrival of this 8-inch-gun cruiser restored to its
narrow balance a doubtful situation.

It had been most exciting to follow the drama of this brilliant
action from the Admiralty War Room, where I spent a large
part of the 13th. Our anxieties did not end with the day. Mr.
Chamberlain was at that time in France on a visit to the Army.
On the 17th I wrote to him:

December 17, 1939

If the *Spee* breaks out as she may do to-night we hope to renew the
action of the 13th with the *Cumberland*, an eight 8-inch-gun ship, in

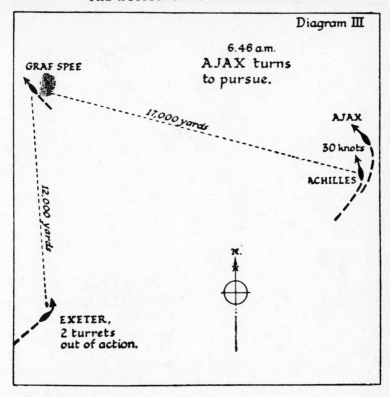

Diagram III

6.46 a.m.
AJAX turns
to pursue.

GRAF SPEE

17,000 yards

AJAX

30 knots

ACHILLES

12,000 yards

EXETER,
2 turrets
out of action.

N.

the place of the six-gun *Exeter*. The *Spee* knows now that *Renown* and *Ark Royal* are oiling at Rio, so this is her best chance. The *Dorsetshire* and *Shropshire*, who are coming across from the Cape, are still three and four days away respectively. It is fortunate that the *Cumberland* was handy at the Falklands, as *Exeter* was heavily damaged. She was hit over a hundred times, one turret smashed, three guns knocked out, and sixty officers and men killed and twenty wounded. Indeed, the *Exeter* fought one of the finest and most resolute actions against superior range and metal on record. Every conceivable precaution has been taken to prevent the *Spee* slipping out unobserved, and I have told Harwood [who is now an Admiral and a K.C.B.] that he is free to attack her anywhere outside the three-mile limit. We should prefer however that she should be interned, as this will be less creditable to the German Navy than being sunk in action. Moreover, a battle of this kind is full of hazard, and needless bloodshed must never be sought.

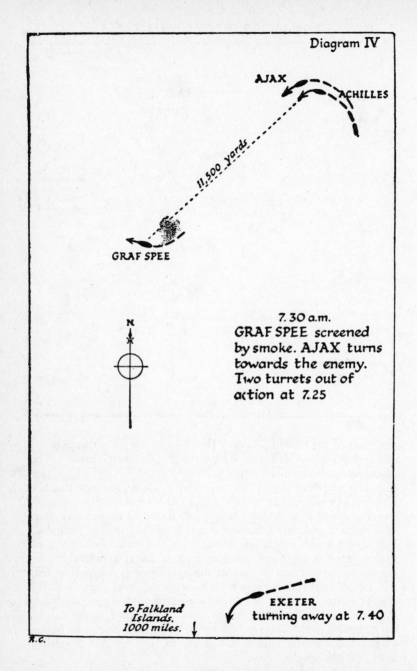

Diagram IV

AJAX

ACHILLES

11,500 Yards

GRAF SPEE

N

7.30 a.m.
GRAF SPEE screened
by smoke. AJAX turns
towards the enemy.
Two turrets out of
action at 7.25

EXETER
turning away at 7.40

To Falkland
Islands.
1000 miles.

R.C.

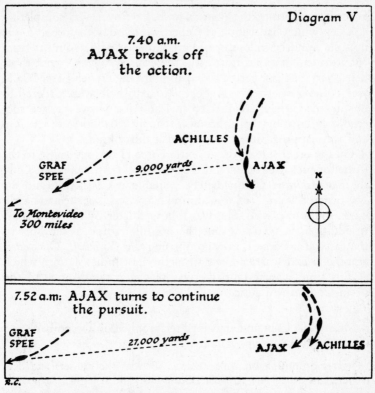

Diagram V

7.40 a.m.
AJAX breaks off
the action.

ACHILLES

GRAF
SPEE

9,000 yards - - - - - - AJAX

N

To Montevideo
300 miles

7.52 a.m: AJAX turns to continue
the pursuit.

GRAF
SPEE

27,000 yards - - - - - -

AJAX   ACHILLES

R.C.

The whole of the Canadians came in safely this morning under the protection of the main fleet, and [are] being welcomed by Anthony, Massey, and I trust a good part of the people of Greenock and Glasgow. We plan to give them a cordial reception. They are to go to Aldershot, where no doubt you will go and see them presently.

There have been ten air attacks to-day on individual ships along the East Coast from Wick to Dover, and some of the merchant ships have been machine-gunned out of pure spite, some of our people being hit on their decks.

I am sure you must be having a most interesting time at the Front, and I expect you will find that change is the best kind of rest.

From the moment when we heard that action was joined we instantly ordered powerful forces to concentrate off Montevideo, but our hunting groups were naturally widely dispersed and none was within two thousand miles of the scene. In the north

Force K, comprising the *Renown* and *Ark Royal*, was completing a sweep which had begun at Capetown ten days before and was now six hundred miles east of Pernambuco and 2,500 miles from Montevideo. Farther north still the cruiser *Neptune*, with three destroyers, had just parted company with the French Force X and was coming south to join Force K. All these were ordered to Montevideo; they had first to fuel at Rio. However, we succeeded in creating the impression that they had already left Rio and were approaching Montevideo at thirty knots.

On the other side of the Atlantic Force H was returning to the Cape for fuel after an extended sweep up the African coast. Only the *Dorsetshire* was immediately available at Capetown, and she was ordered at once to join Admiral Harwood, but she had nearly 4,000 miles to travel. She was followed later by the *Shropshire*. In addition, to guard against the possible escape of the *Spee* to the eastward, Force I, now comprising the *Cornwall*, *Gloucester*, and the aircraft-carrier *Eagle* from the East Indies station, which at this time was at Durban, was placed at the disposal of the C.-in-C. South Atlantic.

* * * * *

Meanwhile Captain Langsdorff telegraphed on December 16 to the German Admiralty as follows:

Strategic position off Montevideo. Besides the cruisers and destroyers, *Ark Royal* and *Renown*. Close blockade at night; escape into open sea and break-through to home waters hopeless. . . . Request decision on whether the ship should be scuttled in spite of insufficient depth in the estuary of the Plate, or whether internment is to be preferred.

At a conference presided over by the Fuehrer, at which Raeder and Jodl were present, the following answer was decided on:

Attempt by all means to extend the time in neutral waters. . . . Fight your way through to Buenos Aires if possible. No internment in Uruguay. Attempt effective destruction if ship is scuttled.

As the German envoy in Montevideo reported later that further attempts to extend the time-limit of seventy-two hours were fruitless, these orders were confirmed by the German Supreme Command.

Accordingly during the afternoon of the 17th the *Spee* trans-

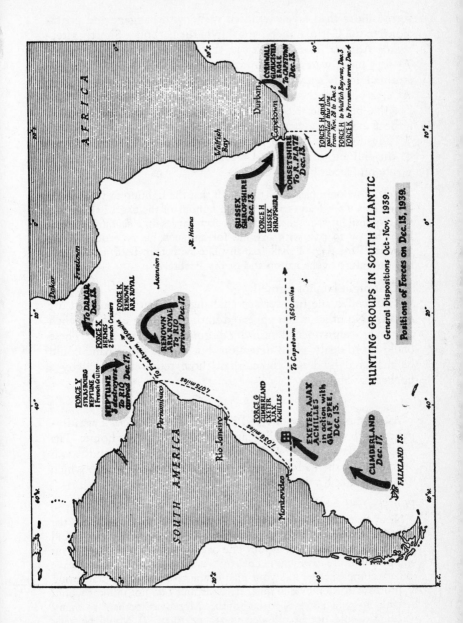

HUNTING GROUPS IN SOUTH ATLANTIC

General Dispositions Oct-Nov, 1939.

Positions of Forces on Dec.13, 1939.

AFRICA

CORNWALL
GLOUCESTER
EAGLE
To CAPETOWN
Dec.13

Durban

Capetown

FORCES H and K
patrolled this line
from Nov. 28 to Dec. 2
FORCE H to Walfish Bay arr., Dec.3
FORCE K to Pernambuco arriv., Dec.4

Walfish
Bay

DORSETSHIRE
To R. PLATE
Dec.13

SUSSEX
SHROPSHIRE
Dec.13

FORCE H
SUSSEX
SHROPSHIRE

R. Helena

Ascension I.

Freetown

Dakar

To DAKAR
Dec.13

FORCE X
HERMES
2 French Cruisers

FORCE K
RENOWN
ARK ROYAL

RENOWN
ARK ROYAL
To RIO
arrived Dec.17

To Capetown   3,650 miles

FORCE Y
STRASBOURG
NEPTUNE
1 French Cruiser

NEPTUNE
To RIO
arrived Dec.17

Pernambuco

1,550 miles to Freetown

1,035 miles

1,075 miles

Rio Janeiro

SOUTH AMERICA

Montevideo

FORCE G
CUMBERLAND
EXETER
AJAX
ACHILLES

EXETER, AJAX
ACHILLES
in action with
GRAF SPEE,
Dec.13.

CUMBERLAND
Dec.17

FALKLAND IS.

ferred more than seven hundred men, with baggage and provisions, to the German merchant ship in the harbour. Shortly afterwards Admiral Harwood learnt that she was weighing anchor. At 6.15 p.m., watched by immense crowds, she left harbour and steamed slowly seawards, awaited hungrily by the British cruisers. At 8.54 p.m., as the sun sank, the *Ajax's* aircraft reported: "*Graf Spee* has blown herself up." The *Renown* and *Ark Royal* were still a thousand miles away.

Langsdorff was broken-hearted by the loss of his ship. In spite of the full authority he had received from his Government, he wrote on December 19:

I can now only prove by my death that the fighting services of the Third Reich are ready to die for the honour of the flag. I alone bear the responsibility for scuttling the pocket-battleship *Admiral Graf Spee.* I am happy to pay with my life for any possible reflection on the honour of the flag. I shall face my fate with firm faith in the cause and the future of the nation and of my Fuehrer.

That night he shot himself.

Thus ended the first surface challenge to British trade on the oceans. No other raider appeared until the spring of 1940, when a new campaign opened, utilising disguised merchant ships. These could more easily avoid detection, but on the other hand could be mastered by lesser forces than those required to destroy a pocket-battleship.

* * * * *

As soon as the news arrived of the end of the *Spee* I was impatient to bring our widely-scattered hunting groups home. The *Spee's* auxiliary, the *Altmark*, was however still afloat, and it was believed that she had on board the crews of the nine ships which had been sunk by the raider.

First Sea Lord   17.XII.39
Now that the South Atlantic is practically clear except for the *Altmark*, it seems of high importance to bring home the *Renown* and *Ark Royal*, together with at least one of the 8-inch-gun cruisers. This will give us more easement in convoy work and enable refits and leave to be accomplished. I like your plan of the two small ships anchoring to-morrow in Montevideo inner harbour, but I do not think it would be right to send Force K so far south. Moreover, perhaps so many warships would not be allowed in at one time. It would be very

convenient if, as you proposed, *Neptune* relieved *Ajax* as soon as the triumphal entry into [Montevideo harbour] is over; and it would be very good if all the returning forces could scrub and search the South Atlantic on their way home for the *Altmark*. I feel that we ought to bring home all that are not absolutely needed. The Northern Patrol will require constant support in two, or better still three, reliefs from the Clyde as long as we stay there. I agree with Captain Tennant that the German Admiralty will be most anxious to do something to get their name back.

Perhaps you will let me know what you think about these ideas.

I was also most anxious about the *Exeter*, and could not accept the proposals made to me to leave her unrepaired in the Falkland Islands till the end of the war.

First Sea Lord, Controller, and others   17.XII.39

This preliminary report of damage to *Exeter* shows the tremendous fire to which she was exposed and the determination with which she was fought. It also reflects high credit on the Constructors' Department that she should have been able to stand up to such a prolonged and severe battering. This story will have to be told as soon as possible, omitting anything undesirable [*i.e.*, what the enemy should not know].

What is proposed about repair? What can be done at the Falklands? I presume she will be patched up sufficiently to come home for long refit.

First Sea Lord, D.C.N.S., Controller   23.XII.39

We ought not readily to accept the non-repair during the war of *Exeter*. She should be strengthened and strutted internally as far as possible, and should transfer her ammunition, or the bulk of it, to some merchant ship or tender. Perhaps she might be filled up in part with barrels or empty oil-drums, and come home with reduced crew under escort either to the Mediterranean or to one of our dockyards. If nothing can be done with her then, she should be stripped of all useful guns and appliances, which can be transferred to new construction.

The above indicates only my general view. Perhaps you will let me know how it can be implemented.

Controller and First Sea Lord   29.XII.39

I have not seen the answer to the telegram from the Rear-Admiral South America about its not being worth while to repair *Exeter*, on which I minuted in the contrary sense. How does this matter now stand? I gathered from you verbally that we were all in agreement she should come home and be thoroughly repaired, and that this need not take so long as the R.A. thought.

What is going to happen to *Exeter* now? How is she going to be brought home, in what condition, and when? We cannot leave her at the Falklands, where either she will be in danger or some valuable ship will be tethered to look after her. I shall be glad to know what is proposed.

My view prevailed. The *Exeter* reached this country safely. I had the honour to pay my tribute to her brave officers and men from her shattered deck in Plymouth Harbour. She was preserved for over two years of distinguished service, until she perished under Japanese guns in the forlorn battle of the Straits of Sunda in 1942.

* * * * *

The effects of the action off the Plate gave intense joy to the British nation and enhanced our prestige throughout the world. The spectacle of the three smaller British ships unhesitatingly attacking and putting to flight their far more heavily gunned and armoured antagonist was everywhere admired. It was contrasted with the disastrous episode of the escape of the *Goeben* in the Straits of Otranto in August 1914. In justice to the Admiral of those days it must be remembered that all Commodore Harwood's ships were faster than the *Spee*, and all except one of Admiral Troubridge's squadron in 1914 were slower than the *Goeben*. Nevertheless the impression was exhilarating, and lightened the dreary and oppressive winter through which we were passing.

The Soviet Government were not pleased with us at this time, and their comment on December 31, 1939, in the *Red Fleet* is an example of their factual reporting:

Nobody would dare to say that the loss of a German battleship is a brilliant victory for the British Fleet. This is rather a demonstration, unprecedented in history, of the impotence of the British. Upon the morning of December 13 the battleship started an artillery duel with the *Exeter*, and within a few minutes obliged the cruiser to withdraw from the action. According to the latest information the *Exeter* sank near the Argentine coast, *en route* for the Falkland Islands.

* * * * *

On December 23 the American Republics made a formal protest to Britain, France, and Germany about the action off the

River Plate, which they claimed to be a violation of the American Security Zone. It also happened about this time that two German merchant ships were intercepted by our cruisers near the coast of the United States. One of these, the liner *Columbus*, of 32,000 tons, was scuttled and survivors were rescued by an American cruiser; the other escaped into territorial waters in Florida. President Roosevelt reluctantly complained about these vexations near the coast of the Western Hemisphere, and in my reply I took the opportunity of stressing the advantages which our action off the Plate had brought to all the South American Republics. Their trade had been hampered by the activities of the German raider and their ports had been used for his supply ships and as information centres. By the laws of war the raider had been entitled to capture all merchant ships trading with us in the South Atlantic, or to sink them after providing for their crews; and this had inflicted grave injury on American commercial interests, particularly in the Argentine. The South American Republics should greet the action off the Plate as a deliverance from all this annoyance. The whole of the South Atlantic was now clear, and might perhaps remain clear, of warlike operations. This relief should be highly valued by the South American States, who might now in practice enjoy for a long period the advantages of a Security Zone of three thousand, rather than three hundred, miles.

I could not forbear from adding that the Royal Navy was carrying a very heavy burden in enforcing respect for international law at sea. The presence of even a single raider in the North Atlantic called for the employment of half our battle-fleet to give sure protection to the world's commerce. The unlimited laying of magnetic mines by the enemy was adding to the strain upon our flotillas and small craft. If we should break under this strain the South American Republics would soon have many worse worries than the sound of one day's distant seaward cannonade; and in quite a short time the United States would also face more direct cares. I therefore felt entitled to ask that full consideration should be given to the burden which we were carrying at this crucial period, and that the best construction should be placed on action which was indispensable if the war was to be ended within reasonable time and in the right way.

CHAPTER XXX

SCANDINAVIA. FINLAND

The Norway Peninsula – Swedish Iron Ore – Neutrality and the Norwegian Corridor – An Error Corrected – Behind the German Veil – Admiral Von Raeder and Herr Rosenberg – Vidkun Quisling – Hitler's Decision, December 14, 1939 – Soviet Action against the Baltic States – Stalin's Demands upon Finland – The Russians Declare War on Finland, November 28, 1939 – Gallant Finnish Resistance – The Soviet Failure and Rebuff – World-wide Satisfaction – Aid to Finland and Norwegian and Swedish Neutrality – The Case for Mining the "Leads" – The Moral Issue.

*T*HE thousand-mile-long peninsula stretching from the mouth of the Baltic to the Arctic Circle had an immense strategic significance. The Norwegian mountains run into the ocean in a continuous fringe of islands. Between these islands and the mainland there was a corridor in territorial waters through which Germany could communicate with the outer sea to the grievous injury of our blockade. German war industry was mainly based upon supplies of Swedish iron ore, which in the summer were drawn from the Swedish port of Luleå, at the head of the Gulf of Bothnia, and in the winter, when this was frozen, from Narvik, on the west coast of Norway. To respect the corridor would be to allow the whole of this traffic to proceed under the shield of neutrality in the face of our superior sea-power. The Admiralty Staff were seriously perturbed at this important advantage being presented to Germany, and at the earliest opportunity I raised the issue in the Cabinet.

My recollection of the previous war was that the British and American Governments had had no scruples about mining the "Leads", as these sheltered waters were called. The great mine barrage which was laid in 1917–18 across the North Sea from

Scotland to Norway could not have been fully effective if German commerce and German U-boats had only to slip round the end of it unmolested. I found however that neither of the Allied Fleets had laid any minefields in Norwegian territorial waters. Their admirals had complained that the barrage, on which enormous quantities of labour and money had been spent, would be ineffective unless this corridor was closed, and all the Allied Governments had therefore put the strongest pressure on Norway to close it themselves. The immense barrage took a long time to lay, and by the time it was finished there was not much doubt how the war would end or that Germany no longer possessed the power to invade Scandinavia. It was not however till the end of September 1918 that the Norwegian Government were persuaded to take action. Before they actually carried out their undertaking the war came to an end.

When eventually I presented this case in the House of Commons, in April 1940, I said:

During the last war, when we were associated with the United States, the Allies felt themselves so deeply injured by this covered way, then being used especially for U-boats setting out on their marauding expeditions, that the British, French, and United States Governments together induced the Norwegians to [undertake to] lay a minefield in their territorial waters across the covered way in order to prevent the abuse by U-boats of this channel. It was only natural that the Admiralty since this war began should have brought this precedent—although it is not exactly on all fours and there are some differences—this modern and highly respectable precedent, to the notice of His Majesty's Government, and should have urged that we should be allowed to lay a minefield of our own in Norwegian territorial waters in order to compel this traffic which was passing in and out to Germany to come out into the open sea and take a chance of being brought into the Contraband Control or being captured as enemy prize by our blockading squadrons and flotillas. It was only natural and it was only right that His Majesty's Government should have been long reluctant to incur the reproach of even a technical violation of international law.

They certainly were long in reaching a decision.

At first the reception of my case was favourable. All my colleagues were deeply impressed with the evil; but strict respect for the neutrality of small States was a principle of conduct to which we all adhered.

First Lord to First Sea Lord and others   19.IX.39

I brought to the notice of the Cabinet this morning the importance of stopping the Norwegian transportation of Swedish iron ore from Narvik, which will begin as soon as the ice forms in the Gulf of Bothnia. I pointed out that we had laid a minefield across the 3-mile limit in Norwegian territorial waters in 1918, with the approval and co-operation of the United States. I suggested that we should repeat this process very shortly. [This, as is explained above, was not an accurate statement, and I was soon apprised of the fact.] The Cabinet, including the Foreign Secretary, appeared strongly favourable to this action.

It is therefore necessary to take all steps to prepare it.

1. The negotiations with the Norwegians for the chartering of their tonnage must be got out of the way first.

2. The Board of Trade would have to make arrangements with Sweden to buy the ore in question, as it is far from our wish to quarrel with the Swedes.

3. The Foreign Office should be made acquainted with our proposals, and the whole story of Anglo-American action in 1918 must be carefully set forth, together with a reasoned case.

4. The operation itself should be studied by the Admiralty Staff concerned. The Economic Warfare Department should be informed as and when necessary.

Pray let me be continually informed of the progress of this plan, which is of the highest importance in crippling the enemy's war industry.

A further Cabinet decision will be necessary when all is in readiness.

On the 29th, at the invitation of my colleagues, and after the whole subject had been minutely examined at the Admiralty, I drafted a paper for the Cabinet upon this subject, and on the chartering of neutral tonnage, which was linked with it.

NORWAY AND SWEDEN

Memorandum by the First Lord of the Admiralty

Chartering Norwegian Tonnage   *September 29, 1939*

1. The Norwegian Delegation is approaching, and in a few days the President of the Board of Trade hopes to make a bargain with them by which he charters all their spare tonnage, the bulk of which consists of tankers.

The Admiralty consider the chartering of this tonnage most important, and Lord Chatfield has written strongly urging it upon them.

German Supplies of Iron Ore from Narvik

2. At the end of November the Gulf of Bothnia normally freezes, so that Swedish iron ore can be sent to Germany only through Oxelosund, in the Baltic, or from Narvik, at the north of Norway. Oxelosund can export only about one-fifth of the weight of ore Germany requires from Sweden. In winter normally the main trade is from Narvik, whence ships can pass down the west coast of Norway, and make the whole voyage to Germany without leaving territorial waters until inside the Skagerrak.

It must be understood that an adequate supply of Swedish iron ore is vital to Germany, and the interception or prevention of these Narvik supplies during the winter months, *i.e.*, from October to the end of April, will greatly reduce her power of resistance. For the first three weeks of the war no iron ore ships left Narvik owing to the reluctance of crews to sail and other causes outside our control. Should this satisfactory state of affairs continue, no special action would be demanded from the Admiralty. Furthermore, negotiations are proceeding with the Swedish Government which in themselves may effectively reduce the supplies of Scandinavian ore to Germany.

Should however the supplies from Narvik to Germany start moving again, more drastic action will be needed.

Relations with Sweden

3. Our relations with Sweden require careful consideration. Germany acts upon Sweden by threats. Our sea-power gives us also powerful weapons, which, if need be, we must use to ration Sweden. Nevertheless, it should be proposed, as part of the policy outlined in paragraph 2, to assist the Swedes so far as possible to dispose of their ore in exchange for our coal; and, should this not suffice, to indemnify them, partly at least, by other means. This is the next step.

Charter and Insurance of all available Neutral Tonnage

4. The above considerations lead to a wider proposal. Ought we not to secure the control, by charter or otherwise, of all the free neutral shipping we can obtain, as well as the Norwegian, and thus give the Allies power to regulate the greater part of the sea transport of the world and recharter it, profitably, to those who act as we wish?

And ought we not to extend to neutral shipping not under our direct control the benefit of our convoy system?

The results so far achieved by the Royal Navy against the U-boat attack seem, in the opinion of the Admiralty, to justify the adoption of this latter course. This would mean that we should offer safe convoy to all vessels of all countries traversing our sea routes, provided they conform to our rules of contraband and pay the necessary premiums in foreign devisen. They would therefore be able to contract themselves out of the war risk, and with the success of our anti-U-boat campaign we may well hope to make a profit to offset its heavy expense. Thus not only vessels owned by us or controlled by us, but independent neutral ships, would all come to enjoy the British protection on the high seas, or be indemnified in case of accidents. It is not believed at the Admiralty that this is beyond our strength. Had some such scheme for the chartering and insurance of neutral shipping been in force from the early days of the last war, there is little doubt that it would have proved a highly profitable speculation. In this war it might well prove to be the foundation of a League of Free Maritime Nations to which it was profitable to belong.

5. It is therefore asked that the Cabinet, if they approve in principle of these four main objectives, should remit the question to the various departments concerned in order that detailed plans may be made for prompt action.

Before circulating this paper to the Cabinet and raising the issue there, I called upon the Admiralty Staff for a thorough re-check of the whole position.

First Lord to the Assistant Chief of the Naval Staff 29.IX.39

Please reconvene the meeting on iron ore we held on Thursday tomorrow morning, while Cabinet is sitting, in order to consider the draft print which I have made. It is no use my asking the Cabinet to take the drastic action suggested against a neutral country unless the results are in the first order of importance.

I am told that there are hardly any German or Swedish ships trying to take ore south from Narvik. Also that the Germans have been accumulating ore by sea at Oxelosund against the freezing up, and so will be able to bring good supplies down the Baltic via the Kiel Canal to the Ruhr during the winter months. Are these statements true? It would be very unpleasant if I went into action on mining the Norwegian territorial waters and was answered that it would not do the trick.

At the same time, assuming that the west coast traffic of Norway in ore is a really important factor worth making an exertion to stop, at what point would you stop it?

Pray explore in detail the coast and let me know the point. Clearly it should be north at any rate of Bergen, thus leaving the southern part of the West Norwegian coast open for any traffic that may come from Norway or out of the Baltic in the Norwegian convoy across to us. All this has to be more explored before I can present my case to the Cabinet. I shall not attempt to do so until Monday or Tuesday.

When all was agreed and settled at the Admiralty I brought the matter a second time before the Cabinet. Again there was general agreement upon the need; but I was unable to obtain assent to action. The Foreign Office arguments about neutrality were weighty, and I could not prevail. I continued, as will be seen, to press my point by every means and on all occasions. It was not however until April 1940 that the decision that I asked for in September 1939 was taken. By that time it was too late.

*　　*　　*　　*　　*

Almost at this very moment, as we now know, German eyes were turned in the same direction. On October 3 Admiral Raeder, Chief of the Naval Staff, submitted a proposal to Hitler headed "Gaining of Bases in Norway." He asked, "That the Fuehrer be informed as soon as possible of the opinions of the Naval War Staff on the possibilities of extending the operational base to the north. It must be ascertained whether it is possible to gain bases in Norway under the combined pressure of Russia and Germany, with the aim of improving our strategic and operational position." He framed therefore a series of notes, which he placed before Hitler on October 10. "In these notes," he wrote, "I stressed the disadvantages which an occupation of Norway by the British would have for us: the control of the approaches to the Baltic, the outflanking of our naval operations and of our air attacks on Britain, the end of our pressure on Sweden. I also stressed the advantages for us of the occupation of the Norwegian coast: outlet to the North Atlantic, no possibility of a British mine barrier, as in the year 1917–18. . . . The Fuehrer saw at once the significance of the Norwegian problem; he asked me to leave the notes, and stated that he wished to consider the question himself."

Rosenberg, the foreign affairs expert of the Nazi Party, and in charge of a special bureau to deal with propaganda activities in foreign countries, shared the Admiral's view. He dreamed of "converting Scandinavia to the idea of a Nordic community embracing the northern peoples under the natural leadership of Germany". Early in 1939 he thought he had discovered an instrument in the extreme Nationalist Party in Norway, which was led by a former Norwegian Minister of War named Vidkun Quisling. Contacts were established, and Quisling's activity was

linked with the plans of the German Naval Staff through Rosenberg's organisation and the German Naval Attaché in Oslo.

Quisling and his assistant, Hagelin, went to Berlin on December 14, and were taken by Raeder to Hitler, to discuss a political
stroke in Norway. Quisling arrived with a detailed plan. Hitler,
careful of secrecy, affected reluctance to increase his commitments, and said he would prefer a neutral Scandinavia. Nevertheless, according to Raeder, it was on this very day that he gave the
order to the Supreme Command to prepare for a Norwegian
Operation.

Of all this we of course knew nothing.

★   ★   ★   ★   ★

Meanwhile the Scandinavian peninsula became the scene of an
unexpected conflict which aroused strong feeling in Britain and
France and powerfully affected the discussion about Norway.
As soon as Germany was involved in war with Great Britain and
France, Soviet Russia, in the spirit of her pact with Germany,
proceeded to block the lines of entry into the Soviet Union from
the west. One passage led from East Prussia through the Baltic
States; another led across the waters of the Gulf of Finland; the
third route was through Finland itself and across the Karelian
Isthmus to a point where the Finnish frontier was only twenty
miles from the suburbs of Leningrad. The Soviets had not forgotten the dangers which Leningrad had faced in 1919. Even the
White Russian Government of Kolchak had informed the Peace
Conference in Paris that bases in the Baltic States and Finland
were a necessary protection for the Russian capital. Stalin had
used the same language to the British and French Missions in the
summer of 1939; and we have seen in earlier chapters how the
natural fears of these small States had been an obstacle to an Anglo-
French Alliance with Russia, and had paved the way for the
Molotov-Ribbentrop agreement.

Stalin had wasted no time. On September 24 the Esthonian
Foreign Minister had been called to Moscow, and four days later
his Government signed a Pact of Mutual Assistance which gave
the Russians the right to garrison key bases in Esthonia. By
October 21 the Red Army and Air Force were installed. The
same procedure was used simultaneously in Latvia, and Soviet
garrisons also appeared in Lithuania. Thus the southern road to

Leningrad and half the Gulf of Finland had been swiftly barred against potential German ambitions by the armed forces of the Soviets. There remained only the approach through Finland.

Early in October Mr. Paasikivi, one of the Finnish statesmen who had signed the peace of 1921 with the Soviet Union, went to Moscow. The Soviet demands were sweeping: the Finnish frontier on the Karelian Isthmus must be moved back a considerable distance so as to remove Leningrad from the range of hostile artillery. The cession of certain Finnish islands in the Gulf of Finland; the lease of the Rybathy Peninsula, together with Finland's only ice-free port in the Arctic Sea, Petsamo; and, above all, the leasing of the port of Hango, at the entrance of the Gulf of Finland, as a Russian naval and air base, completed the Soviet requirements. The Finns were prepared to make concessions on every point except the last. With the keys of the Gulf in Russian hands the strategic and national security of Finland seemed to them to vanish. The negotiations broke down on November 13, and the Finnish Government began to mobilise, and strengthen their troops on the Karelian frontier. On November 28 Molotov denounced the Non-Aggression Pact between Finland and Russia; two days later the Russians attacked at eight points along Finland's thousand-mile frontier, and on the same morning the capital, Helsingfors, was bombed by the Red Air Force.

The brunt of the Russian attack fell at first upon the frontier defences of the Finns in the Karelian Isthmus. These comprised a fortified zone about twenty miles in depth running north and south through forest country, deep in snow. This was called the "Mannerheim Line", after the Finnish Commander-in-Chief and saviour of Finland from Bolshevik subjugation in 1917. The indignation excited in Britain, France, and even more vehemently in the United States, at the unprovoked attack by the enormous Soviet Power upon a small, spirited, and highly-civilised nation was soon followed by astonishment and relief. The early weeks of fighting brought no success to the Soviet forces, which in the first instance were drawn almost entirely from the Leningrad garrison. The Finnish Army, whose total fighting strength was only about 200,000 men, gave a good account of themselves. The Russian tanks were encountered with audacity and a new type of hand-grenade, soon nicknamed "the Molotov Cocktail".

It is probable that the Soviet Government had counted on a

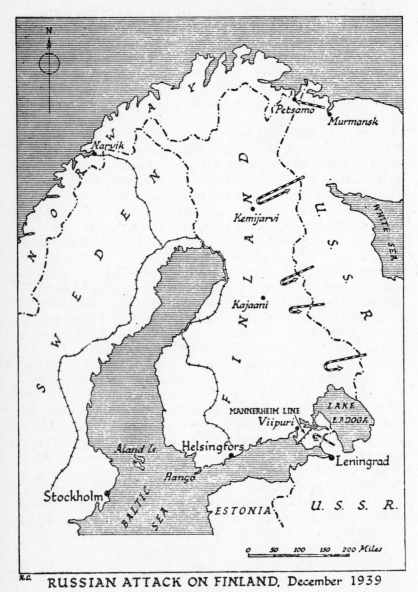

RUSSIAN ATTACK ON FINLAND, December 1939

walk-over. Their early air raids on Helsingfors and elsewhere, though not on a heavy scale, were expected to strike terror. The troops they used at first, though numerically much stronger, were inferior in quality and ill-trained. The effect of the air raids and of the invasion of their land roused the Finns, who rallied to a man against the aggressor and fought with absolute determination and the utmost skill. It is true that the Russian division which carried out the attack on Petsamo had little difficulty in throwing back the 700 Finns in that area. But the attack on the "Waist" of Finland proved disastrous to the invaders. The country here is almost entirely pine forests, gently undulating and at the time covered with a foot of hard snow. The cold was intense. The Finns were well equipped with skis and warm clothing, of which the Russians had neither. Moreover, the Finns proved themselves aggressive individual fighters, highly trained in reconnaissance and forest warfare. The Russians relied in vain on numbers and heavier weapons. All along this front the Finnish frontier posts withdrew slowly down the roads, followed by the Russian columns. After these had penetrated about thirty miles they were set upon by the Finns. Held in front at Finnish defence lines constructed in the forests, violently attacked in flank by day and

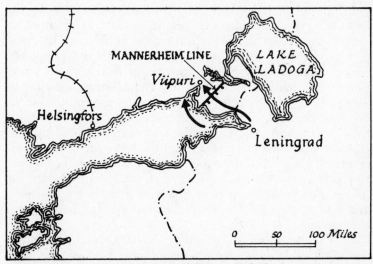

BREAKING THE MANNERHEIM LINE, March 1940

night, their communications severed behind them, the columns were cut to pieces, or, if lucky, got back after heavy loss whence they came. By the end of December the whole Russian plan for driving in across the "Waist" had broken down.

Meanwhile the attacks against the Mannerheim Line in the Karelian Isthmus fared no better. North of Lake Ladoga a turning movement attempted by about two Soviet divisions met the same fate as the operations farther north. Against the Line itself a series of mass attacks by nearly twelve divisions was launched in early December, and continued throughout the month. The Russian artillery bombardments were inadequate; their tanks were mostly light, and a succession of frontal attacks were repulsed with heavy losses and no gains. By the end of the year failure all along the front convinced the Soviet Government that they had to deal with a very different enemy from what they had expected. They determined upon a major effort. Realising that in the forest warfare of the north they could not overcome by mere weight of numbers the superior tactics and training of the Finns, they decided to concentrate on piercing the Mannerheim Line by methods of siege warfare in which the power of massed heavy artillery and heavy tanks could be brought into full play. This required preparation on a large scale, and from the end of the year fighting died down all along the Finnish front, leaving the Finns so far victorious over their mighty assailant. This surprising event was received with equal satisfaction in all countries, belligerent or neutral, throughout the world. It was a pretty bad advertisement for the Soviet Army. In British circles many people congratulated themselves that we had not gone out of our way to bring the Soviets in on our side, and preened themselves on their foresight. The conclusion was drawn too hastily that the Russian Army had been ruined by the purge, and that the inherent rottenness and degradation of their system of government and society was now proved. It was not only in England that this view was taken. There is no doubt that Hitler and his generals meditated profoundly upon the Finnish exposure, and that it played a potent part in influencing the Fuehrer's thought.

All the resentment felt against the Soviet Government for the Ribbentrop-Molotov pact was fanned into flame by this latest exhibition of brutal bullying and aggression. With this was also mingled scorn for the inefficiency displayed by the Soviet troops and enthusiasm for the gallant Finns. In spite of the Great War which had been declared, there was a keen desire to help the Finns by aircraft and other precious war material and by volunteers from Britain, from the United States, and still more from France. Alike for the munitions supplies and the volunteers there was only one possible route to Finland. The iron ore port of Narvik, with its railroad over the mountains to the Swedish iron mines, acquired a new sentimental if not strategic significance. Its use as a line of supply for the Finnish armies affected the neutrality both of Norway and Sweden. These two States, in equal fear of Germany and Russia, had no aim but to keep out of the wars by which they were encircled and might be engulfed. For them this seemed the only chance of survival. But whereas the British Government were naturally reluctant to commit even a technical infringement of Norwegian territorial waters by laying mines in the Leads for their own advantage against Germany, they moved upon a generous emotion, only indirectly connected with our war problem, towards a far more serious demand upon both Norway and Sweden for the free passage of men and supplies to Finland.

I sympathised ardently with the Finns and supported all proposals for their aid; and I welcomed this new and favourable breeze as a means of achieving the major strategic advantage of cutting off the vital iron ore supplies of Germany. If Narvik was to become a kind of Allied base to supply the Finns, it would certainly be easy to prevent the German ships loading ore at the port and sailing safely down the Leads to Germany. Once Norwegian and Swedish protestations were overborne, for whatever reason, the greater measures would include the less. The Admiralty's eyes were also fixed at this time upon the movements of a large and powerful Russian ice-breaker which was to be sent from Murmansk to Germany, ostensibly for repairs, but much more probably to open the now frozen Baltic port of Luleå for the German ore-ships. I therefore renewed my efforts to win consent to the simple and bloodless operation of mining the Leads, for which a certain precedent from the previous war existed. As the question raises moral issues, I feel it right to set the case in its final form as I made it after prolonged reflection and debate.

NORWAY—IRON ORE TRAFFIC

NOTE BY THE FIRST LORD OF THE ADMIRALTY

16.XII.39

1. The effectual stoppage of the Norwegian ore supplies to Germany ranks as a major offensive operation of war. No other measure is open to us for many months to come which gives so good a chance of abridging the waste and destruction of the conflict, or of perhaps preventing the vast slaughters which will attend the grapple of the main armies.

2. If the advantage is held to outweigh the obvious and serious objections, the whole process of stoppage must be enforced. The ore from Luleå is already stopped by the winter ice, which must not be [allowed to be] broken by the Soviet ice-breaker, should the attempt be made. The ore from Narvik must be stopped by laying successively a series of small minefields in Norwegian territorial waters at the two or three suitable points on the coast, which will force the ships carrying ore to Germany to quit territorial waters and come on to the high seas, where, if German, they will be taken as prize, or, if neutral, be subjected to our contraband control. The ore from Oxelosund, the main ice-free port in the Baltic, must also be prevented from leaving by methods which will be neither diplomatic nor military. All these three ports must be dealt with in various appropriate ways as soon as possible.

3. Thus it is not a question of denying Germany a mere million tons between now and May, but of cutting off her whole winter supply, except the negligible amounts that can be got from Gavle, or other minor ice-free Baltic ports. Germany would therefore undergo a severe deprivation, tending to crisis before the summer. But when the ice melts in the Gulf of Bothnia the abundant supply from Luleå would again be open, and Germany is no doubt planning not only to get as much as she can during the winter, but to make up the whole $9\frac{1}{2}$ million tons which she needs, or even more, between May 1 and December 15, 1940. After this she might hope to organise Russian supplies and be able to wage a very long war.

4. It may well be that, should we reach the month of May with Germany starving for ore for her industries and her munitions, the prevention of the reopening of Luleå may become [for us] a principal naval objective. The laying of a declared minefield, including magnetic mines, off Luleå by British submarines would be one way. There are others. If Germany can be cut from all Swedish ore supplies from now onwards till the end of 1940 a blow will have been struck at her war-making capacity equal to a first-class victory in the field or from

the air, and without any serious sacrifice of life. It might indeed be immediately decisive.

5. To every blow struck in war there is a counter. If you fire at the enemy he will fire back. It is most necessary therefore to face squarely the counter-measures which may be taken by Germany, or constrained by her from Norway or Sweden. As to Norway, there are three pairs of events which are linked together. First, the Germans, conducting war in a cruel and lawless manner, have violated the territorial waters of Norway, sinking without warning or succour a number of British and neutral vessels. To that our response is to lay the minefields mentioned above. It is suggested that Norway, by way of protest, may cancel the valuable agreement we have made with her for chartering her tankers and other shipping. But then she would lose the extremely profitable bargain she has made with us, and this shipping would become valueless to her in view of our contraband control. Her ships would be idle, and her owners impoverished. It would not be in Norwegian interests for her Government to take this step; and interest is a powerful factor. Thirdly, Norway could retaliate by refusing to export to us the aluminium and other war materials which are important to the Air Ministry and the Ministry of Supply. But here again her interests would suffer. Not only would she not receive the valuable gains which this trade brings her, but Great Britain, by denying her bauxite and other indispensable raw materials, could bring the whole industry of Norway, centring upon Oslo and Bergen, to a complete standstill. In short, Norway, by retaliating against us, would be involved in economic and industrial ruin.

6. Norwegian sympathies are on our side, and her future independence from German overlordship hangs upon the victory of the Allies. It is not reasonable to suppose that she will take either of the counter-measures mentioned above (although she may threaten them), unless she is compelled to do so by German brute force.

7. This will certainly be applied to her anyway, and whatever we do, if Germany thinks it her interest to dominate forcibly the Scandinavian peninsula. In that case the war would spread to Norway and Sweden, and with our command of the seas there is no reason why French and British troops should not meet German invaders on Scandinavian soil. At any rate, we can certainly take and hold whatever islands or suitable points on the Norwegian coast we choose. Our northern blockade of Germany would then become absolute. We could, for instance, occupy Narvik and Bergen, and keep them open for our own trade while closing them completely to Germany. It cannot be too strongly emphasised that British control of the Norwegian coast-line is a strategic objective of first-class importance.

It is not therefore seen how, even if retaliation by Germany were to run its full course, we should be worse off for the action now proposed. On the contrary, we have more to gain than lose by a German attack upon Norway or Sweden. This point is capable of more elaboration than is necessary here.

There is no reason why we should not manage to secure a large and long-continued supply of iron ore from Sweden through Narvik while at the same time diverting all supplies of ore from Germany. This must be our aim.

I concluded as follows:

8. The effect of our action against Norway upon world opinion and upon our own reputation must be considered. We have taken up arms in accordance with the principles of the Covenant of the League in order to aid the victims of German aggression. No technical infringement of international law, so long as it is unaccompanied by inhumanity of any kind, can deprive us of the good wishes of neutral countries. No evil effect will be produced upon the greatest of all neutrals, the United States. We have reason to believe that they will handle the matter in the way most calculated to help us. And they are very resourceful.

9. The final tribunal is our own conscience. We are fighting to re-establish the reign of law and to protect the liberties of small countries. Our defeat would mean an age of barbaric violence, and would be fatal not only to ourselves, but to the independent life of every small country in Europe. Acting in the name of the Covenant, and as virtual mandatories of the League and all it stands for, we have a right, and indeed are bound in duty, to abrogate for a space some of the conventions of the very laws we seek to consolidate and reaffirm. Small nations must not tie our hands when we are fighting for their rights and freedom. The letter of the law must not in supreme emergency obstruct those who are charged with its protection and enforcement. It would not be right or rational that the Aggressor Power should gain one set of advantages by tearing up all laws, and another set by sheltering behind the innate respect for law of its opponents. Humanity, rather than legality, must be our guide.

Of all this history must be the judge. We now face events.

* * * * *

My memorandum was considered by the Cabinet on December 22, and I pleaded the case to the best of my ability. I could not obtain any decision for action. Diplomatic protest might be made to Norway about the misuse of her territorial waters by

Germany, and the Chiefs of Staff were instructed to consider the military consequences of commitments on Scandinavian soil. They were authorised to plan for landing a force at Narvik for the sake of Finland, and also a possible German occupation of Southern Norway. But no executive orders could be issued to the Admiralty. In a paper which I circulated on December 24 I summarised the Intelligence reports which showed the possibilities of a Russian design upon Norway. The Soviet were said to have three divisions concentrated at Murmansk preparing for a seaborne expedition. "It may be," I concluded, "that this theatre will become the scene of early activities." This proved only too true; but from a different quarter.

CHAPTER XXXI

A DARK NEW YEAR

The Trance Continues – "Catherine": The Final Phase – Tension with Russia – Mussolini's Misgivings – Mr. Hore-Belisha Leaves the War Office – Impediments to Action – A Twilight Mood in the Factories – The Results in May – Capture of the German Plans Against Belgium – Work and Growth of the British Expeditionary Force – No Armoured Division – Deterioration of the French Army – Communist Intrigues – German Plans for the Invasion of Norway – The Supreme War Council of February 5 – My First Attendance – The "Altmark" Incident – Captain Philip Vian – Rescue of the British Prisoners – Mr. Chamberlain's Effective Defence – Hitler Appoints General von Falkenhorst to Command Against Norway – Norway Before France – German Air Attack on Our East Coast Shipping – Counter-measures – Satisfactory Results of the First Six Months' Sea War – Navy Estimates Speech, February 27, 1940.

*T*HE END of the year 1939 left the war still in its sinister trance. An occasional cannon-shot or reconnoitring patrol alone broke the silence of the Western Front. The armies gaped at each other from behind their rising fortifications across an undisputed "No-man's-land".

There is a certain similarity [I wrote to Pound on Christmas Day] between the position now and at the end of the year 1914. The transition from peace to war has been accomplished. The outer seas, for the moment at any rate, are clear from enemy surface craft. The lines in France are static. But in addition on the sea we have repelled the first U-boat attack, which previously did not begin till February 1915, and we can see our way through the magnetic mine novelty. Moreover, in France the lines run along the frontiers instead of six or seven of the French provinces and Belgium being in the enemy's hands. Thus I feel we may compare the position now very favourably with that of 1914. And also I have the feeling (which may be corrected at any moment) that the Kaiser's Germany was a much tougher customer than Nazi Germany.

This is the best I can do for a Christmas card in these hard times.

I was by now increasingly convinced that there could be no Operation "Catherine" in 1940. "The sending of a superior surface fleet into the Baltic," I wrote to Pound (January 6), "though eminently desirable, is not essential to the seizure and retention of the ironfields. While therefore every preparation to send the Fleet in should continue, and strong efforts should be made, it would be wrong to try it unless we can see our way to maintaining it under air attack, and still more wrong to make the seizure of the ironfields dependent upon the sending of a surface fleet. Let us advance with confidence and see how the naval side develops as events unfold."

And again a week later:

First Lord to First Sea Lord   15.I.40

1. I have carefully considered all the papers you have been good enough to send me in reply to my various minutes about "Catherine". I have come reluctantly but quite definitely to the conclusion that the operation we outlined in the autumn will not be practicable this year. We have not yet obtained sufficient mastery over U-boats, mines, and raiders to enable us to fit for their special duties the many smaller vessels required. The problem of making our ships comparatively secure against air attack has not been solved. The dive-bomber remains a formidable menace. The Rockets [called for secrecy "the U.P. weapon", *i.e.*, unrotated projectile], though progressing rapidly towards the production stage, will not be available in sufficient quantities, even if all goes well, for many months to come. We have not been able so far to give the additional armour protection to our larger ships. The political situation in the Baltic is as baffling as ever. On the other hand, the arrival of the *Bismarck* in September adds greatly to the scale of the surface resistance to be encountered.

2. But the war may well be raging in 1941, and no one can tell what opportunities may present themselves then. I wish therefore that all the preparations of various ships and auxiliaries outlined in your table and marked as "beneficial" should continue as opportunity offers, and that when ships come into the dockyards for repair or refit everything should be done to them which will not delay their return to service. And it would surely be only common prudence, in view

of the attitude of Russia, to go on warming our destroyers for service in winter seas. I am glad to feel that we are agreed in this.

★　　★　　★　　★　　★

So far no ally had espoused our cause. The United States was cooler than in any other period. I persevered in my correspondence with the President, but with little response. The Chancellor of the Exchequer groaned about our dwindling dollar resources. We had already signed a pact of mutual assistance with Turkey, and were considering what aid we could give her from our narrow margins. The stresses created by the Finnish war had worsened our relations, already bad, with the Soviets. Any action we might undertake to help the Finns might lead to war with Russia. The fundamental antagonisms between the Soviet Government and Nazi Germany did not prevent the Kremlin actively aiding by supplies and facilities the development of Hitler's power. Communism in France and any that existed in Britain denounced the "Imperialist-Capitalist" war, and did what they could to hamper work in the munitions factories. They certainly exercised a depressing and subversive influence within the French Army, already wearied by inaction. We continued to court Italy by civilities and favourable contracts, but we could feel no security, or progress towards friendship. Count Ciano was polite to our Ambassador. Mussolini stood aloof.

The Italian Dictator was not however without his own misgivings. On January 3 he wrote a revealing letter to Hitler expressing his distaste for the German agreement with Russia:

No one knows better than I, with forty years' political experience, that policy—particularly a revolutionary policy—has its tactical requirements. I recognised the Soviets in 1924. In 1934 I signed with them a treaty of commerce and friendship. I therefore understood that, *especially as Ribbentrop's forecast about the non-intervention of Britain and France has not come off*, you are obliged to avoid the Second Front. You have had to pay for this in that Russia has, without striking a blow, been the great profiteer in the war in Poland and the Baltic.

But I who was born a revolutionary and have not modified my revolutionary mentality tell you that you cannot permanently sacrifice the principles of *your* revolution to the tactical requirements of a given moment. . . . I have also the definite duty to add that a further step in the relations with Moscow would have catastrophic repercussions in Italy, where the unanimity of anti-Bolshevist feeling is absolute, granite-hard, and unbreakable. Permit me to think that this will not happen. The solution of your *Lebensraum* is in Russia, and nowhere else. . . . The day when we shall have demolished Bolshevism we shall have kept faith with both our revolutions. Then it will be the turn of the great democracies, who will not be able to survive the cancer which gnaws them. . . .

* * * * *

On January 6 I again visited France, to explain my two mechanical projects, Cultivator No. 6 and the fluvial mine (Operation "Royal Marine")* to the French High Command. In the morning, before I left, the Prime Minister sent for me and told me he had decided to make a change at the War Office, and that Mr. Hore-Belisha would give place to Mr. Oliver Stanley. Late that night Mr. Hore-Belisha called me on the telephone at our Embassy in Paris and told me what I knew already. I pressed him, but without success, to take one of the other offices which were open to him. The Government was itself in low water at this time, and almost the whole Press of the country declared that a most energetic and live figure had been lost to the Government. He quitted the War Office amid a chorus of newspaper tributes. Parliament does not take its opinion from the newspapers; indeed, it often reacts in the opposite sense. When the House of Commons met a week later Mr. Hore-Belisha had few champions, and refrained from making any statement. I wrote to him as follows:

January 10, 1940

I much regret that our brief association as colleagues has ended. In the last war I went through the same experience as you have suffered, and I know how bitter and painful it is to anyone with his heart in the job. I was not consulted in the changes that were proposed. I was only informed after they had been decided. At the same time, I should fail in candour if I did not let you know that I thought it would have been better if you had gone to the Board of Trade or the Ministry of Information, and I am very sorry that you did not see your way to accept the first of these important offices.

The outstanding achievement of your tenure of the War Office was the passage of conscription in time of peace. You may rest with confidence upon this, and I hope that it will not be long before we are colleagues again, and that this temporary set-back will prove no serious obstacle to your opportunities of serving the country.

* See Appendices O and Q.

497

It was not possible for me to realise my hope until, after the break-up of the National Coalition, I formed the so-called "Caretaker Government" in May 1945. Belisha then became Minister of National Insurance. In the interval he had been one of our severe critics; but I was very glad to be able to bring so able a man back into the Administration.

* * * * *

All January the Finns stood firm, and at the end of the month the growing Russian armies were still held in their positions. The Red Air Force continued to bomb Helsingfors and Viipuri, and the cry from the Finnish Government for aircraft and war materials grew louder. As the Arctic nights shortened the Soviet air offensive would increase, not only upon the towns of Finland, but upon the communications of their armies. Only a trickle of war material and only a few thousand volunteers from the Scandinavian countries had reached Finland so far. A bureau for recruiting was opened in London in January, and several scores of British aircraft were sent to Finland, some direct by air. Nothing in fact of any use was done.

The delays about Narvik continued interminably. Although the Cabinet were prepared to contemplate pressure upon Norway and Sweden to allow aid to pass to Finland, they remained opposed to the much smaller operation of mining the Leads. The first was noble, the second merely tactical. Besides, everyone could see that Norway and Sweden would refuse facilities for aid, so nothing would come of the project anyway.

In my vexation after one of our Cabinets I wrote to a colleague:

January 15, 1940

My disquiet was due mainly to the awful difficulties which our machinery of war-conduct presents to positive action. I see such immense walls of prevention, all built and building, that I wonder whether any plan will have a chance of climbing over them. Just look at the arguments which have had to be surmounted in the seven weeks we have discussed this Narvik operation. First, the objections of the Economic Departments, Supply, Board of Trade, etc. Secondly, the Joint Planning Committee. Thirdly, the Chiefs of Staff Committee. Fourthly, the insidious argument, "Don't spoil the big plan for the sake of the small," when there is really very little chance of the big plan being resolutely attempted. Fifthly, the juridical and moral objections, all gradually worn down. Sixthly, the attitude of neutrals, and above all the United States. But see how well the United States have responded to our *démarche!* Seventhly, the Cabinet itself, with its many angles of criticism. Eighthly, when all this has been smoothed out, the French have to be consulted. Finally, the Dominions and their consciences have to be squared, they not having gone through the process by which opinion has advanced at home. All this makes me feel that under the present arrangements we shall be reduced to waiting upon the terrible attacks of the enemy, against which it is impossible to prepare in every quarter simultaneously without fatal dissipation of strength.

I have two or three projects moving forward, but all, I fear, will succumb before the tremendous array of negative arguments and forces. Pardon me, therefore, if I showed distress. One thing is absolutely certain, namely, that victory will never be found by taking the line of least resistance.

However, all this Narvik story is for the moment put on one side by the threat to the Low Countries. If this materialises the position will have to be studied in the light of entirely new events. . . . Should a great battle engage in the Low Countries the effects upon Norway and Sweden may well be decisive. Even if the battle ends only in a stalemate they may feel far more free, and to us a diversion may become even more needful.

* * * * *

There were other causes for uneasiness. Progress in converting our industries to war production was not up to the pace required. In a speech at Manchester on January 27 I urged the immense importance of expanding our labour supply and of bringing great numbers of women into industry to replace the men taken for the armed forces and to augment our strength. I continued:

We have to make a huge expansion, especially of those capable of performing skilled or semi-skilled operations. Here we must specially count for aid and guidance upon our Labour colleagues and trade union leaders. I can speak with some knowledge about this, having presided over the former Ministry of Munitions in its culminating phase. Millions of new workers will be needed, and more than a million women must come boldly forward into our war industries—into the shell plants, the munitions works, and into the aircraft factories. Without this expansion of labour and without allowing the women of Britain to enter the struggle as they desire to do, we should fail utterly to bear our fair share of the burden which France and Britain have jointly assumed.

Little was however done, and the sense of extreme emergency seemed lacking. There was a "twilight" mood in the ranks of Labour and of those who directed production as well as in the military operations. It was not till the beginning of May that a survey of employment in the engineering, motor, and aircraft group of industries which was presented to the Cabinet revealed the facts in an indisputable form. This paper was searchingly examined by my statistical department under Professor Lindemann. In spite of the distractions and excitements of the Norwegian hurly-burly then in progress, I found time to address the following note to my colleagues:

NOTE BY THE FIRST LORD OF THE ADMIRALTY

May 4, 1940

This report suggests that in this fundamental group, at any rate, we have hardly begun to organise man-power for the production of munitions.

In [previous papers] it was estimated that a very large expansion, amounting to 71.5 per cent. of the number engaged in the metal industry, would be needed in the first year of war. Actually the engineering, motor, and aircraft group, which covers three-fifths of the metal industry and which is discussed in this survey, has only expanded by 11.1 per cent. (122,000) between June 1939 and April 1940. This is less than one-sixth of the expansion stated to be required. Without any Government intervention, by the mere improvement of trade, the number increased as quickly as this in the year 1936–37.

Although 350,000 boys leave school each year, there is an increase of only 25,000 in the number of males under 21 employed in this group. Moreover, the proportion of women and young persons has only increased from 26.6 per cent. to 27.6 per cent. In the engineering, motor, and aircraft group we now have only one woman for every twelve men. During the last war the ratio of women to men in the metal industries increased from one woman for every ten men to one woman for every three men. In the first year of the last war, July 1914 to July 1915, the new workers drafted into the metal industries amounted to 20 per cent. of those already there. In the group under survey, which may fairly be taken as typical of the whole metal industry, only 11 per cent. have been added in the last ten months.

Admiralty establishments, in which employment has been increased by nearly 27 per cent., have not been considered here, as no figures of the different types of labour are given.

* * * * *

On January 10 anxieties about the Western Front received confirmation. A German staff major of the 7th Air Division had been ordered to take some documents to headquarters in Cologne. He missed his train and decided to fly. His machine overshot the mark and made a forced landing in Belgium, where Belgian troops arrested him and impounded his papers, which he tried desperately to destroy. These contained the entire and actual scheme for the invasion of Belgium, Holland, and France on which Hitler had resolved. Shortly the German major was released to explain matters to his superiors. I was told about all this at the time, and it seemed to me incredible that the Belgians would not make a plan to invite us in. But they did nothing about it. It was argued in all three countries concerned that probably it was a plant. But this could not be true. There could be no sense in the Germans trying to make the Belgians believe that they were going to attack them in the near future. This might make them do the very last thing the Germans wanted, namely, make a plan with the French and British Armies to come forward privily and quickly one fine night. I therefore believed in the impending attack.

On January 13 Admiral Keyes telephoned to me that the King of the Belgians might be able to persuade his Ministers to invite French and British troops into Belgium "at once" if we would agree to give certain far-reaching guarantees. "At once" was taken by us to mean immediately and not "as soon as Germany invades". The War Cabinet decided to reply that we could give no guarantees other than those implicit in a military alliance, and that the invitation to enter Belgium must be given soon enough to enable the Allied troops to forestall a German invasion, which the Belgian Government apparently thought was imminent. Admiral Keyes telegraphed on January 15 that the King thought this reply would have a very bad effect if he communicated it to his Government, that if Allied troops entered "at once" Belgium and Holland would be immediately involved in war, and that it would be better that the onus of breaking Belgian neutrality should rest on Germany. A similar reply was given by the Belgian Government to M. Daladier, and the French Ambassador in London also told us that the Belgian Government thought that if Germany were left to commit an act of aggression Anglo-French help would "acquire a moral character" which "would increase the chance of success".

Thus the Belgian King and his Army staff merely waited, hoping that all would turn out well. In spite of the German major's papers no further action of any kind was taken by the Allies or the threatened States. Hitler, on the other hand, as we know, summoned Goering to his presence, and on being told that the captured papers were in fact the complete plans for invasion, ordered, after venting his anger, new variants to be prepared.

It was thus clear at the beginning of 1940 that Hitler had a detailed plan involving both Belgium and Holland for the invasion of France. Should this begin at any moment, General Gamelin's Plan D would be put in operation, including the movement of the Seventh French Army and the British Army. Plan D had been worked out in exact detail, and required only one single word to set it in motion. This course, though deprecated at the outset of the war by the British Chiefs of Staff, had been definitely and formally confirmed in Paris on November 17, 1939. On this basis the Allies awaited the impending shock, and Hitler the campaigning season, for which the weather might well be favourable from April onwards.

During the winter and spring the B.E.F. were extremely busy setting themselves to rights, fortifying their line and preparing for war, whether offensive or defensive. From the highest rank to the lowest all were hard at it, and the good showing that they eventually made was due largely to the full use made of the opportunities provided during the winter. The British was a a far better army at the end of the "Twilight War". It was also larger. The 42nd and 44th Divisions arrived in France in March, and went on to the frontier line in the latter half of April 1940. In that month there also arrived the 12th, 23rd, and 46th Divisions. These were sent to complete their training in France and to augment the labour force for all the work in hand. They were short even of the ordinary unit weapons and equipment, and had no artillery. Nevertheless they were inevitably drawn into the fighting when it began, and acquitted themselves well.

The awful gap, reflecting on our pre-war arrangements, was *the absence of even one armoured division in the British Expeditionary Force*. Britain, the cradle of the tank in all its variants, had between the wars so far neglected the development of this weapon, soon to dominate the battlefields, that eight months after the declaration of war our small but good Army had only with it, when the hour of trial arrived, the 1st Army Tank Brigade, comprising 17 light tanks and 100 "Infantry" tanks. Only 23 of the latter carried even the 2-pdr. gun, the rest machine-guns only. There were also seven cavalry and Yeomanry regiments equipped with carriers and light tanks which were in process of being formed into two light armoured brigades. Apart from the lack of armour, the progress in the efficiency of the B.E.F. was marked.

*   *   *   *   *

Developments on the French front were less satisfactory. In a great national conscript force the mood of the people is closely reflected in its army, the more so when that army is quartered in the homeland and contacts are close. It cannot be said that France in 1939–40 viewed the war with uprising spirit, or even with much confidence. The restless internal politics of the past decade had bred disunity and discontents. Important elements, in reaction to growing Communism, had swung towards Fascism, lending a ready ear to Goebbels' skilful propaganda and passing it on in gossip and rumour. So also in the Army the disintegrating influences of both Communism and Fascism were at work; the long winter months of waiting gave time and opportunity for the poisons to be established.

Very many factors go to the building up of sound morale in an army, but one of the greatest is that the men be fully employed at useful and interesting work. Idleness is a dangerous breeding-ground. Throughout the winter there were many tasks that needed doing: training demanded continuous attention; defences were far from satisfactory or complete—even the Maginot Line lacked many supplementary field works; physical fitness demands exercise. Yet visitors to the French front were often struck by the prevailing atmosphere of calm aloofness, by the seemingly poor quality of the work in hand, by the lack of visible activity of any kind. The emptiness of the roads behind the line was in great contrast to the continual coming and going which extended for miles behind the British sector.

There can be no doubt that the quality of the French Army was allowed to deteriorate during the winter, and that it would have fought better in the autumn than in the spring. Soon it was to be stunned by the swiftness and violence of the German assault. It was not until the last phases of that brief campaign that the true fighting qualities of the French soldier rose uppermost in defence of his country against the age-long enemy. But then it was too late.

*   *   *   *   *

Meanwhile the German plans for a direct assault on Norway and a lightning occupation of Denmark also were advancing. General Keitel drew up a memorandum on this subject on January 27, 1940:

The Fuehrer and Supreme Commander of the Armed Forces wishes that Study N should be further worked on under my direct and personal guidance, and in the closest conjunction with the general war policy. For these reasons the Fuehrer has commissioned me to take over the direction of further preparations.

The detailed planning for this operation proceeded through the normal channels.

*   *   *   *   *

In early February, when the Prime Minister was going to the Supreme War Council in Paris, he invited me for the first time to go with him. I suggested that we should go by sea, which I could arrange; so we all sailed from Dover in a destroyer, and reached Paris in time for a meeting in the evening. On the way over Mr. Chamberlain showed me the reply he had given to the peace suggestions which Mr. Sumner Welles had gathered. This struck me favourably, and when I had read it in his presence I said to him: "I am proud to serve in your Government." He seemed pleased at this.

The main subject of discussion on February 5 was "Aid to Finland", and plans were approved to prepare three or four divisions and persuade Norway and Sweden to let us send supplies and reinforcements to the Finns, and incidentally to get control of the Gällivare ore-field. As might be expected, the Swedes did not agree to this, and, though extensive preparations were made, the whole project fell to the ground. Mr. Chamberlain conducted the proceedings himself on our behalf, and only minor interventions were made by the various British Ministers attending. I am not recorded as having said a word.

The next day, when we came to re-cross the Channel, an amusing incident occurred. We sighted a floating mine. So I said to the captain, "Let's blow it up by gunfire." It burst with a good bang, and a large piece of wreckage sailed over towards us and seemed for an instant as if it were going to settle on the bridge, where all the politicians and some of the other swells were clustered. However, it landed on the forecastle, which was happily bare, and no one was hurt. Thus everything passed off pleasantly. From this time onwards I was invited by the Prime Minister to accompany him, with others, to the meetings of the Supreme War Council. But I could not provide an equal entertainment each time.

★　★　★　★　★

The Council decided that it was of the first importance that Finland should be saved; that she could not hold out after the spring without reinforcements of thirty to forty thousand trained men; that the present stream of heterogeneous volunteers was not sufficient; and that the destruction of Finland would be a major defeat for the Allies. It was therefore necessary to send Allied troops either through Petsamo or through Narvik and/or other Norwegian ports. The operation through Narvik was preferred, as it would enable the Allies to "kill two birds with one stone" (*i.e.*, help Finland and cut off the iron ore). Two British divisions due to start for France in February should be retained in England and prepared for fighting in Norway. Meanwhile every effort should be made to procure the assent and if possible the co-operation of the Norwegians and Swedes. The issue of what to do if Norway and Sweden refused, as seemed probable, was never faced.

A vivid episode now sharpened everything in Scandinavia. The reader will remember my concern that the *Altmark*, the auxiliary of the *Spee*, should be captured. This vessel was also a floating prison for the crews of our sunk merchant-ships. British captives released by Captain Langsdorff according to international law in Montevideo harbour told us that nearly three hundred British merchant seamen were on board the *Altmark*. This vessel hid in the South Atlantic for nearly two months, and then, hoping that the search had died down, her captain made a bid to return to Germany. Luck and the weather favoured her, and not until February 14, after passing between Iceland and the Faroes, was she sighted by our aircraft in Norwegian territorial waters.

First Lord to First Sea Lord 16.2.40

On the position as reported to me this morning, it would seem that the cruiser and destroyers should sweep northward during the day up the coast of Norway, not hesitating to arrest *Altmark* in territorial waters should she be found. This ship is violating neutrality in carrying British prisoners of war to Germany. Surely another cruiser or two should be sent to rummage the Skagerrak to-night? The *Altmark* must be regarded as an invaluable trophy.

In the words of an Admiralty communiqué, "certain of His Majesty's ships which were conveniently disposed were set in motion". A destroyer flotilla, under the command of Captain Philip Vian, of H.M.S. *Cossack*, intercepted the *Altmark*, but did not immediately molest her. She took refuge in Jösing Fiord, a narrow inlet about a mile and a half long surrounded by high snow-clad cliffs. Two British destroyers were told to board her for examination. At the entrance to the fiord they were met by two Norwegian gunboats, who informed them that the ship was unarmed, had been examined the previous day, and had received permission to proceed to Germany, making use of Norwegian territorial waters. Our destroyers thereupon withdrew.

When this information reached the Admiralty I intervened, and, with the concurrence of the Foreign Secretary, ordered our ships to enter the fiord. I did not often act so directly; but I now sent Captain Vian the following order:

February 16, 1940, 5.25 p.m.

Unless Norwegian torpedo-boat undertakes to convoy *Altmark* to Bergen with a joint Anglo-Norwegian guard on board, and a joint escort, you should board *Altmark*, liberate the prisoners, and take possession of the ship pending further instructions. If Norwegian torpedo-boat interferes, you should warn her to stand off. If she fires upon you, you should not reply unless attack is serious, in which case you should defend yourself, using no more force than is necessary, and ceasing fire when she desists.

Vian did the rest. That night in the *Cossack* with searchlights burning he entered the fiord through the icefloes. He first went on board the Norwegian gunboat *Kjell* and requested that the *Altmark* should be taken to Bergen under a joint escort, for inquiry according to international law. The Norwegian captain repeated his assurance that the *Altmark* had been twice searched, that she was unarmed, and that no British prisoners had been found. Vian then stated that he was going to board her, and invited the Norwegian officer to join him. This offer was eventually declined.

Meanwhile the *Altmark* got under way, and in trying to ram the *Cossack* ran herself aground. The *Cossack* forced her way alongside and a boarding party sprang across, after grappling the two ships together. A sharp hand-to-hand fight followed, in which four Germans were killed and five wounded; part of the crew fled ashore and the rest surrendered. The search began for the British prisoners. They were soon found in their hundreds, battened down, locked in storerooms, and even in an empty oil-tank. Then came the cry, "The Navy's here!" The doors were broken in and the captives rushed on deck. Altogether 299 prisoners were released and transferred to our destroyers. It was also found that the *Altmark* carried two pom-poms and four machine-guns, and that, despite having been boarded twice by the Norwegians, she had not been searched. The Norwegian gunboats remained passive observers throughout. By midnight Vian was clear of the fiord, and making for the Forth.

Admiral Pound and I sat up together in some anxiety in the Admiralty War Room. I had put a good screw on the Foreign Office, and was fully aware of the technical gravity of the measures taken. To judge them fairly it must be remembered that up to that date Germany had sunk 218,000 tons of Scandinavian shipping, with a loss of 555 Scandinavian lives. But what mattered at home and in the Cabinet was whether British prisoners were found on board or not. We were delighted when at three o'clock in the morning news came that three hundred had been found and rescued. This was a dominating fact.

On the assumption that the prisoners were in a pitiable condition from starvation and confinement, we directed ambulances, doctors, the Press, and photographers to the port of Leith to receive them. As however it appeared that they were in good health, had been well looked after on the destroyers, and came ashore in a hearty condition, no publicity was given to this aspect. Their rescue and Captain Vian's conduct aroused a wave of enthusiasm in Britain almost equal to that which followed the sinking of the *Graf Spee*. Both these events strengthened my hand and the prestige of the Admiralty. "The Navy's here!" was passed from lip to lip.

Every allowance must be made for the behaviour of the Norwegian Government, which was of course quivering under the German terror and exploiting our forbearance. They protested vehemently against the entry of their territorial waters. Mr. Chamberlain's speech in the House of Commons contained the essence of the British reply:

According to the views expressed by Professor Koht [the Norwegian Foreign Minister], the Norwegian Government see no objection to the use of Norwegian territorial waters for hundreds of miles by a German warship for the purpose of escaping capture on the high seas and of conveying British prisoners to a German prison camp. Such a doctrine is at variance with international law as His Majesty's Government understand it. It would in their view legalise the abuse by German warships of neutral waters and create a position which His Majesty's Government could in no circumstances accept.

$$* \quad * \quad * \quad * \quad *$$

Hitler's decision to invade Norway had, as we have seen, been taken on December 14, and the staff work was proceeding under Keitel. The incident of the *Altmark* no doubt gave a spur to action. At Keitel's suggestion, on February 20 Hitler summoned urgently to Berlin General von Falkenhorst, who was at that time in command of an Army Corps at Coblenz. Falkenhorst had taken part in the German campaign in Finland in 1918, and upon this subject the interview with the Fuehrer opened. The General described the conversation at the Nuremberg trials.

Hitler reminded me of my experience in Finland, and said to me, "Sit down and tell me what you did." After a moment the Fuehrer interrupted me. He led me to a table covered with maps. "I have a similar thing in mind," he said: "the occupation of Norway; because I am informed that the English intend to land there, and I want to be there before them."

Then, marching up and down, he expounded to me his reasons. "The occupation of Norway by the British would be a strategic turning movement which would lead them into the Baltic, where we have neither troops nor coastal fortifications. The success which we have gained in the East and which we are going to win in the West would be annihilated, because the enemy would find himself in a position to advance on Berlin and to break the backbone of our two fronts. In the second and third place, the conquest of Norway will ensure the liberty of movement of our Fleet in the Bay of Wilhelmshaven, and will protect our imports of Swedish ore." . . . Finally he said to me, "I appoint you to the command of the expedition."

That afternoon Falkenhorst was summoned again to the Chancellery to discuss with Hitler, Keitel, and Jodl the detailed operational plans for the Norwegian expedition. The question of priorities was of supreme importance. Would Hitler commit himself in Norway before or after the execution of "Case Yellow" —the attack on France? On March 1 he made his decision: Norway was to come first. The entry in Jodl's diary for March 3 reads: "The Fuehrer decides to carry out 'Weser Exercise' before 'Case Yellow', with a few days' interval."

★   ★   ★   ★   ★

A vexatious air attack had recently begun on our shipping all along the East Coast. Besides ocean-going vessels destined for the large ports there were on any given day about 320 ships of between 500 and 2,000 tons either at sea or in harbour on the coast, many engaged in coal transport to London and the south. Only a few of these small vessels had yet been provided with an anti-aircraft gun, and the enemy aircraft therefore concentrated upon this easy prey. They even attacked the lightships. These faithful servants of the seamen, moored in exposed positions near the shoals along our coasts, were of use to all, even the marauding U-boat itself, and had never been touched in any previous war. Several were now sunk or damaged, the worst case being off the Humber, where a fierce machine-gun attack killed eight out of the lightship's crew of nine.

As a defence against air attack the convoy system proved as effective as it had against the U-boats, but everything was now done to find some kind of weapon for each ship. In our dearth of ack-ack guns all sorts of contrivances were used. Even a life-saving rocket brought down an air bandit. The spare machine-guns from the Home Fleet were distributed to British and Allied merchant ships on the East Coast with naval gunners. These men and their weapons were shifted from ship to ship for each voyage through the danger zone. By the end of February the Army was able to help, and thus began an organisation later known as the Maritime Royal Artillery. At the height of the war in 1944 more than 38,000 officers and men from the regular forces were employed in this task, of which 14,000 were found by the Army. Over considerable sections of the East Coast convoy route air fighter protection from the nearest airfields could soon be given on call. Thus the efforts of all three Services were combined. An increasing toll was taken of the raiders. Shooting up ordinary defenceless shipping of all countries turned out to be more costly than had been expected, and the attacks diminished.

Not all the horizon was dark. In the outer seas there had been no further signs of raider activity since the destruction of the *Graf Spee* in December, and the work of sweeping German shipping from the seas continued. During February six German ships left Spain in an attempt to reach Germany. Only one succeeded; of the remainder three were captured, one scuttled herself, and one was wrecked in Norway. Seven other German ships attempting to run the blockade were intercepted by our patrols during February and March. All except one of these were scuttled by their captains. Altogether by the beginning of April 1940 seventy-one ships, of 340,000 tons, had been lost to the Germans by capture or scuttling, while 215 German ships still remained cooped in neutral ports. Finding our merchant ships armed, the U-boats had abandoned the gun for the torpedo. Their next descent had been from the torpedo to the lowest form of warfare—the undeclared mine. We have seen how the magnetic mine attack had been met and mastered. Nevertheless more than half our losses in January were from this cause, and more than two-thirds of the total fell on neutrals.

On the Navy Estimates at the end of February I reviewed the salient features of the war at sea. The Germans, I surmised, had lost half the U-boats with which they had entered the war. Contrary to expectation, few new ones had yet made their appearance. Actually, as we now know, sixteen U-boats had been sunk and nine added up to the end of February. The enemy's main effort had not yet developed. Our programme of small-ship building, both in the form of escort vessels and in replacement of merchant ships, was very large. The Admiralty had taken over control of merchant ship-building, and Sir James Lithgow, the Glasgow ship-builder, had joined the Board for this purpose. In the first six months of this new war, after making allowance for gains through new construction and transfers from foreign flags, our net loss had been less than 200,000 tons, compared with

450,000 tons in the single deadly month of April 1917.* Meanwhile we had continued to capture more cargoes in tonnage destined for the enemy than we had lost ourselves.

Each month [I said in ending my speech] there has been a steady improvement in imports. In January the Navy carried safely into British harbours, despite U-boats and mines and the winter gales and fog, considerably more than four-fifths of the peace-time average for the three preceding years. . . . When we consider the great number of British ships which have been withdrawn for naval service or for the transport of our armies across the Channel or of troop convoys across the globe, there is nothing in these results, to put it mildly, which should cause despondency or alarm.

* See Appendix P.

CHAPTER XXXII

BEFORE THE STORM

March 1940

*The Fleet Returns to Scapa Flow – Our Voyage through the Minches –
"Mines Reported in the Fairway" – An Air Alarm – Improvements at
Scapa – Hitler's Plans as Now Known – Desperate Plight of Finland –
M. Daladier's Vain Efforts – The Russo-Finnish Armistice Terms –
New Dangers in Scandinavia – Operation "Royal Marine" – The
Fluvial Mines Ready – M. Daladier's Opposition – The Fall of the
Daladier Government – My Letter to the New Premier, M. Reynaud –
Meeting of Supreme War Council, March 28 – Mr. Chamberlain's
Survey – Decision to Mine the Norwegian Leads at Last – Seven
Months' Delay – Various Offensive Proposals and Devices – Mr.
Chamberlain's Speech of April 5, 1940 – Signs of Impending German
Action.*

MARCH 12 was the long-desired date for the reoccupation and use of Scapa as the main base of the Home Fleet. I thought I would give myself the treat of being present on this occasion in our naval affairs, and embarked accordingly in Admiral Forbes' flagship at the Clyde.

The fleet comprised five capital ships, a cruiser squadron, and perhaps a score of destroyers. The twenty-hour voyage lay through the Minches. We were to pass the Northern Straits at dawn and reach Scapa about noon. The *Hood* and other ships from Rosyth, moving up the East Coast, would be there some hours before us. The navigation of the Minches is intricate, and the northern exit barely a mile wide. On every side are rocky shores and reefs, and three U-boats were reported in these enclosed waters. We had to proceed at high speed and by zigzag. All the usual peace-time lights were out. This was therefore a task in navigation which the Navy keenly appreciated. However, just as we were about to start after luncheon the Master of the Fleet, navigating officer of the flagship, on whom the prime direct responsibility lay, was suddenly stricken by influenza. So a very young-looking lieutenant who was his assistant came up on to the bridge to take charge of the movement of the fleet. I was struck by this officer, who without any notice had to undertake so serious a task, requiring such perfect science, accuracy, and judgment. His composure did not entirely conceal his satisfaction.

I had many things to discuss with the Commander-in-Chief, and it was not until after midnight that I went up on to the bridge. All was velvet black. The air was clear, but no stars were to be seen, and there was no moon. The great ship ploughed along at about 16 knots. One could just see astern the dark mass of the following battleship. Here were nearly thirty vessels steaming in company and moving in order with no lights of any kind except their tiny stern-lights, and constantly changing course in accordance with the prescribed anti-U-boat ritual. It was five hours since they had had any observation of the land or the heavens. Presently the Admiral joined me, and I said to him, "Here is one of the things I should be very sorry to be made responsible for carrying out. How are you going to make sure you will hit the narrow exit from the Minches at daylight?" "What would you do, sir," he said, "if you were at this moment the only person who could give an order?" I replied at once, "I should anchor and wait till morning. 'Anchor, Hardy,' as Nelson said." But the Admiral answered, "We have nearly a hundred fathoms of water beneath us now." I had of course complete confidence, gained over many years, in the Navy, and I only tell this tale to bring home to the general reader the marvellous skill and precision with which what seem to landsmen to be impossible feats of this kind are performed as a matter of course when necessary.

It was eight o'clock before I woke, and we were in the broad waters north of the Minches, steering round the western extremity of Scotland towards Scapa Flow. We were perhaps half an hour's steaming from the entrance to Scapa when a signal reached us saying that several German aircraft had dropped mines in the main entrance we were about to use. Admiral Forbes thereupon decided that he must stand out to the westward for twenty-four hours until the channel had been reported clear, and on this the whole fleet began to change its course. "I can easily put you ashore in a destroyer if you care to tranship," he said. "The *Hood* is already in harbour and can look after you." As I had snatched these three days from London with difficulty, I accepted this offer. Our baggage was rapidly brought on deck, the flagship reduced her speed to three or four knots, and a cutter manned by twelve men in their lifebelts was lowered from the davits. My small party was already in it, and I was taking leave of the Admiral, when an air-raid alarm sounded, and the whole ship flashed into activity as all the ack-ack batteries were manned and other measures taken.

I was worried that the ship should have had to slow down in waters where we knew there were U-boats, but the Admiral said it was quite all right, and pointed to five destroyers which were circling round her at high speed, while a sixth waited for us. We were a quarter of an hour rowing across the mile that separated us from our destroyer. It was like in the olden times, except that the sailors had not so much practice with the oars. The flagship had already regained her speed and was steaming off after the rest of the fleet before we climbed on board. All the officers were at their action stations on the destroyer, and we were welcomed by the surgeon, who took us into the wardroom, where all the instruments of his profession were laid out on the table ready for accidents. But no air raid occurred, and we immediately proceeded at high speed into Scapa. We entered through Switha Sound, which is a small and subsidiary channel and was not affected by the mine-dropping. "This is the tradesmen's entrance," said Thompson, my Flag Commander. It was in fact the one assigned to the storeships. "It's the only one," said the destroyer lieutenant stiffly, "that the flotillas are allowed to use." To make everything go well I asked him if he could remember Kipling's poem about

> "Mines reported in the fairway,
> Warn all traffic and detain.
> 'Sent up . . ."

and here I let him carry on, which he did correctly:

> "*Unity, Claribel, Assyrian, Stormcock,* and *Golden Gain.*"*

* Quoted from "Mine Sweepers", from *Sea Warfare*, by permission of Mrs. Bambridge and Messrs. Macmillan & Co., Ltd.

We soon found our way to the *Hood*, where Admiral Whitworth received us, having gathered most of his captains, and I passed a pleasant night on board before the long round of inspections which filled the next day. This was the last time I set foot upon the *Hood*, although she had nearly two years of war service to perform before her destruction by the *Bismarck* in 1941.

More than six months of constant exertion and the highest priorities had repaired the peace-time neglect at Scapa. The three main entrances were defended with booms and mines, and three additional blockships, among others, had already been placed in Kirk Sound, through which Prien's U-boat had slipped to destroy the *Royal Oak*. Many more blockships were yet to come. A large garrison guarded the base and the still-growing batteries. We had planned for over 120 ack-ack guns, with numerous searchlights and a balloon barrage, to command the air over the fleet anchorage. Not all these measures were yet complete, but the air defences were already formidable. Many small craft patrolled the approaches in ceaseless activity, and two or three squadrons of Hurricane fighters from the airfields in Caithness could be guided to an assailant in darkness or daylight by one of the finest Radar installations then in existence. At last the Home Fleet had a home. It was the famous home from which in the previous war the Royal Navy had ruled the seas.

* * * * *

Although, as we now know, May 10 was already chosen for the invasion of France and the Low Countries, Hitler had not yet fixed the actual date of the prior Norway onslaught. Much was to precede it. On March 14 Jodl wrote in his diary:

The English keep vigil in the North Sea with fifteen to sixteen submarines; doubtful whether reason to safeguard own operations or prevent operations by Germans. Fuehrer has not yet decided what reason to give for Weser Exercise.

There was a hum of activity in the planning sections of the German war machine. Preparations both for the attack on Norway and the invasion of France continued simultaneously and efficiently. On March 20 Falkenhorst reported that his side of the "Weser" operation plan was ready. The Fuehrer held a Military Conference on the afternoon of March 16, and D-Day was provisionally fixed, apparently for April 9. Admiral Raeder reported to the conference:

. . . In my opinion the danger of a British landing in Norway is no longer acute at present. . . . The question of what the British will do in the North in the near future can be answered as follows: They will make further attempts to disrupt German trade in neutral waters and to cause incidents in order perhaps to create a pretext for action against Norway. One object has been and still is to cut off Germany's imports from Narvik. These will be cut off at least for a time however, even if the Weser operation is carried out.

Sooner or later Germany will be faced with the necessity of carrying out the Weser operation. Therefore it is advisable to do so as soon as possible, by April 15 at the latest, since after that date the nights are too short; there will be a new moon on April 7. The operational possibilities of the Navy will be restricted too much if the Weser operation is postponed any longer. The submarines can remain in position only for two to three weeks more. Weather of the type favourable for operation "Gelb" [Yellow] is not to be waited for in the case of the Weser operation; overcast, foggy weather is more satisfactory for the latter. The general state of preparedness of the naval forces and ships is at present good.

*　　*　　*　　*　　*

From the beginning of the year the Soviets had brought their main power to bear on the Finns. They redoubled their efforts to pierce the Mannerheim Line before the melting of the snows. Alas, this year the spring and its thaw, on which the hard-pressed Finns based their hopes, came nearly six weeks late. The great Soviet offensive on the isthmus, which was to last forty-two days, opened on February 1, combined with heavy air bombing of base depots and railway junctions behind the lines. Ten days of heavy bombardment from Soviet guns, massed wheel to wheel, heralded the main infantry attack. After a fortnight's fighting the line was breached. The air attacks on the key fort and base of Viipuri increased in intensity. By the end of the month the Mannerheim defence system had been disorganised and the Russians were able to concentrate against the Gulf of Viipuri. The Finns were short of ammunition and their troops exhausted.

The honourable correctitude which had deprived us of any strategic initiative equally hampered all effective measures for sending munitions to Finland. We had been able so far only to send from our own scanty store contributions insignificant to the

Finns. In France however a warmer and deeper sentiment prevailed, and this was strongly fostered by M. Daladier. On March 2, without consulting the British Government, he agreed to send fifty thousand volunteers and a hundred bombers to Finland. We could certainly not act on this scale, and in view of the documents found on the German major in Belgium, and of the ceaseless Intelligence reports of the steady massing of German troops on the Western Front, it went far beyond what prudence would allow. However, it was agreed to send fifty British bombers. On March 12 the Cabinet again decided to revive the plans for military landings at Narvik and Trondheim, to be followed at Stavanger and Bergen, as a part of the extended help to Finland into which we had been drawn by the French. These plans were to be available for action on March 20, although the need of Norwegian and Swedish permission had not been met. Meanwhile on March 7 Mr. Paasikivi had gone again to Moscow, this time to discuss armistice terms. On the 12th the Russian terms were accepted by the Finns. All our plans for military landings were again shelved, and the forces which were being collected were to some extent dispersed. The two divisions which had been held back in England were now allowed to proceed to France, and our striking power towards Norway was reduced to eleven battalions.

*　*　*　*　*

Meanwhile Operation "Royal Marine" had ripened. Five months of intensive effort with Admiralty priorities behind it had brought its punctual fruition. Admiral FitzGerald and his trained detachments of British naval officers and marines, each man aflame with the idea of a novel stroke in the war, were established on the upper reaches of the Rhine, ready to strike when permission could be obtained. My detailed explanation of the plan will be found in Appendix Q. In March all preparations were perfected, and I at length appealed both to my colleagues and to the French. The War Cabinet were very ready to let me begin this carefully-prepared offensive plan, and left it to me, with Foreign Office support, to do what I could with the French. In all their wars and troubles in my lifetime I have been bound up with the French, and I believed that they would do as much for me as for any other foreigner alive. But in this phase of the Twilight War I could not move them. When I pressed very hard they used a method of refusal which I never met before or since. M. Daladier told me with an air of exceptional formality that "The President of the Republic himself had intervened, and that no aggressive action must be taken which might only draw reprisals upon France." This idea of not irritating the enemy did not commend itself to me. Hitler had done his best to strangle our commerce by the indiscriminate mining of our harbours. We had beaten him by defensive means alone. Good, decent, civilised people, it appeared, must never themselves strike till after they have been struck dead. In these days the fearful German volcano and all its subterranean fires drew near to their explosion-point. There were still months of pretended war. On the one side endless discussions about trivial points, no decisions taken, or if taken rescinded, and the rule "Don't be unkind to the enemy; you will only make him angry." On the other, doom preparing—a vast machine grinding forward ready to break upon us!

* * * * *

The military collapse of Finland led to further repercussions. On March 18 Hitler met Mussolini at the Brenner Pass. Hitler deliberately gave the impression to his Italian host that there was no question of Germany launching a land offensive in the West. On the 19th Mr. Chamberlain spoke in the House of Commons. In view of growing criticism he reviewed in some detail the story of British aid to Finland. He rightly emphasised that our main consideration had been the desire to respect the neutrality of Norway and Sweden, and he also defended the Government for not being hustled into attempts to succour the Finns, which had offered little chance of success. The defeat of Finland was fatal to the Daladier Government, whose chief had taken such marked, if tardy, action, and who had personally given disproportionate prominence to this part of our anxieties. On March 21 a new Cabinet was formed, under M. Reynaud, pledged to an increasingly vigorous conduct of the war.

My relations with M. Reynaud stood on a different footing from any I had established with M. Daladier. Reynaud, Mandel, and I had felt the same emotions about Munich. Daladier had been on the other side. I therefore welcomed the change in the French Government, and I also hoped that my fluvial mines would now have a better chance of acceptance.

Mr. Churchill to M. Reynaud   *March 22, 1940*

I cannot tell you how glad I am that all has been accomplished so successfully and speedily, and especially that Daladier has been rallied to your Cabinet. This is much admired over here, and also Blum's self-effacing behaviour.

I rejoice that you are at the helm, and that Mandel is with you, and I look forward to the very closest and most active co-operation between our two Governments. I share, as you know, all the anxieties you expressed to me the other night about the general course of the war, and the need for strenuous and drastic measures; but I little thought when we spoke that events would so soon take a decisive turn for you. We have thought so much alike during the last three or four years that I am most hopeful that the closest understanding will prevail, and that I may contribute to it.

I now send you the letter which I wrote to Gamelin upon the bus'ness which brought me to Paris last week, and I beg you to give the project your immediate sympathetic consideration. Both the Prime Minister and Lord Halifax have become very keen upon this operation ["Royal Marine"], and we were all three about to press it strongly upon your predecessor. It seems a great pity to lose this valuable time. I have now upwards of 6,000 mines ready and moving forward in an endless flow—alas, only on land—and of course there is always danger of secrecy being lost when delays occur.

I look forward to an early meeting of the Supreme Council, where I trust concerted action may be arranged between French and English *colleagues*—for that is what we are.

Pray give my kind regards to Mandel, and believe me, with the warmest wishes for your success, in which our common safety is deeply involved . . .

The French Ministers came to London for a meeting of the Supreme War Council on March 28. Mr. Chamberlain opened with a full and clear description of the scene as he saw it. To my great satisfaction he said his first proposal was that "a certain operation, generally known as the 'Royal Marine', should be put into operation immediately." He described how this project would be carried out, and stated that stocks had been accumulated for effective and continuous execution. There would be complete surprise. The operation would take place in that part of the Rhine used almost exclusively for military purposes. No similar operation had ever been carried out before, nor had equipment previously been designed capable of taking advantage of river conditions and working successfully against the barrages and types of craft found in rivers. Finally, owing to the design of the weapons, neutral waters would not be affected. The British anticipated that this attack would create the utmost consternation and confusion. It was well known that no people were more thorough than the Germans in preparation and planning; but equally no people could be more completely upset when their plans miscarried. They could not improvise. Again, the war had found the German railways in a precarious state, and therefore their dependence on their inland waterways had increased. In addition to the floating mines other weapons had been designed to be dropped from aircraft in canals within Germany itself, where there was no current. He urged that surprise depended upon speed. Secrecy would be endangered by delay, and the river conditions were about to be particularly favourable. As to German retaliation, if Germany thought it worth while to bomb French or British cities she would not wait for a pretext. Everything was ready. It was only necessary for the French High Command to give the order.

He then said that Germany had two weaknesses: her supplies of iron ore and of oil. The main sources of supply of these were situated at the opposite ends of Europe. The iron ore came from the North. He unfolded with precision the case for intercepting the German iron ore supplies from Sweden. He dealt also with the Roumanian and Baku oilfields, which ought to be denied to Germany, if possible by diplomacy. I listened to this powerful argument with increasing pleasure. I had not realised how fully Mr. Chamberlain and I were agreed.

M. Reynaud spoke of the impact of German propaganda upon French morale. The German radio blared each night that the Reich had no quarrel with France; that the origin of the war was to be found in the blank cheque given by Britain to Poland; that France had been draggged into war at the heels of the British, and even that she was not in a position to sustain the struggle. Goebbels' policy towards France seemed to be to let the war run on at the present reduced tempo, counting upon growing discouragement among the five million Frenchmen now called up and upon the emergence of a French Government willing to come to compromise terms with Germany at the expense of Great Britain.

The question, he said, was widely asked in France, "How can

the Allies win the war?" The number of divisions, "despite British efforts", was increasing faster on the German side than on ours. When therefore could we hope to secure that superiority in man-power required for successful action in the West? We had no knowledge of what was going on in Germany in material equip-ment. There was a general feeling in France that the war had reached a deadlock, and that Germany had only to wait. Unless some action were taken to cut the enemy's supply of oil and other raw material "the feeling might grow that blockade was not a weapon strong enough to secure victory for the Allied cause". About the operation "Royal Marine" he said that, though good in itself, it could not be decisive, and that any reprisals would fall upon France. However, if other things were settled he would make a special effort to secure French concurrence. He was far more responsive about cutting off supplies of Swedish iron ore, and he stated that there was an exact relation between the supplies of Swedish iron ore to Germany and the output of the German iron and steel industry. His conclusion was that the Allies should lay mines in the territorial waters along the Norwegian coast and later obstruct by similar action ore being carried from the port of Luleå to Germany. He emphasised the importance of hampering German supplies of Roumanian oil.

It was at last decided that, after addressing communications in general terms to Norway and Sweden, we should lay minefields in Norwegian territorial waters on April 5, and that, *subject to the concurrence of the French War Committee*, "Royal Marine" should be begun by launching the fluvial mines in the Rhine on April 4, and on April 15 upon the German canals from the air. It was also agreed that if Germany invaded Belgium the Allies should im-mediately move into that country without waiting for a formal invitation; and that if Germany invaded Holland, and Belgium did not go to her assistance, the Allies should consider themselves free to enter Belgium for the purpose of helping Holland.

Finally, as an obvious point on which all were at one, the com-muniqué stated that the British and French Governments had agreed on the following solemn declaration:

That during the present war they would neither negotiate nor conclude an armistice or treaty of peace except by mutual agreement.

This pact later acquired high importance.

*   *   *   *   *

On April 3 the British Cabinet implemented the resolve of the Supreme War Council, and the Admiralty was authorised to mine the Norwegian Leads on April 8. I called the actual mining operation "Wilfred", because by itself it was so small and innocent. As our mining of Norwegian waters might provoke a German retort, it was also agreed that a British brigade and a French contingent should be sent to Narvik to clear the port and advance to the Swedish frontier. Other forces should be dispatched to Stavanger, Bergen, and Trondheim, in order to deny these bases to the enemy.

It is worth while looking back on the stages by which at last the decision to mine the Leads was reached.* I had asked for it on September 29, 1939. Nothing relevant had altered in the meanwhile. The moral and technical objections on the score of neutrality, the possibility of German retaliation against Norway, the importance of stopping the flow of iron ore from Narvik to Germany, the effect on neutral and world-wide opinion—all were exactly the same. But at last the Supreme War Council was convinced, and at last the War Cabinet were reconciled to the scheme, and indeed resolved upon it. Once had they given consent and withdrawn it. Then their mind had been overlaid by the complications of the Finnish war. On sixty days "Aid to Finland" had been part of the Cabinet agenda. Nothing had come of it all. Finland had been crushed into submission by Russia. Now after all this vain boggling, hesitation, changes of policy, arguments between good and worthy people unending, we had at last reached the simple point on which action had been demanded seven months before. But in war seven months is a long time Now Hitler was ready, and ready with a far more

* September 29, 1939. First Lord calls attention of the Cabinet to the value of Swedish iron ore to the German economy.
November 27, 1939. First Lord addresses a minute to the First Sea Lord asking for examination of proposal to mine the Leads.
December 15, 1939 First Lord raises in Cabinet the question of iron ore shipments to Germany.
December 16, 1939. Circulation of detailed memorandum on the subject to the Cabinet.
December 22, 1939. Memorandum considered by the Cabinet.
February 5, 1940. Detailed discussion of issue in connection with aid to Finland at Supreme War Council in Paris (W.S.C. present)
February 19, 1940. Renewed discussion of mining of Leads in British Cabinet. Admiralty authorised to make preparations.
February 29, 1940. Authorisation cancelled.
March 28, 1940. Resolution of Supreme War Council that minefields should be laid.
April 3, 1940. Final decision taken by British Cabinet.
April 8, 1940. The minefields laid.

powerful and well-prepared plan. One can hardly find a more perfect example of the impotence and fatuity of waging war by committee, or rather by groups of committees. It fell to my lot in the weeks which followed to bear much of the burden and some of the odium of the ill-starred Norwegian campaign, the course of which will presently be described. Had I been allowed to act with freedom and design when I first demanded permission a far more agreeable conclusion might have been reached in this key theatre, with favourable consequences in every direction. But now all was to be disaster.

> He that will not when he may,
> When he will he shall have nay.

* * * * *

It may here be right to set forth the various offensive proposals and devices which in my subordinate position I put forward during the Twilight War. The first was the entry and domination of the Baltic, which was the sovereign plan if it were possible. It was vetoed by the growing realisation of air-power. The second was the creation of a Close Action squadron of naval tortoises not too much afraid of the air-bomb or torpedo, by the reconstruction of the *Royal Sovereign* class of battleships. This fell by the way through the movement of the war and the priorities which had to be given to aircraft-carriers. The third was the simple tactical operation of laying mines in the Norwegian Leads to cut off the vital German iron ore supplies. Fourthly comes "Cultivator No. 6" (Appendix O)—namely, a long-term means for breaking the deadlock on the French front without a repetition of the slaughter of the previous war. This was superseded by the onrush of German armour turning our own invention of tanks to our undoing, and proving the ascendancy of the offensive in this new war. The fifth was the operation "Royal Marine"— namely, the paralysing of traffic on the Rhine by the dropping and discharge of fluvial mines. This played its limited part and proved its virtue from the moment when it was permitted. It was however swept away in the general collapse of the French resistance. In any case it required prolonged application to cause major injury to the enemy.

To sum up: In the war of armies on the ground I was under the thrall of defensive fire-power. On the sea I strove persistently within my sphere to assert the initiative against the enemy as a relief from the terrible ordeal of presenting our enormous target of sea commerce to his attack. But in this prolonged trance of the Twilight or "Phoney" War, as it was commonly called in the United States, neither France nor Britain was capable of meeting the German vengeance thrust. It was only after France had been flattened out that Britain, thanks to her island advantage, developed out of the pangs of defeat and the menace of annihilation a national resolve equal to that of Germany.

★　★　★　★　★

Ominous items of news of varying credibility now began to come in. At the meeting of the War Cabinet on April 3 the Secretary of State for War told us that a report had been received at the War Office that the Germans had been collecting strong forces of troops at Rostock with the intention of taking Scandinavia if necessary. The Foreign Secretary said that the news from Stockholm tended to confirm this report. According to the Swedish Legation in Berlin, 200,000 tons of German shipping were now concentrated at Stettin and Swinemünde, with troops on board which rumour placed at 400,000. It was suggested that these forces were in readiness to deliver a counter-stroke against a possible attack by us upon Narvik or other Norwegian ports, about which the Germans were said to be still nervous.

Soon we learnt that the French War Committee would not agree to the launching of "Royal Marine". They were in favour of mining the Norwegian Leads, but opposed to anything that might draw retaliation on France. Through the French Ambassador Reynaud expressed his regret. Mr. Chamberlain, who was much inclined to aggressive action of some kind at this stage, was vexed at this refusal, and in a conversation with M. Corbin he linked the two operations together. The British would cut off the ore supplies of Germany as the French desired, provided that at the same time the French allowed us to retaliate by means of "Royal Marine" for all the injuries we had suffered and were enduring from the magnetic mine. Keen as I was on "Royal Marine", I had not expected him to go so far as this. Both operations were methods of making offensive war upon the enemy, and bringing to an end the twilight period, from the prolongation of which I now believed Germany was the gainer.

However, if a few days would enable us to bring the French into agreement upon the punctual execution of the two projects, I was agreeable to postponing "Wilfred" for a few days.

The Prime Minister was so favourable to my views at this juncture that we seemed almost to think as one. He asked me to go over to Paris and see what I could do to persuade M. Daladier, who was evidently the stumbling-block. I met M. Reynaud and several others of his Ministers at dinner on the night of the 4th at the British Embassy, and we seemed in pretty good agreement. Daladier had been invited to attend, but professed a previous engagement. It was arranged that I should see him the next morning. While meaning to do my utmost to persuade Daladier, I asked permission from the Cabinet to make it clear that we would go forward with "Wilfred" even if "Royal Marine" was vetoed.

I visited Daladier at the Rue St. Dominique at noon on the 5th, and had a serious talk with him. I commented on his absence from our dinner the night before. He pleaded his previous engagement. It was evident to me that a considerable gulf existed between the new and the former Premier. Daladier argued that in three months' time the French aviation would be sufficiently improved for the necessary measures to be taken to meet German reactions to "Royal Marine". For this he was prepared to give a firm date in writing. He made a strong case about the defenceless French factories. Finally he assured me that the period of political crises in France was over, and that he would work in harmony with M. Reynaud. On this we parted.

I reported by telephone to the War Cabinet, who were agreed that "Wilfred" should go forward notwithstanding the French refusal of "Royal Marine", but wished this to be the subject of a formal communication. At their meeting on April 5 the Foreign Secretary was instructed to inform the French Government that notwithstanding the great importance we had throughout attached to carrying out the "Royal Marine" operation at an early date, and simultaneously with the proposed operation in Norwegian territorial waters, we were nevertheless prepared as a concession to their wishes to proceed with the latter alone. The date was thus finally fixed for April 8.

*　　*　　*　　*　　*

On Thursday, April 4, 1940, the Prime Minister addressed the Central Council of the National Union of Conservative and Unionist Associations in a spirit of unusual optimism:

After seven months of war I feel ten times as confident of victory as I did at the beginning. . . . I feel that during the seven months our relative position towards the enemy has become a great deal stronger than it was.

Consider the difference between the ways of a country like Germany and our own. Long before the war Germany was making preparations for it. She was increasing her armed forces on land and in the air with feverish haste, she was devoting all her resources to turning out arms and equipment and to building up huge reserves of stocks; in fact, she was turning herself into a fully armed camp. On the other hand, we, a peaceful nation, were carrying on with our peaceful pursuits. It is true that we had been driven by what was going on in Germany to begin to build up again those defences which we had so long left in abeyance, but we postponed as long as any hope of peace remained— we continually postponed—those drastic measures which were necessary if we were to put the country on a war footing.

The result was that when war did break out German preparations were far ahead of our own, and it was natural then to expect that the enemy would take advantage of his initial superiority to make an endeavour to overwhelm us and France before we had time to make good our deficiencies. Is it not a very extraordinary thing that no such attempt was made? Whatever may be the reason—whether it was that Hitler thought he might get away with what he had got without fighting for it, or whether it was that after all the preparations were not sufficiently complete—however, one thing is certain: he missed the bus.

And so the seven months that we have had have enabled us to make good and remove our weaknesses, to consolidate, and to tune up every arm, offensive and defensive, and so enormously to add to our fighting strength that we can face the future with a calm and steady mind whatever it brings.

Perhaps you may say, "Yes, but has not the enemy too been busy?" I have not the slightest doubt he has. I would be the last to underrate the [his] strength or determination to use that strength without scruple and without mercy if he thinks he can do so without getting his blows returned with interest. I grant that. But I say this too: the very completeness of his preparations has left him very little margin of strength still to call upon.

This proved an ill-judged utterance. Its main assumption that we and the French were relatively stronger than at the beginning of the war was not reasonable. As has been previously explained, the Germans were now in the fourth year of vehement munitions manufacture, whereas we were at a much earlier stage, probably comparable in fruitfulness to the second year. Moreover, with every month that had passed the German Army, now four years old, was becoming a mature and perfected weapon, and the former advantage of the French Army in training and cohesion was steadily passing away. The Prime Minister showed no premonition that we were on the eve of great events, whereas it seemed almost certain to me that the land war was about to begin. Above all, the expression "Hitler missed the bus" was unlucky.

All lay in suspense. The various minor expedients I had been able to suggest had gained acceptance; but nothing of a major character had been done by either side. Our plans, such as they were, rested upon enforcing the blockade by the mining of the Norwegian corridor in the north and by hampering German oil supplies from the south-east. Complete immobility and silence reigned behind the German front. Suddenly the passive or small-scale policy of the Allies was swept away by a cataract of violent surprises. We were to learn what total war means.

CHAPTER XXXIII

THE CLASH AT SEA

April 1940

Lord Chatfield's Retirement – The Prime Minister Invites Me to Preside over the Military Co-ordination Committee – An Awkward Arrangement – "Wilfred" – Oslo – The German Seizure of Norway – Tragedy of Neutrality – All the Fleets at Sea – The "Glowworm" – The "Renown" Engages the "Scharnhorst" and "Gneisenau" – The Home Fleet off Bergen – Action by British Submarines – Warburton-Lee's Flotilla at Narvik – Supreme War Council Meets in London, April 9 – Its Conclusions – My Minute to the First Sea Lord, April 10 – Anger in England – Debate in Parliament, April 11 – The "Warspite" and her Flotilla exterminate the German Destroyers at Narvik – Letter from the King.

BEFORE resuming the narrative I must explain the alterations in my position which occurred during the month of April 1940.

Lord Chatfield's office as Minister for the Co-ordination of Defence had become redundant, and on the 3rd Mr. Chamberlain accepted his resignation, which he proffered freely. On the 4th a statement was issued from No. 10 Downing Street that it was not proposed to fill the vacant post, but that arrangements were being made for the First Lord of the Admiralty, as the senior Service Minister concerned, to preside over the Military Co-ordination Committee. Accordingly I took the chair at its meetings, which were held daily, and sometimes twice daily, from the 8th to 15th of April. I had therefore an exceptional measure of responsibility, but no power of effective direction. Among the other Service Ministers who were also members of the War Cabinet I was "first among equals". I had however no power to

take or to enforce decisions. I had to carry with me both the
Service Ministers and their professional chiefs. Thus many
important and able men had a right and duty to express their
views on the swiftly-changing phases of the battle—for battle it
was—which now began.

The Chiefs of Staff sat daily together after discussing the whole
situation with their respective Ministers. They then arrived at
their own decisions, which obviously became of dominant importance. I learned about these either from the First Sea Lord,
who kept nothing from me, or by the various memoranda or
aide-mémoires which the Chiefs of Staff Committee issued. If I
wished to question any of these opinions I could of course raise
them in the first instance at my Co-ordinating Committee, where
the Chiefs of Staff, supported by their departmental Ministers,
whom they had usually carried along with them, were all present
as individual members. There was a copious flow of polite conversation, at the end of which a tactful report was drawn up by
the secretary in attendance and checked by the three Service
departments to make sure there were no discrepancies. Thus we
had arrived at those broad, happy uplands where everything is
settled for the greatest good of the greatest number by the
common sense of most after the consultation of all. But in war
of the kind we were now to feel the conditions were different.
Alas, I must write it: the actual conflict had to be more like one
ruffian bashing the other on the snout with a club, a hammer, or
something better. All this is deplorable, and it is one of the many
good reasons for avoiding war, and having everything settled by
agreement in a friendly manner, with full consideration for the
rights of minorities and the faithful recording of dissentient
opinions.

The Defence Committee of the War Cabinet sat almost every
day to discuss the reports of the Military Co-ordination Committee and those of the Chiefs of Staff; and their conclusions or
divergences were again referred to frequent Cabinets. All had to
be explained and re-explained; and by the time this process was
completed the whole scene had often changed. At the Admiralty,
which is of necessity in war-time a battle headquarters, decisions
affecting the Fleet were taken on the instant, and only in the
gravest cases referred to the Prime Minister, who supported us
on every occasion. Where the action of the other Services was

involved the procedure could not possibly keep pace with events. However, at the beginning of the Norway campaign the Admiralty in the nature of things had three-quarters of the executive business in its own hands.

I do not pretend that, whatever my powers, I should have been able to take better decisions or reach good solutions of the problems with which we were now confronted. The impact of the events about to be described was so violent and the conditions so chaotic that I soon perceived that only the authority of the Prime Minister could reign over the Military Co-ordination Committee. Accordingly on the 15th I requested Mr. Chamberlain to take the chair, and he presided at practically every one of our subsequent meetings during the campaign in Norway. He and I continued in close agreement, and he gave his supreme authority to the views which I expressed. I was most intimately involved in the conduct of the unhappy effort to rescue Norway when it was already too late. The change in chairmanship was announced to Parliament by the Prime Minister in reply to a question as follows:

I have agreed at the request of the First Lord of the Admiralty to take the chair myself at the meetings of the Co-ordination Committee when matters of exceptional importance relating to the general conduct of the war are under discussion.

Loyalty and goodwill were forthcoming from all concerned. Nevertheless both the Prime Minister and I were acutely conscious of the formlessness of our system, especially when in contact with the surprising course of events. Although the Admiralty was at this time inevitably the prime mover, obvious objections could be raised to an organisation in which one of the Service Ministers attempted to concert all the operations of the other Services, while at the same time managing the whole business of the Admiralty and having a special responsibility for the naval movements. These difficulties were not removed by the fact that the Prime Minister himself took the chair and backed me up. But while one stroke of misfortune after another, the results of want of means or of indifferent management, fell upon us almost daily, I nevertheless continued to hold my position in this fluid, friendly, but unfocused circle.

* * * * *

On the evening of Friday, April 5, the German Minister in Oslo invited distinguished guests, including members of the Government, to a film show at the Legation. The film depicted the German conquest of Poland, and culminated in a crescendo of horror scenes during the German bombing of Warsaw. The caption read: "For this they could thank their English and French friends." The party broke up in silence and dismay. The Norwegian Government was however chiefly concerned with the activities of the British. Between 4.30 and 5 a.m. on April 8 four British destroyers laid our minefield off the entrance to West Fiord, the channel to the port of Narvik. At 5 a.m. the news was broadcast from London, and at 5.30 a note from His Majesty's Government was handed to the Norwegian Foreign Minister. The morning in Oslo was spent in drafting protests to London. But later that afternoon the Admiralty informed the Norwegian Legation in London that German warships had been sighted off the Norwegian coast proceeding northwards, and presumably bound for Narvik. About the same time reports reached the Norwegian capital that a German troopship, the *Rio de Janeiro*, had been sunk off the south coast of Norway by the Polish submarine *Orzel*, that large numbers of German soldiers had been rescued by the local fishermen, and that they said they were bound for Bergen to help the Norwegians defend their country against the British and French. More was to come. Germany had broken into Denmark, but the news did not reach Norway until after she herself was invaded. Thus she received no formal warning. Denmark was easily overrun after a resistance in which a few faithful soldiers were killed.

That night German warships approached Oslo. The outer batteries opened fire. The Norwegian defending force consisted of a mine-layer, the *Olav Tryggvason*, and two minesweepers. After dawn two German minesweepers entered the mouth of the fiord to disembark troops in the neighbourhood of the shore batteries. One was sunk by the *Olav Tryggvason*, but the German troops were landed and the batteries taken. The gallant mine-layer however held off two German destroyers at the mouth of the fiord and damaged the cruiser *Emden*. An armed Norwegian whaler mounting a single gun also went into action at once and without special orders against the invaders. Her gun was smashed and the commander had both legs shot off. To avoid unnerving his men, he rolled himself overboard and died nobly. The main German force, led by the heavy cruiser *Bluecher*, now entered the fiord, making for the narrows defended by the fortress of Oscarsborg. The Norwegian batteries opened, and two torpedoes fired from the shore at 500 yards scored a decisive strike. The *Bluecher* sank rapidly, taking with her the senior officers of the German administrative staff and detachments of the Gestapo. The other German ships, including the *Luetzow*, retired. The damaged *Emden* took no further part in the fighting at sea. Oslo was ultimately taken, not from the sea, but by troop-carrying aeroplanes and by landings in the fiord.

Hitler's plan immediately flashed into its full scope. German forces descended at Kristiansand, at Stavanger, and to the north at Bergen and Trondheim.

The most daring stroke was at Narvik. For a week supposedly empty German ore-ships returning to that port in the ordinary course had been moving up the corridor sanctioned by Norwegian neutrality, filled with supplies and ammunition. Ten German destroyers, each carrying two hundred soldiers and supported by the *Scharnhorst* and *Gneisenau*, had left Germany some days before, and reached Narvik early on the 9th.

Two Norwegian warships, *Norge* and *Eidsvold*, lay in the fiord. They were prepared to fight to the last. At dawn destroyers were sighted approaching the harbour at high speed, but in the prevailing snow-squalls their identity was not at first established. Soon a German officer appeared in a motor launch and demanded the surrender of the *Eidsvold*. On receiving from the commanding officer the curt reply, "I attack," he withdrew, but almost at once the ship was destroyed with nearly all hands by a volley of torpedoes. Meanwhile the *Norge* opened fire, but in a few minutes she too was torpedoed and sank instantly.

In this gallant but hopeless resistance 287 Norwegian seamen perished, less than a hundred being saved from the two ships. Thereafter the capture of Narvik was easy. It was a strategic key —for ever to be denied us.

* * * * *

Surprise, ruthlessness, and precision were the characteristics of the onslaught upon innocent and naked Norway. Nowhere did the initial landing forces exceed two thousand men. Seven army divisions were employed, embarking principally from Hamburg and Bremen and for the follow-up from Stettin and Danzig. Three divisions were used in the assault phase, and four supported them through Oslo and Trondheim. Eight hundred operational aircraft and 250 to 300 transport planes were the salient and vital feature of the design. Within forty-eight hours all the main ports of Norway were in the German grip.

*   *   *   *   *

On the night of Sunday, the 7th, our air reconnaissance reported that a German fleet, consisting of a battle-cruiser, two light cruisers, fourteen destroyers, and another ship, probably a transport, had been seen the day before moving towards the Naze across the mouth of the Skaggerak. We found it hard at the Admiralty to believe that this force was going to Narvik. In spite of a report from Copenhagen that Hitler meant to seize that port, it was thought by the Naval Staff that the German ships would probably turn back into the Skaggerak. Nevertheless the following movement was at once ordered. The Home Fleet, comprising *Rodney*, *Repulse*, *Valiant*, two cruisers, and ten destroyers, was already under steam, and left Scapa at 8.30 p.m. on April 7; the Second Cruiser Squadron, of two cruisers and fifteen destroyers, started from Rosyth at 10 p.m. on the same night. The First Cruiser Squadron, which had been embarking troops at Rosyth for the possible occupation of Norwegian ports in the event of a German attack, was ordered to march the soldiers ashore, even without their equipment, and join the Fleet at sea at the earliest moment. The cruiser *Aurora* and six destroyers similarly engaged in the Clyde were ordered to Scapa. All these decisive steps were concerted with the Commander-in-Chief. In short everything available was ordered out, on the assumption—which we had by no means accepted—that a major emergency had come. At the same time the mine-laying operation off Narvik, by four destroyers, was in progress, covered by the battle-cruiser *Renown*, the cruiser *Birmingham*, and eight destroyers.

When the War Cabinet met on Monday morning I reported that the minefields in the West Fiord had been laid between 4.30 and 5 a.m. I also explained in detail that all our fleets were at sea. But by now we had assurance that the main German naval force was undoubtedly making towards Narvik. On the way to lay the minefield "Wilfred" one of our destroyers, the *Glowworm*, having lost a man overboard during the night, stopped behind to search for him and became separated from the rest of the force. At 8.30 a.m. on the 8th the *Glowworm* had reported herself engaged with an enemy destroyer about 150 miles south-west of West Fiord. Shortly afterwards she had reported seeing another destroyer ahead of her, and later that she was engaging a superior force. After 9.45 she had become silent, since when nothing had been heard from her. On this it was calculated that the German forces, unless intercepted, could reach Narvik about 10 p.m. that night. They would, we hoped, be engaged by the *Renown* and the *Birmingham* and their destroyers. An action might therefore take place very shortly. "It is impossible," I said, "to forecast the hazards of war, but such an action should not be on terms unfavourable for us." Moreover, the Commander-in-Chief with the whole Home Fleet would be approaching the scene from the south. He would now be about opposite Statland. He was fully informed on all points known to us, though naturally he was remaining silent. The Germans knew that the Fleet was at sea, since a U-boat near the Orkneys had been heard to transmit a long message as the Fleet left Scapa. Meanwhile the Second Cruiser Squadron, off Aberdeen, moving north, had reported that it was being shadowed by aircraft and expected to be attacked about noon. All possible measures were being taken by the Navy and the R.A.F. to bring fighters to the scene. No aircraft-carriers were available, but flying-boats were working. The weather was thick in places, but believed to be better in the north and improving.

The War Cabinet took note of my statement and invited me to pass on to the Norwegian naval authorities the information we had received about German naval movements. On the whole the opinion was that Hitler's aim was Narvik.

On April 9 Mr. Chamberlain summoned us to a War Cabinet at 8.30 a.m., when the facts, as then known to us, about the German invasion of Norway and Denmark were discussed. The War Cabinet agreed that I should authorise the Commander-in-Chief of the Home Fleet to take all possible steps to clear Bergen and Trondheim of enemy forces, and that the Chiefs of Staff should set on foot preparations for military expeditions to recapture both those places and to occupy Narvik. These expeditions should not however move until the naval situation had been cleared up.

★　★　★　★　★

Since the war we have learned from German records what happened to the *Glowworm*. Early on the morning of Monday, the 8th, she encountered first one and then a second enemy destroyer. A running fight ensued in a heavy sea until the cruiser *Hipper* appeared on the scene. When the *Hipper* opened fire the *Glowworm* retired behind a smoke-screen. The *Hipper*, pressing on through the smoke, presently emerged to find the British destroyer very close and coming straight for her at full speed. There was no time for the *Hipper* to avoid the impact, and the *Glowworm* rammed her 10,000-ton adversary, tearing a hole forty metres wide in her side. She then fell away crippled and blazing. A few minutes later she blew up. The *Hipper* picked up forty survivors; her gallant captain was being hauled to safety when he fell back exhausted from the cruiser's deck and was lost. Thus the *Glowworm's* light was quenched, but her captain, Lieutenant-Commander Gerard Roope, was awarded the Victoria Cross posthumously, and the story will long be remembered.

When the *Glowworm's* signals ceased abruptly we had good hopes of bringing to action the main German forces which had ventured so far. During Monday we had a superior force on either side of them. Calculations of the sea areas to be swept gave prospects of contact, and any contact meant concentration upon them. We did not then know that the *Hipper* was escorting German forces to Trondheim. She entered Trondheim that night, but the *Glowworm* had put this powerful vessel out of action for a month.

Vice-Admiral Whitworth, in the *Renown*, on receiving *Glowworm's* signals first steered south, hoping to intercept the enemy, but on later information and Admiralty instructions he decided to cover the approaches to Narvik. Tuesday, the 9th, was a tempestuous day, with the seas running high under furious gales and snow-storms. At early dawn the *Renown* sighted two darkened ships some 50 miles to seaward of West Fiord. These were the *Scharnhorst* and *Gneisenau*, who had just completed the task of escorting their expedition to Narvik, but at the time it was believed that only one of the two was a battle-cruiser. The *Renown* opened fire first at 18,000 yards, and soon hit the *Gneisenau*,

destroying her main gun-control equipment and for a time causing her to stop firing. Her consort screened her with smoke. Both
ships then turned away to the north, and the action became a
chase. Meanwhile the *Renown* had received two hits, but these
caused little damage, and presently she scored a second and later
a third hit on the *Gneisenau*. In the heavy seas the *Renown* drove
forward at full speed, but soon had to reduce to 20 knots. Amid
intermittent snow-squalls and German smoke-screens the fire on
both sides became ineffective. Although the *Renown* strained herself to the utmost in trying to overhaul the German ships, they at
last drew away out of sight to the northward.

★   ★   ★   ★   ★

Meanwhile on the morning of April 9 Admiral Forbes with the
main fleet was abreast of Bergen. At 6.20 a.m. he asked the
Admiralty for news of the German strength there, as he intended
to send in a force of cruisers and destroyers under Vice-Admiral
Layton to attack any German ships they might find. The
Admiralty had the same idea, and at 8.20 made him the following signal:

Prepare plans for attacking German warships and transports in
Bergen and for controlling the approaches to the port on the supposi-
tion that defences are still in hands of Norwegians. Similar plans as
regards Trondheim should also be prepared if you have sufficient
forces for both.

The Admiralty sanctioned Admiral Forbes' plan for attacking
Bergen, but later warned him that he must no longer count on
the defences being friendly. To avoid dispersion, the attack on
Trondheim was postponed until the German battle-cruisers
should be found. At about 11.30 four cruisers and seven destroyers, under the Vice-Admiral, started for Bergen, eighty miles
away, making only 16 knots against a head wind and a rough
sea. Presently aircraft reported two cruisers in Bergen instead of
one. With only seven destroyers the prospects of success were
distinctly reduced, unless our cruisers went in too. The First Sea
Lord thought the risk to these vessels, both from mines and the
air, excessive. He consulted me on my return from the Cabinet
meeting, and after reading the signals which had passed during
the morning, and a brief discussion in the War Room, I concurred
in his view. We therefore cancelled the attack. Looking back

on this affair, I consider that the Admiralty kept too close a control upon the Commander-in-Chief, and after learning his original intention to force the passage into Bergen we should have confined ourselves to sending him information.

That afternoon strong air attacks were made on the Fleet, chiefly against Vice-Admiral Layton's ships. The destroyer *Gurkha* was sunk, and the cruisers *Southampton* and *Glasgow* damaged by near misses. In addition the flagship *Rodney* was hit, but her strong deck-armour prevented serious damage.

When the cruiser attack on Bergen was cancelled Admiral Forbes proposed to use torpedo-carrying naval aircraft from the carrier *Furious* at dusk on April 10. The Admiralty agreed, and also arranged attacks by R.A.F. bombers on the evening of the 9th and by naval aircraft from Hatston (Orkney) on the morning of the 10th. Meanwhile our cruisers and destroyers continued to blockade the approaches. The air attacks were successful, and the cruiser *Koenigsberg* was sunk by three bombs from naval aircraft. The *Furious* was now diverted to Trondheim, where our air patrols reported two enemy cruisers and two destroyers. Eighteen aircraft attacked at dawn on the 11th, but found only two destroyers and a submarine, besides merchant ships. Unluckily the wounded *Hipper* had left during the night, no cruisers were found, and the attack on the two German destroyers failed because our torpedoes grounded in shallow water before reaching their targets.

Meanwhile our submarines were active in the Skagerrak and Kattegat. On the night of the 8th they had sighted and attacked enemy ships northward bound from the Baltic, but without success. However, on the 9th the *Truant* sank the cruiser *Karlsruhe* off Kristiansand, and the following night the *Spearfish* torpedoed the pocket-battleship *Luetzow* returning from Oslo. Besides these successes submarines accounted for at least nine enemy transport and supply ships, with heavy loss of life, during the first week of this campaign. Our own losses were severe, and three British submarines perished during April in the heavily-defended approaches to the Baltic.

*　*　*　*　*

On the morning of the 9th the situation at Narvik was obscure. Hoping to forestall a German seizure of the port, the Commander-in-Chief directed Captain Warburton-Lee, commanding our destroyers, to enter the fiord and prevent any landing. Meanwhile the Admiralty transmitted a Press report to him indicating that one ship had already entered the port and landed a small force. The message went on:

Proceed to Narvik and sink or capture enemy ship. It is at your discretion to land forces, if you think you can recapture Narvik from number of enemy present.

Accordingly Captain Warburton-Lee, with the five destroyers of his own flotilla, *Hardy*, *Hunter*, *Havock*, *Hotspur*, and *Hostile*, entered West Fiord. He was told by Norwegian pilots at Tranoy that six ships larger than his own and a U-boat had passed in and that the entrance to the harbour was mined. He signalled this information and added: "Intend attacking at dawn". Admiral Whitworth, who received the signals, considered whether he might stiffen the attacking forces from his own now augmented squadron, but the time seemed too short and he felt that intervention by him at this stage might cause delay. In fact, we in the Admiralty were not prepared to risk the *Renown*—one of our only three battle-cruisers—in such an enterprise. The last Admiralty message passed to Captain Warburton-Lee was as follows:

Norwegian coast defence ships may be in German hands: you alone can judge whether in these circumstances attack should be made. Shall support whatever decision you take.

His reply was:

Going into action.

In the mist and snowstorms of April 10 the five British destroyers steamed up the fiord, and at dawn stood off Narvik. Inside the harbour were five enemy destroyers. In the first attack the *Hardy* torpedoed the ship bearing the pennant of the German Commodore, who was killed; another destroyer was sunk by two torpedoes, and the remaining three were so smothered by gunfire that they could offer no effective resistance. There were also in the harbour twenty-three merchant ships of various nations, including five British; six German were destroyed. Only three of our five destroyers had hitherto attacked. The *Hotspur* and *Hostile* had been left in reserve to guard against any shore batteries or against fresh German ships approaching. They now joined in a second attack, and the *Hotspur* sank two more merchantmen with torpedoes. Captain Warburton-Lee's ships were unscathed; the enemy's fire was apparently silenced, and after an hour's fighting no ships had come out from any of the inlets against him.

But now fortune turned. As he was coming back from a third attack Captain Warburton-Lee sighted three fresh ships approaching from Herjangs Fiord. They showed no signs of wishing to close the range, and action began at 7,000 yards. Suddenly out of the mist ahead appeared two more warships. They were not, as was at first hoped, British reinforcements, but German destroyers which had been anchored in Ballangen Fiord. Soon the heavier guns of the German ships began to tell; the bridge of the *Hardy* was shattered, Warburton-Lee mortally stricken, and all his officers and companions killed or wounded except Lieutenant Stanning, his secretary, who took the wheel. A shell then exploded in the engine-room, and under heavy fire the destroyer was beached. The last signal from the *Hardy's* captain to his flotilla was "Continue to engage the enemy."

Meanwhile the *Hunter* had been sunk, and the *Hotspur* and the *Hostile*, which were both damaged, with the *Havock*, made for the open sea. The enemy who had barred their passage was by now in no condition to stop them. Half an hour later they encountered a large ship coming in from the sea, which proved to be the *Rauenfels*, carrying the German reserve ammunition. She was fired upon by the *Havock*, and soon blew up. The survivors of the *Hardy* struggled ashore with the body of their commander, who was awarded posthumously the Victoria Cross. He and they had left their mark on the enemy and in our naval records.

*　　*　　*　　*　　*

On the 9th MM. Reynaud and Daladier, with Admiral Darlan, flew over to London, and in the afternoon a Supreme War Council meeting was held to deal with what they called "the German action in consequence of the laying of mines within Norwegian territorial waters". Mr. Chamberlain at once pointed out that the enemy's measures had certainly been planned in advance and quite independently of ours. Even at that date this was obvious. M. Reynaud informed us that the French War

Committee, presided over by the President, had that morning decided in principle on moving forward into Belgium should the Germans attack. The addition, he said, of eighteen to twenty Belgian divisions, besides the shortening of the front, would to all intents and purposes wipe out the German preponderance in the West. The French would be prepared to connect such an operation with the laying of the fluvial mines in the Rhine. He added that his reports from Belgium and Holland indicated the imminence of a German attack on the Low Countries; some said days, some said hours.

On the question of the military expedition to Norway, the Secretary of State for War reminded the Council that the two British divisions originally assembled for assistance to Finland had since been sent to France. There were only eleven battalions available in the United Kingdom. Two of these were sailing that night. The rest, for various reasons, would not be ready to sail for three or four days or more.

The Council agreed that strong forces should be sent where possible to ports on the Norwegian seaboard, and joint plans were made. A French Alpine division was ordered to embark within two or three days. We were able to provide two British battalions that night, a further five battalions within three days, and four more within fourteen days—eleven in all. Any additional British forces for Scandinavia would have to be withdrawn from France. Suitable measures were to be taken to occupy the Faroe Islands, and assurances of protection would be given to Iceland. Naval arrangements were concerted in the Mediterranean in the event of Italian intervention. It was also decided that urgent representations should be made to the Belgian Government to invite the Allied armies to move forward into Belgium. Finally, it was confirmed that if Germany made an attack in the West or entered Belgium "Royal Marine" should be carried out.

★   ★   ★   ★   ★

I was far from content with what had happened so far in Norway. I wrote to Admiral Pound:

10.IV.40

The Germans have succeeded in occupying all the ports on the Norwegian coast, including Narvik, and large-scale operations will be required to turn them out of any of them. Norwegian neutrality and our respect for it have made it impossible to prevent this ruthless *coup*. It is now necessary to take a new view. We must put up with the disadvantage of closer air attack on our northern bases. We must seal up Bergen with a watchful minefield, and concentrate on Narvik, for which long and severe fighting will be required.

It is immediately necessary to obtain one or two fuelling bases on the Norwegian coast, and a wide choice presents itself. This is being studied by the Staff. The advantage of our having a base, even improvised, on the Norwegian coast is very great, and now that the enemy have bases there we cannot carry on without it. The Naval Staff are selecting various alternatives which are suitable anchorages capable of defence, and without communications with the interior. Unless we have this quite soon we cannot compete with the Germans in their new position.

We must also take our advantages in the Faroes.

Narvik must be fought for. Although we have been completely outwitted, there is no reason to suppose that prolonged and serious fighting in this area will not impose a greater drain on the enemy than on ourselves.

For three days we were deluged with reports and rumours from neutral countries and triumphant claims by Germany of the losses they had inflicted on the British Navy, and of their master-stroke in seizing Norway in the teeth of our superior naval power. It was obvious that Britain had been forestalled, surprised, and, as I had written to the First Sea Lord, outwitted. Anger swept the country, and the brunt fell upon the Admiralty. On Thursday, the 11th, I had to face a disturbed and indignant House of Commons. I followed the method I have always found most effective on such occasions, of giving a calm, unhurried factual narrative of events in their sequence, laying full emphasis upon ugly truths. I explained for the first time in public the disadvantage we had suffered since the beginning of the war by German's abuse of the Norwegian corridor, or "covered way", and how we had at last overcome the scruple which "caused us injury at the same time that it did us honour".

It is not the slightest use blaming the Allies for not being able to give substantial help and protection to neutral countries if we are held at arm's-length until these neutrals are actually attacked on a scientifically-prepared plan by Germany. The strict observance of neutrality by Norway has been a contributory cause to the sufferings to which she is now exposed and to the limits of the aid which we can give her.

I trust this fact will be meditated upon by other countries who may to-morrow, or a week hence, *or a month hence*, find themselves the victims of an equally elaborately worked out staff plan for their destruction and enslavement.

I described the recent reoccupation by our Fleet of Scapa Flow, and the instant movement we had made to intercept the German forces in the North, and how the enemy were in fact caught between two superior forces.

However, they got away. . . . You may look at the map and see flags stuck in at different points and consider that the results will be certain, but when you get out on the sea, with its vast distances, its storms and mists, and with night coming on, and all the uncertainties which exist, you cannot possibly expect that the kind of conditions which would be appropriate to the movements of armies have any application to the haphazard conditions of war at sea. . . . When we speak of the command of the seas it does not mean command of every part of the sea at the same moment, or at every moment. It only means that we can make our will prevail ultimately in any part of the seas which may be selected for operations, and thus indirectly make our will prevail in every part of the sea. Anything more foolish than to suppose that the life and strength of the Royal Navy should have been expended in ceaselessly patrolling up and down the Norwegian and Danish coasts as a target for the U-boats on the chance that Hitler would launch a blow like this cannot be imagined.

The House listened with growing acceptance to the account, of which the news had just reached me, of Tuesday's brush between the *Renown* and the enemy, of the air attack on the British fleet off Bergen, and especially of Warburton-Lee's incursion and action at Narvik. At the end I said:

Everyone must recognise the extraordinary and reckless gambling which has flung the whole German Fleet out upon the savage seas of war, as if it were a mere counter to be cast away for a particular operation. . . . This very recklessness makes me feel that these costly operations may be only the prelude to far larger events which impend on land. We have probably arrived now at the first main clinch of the war.

After an hour and a half the House seemed to be very much less estranged. A little later there would have been more to tell.

* * * * *

By the morning of April 10 the *Warspite* had joined the Commander-in-Chief, who was proceeding towards Narvik. On learning about Captain Warburton-Lee's attack at dawn we resolved to try again. The cruiser *Penelope*, with destroyer support, was ordered to attack, "if in the light of experience this morning you consider it a justifiable operation". But while the signals were passing, *Penelope*, in searching for enemy transports reported off Bodo, ran ashore. The next day (12th) a dive-bombing attack on enemy ships in Narvik harbour was made from the *Furious*. The attack was pressed home in terrible weather and low visibility, and four hits on destroyers were claimed for the loss of two aircraft. This was not enough. We wanted Narvik very much, and were determined at least to clear it of the German Navy. The climax was now at hand.

The precious *Renown* was kept out of it. Admiral Whitworth shifted his flag to the *Warspite* at sea, and at noon on the 13th he entered the fiord, escorted by nine destroyers and by dive-bombers from the *Furious*. There were no minefields; but a U-boat was driven off by the destroyers, and a second sunk by the *Warspite's* own "Swordfish" aircraft, which also detected a German destroyer lurking in an inlet to launch her torpedoes on the battleship from this ambush. The hostile destroyer was quickly overwhelmed. At 1.30 p.m., when our ships were through the Narrows and a dozen miles from Narvik, five enemy destroyers appeared ahead in the haze. At once a fierce fight began, with all ships on both sides firing and manoeuvring rapidly. The *Warspite* found no shore batteries to attack, and intervened in deadly fashion in the destroyer fight. The thunder of her 15-inch guns reverberated among the surrounding mountains like the voice of doom. The enemy, heavily over-matched, retreated, and the action broke up into separate combats. Some of our ships went into Narvik harbour to complete the task of destruction there; others, led by the *Eskimo*, pursued three Germans who sought refuge in the head-waters of Römbaks Fiord and annihilated them there. The bows of the *Eskimo* were blown off by a torpedo; but in this second sea-fight off Narvik the eight enemy destroyers which had survived Warburton-Lee's attack were all sunk or wrecked without the loss of a single British ship.

When the action was over Admiral Whitworth thought of

throwing a landing party of seamen and marines ashore to occupy the town, where there seemed for the moment to be no opposition. Unless the fire of the *Warspite* could dominate the scene, an inevitable counter-attack by a greatly superior number of German soldiers must be expected. With the risk from the air and from U-boats he did not feel justified in exposing this fine ship so long. His decision was endorsed when a dozen German aircraft appeared at 6 p.m. Accordingly he withdrew early next morning, after embarking the wounded from the destroyers. "My impression," he said, "is that the enemy forces in Narvik were thoroughly frightened as a result of to-day's action. I recommend that the town be occupied without delay by the main landing force." Two destroyers were left off the port to watch events, and one of these rescued the survivors of the *Hardy*, who had meanwhile maintained themselves on shore.

* * * * *

His Majesty, whose naval instincts were powerfully stirred by this clash of the British and German Navies in Northern waters, wrote me the following encouraging letter:

BUCKINGHAM PALACE
April 12, 1940

My dear Mr. Churchill,

I have been wanting to have a talk with you about the recent striking events in the North Sea, which, as a sailor, I have naturally followed with the keenest interest, but I have purposely refrained from taking up any of your time as I know what a great strain has been placed upon you by your increased responsibilities as Chairman of the Co-ordination Committee. I shall however ask you to come and see me as soon as there is a lull. In the meantime I would like to congratulate you on the splendid way in which, under your direction, the Navy is countering the German move against Scandinavia. I also beg of you to take care of yourself and get as much rest as you possibly can in these critical days.

Believe me,
Yours very sincerely,
GEORGE R.I.

CHAPTER XXXIV

NARVIK

Hitler's Outrage on Norway – Long-Prepared Treachery – Norwegian Resistance – Appeal to the Allies – The Position of Sweden – The Narvik Expedition – Instructions to General Mackesy – And to Lord Cork – Question of a Direct Assault – General Mackesy Adverse – My Desire to Concentrate on Narvik and to Attempt to Storm it – War Cabinet Conclusions of April 13 – The Trondheim Project Mooted – Disappointing News from Narvik – My Note to the Military Co-ordination Committee of April 17 – Our Telegram to the Naval and Military Commanders – Deadlock at Narvik.

*F*OR MANY generations Norway, with its homely, rugged population engaged in trade, shipping, fishing, and agriculture, had stood outside the turmoil of world politics. Far off were the days when the Vikings had sallied forth to conquer or ravage a large part of the then known world. The Hundred Years War, the Thirty Years War, the wars of William III and Marlborough, the Napoleonic convulsion, and later conflicts, had left Norway, though separated from Denmark, otherwise unmoved and unscathed. A large proportion of the people had hitherto thought of neutrality and neutrality alone. A tiny army and a population with no desires except to live peaceably in their own mountainous and semi-Arctic country now fell victims to the new German aggression.

It had been the policy of Germany for many years to profess cordial sympathy and friendship for Norway. After the previous war some thousands of German children had found food and shelter with the Norwegians. These had now grown up in Germany, and many of them were ardent Nazis. There was also the Major Quisling, who with a handful of young men had aped and reproduced in Norway on an insignificant scale the Fascist movement. For some years past Nordic meetings had been arranged in Germany to which large numbers of Norwegians had been invited. German lecturers, actors, singers, and men of science had visited Norway in the promotion of a common culture. All this had been woven into the texture of the Hitlerite military plan, and a widely-scattered internal pro-German conspiracy set on foot. In this every member of the German diplomatic or consular service, every German purchasing agency, played its part under directions from the German Legation in Oslo. The deed of infamy and treachery now performed may take its place with the Sicilian Vespers and the massacre of St. Bartholomew. The President of the Norwegian Parliament, Carl Hambro, has written:

In the case of Poland and later in those of Holland and Belgium notes had been exchanged, ultimata had been presented. In the case of Norway the Germans under the mask of friendship tried to extinguish the nation in one dark night, silently, murderously, without any declaration of war, without any warning given. What stupefied the Norwegians more than the act of aggression itself was the national realisation that a great Power, for years professing its friendship, suddenly appeared a deadly enemy, and that men and women with whom one had had intimate business or professional relations, who had been cordially welcomed in one's home, were spies and agents of destruction. More than by the violation of treaties and every international obligation, the people of Norway were dazed to find that for years their German friends had been elaborating the most detailed plans for the invasion and subsequent enslaving of their country.[*]

The King, the Government, the Army, and the people, as soon as they realised what was happening, flamed into furious anger. But it was all too late. German infiltration and propaganda had hitherto clouded their vision, and now sapped their powers of resistance. Major Quisling presented himself at the radio, now in German hands, as the pro-German ruler of the conquered land. Almost all Norwegian officials refused to serve him. The Army was mobilised, and at once began, under General Ruge, to fight the invaders pressing northwards from Oslo. Patriots who could find arms took to the mountains and the forests. The King, the Ministry, and the Parliament withdrew first to Hamar, a hundred miles from Oslo. They were hotly pursued by German

[*] Carl J. Hambro, *I Saw it Happen in Norway*, p. 23.

armoured cars, and ferocious attempts were made to exterminate
them by bombing and machine-gunning from the air. They
continued however to issue proclamations to the whole country
urging the most strenuous resistance. The rest of the population
was overpowered and terrorised by bloody examples into stupefied or sullen submission. The peninsula of Norway is nearly a
thousand miles long. It is sparsely inhabited, and roads and railways are few, especially to the northward. The rapidity with
which Hitler effected the domination of the country was a remarkable feat of war and policy, and an enduring example of
German thoroughness, wickedness, and brutality.

The Norwegian Government, hitherto in their fear of Germany so frigid to us, now made vehement appeals for succour.
It was from the beginning obviously impossible for us to rescue
Southern Norway. Almost all our trained troops, and many only
half trained, were in France. Our modest but growing Air Force
was fully assigned to supporting the British Expeditionary Force,
to Home Defence, and vigorous training. All our anti-aircraft
guns were demanded ten times over for vulnerable points of the
highest importance. Still, we felt bound to do our utmost to go to
their aid, even at violent derangement of our own preparations
and interests. Narvik, it seemed, could certainly be seized and
defended with benefit to the whole Allied cause. Here the King
of Norway might fly his flag unconquered. Trondheim might be
fought for, at any rate as a means of delaying the northward
advance of the invader until Narvik could be regained and made
the base of an army. This, it seemed, could be maintained from
the sea at a strength superior to anything which could be brought
against it by land through five hundred miles of mountain
country. The Cabinet heartily approved all possible measures for
the rescue and defence of Narvik and Trondheim. The troops
which had been released from the Finnish project, and a nucleus
kept in hand for Narvik, could soon be ready. They lacked aircraft, anti-aircraft guns, anti-tank guns, tanks, transport, and
training. The whole of Northern Norway was covered with
snow to depths which none of our soldiers had ever seen, felt, or
imagined. There were neither snow-shoes nor skis—still less
skiers. We must do our best. Thus began this ramshackle campaign.

★　　★　　★　　★　　★

There was every reason to believe that Sweden would be the next victim of Germany or Russia, or perhaps even of both. If Sweden came to the aid of her agonised neighbour the military situation would be for the time being transformed. The Swedes had a good army. They could enter Norway easily. They could be at Trondheim in force before the Germans. We could join them there. But what would be the fate of Sweden in the months that followed? Hitler's vengeance would lay them low, and the Bear would maul them from the East. On the other hand, the Swedes could purchase neutrality by supplying the Germans with all the iron ore they wanted throughout the approaching summer. For Sweden the choice was a profitable neutrality or subjugation. She could not be blamed because she did not view the issue from the standpoint of our unready but now eager Island.

After the Cabinet on the morning of April 11 I wrote the following minute, which the sacrifices we were making for the rights of small States and the Law of Nations may justify:

Prime Minister
Foreign Secretary

I am not entirely satisfied with the result of the discussion this morning, or with my contribution to it. What we want is that Sweden should not remain neutral, but declare war on Germany. What we do not want is either to provide the three divisions which we dangled to procure the Finland project, or to keep her fully supplied with food as long as the war lasts, or to bomb Berlin, etc., if Stockholm is bombed. These stakes are more than it is worth while paying at the present time. On the other hand, we should do everything to encourage her into the war by general assurances that we will give all the help we can, that our troops will be active in the Scandinavian peninsula, that we will make common cause with her as good allies, and will not make peace without her, or till she is righted. Have we given this impulse to the Anglo-French mission? If not, there is still time to do it. Moreover, our diplomacy should be active at Stockholm.

It must be remembered that Sweden will say "Thank you for nothing" about any offers on our part to defend the Gällivare ironfield. She can easily do this herself. Her trouble is to the south, where we can do but little. Still, it will be something to assure her that we intend to open the Narvik route to Sweden from the Atlantic by main force as soon as possible, and also that we propose to clean up the German lodgments on the Norwegian coast *seriatim*, thus opening other channels.

If the great battle opens in Flanders the Germans will not have much to spare for Scandinavia, and if, on the other hand, the Germans do not attack in the West we can afford to send troops to Scandinavia in proportion as German divisions are withdrawn from the Western Front. It seems to me we must not throw cold water on the French idea of trying to induce the Swedes to enter the war. It would be disastrous if they remained neutral and bought Germany off with ore from Gällivare, down the Gulf of Bothnia.

I must apologise for not having sufficiently gripped this issue in my mind this morning, but I only came in after the discussion had begun, and did not address myself properly to it.

There was justice in the Foreign Secretary's reply, by which I was convinced. He said that the Prime Minister and he agreed with my general view, but doubted the method I favoured of approaching Sweden.

April 11, 1940

From all the information that we have from Swedish sources that are friendly to the Allies, it appears that any representations that can be readily translated in their mind into an attempt by us to drag them into the war will be likely to have an effect opposite to that which we want. Their immediate reaction would be that we were endeavouring to get them to do what, until we have established a position in one or more of the Norwegian ports, we were unable or unwilling to do ourselves. And accordingly the result would do us more harm than good.

* * * * *

It was easy to regather at short notice the small forces for a Narvik expedition which had been dispersed a few days earlier. One British brigade and its ancillary troops began to embark immediately, and the first convoy sailed for Narvik on April 12. This was to be followed in a week or two by three battalions of Chasseurs Alpins and other French troops. There were also Norwegian forces north of Narvik which would help our landings. Major-General Mackesy had been selected on April 5 to command any expedition which might be sent to Narvik. His instructions were couched in a form appropriate to the case of a friendly neutral Power from whom some facilities are required. They contained among their appendices the following reference to bombardment:

It is clearly illegal to bombard a populated area in the hope of hitting a legitimate target which is known to be in the area but which cannot be precisely located and identified.

In the face of the German onslaught new and stiffer instructions were issued to the General on the 10th. They gave him more latitude, but did not cancel this particular injunction. Their substance was as follows:

His Majesty's Government and the Government of the French Republic have decided to send a Field Force to initiate operations against Germany in Northern Norway. The object of the force will be to eject the Germans from the Narvik area and establish control of Narvik itself. . . . Your initial task will be to establish your force at Harstad, ensure the co-operation of Norwegian forces that may be there, and obtain the information necessary to enable you to plan your further operations. It is not intended that you should land in the face of opposition. You may however be faced with opposition owing to mistaken identity; you will therefore take such steps as are suitable to establish the nationality of your force before abandoning the attempt. The decision whether to land or not will be taken by the senior naval officer in consultation with you. If landing is impossible at Harstad some other suitable locality should be tried. A landing must be carried out when you have sufficient troops.

At the same time a personal letter from General Ironside, the C.I.G.S., was given to General Mackesy, which included the remark:

You may have a chance of taking advantage of naval action, and should do so if you can. Boldness is required.

This struck a somewhat different note from the formal instructions.

My contacts with Lord Cork and Orrery had become intimate in the long months during which the active discussions of Baltic strategy had proceeded. In spite of some differences of view about "Catherine", his relations with the First Sea Lord were good. I was fully conscious from long and hard experience of the difference between pushing things audaciously on paper so as to get them explored and tested—the processes of mental reconnaissance-in-force—and actually doing them or getting them done. Admiral Pound and I were both agreed from slightly different angles that Lord Cork should command the naval forces in this amphibious adventure in the North. We both urged him not to hesitate to run risks, but to strike hard to seize Narvik. As we were all agreed and could talk things over together, we left him exceptional discretion and did not give him any written orders. He knew exactly what we wanted. In his dispatch he says, " My impression on leaving London was quite clear that it was desired by His Majesty's Government to turn the enemy out of Narvik at the earliest possible moment, and that I was to act with all promptitude in order to attain this result."

Our Staff work at this time had not been tempered by war experience, nor was the action of the Service departments concerted except by the meetings of the Military Co-ordination Committee, over which I had just begun to preside. Neither I, as chairman of the Committee, nor the Admiralty were made acquainted with the War Office instructions to General Mackesy, and as the Admiralty directions had been given orally to Lord Cork there was no written text to communicate to the War Office. The instructions of the two departments, although animated by the same purpose, were somewhat different in tone and emphasis; and this may have helped to cause the divergences which presently developed between the military and naval commanders.

Lord Cork sailed from Rosyth at high speed in the *Aurora* on the night of April 12.* He had intended to meet General Mackesy at Harstad, a small port on the island of Hinnoy, in Vaags Fiord, which, although sixty miles from Narvik, had been selected as the military base. However, on the 14th he received a signal from Admiral Whitworth in the *Warspite*, who had exterminated all the German destroyers and supply ships the day before, saying, "I am convinced that Narvik can be taken by direct assault now without fear of meeting serious opposition on landing. I consider that the main landing force need only be small. . . ." Lord Cork therefore diverted the *Aurora* to Skjel Fiord, in the Lofoten Islands, flanking the approach to Narvik, and sent a message ordering the *Southampton* to join him there. His intention was to organise a force for an immediate assault, consisting of two companies of the Scots Guards who had been embarked in the *Southampton*, and a force of seamen and marines from the *Warspite* and other ships already in Skjel Fiord. He could not however get in touch with the *Southampton* except, after some delay, through the Admiralty, whose reply contained the following sentence: "We think it imperative that you and the General should be together and act together and that no attack should be

* A sketch map of the Narvik operations will be found on page 591.

made except in concert." He therefore left Skjel Fiord for Harstad, and led the convoy carrying the 24th Brigade into harbour there on the morning of the 15th. His escorting destroyers sank U.49, which was prowling near by.

Lord Cork now urged General Mackesy to take advantage of the destruction of all the German naval force and to make a direct attack on Narvik as soon as possible, but the General replied that the harbour was strongly held by the enemy with machine-gun posts. He also pointed out that his transports had not been loaded for an assault, but only for an unopposed landing. He opened his headquarters at the hotel in Harstad, and his troops began to land thereabouts. The next day he stated that, on the information available, landing at Narvik was not possible, nor would naval bombardment make it so. Lord Cork considered that with the help of overwhelming gun-fire troops could be landed in Narvik with little loss; but the General did not agree, and could find some cover in his instructions. From the Admiralty we urged an immediate assault. A deadlock arose between the military and naval chiefs.

At this time the weather greatly worsened, and dense falls of snow seemed to paralyse all movement by our troops, unequipped and untrained for such conditions. Meanwhile the Germans in Narvik held our ever-growing forces at bay with their machine-guns. Here was a serious and unexpected check.

* * * * *

Most of the business of our improvised campaign passed through my hands, and I prefer to record it as far as possible in my own words at the time. The Prime Minister had a strong desire, shared by the War Cabinet, to occupy Trondheim as well as Narvik. This Operation "Maurice", as it was called, promised to be a big undertaking. According to the records of our Military Co-ordination Committee of April 13, I was

very apprehensive of any proposals which might tend to weaken our intention to seize Narvik. Nothing must be allowed to deflect us from making the capture of this place as certain as possible. Our plans against Narvik had been very carefully laid, and there seemed every chance that they would be successful if they were allowed to proceed without being tampered with. Trondheim was, on the other hand, a much more speculative affair, and I deprecated any suggestion which might lead to the diversion of the Chasseurs Alpins until we had definitely established ourselves at Narvik. Otherwise we might find ourselves committed to a number of ineffectual operations along the Norwegian coast, none of which would succeed.

At the same time consideration had already been given to the Trondheim area, and plans were being made to secure landing-points in case a larger-scale action should be needed. A small landing of naval forces would take place at Namsos that afternoon. The Chief of the Imperial General Staff had collected a force of five battalions, two of which would be ready to land on the Norwegian coast on April 16, and three more on April 21 if desired. The actual points at which landings were to be made would be decided that night.

General Mackesy's original orders had been that, after landing at Narvik, he should push rapidly on to the Gällivare ore-field. He has now been told to go no farther than the Swedish frontier, since, if Sweden were friendly, there need be no fear for the ore-fields, and if hostile the difficulties of occupying them would be too great.

I also said that:

It might be necessary to proceed to invest the German forces in Narvik. But we should not allow the operation to degenerate into an investment except after a very determined battle. On this understanding I was willing to send a telegram to the French saying that we hoped and thought that we should be successful in seizing Narvik by a *coup-de-main*. We should explain that this had been made easier by a change in the orders, which did not now require the expedition to go beyond the Swedish frontier.

It was decided by the War Cabinet to attempt both the Narvik and Trondheim operations. The Secretary of State for War, with foresight, warned us that reinforcements for Norway might soon be required from our Army in France, and suggested that we should address the French on the point at a very early date. I agreed with this, but thought it premature to approach the French for a day or two. This was accepted. The War Cabinet approved a proposal to inform the Swedish and Norwegian Governments that we intended to recapture both Trondheim and Narvik; that we recognised the supreme importance of Trondheim as a strategic centre, but that it was important to secure Narvik as a naval base. We added that we had no intention that our forces should proceed over the Swedish frontier. We were at the same time to invite the French Government to

give us liberty to use the Chasseurs Alpins for operations elsewhere than at Narvik, telling them what we were saying to the
Swedish and Norwegian Governments. Neither I nor Mr.
Stanley liked the dispersion of our forces. We were still inclined
to concentrate all on Narvik, except for diversions elsewhere. But
we deferred to the general view, for which there was no lack of
good reasons.

*   *   *   *   *

On the night of the 16th-17th disappointing news arrived from
Narvik. General Mackesy had, it appeared, no intention of trying
to seize the town by an immediate assault protected by the close-
range bombardment of the Fleet, and Lord Cork could not move
him. I stated the position to my Committee as it then appeared.

April 17

1. Lord Cork's telegram shows that General Mackesy proposes to
take two unoccupied positions on the approaches to Narvik and to
hold on there until the snow melts, perhaps at the end of the month.
The General expects that the first demi-brigade of Chasseurs Alpins
will be sent to him, which it certainly will not be. This policy means
that we shall be held up in front of Narvik for several weeks. Meanwhile the Germans will proclaim that we are brought to a standstill
and that Narvik is still in their possession. The effects of this will be
damaging both upon Norwegians and neutrals. Moreover, the
German fortification of Narvik will continue, requiring a greater
effort when the time comes. This information is at once unexpected
and disagreeable. One of the best Regular brigades in the Army will
be wasting away, losing men by sickness, and playing no part. It is
for consideration whether a telegram on the following lines should not
be sent to Lord Cork and General Mackesy:

"Your proposals involve damaging deadlock at Narvik and the
neutralisation of one of our best brigades. We cannot send you the
Chasseurs Alpins. The *Warspite* will be needed elsewhere in two or
three days. Full consideration should therefore be given by you to
an assault upon Narvik, covered by the *Warspite* and the destroyers,
which might also operate at Römbaks Fiord. The capture of the port
and town would be an important success. We should like to receive
from you the reasons why this is not possible, and your estimate of
the degree of resistance to be expected on the waterfront. Matter
most urgent."

2. The second point which requires decision is whether the Chasseurs
Alpins shall go straight on to join General Carton de Wiart at or beyond Namsos, or whether, as is easy, they should be held back at

Scapa and used for the Trondheim operation on the 22nd or 23rd, together with other troops available for this main attack.

3. Two battalions of the 146th Brigade will, it is hoped, have been landed before dawn to-day at Namsos and Bandsund. The 3rd Battalion, in the *Chrobry*, will make a dangerous voyage to-morrow to Namsos, arriving, if all is well, about dusk, and landing. The anchorage of Lillejonas was bombed all the afternoon without the two transports being hit, and the large 18,000-tonner is now returning empty to Scapa Flow. If the leading Chasseurs Alpins are to be used at Namsos they must go there direct instead of making rendezvous at Lillejonas.

4. The question of whether the forces now available for the main attack on Trondheim are adequate must also be decided to-day. The two Guards battalions that were to be mobilised, *i.e.*, equipped, cannot be ready in time. The two French Foreign Legion battalions cannot arrive in time. A Regular brigade from France can however be ready to sail from Rosyth on the 20th. The first and second demi-brigades of the Chasseurs Alpins can also be in time. A thousand Canadians have been made available. There is also a brigade of Territorials. Is this enough to prevail over the Germans in Trondheim? The dangers of delay are very great and need not be restated.

5. Admiral Holland leaves to-night to meet the Commander-in-Chief Home Fleet on his return to Scapa on the 18th, and he must carry with him full and clear decisions. It may be taken as certain that the Navy will cheerfully undertake to carry troops to Trondheim.

6. It is probable that fighting will take place to-night and to-morrow morning for the possession of Andalsnes. We hope to have landed an advance party from the cruiser *Calcutta*, and are moving sufficient cruisers to meet a possible attack by five enemy destroyers at dawn.

7. The naval bombardments of Stavanger aerodrome will begin at dawn [to-day].

The Committee agreed to the telegram, which was accordingly sent. It produced no effect. It must remain a matter of opinion whether such an assault would have succeeded. It involved no marches through the snow, but, on the other hand, landings from open boats both in Narvik harbour and in Römbaks Fiord, under machine-gun fire. I counted upon the effect of close-range bombardment by the tremendous ship's batteries, which would blast the waterfronts and cover with smoke and clouds of snow and earth the whole of the German machine-gun posts. Suitable high-explosive shells had been provided by the Admiralty both for the battleship and the destroyers. Certainly Lord Cork, on the spot and able to measure the character of the bombardment, was strongly in favour of making the attempt. We had over four thousand of our best Regular troops, including the Guards Brigade and Marines, who, once they set foot on shore, would become intermingled at close quarters with the German defenders, whose regular troops, apart from the crews rescued from the sunken destroyers, we estimated, correctly as we now know, at no more than half their number. This would have been considered a fair proposition on the Western Front in the previous war, and no new factors were at work here. Later on in this war scores of such assaults were made and often succeeded. Moreover, the orders sent to the commanders were of such a clear and imperative character, and so evidently contemplated heavy losses, that they should have been obeyed. The responsibility for a bloody repulse would fall exclusively on the home authorities, and very directly upon me. I was content that this should be so; but nothing I or my colleagues or Cork could do or say produced the slightest effect on the General. He was resolved to wait till the snow melted. As for the bombardment, he could point to the paragraph in his instructions against endangering the civil population. When we contrast this spirit with the absolutely reckless gambling in lives and ships and the almost frenzied vigour, based upon long and profound calculations, which had gained the Germans their brilliant success, the disadvantages under which we lay in waging this campaign are obvious.

CHAPTER XXXV

TRONDHEIM

*T*RONDHEIM, if it were within our strength, was of course the key to any considerable operations in Central Norway. To gain it meant a safe harbour with quays and docks upon which an army of 50,000 men or more could be built up and based. Near by was an air-field from which several fighter squadrons could work. The possession of Trondheim would open direct railway contact with Sweden, and greatly improve the chances of Swedish intervention or the degree of mutual aid possible if Sweden were herself attacked. From Trondheim alone the northward advance of the German invasion from Oslo could be securely barred. On the broadest grounds of policy and strategy it would be good for the Allies to fight Hitler on the largest possible scale in Central Norway, if that was where he wanted to go. Narvik, far away to the north, could be stormed or reduced at leisure and would all the while be protected. We had the effective command of the sea. As to the air, if we could establish ourselves firmly on Norwegian airfields we should not hesitate to fight the German Air Force there to any extent which the severely limiting conditions allowed to either side.

All these reasons had simultaneously convinced the French War Council, the British War Cabinet, and most of their advisers. The British and French Prime Ministers were at one. General Gamelin was willing to withdraw French or release British divisions from France for Norway to the same extent that the Germans diverted their forces thither. He evidently welcomed a prolonged battle on a large scale south of Trondheim, where the ground was almost everywhere favourable to defence. It seemed that we could certainly bring forces and supplies to the scene across the open sea and through Trondheim far quicker than the Germans could fight their way up the single road and railway-line from Oslo, both of which might be cut behind them by bombs or parties dropped from the air. The only question was, could we take Trondheim in time? Could we get there before the main enemy army arrived from the south? and for this purpose could we obtain even a passing relief from their present unchallenged air domination?

There was a surge of opinion in favour of Trondheim which extended far beyond Cabinet circles. The advantages were so obvious that all could see them. The public, the clubs, the newspapers and their military correspondents had for some days past been discussing such a policy freely. My great friend Admiral of the Fleet Sir Roger Keyes, champion of forcing the Dardanelles, hero and victor of Zeebrugge, passionately longed to lead the Fleet or any portion of it past the batteries into the Trondheim Fiord and storm the town by landings from the sea. The appointment of Lord Cork, also an Admiral of the Fleet, to command the naval operations at Narvik although he was senior to the Commander-in-Chief, Admiral Forbes, himself, seemed to remove the difficulties of rank. Admirals of the Fleet are always on the active list, and Keyes had many contacts at the Admiralty. He spoke and wrote to me repeatedly with vehemence, reminding me of the Dardanelles and how easily the straits could have been forced if we had not been stopped by timid obstructionists. I also pondered a good deal upon the lessons of the Dardanelles. Certainly the Trondheim batteries and any minefields that might have been laid were trivial compared with those we had then had to face. On the other hand, there was the aeroplane, capable of dropping its bombs on the unprotected decks of the very few great ships which now constituted the naval power of Britain on the oceans.

At the Admiralty the First Sea Lord and the Naval Staff generally did not shrink from the venture. On April 13 the

Admiralty had officially informed the Commander-in-Chief of the Supreme Council's decision to allot troops for the capture of Trondheim, and had raised with him in a positive manner the question whether the Home Fleet should not force the passage.

Do you consider [the message ran] that the shore batteries could be either destroyed or dominated to such an extent as to permit transports to enter? If so, how many ships and what type would you propose?

On this Admiral Forbes asked for details about the Trondheim defences. He agreed that the shore batteries might be destroyed or dominated in daylight by battleships, if provided with suitable ammunition. None was carried at that moment in Home Fleet ships. The first and most important task, he said, was to protect troopships from heavy air attack over the thirty-miles approach through narrow waters, and the next to carry out an opposed landing of which ample warning had been given. In the circumstances he did not consider the operation feasible.

The Naval Staff persisted in their view, and the Admiralty, with my earnest agreement, replied on April 15 as follows:

We still think that the operation described should be further studied. It could not take place for seven days, which would be devoted to careful preparation. Danger from air not appreciably less wherever these large troopships are brought into the danger zone. Our idea would be that in addition to R.A.F. bombing of Stavanger aerodrome *Suffolk* should bombard with high explosive at dawn, hoping thereby to put the aerodrome out of business. The aerodrome at Trondheim could be dealt with by Fleet Air Arm bombers and subsequently by bombardment. High-explosive shells for 15-inch guns have been ordered to Rosyth. *Furious* and First Cruiser Squadron would be required for this operation. Pray therefore consider this important project further.

Admiral Forbes, although not fully convinced of its soundness, therefore addressed himself to the project in an increasingly favourable mood. In a further reply he said that he did not anticipate great difficulties from the naval side, except that he could not provide air defence for the transports while carrying out the landing. The naval force required would be the *Valiant* and *Renown* to give air defence to the *Glorious*, the *Warspite* to bombard, at least four ack-ack cruisers, and about twenty destroyers.

★　★　★　★　★

While plans for the frontal attack on Trondheim from the sea were being advanced with all speed, two subsidiary landings were already in progress designed to envelop the town from the landward side. Of these the first was a hundred miles to the north, at Namsos, where Major-General Carton de Wiart, V.C., had been chosen to command the troops, with orders "to secure the Trondheim area". He was informed that the Navy were making a preliminary lodgment with a party about three hundred strong in order to take and hold points for his disembarkation. The idea was that two infantry brigades and a light division of Chasseurs Alpins should land hereabouts in conjunction with the main attack by the Navy upon Trondheim, Operation "Hammer". For this purpose the 146th Brigade and the Chasseurs Alpins were being diverted from Narvik. Carton de Wiart started forthwith in a flying-boat, and reached Namsos under heavy air attack on the evening of the 15th. His staff officer was wounded, but he took effective charge on the spot. The second landing was at Andalsnes, about a hundred and fifty miles by road to the south-west of Trondheim. Here also the Navy had made a lodgment, and on April 18 Brigadier Morgan with a military force arrived and took command. Lieutenant-General Massy was appointed Commander-in-Chief of all the forces operating in Central Norway. This officer had to exercise his command from the War Office because there was as yet no place for his headquarters on the other side.

*　　*　　*　　*　　*

On the 15th I reported that all these plans were being developed, but the difficulties were serious. Namsos was under four feet of snow and offered no concealment from the air. The enemy enjoyed complete air mastery, and we had neither anti-aircraft guns nor any airfield from which protecting squadrons might operate. Admiral Forbes had not, I said, at first been very keen on forcing his way into Trondheim because of the risk of air attack. It was of course of first importance that the Royal Air Force should continue to harass the Stavanger airfield, by which the enemy aeroplanes were passing northwards. The *Suffolk* would bombard the Stavanger airfield with her 8-inch guns on April 17. This was approved and the bombardment took place as planned. Some damage was done to the airfield, but during her withdrawal the *Suffolk* was continuously bombed for seven hours. She was heavily hit, and reached Scapa Flow the following day with her quarterdeck awash.

*   *   *   *   *

The Secretary of State for War had now to nominate a Military Commander. The auspices were unfavourable. Colonel Stanley's first choice fell upon Major-General Hotblack, who was highly reputed, and on April 17 he was briefed for his task at a meeting of the Chiefs of Staff held in the Admiralty. That night at 12.30 a.m. he had a fit on the Duke of York's Steps, and was picked up unconscious some time later. He had luckily left all his papers with his staff, who were working on them. The next morning Brigadier Berney-Ficklin was appointed to succeed Hotblack. He too was briefed, and started by train for Edinburgh. On April 19 he and his staff left by air for Scapa. They crashed on the airfield at Kirkwall, and the pilot was seriously injured. Every day counted.

On April 17 I explained in outline to the Supreme War Council the plan which the staffs were making for the landing at Trondheim. The forces immediately available were one Regular brigade from France (2,500 strong), 1,000 Canadians, and about 1,000 men of a Territorial brigade as a reserve. The Military Co-ordination Committee had been advised that the forces available were adequate and that the risks, although very considerable, were justified. The operation would be supported by the full strength of the Fleet, and two carriers would be available, with a total of about 100 aircraft, including 45 fighters. The provisional date for the landing was April 22. The second demi-brigade of Chasseurs Alpins would not reach Trondheim until April 25, when it was hoped they would be able to disembark at the quays at Trondheim.

Asked whether the Chiefs of Staff were in agreement with the plans as outlined, the Chief of the Air Staff said on their behalf and in their presence that they were. The operation was of course attended by considerable risks, but these were worth running. The Prime Minister agreed with this view, and emphasised the importance of air co-operation. The War Cabinet gave cordial approval to the enterprise. I did my best to have it carried out.

Up to this point all the staffs and their chiefs had seemed resolved upon the central thrust at Trondheim. Admiral Forbes was actively preparing to strike, and there seemed no reason why the date of the 22nd should not be kept. Although Narvik was my pet, I threw myself with increasing confidence into this daring adventure, and was willing that the Fleet should risk the petty batteries at the entrance to the fiord, the possible minefields, and, most serious, the air. The ships carried what was in those days very powerful anti-aircraft armament. A group of ships had a combined overhead fire-power which few aircraft would care to encounter at a distance where bombing would be accurate. I must here explain that the power of an air force is terrific when there is nothing to oppose it. The pilots can fly as low as they please, and are often safer fifty feet off the ground than high up. They can cast their bombs with precision and use their machine-guns on troops with no more risk than that of a lucky rifle-bullet. These hard conditions had to be faced by our small expeditions at Namsos and Andalsnes, but the Fleet, with its A.A. batteries and a hundred seaborne aeroplanes, might well be superior during the actual operation to any air-power the enemy could bring. If Trondheim were taken, the neighbouring airfield of Vaernes would be in our hands, and in a few days we could have not only a considerable garrison in the town but also several fighter squadrons of the R.A.F. in action. Left to myself, I would have stuck to my first love, Narvik; but, serving as I did a loyal chief and friendly Cabinet, I now looked forward to this exciting enterprise to which so many staid and cautious Ministers had given their strong adherence, and which seemed to find much favour with the Naval Staff and indeed among all our experts. Such was the position on the 17th.

Meanwhile I felt that we should do our utmost to keep the King of Norway and his advisers informed of our plans by sending him an officer who understood the Norwegian scene and could speak with authority. Admiral Sir Edward Evans was well suited to this task, and was sent to Norway by air through Stockholm to make contact with the King at his headquarters. There he was to do everything possible to aid the Norwegian Government in their resistance and explain the measures which the British Government were taking to assist them. From April 22 he was for some days in consultation with the King and the principal Norwegian authorities, helping them to understand both our plans and our difficulties.

★　★　★　★　★

During the 18th a vehement and decisive change in the opinions of the Chiefs of Staff and of the Admiralty occurred. This change was brought about first by increasing realisation of the magnitude of the naval stake in hazarding so many of our finest capital ships, and also by War Office arguments that even if the Fleet got in and got out again the opposed landing of the troops in the face of the German air-power would be perilous. On the other hand, the landings which were already being successfully carried out both north and south of Trondheim seemed to all these authorities to offer a far less dangerous solution. The Chiefs of Staff drew up a long paper opposing Operation "Hammer".

This began with a reminder that a combined operation involving an opposed landing was one of the most difficult and hazardous operations of war. The Chiefs of Staff had always realised that this particular operation would involve very serious risks; for, owing to the urgency of the situation, there had not been time for the detailed and meticulous preparation which should have been given to an operation of this character, and as there had been no reconnaissance or air photographs the plan had been worked out from maps and charts. The plan had the further disadvantage that it would involve concentrating almost the whole of the Home Fleet in an area where it could be subjected to heavy attack from the air. There were also new factors in the situation which should be taken into account. We had seized the landing places at Namsos and Andalsnes and established forces ashore there; there were reliable reports that the Germans were improving the defences at Trondheim, and reports of our intentions to make a direct landing at Trondheim had appeared in the Press. On reconsidering the original project in the light of these new factors the Chiefs of Staff unanimously recommended a change of plan.

They still thought it essential that we should seize Trondheim and use it as a base for subsequent operations in Scandinavia; but they urged that, instead of the direct frontal assault, we should take advantage of our unexpected success in landing forces at Namsos and Andalsnes and develop a pincers movement on

Trondheim from north and south. By this means, they declared, we could turn a venture which was attended by grave hazards into an operation which could achieve the same results with much less risk. By this change of plan the Press reports of our intentions could also be turned to our advantage; for by judicious leakages we could hope to leave the enemy under the impression that we still intended to persist in our original plan. The Chiefs of Staff therefore recommended that we should push in the maximum forces possible at Namsos and Andalsnes, seize control of the road and rail communications running through Dombas, and envelop Trondheim from the north and south. Shortly before the main landings at Namsos and Andalsnes the outer forts at Trondheim should be bombarded from the sea with a view to leading the enemy to suppose that a direct assault was due to take place. We should thus invest Trondheim by land and blockade it by sea; and although its capture would take longer than originally contemplated, our main forces might be put ashore at a slightly earlier date. Finally, the Chiefs of Staff pointed out that such an enveloping operation, as opposed to a direct assault, would release a large number of valuable units of the Fleet for operations in other areas, *e.g.*, at Narvik. These powerful recommendations were put forward with the authority not only of the three Chiefs of Staff, but of their three able deputies, including Admiral Tom Phillips and Sir John Dill, newly appointed.

No more decisive stopper on a positive amphibious plan can be imagined, nor have I seen a Government or Minister who would have overridden it. Under the prevailing arrangement the Chiefs of Staff worked as a separate and largely independent body, without guidance or direction from the Prime Minister or any effective representative of the supreme executive power. Moreover, the leaders of the three Services had not yet got the conception of the war as a whole, and were influenced unduly by the departmental outlook of their own Services. They met together, after talking things over with their respective Ministers, and issued *aide-mémoires* or memoranda which carried enormous weight. Here was the fatal weakness of our system of conducting war at this time.

When I became aware of this right-about-turn I was indignant, and questioned searchingly the officers concerned. It was soon plain to me that all professional opinion was now adverse to the operation which only a few days before it had spontaneously espoused. Of course there was at hand, in passionate ardour for action and glory, Sir Roger Keyes. He was scornful of these belated fears and second thoughts. He volunteered to lead a handful of older ships with the necessary transports into Trondheim Fiord, land the troops, and storm the place, before the Germans got any stronger. Roger Keyes had formidable credentials of achievement. In him there burned a flame. It was suggested in the May debates that "the iron of the Dardanelles had entered into my soul", meaning that on account of my downfall on that occasion I had no longer the capacity to dare; but this was really not true. The difficulties of acting from a subordinate position in the violent manner required are of the first magnitude.

Moreover, the personal relations of the high naval figures involved were peculiar. Roger Keyes, like Lord Cork, was senior to the Commander-in-Chief and the First Sea Lord. Admiral Pound had been for two years Keyes' Staff Officer in the Mediterranean. For me to take Roger Keyes' advice against his would have entailed his resignation, and Admiral Forbes might well have asked to be relieved of his command. It was certainly not my duty in the position I held to confront the Prime Minister and my War Cabinet colleagues with these personal dramas at such a time, and upon an operation which, for all its attractiveness and interest, was essentially minor even in relation to the Norwegian campaign, to say nothing of the general war. I therefore had no doubt that we must accept the Staff view in spite of their change of mind and the obvious objections that could be raised against their mutilated plan.

I accordingly submitted to the abandonment of "Hammer". I reported the facts to the Prime Minister on the afternoon of the 18th, and though bitterly disappointed he, like me, had no choice but to accept the new position. In war, as in life, it is often necessary, when some cherished scheme has failed, to take up the best alternative open, and if so it is folly not to work for it with all your might. I therefore turned my guns round too. I reported in writing to the Co-ordinating Committee on April 19 as follows:

1. The considerable advance made by Carton de Wiart, the very easy landings we have had at Andalsnes and other ports in this southern

fiord, the indiscretions of the Press, pointing to a storm of Trondheim, and the very heavy naval forces required for this operation called "Hammer", with the undoubted major risk of keeping so many valuable ships so many hours under close air attack, have led the Chiefs of Staff and their deputies to advise that there should be a complete alteration of the emphasis between the two pincers attacks and the centre attack; in the following sense: that the main weight should be thrown into the northern and southern pincers, and that the central attack on Trondheim should be reduced to a demonstration.

2. Owing to the rapidity with which events and opinions have moved, it became necessary to take a decision, of which the Prime Minister had approved, as set out above, and orders are being issued accordingly.

3. It is proposed to encourage the idea that a central attack upon Trondheim is afoot, and to emphasise this by a bombardment by battleships of the outer forts at the suitable moment.

4. Every effort will be made to strengthen Carton de Wiart with artillery, without which his force is not well composed.

5. All the troops we have now under orders for "Hammer" will be shoved in as quickly as possible, mostly in warships, at the various ports of the Romsdal Fiord, to press on to Dombas, and then, some delaying force being sent southward to the Norwegian main front, the bulk will turn north towards Trondheim. There is already one brigade (Morgan's) ashore beyond Andalsnes, with the 600 Marines. The brigade from France and the supporting Territorial brigade will all be thrown in here as quickly as possible. This should enable Dombas to be secured, and the control to be extended to the more easterly of the two Norwegian railways running from Oslo to Trondheim, Storen being a particularly advantageous point. The destination of the second demi-brigade of Chasseurs Alpins, the two battalions of the French Foreign Legion, and the thousand Canadians can for to-day or to-morrow be left open.

6. The position of the Namsos force must be regarded as somewhat hazardous, but its commander is used to taking risks. On the other hand, it is not seen why we cannot bring decisive superiority to bear along the Andalsnes-Dombas railway, and operate as occasion serves beyond that most important point, the object being the isolation of Trondheim and its capture.

7. Although this change of emphasis is to be deprecated on account of its being a change, it must be recognised that we move from a more hazardous to a less hazardous operation, and greatly reduce the strain upon the Navy involved in "Hammer". It would seem that our results would be equally achieved by the safer plan, and it does not follow

that they will be delayed. We can certainly get more men sooner on to Norwegian soil by this method than the other.

8. It is not possible to deprive Narvik of its battleship at the moment when we have urged strenuous action. *Warspite* has therefore been ordered to return [there]. Some further reinforcement will be required for Narvik, which must be studied at once. The Canadians should be considered.

9. At the same time the sweep of the Skagerrak will now become possible, to clear away the enemy anti-submarine craft and aid our submarines.

The next day I explained to the War Cabinet the circumstances in which it had been decided to call off the direct assault on Trondheim, and stated that the new plan which the Prime Minister had approved was broadly to send the whole of the 1st Light Division of Chasseurs Alpins to General Carton de Wiart for his attack on the Trondheim area from the north, and to send the regular brigades from France to reinforce Brigadier Morgan, who had landed at Andalsnes and had pushed on troops to hold Dombas. Another Territorial brigade would be put in on the southern line. It might be possible to push part of this southern force right forward to reinforce the Norwegians on the Oslo front. We had been fortunate in getting all our troops ashore without loss so far (except of the ship carrying all Brigadier Morgan's vehicles), and the present plans provided for the disembarkation of some 25,000 men by the end of the first week in May. The French had offered two more light divisions. The chief limiting factor was the provision of the necessary bases and lines of communication on which the forces were to be maintained. These would be liable to heavy air attack.

The Secretary of State for War then said that the new plan was little less hazardous than the direct attack on Trondheim. Until we had secured the Trondheim aerodrome little could be done to offset the heavy scale of enemy air attack. Nor was it altogether correct to describe the new plan as a "pincers movement" against Trondheim, since while the northern force would bring pressure to bear in the near future, the first task of the southern force must be to secure themselves against a German attack from the south. It might well be a month before any serious move could be made against Trondheim from this direction. This was a sound criticism. General Ironside however strongly supported the new movement, expressing the hope that General Carton de Wiart, who when reinforced by the French would have, he said, quite a large force at his disposal, a large part of which would be highly mobile, might get astride the railway from Trondheim to Sweden. The troops already at Dombas had no guns or transport. They should however be able to hold a defensive position. I then added that the direct assault on Trondheim had been deemed to involve undue risk both to the Fleet and to our landing-parties. If in the course of a successful assault the Fleet were to lose a capital ship by enemy air action this loss would have to be set against the success of the operation. Again, it was obvious that the landing parties might suffer heavy casualties, and General Massy took the view that the stake was out of proportion to the results desired, particularly as these could be obtained by other methods. The Secretary of State for War, having justly pointed out that these other methods offered no sure or satisfactory solution, was content they should be tried. It was evident to us all that we had in fact only a choice of unpleasant courses before us, and also a compulsion to act. The War Cabinet endorsed the transformation of the plan against Trondheim.

I now reverted to Narvik, which seemed at once more important and more feasible since the attack on Trondheim was abandoned, and addressed a note to my Committee as follows:

The importance and urgency of reaching a decision at Narvik can hardly be overrated. If the operations become static the situation will deteriorate for us. When the ice melts in the Gulf of Bothnia, at the latest in a month from now, the Germans may demand of the Swedes free passage for their troops through the ore-field in order to reinforce their people in Narvik, and may also demand control of the ore-field. They might promise Sweden that if she agreed to this in the far North she would be let entirely alone in the rest of the country. Anyhow, we ought to take it for granted that the Germans will try to enter the ore-field and carry succour to the Narvik garrison by force or favour. We have therefore at the outside only a month to spare.

2. In this month we have not only to reduce and capture the town and the landed Germans, but to get up the railway to the Swedish frontier and to secure an effective, well-defended seaplane base on some lake, in order, if we cannot obtain control of the ore-field, to prevent its being worked under German control. It would seem necessary that at least 3,000 [more] good troops should be directed upon Narvik forthwith, and should reach there by the end of the first week in May

at latest. The orders for this should be given now, as nothing will be easier than to divert the troops if in the meanwhile the situation is cleared up. It would be a great administrative advantage if these troops were British, but if this cannot be managed for any reason, could not the leading brigade of the Second French Light Division be directed upon Narvik? There ought to be no undue danger in bringing a big ship into Skjel Fiord or thereabouts.

3. I should be very glad if the Deputy Chief of Naval Staff could consult with an officer of equal standing in the War Office upon how this need can be met, together with ships and times. Failure to take Narvik will be a major disaster, and will carry with it the control by Germany of the ore-field.

The general position as it was viewed at this moment cannot be better stated than in a paper written by General Ismay on April 21.

The object of operations at Narvik is to capture the town and obtain possession of the railway to the Swedish frontier. We should then be in a position to put a force, if necessary, into the Gällivare ore-fields, the possession of which is the main objective of the whole of the operations in Scandinavia.

As soon as the ice melts in Luleå, in about a month's time, we must expect that the Germans will obtain, by threats or force, a passage for their troops, in order that they themselves may secure Gällivare and perhaps go forward and reinforce their troops at Narvik. It is therefore essential that Narvik should be liquidated in about a month.

The object of operations in the Trondheim area is to capture Trondheim, and thereby obtain a base for further operations in Central Norway, and Sweden if necessary. Landings have been made at Namsos on the north of Trondheim and Andalsnes on the south. The intention is that the Namsos force will establish itself astride the railway running eastward from Trondheim, thus encircling the Germans there on the east and north-east. The force landed at Andalsnes has as its first rôle the occupation of a defensive position, in co-operation with the Norwegians at Lillehammer, to block any reinforcement of Trondheim from the main German landing at Oslo. The roads and railways between Oslo and Trondheim have both to be covered. When this has been achieved some troops will work northward and bring pressure to bear on Trondheim from the south.

At the present moment our main attention is directed to the Trondheim area. It is essential to support the Norwegians and ensure that Trondheim is not reinforced. The capture of Narvik is not at the present moment so urgent, but it will become increasingly so as the thaw in the Gulf of Bothnia approaches. If Sweden enters the war Narvik becomes the vital spot.

The operations in Central Norway which are now being undertaken are of an extremely hazardous nature, and we are confronted with serious difficulties. Among these the chief are, first, that the urgent need of coming to the assistance of the Norwegians without delay has forced us to throw ashore hastily-improvised forces—making use of whatever was readily available; secondly, that our entry into Norway is perforce through bases which are inadequate for the maintenance of big formations. The only recognised base in the area is Trondheim, which is in the hands of the enemy. We are making use of Namsos and Andalsnes, which are only minor ports, possessing few, if any, facilities for unloading military stores, and served by poor communications with the interior. Consequently, the landing of mechanical transport, artillery, supplies, and petrol (nothing is obtainable locally) is a matter which, even if we were not hampered in other ways, would present considerable difficulty. Thus, until we succeed in capturing Trondheim the size of the forces which we can maintain in Norway is strictly limited.

Of course it may be said that all Norwegian enterprises, however locally successful, to which we might have committed ourselves would have been swept away by the results of the fearful battle in France which was now so near. Within a month the main Allied armies were to be shattered or driven into the sea. Everything we had would be drawn into the struggle for life. It was therefore lucky for us that we were not able to build up a substantial army and air force round Trondheim. The veils of the future are lifted one by one, and mortals must act from day to day. On the knowledge we had in the middle of April, I remain of the opinion that, having gone so far, we ought to have persisted in carrying out Operation "Hammer" and the threefold attack on Trondheim, on which all had been agreed; but I accept my full share of responsibility for not enforcing this upon our expert advisers when they became so decidedly adverse to it and presented us with serious objections. In that case however it would have been better to abandon the whole enterprise against Trondheim and concentrate all upon Narvik. But for this it was now too late. Many of the troops were ashore, and the Norwegians crying for help.

CHAPTER XXXVI

FRUSTRATION IN NORWAY

Lord Cork Appointed to the Supreme Command at Narvik – His Letter to Me – General Mackesy's Protest against Bombardment – The Cabinet's Reply – The Eighth Meeting of the Supreme War Council, April 22 – German and Allied Strength on Land and in the Air – The Scandinavian Tangle – Decisions upon Trondheim and Narvik – A Further Change in Control – Directive of May 1 – The Trondheim Operation – The Namsos Failure – Paget in the Andalsnes Excursion – Decision of the War Cabinet to Evacuate Central Norway – The Mosjoen Fiasco – My Report of May 4 – Gubbins' Force – The German Northward Advance – German Superiority in Method and Quality.

O N APRIL 20 I had procured agreement to the appointment of Lord Cork as sole commander of the naval, military, and air forces in the Narvik areas, thus bringing General Mackesy directly under his authority. There was never any doubt of Lord Cork's vigorously offensive spirit. He realised acutely the danger of delay; but the physical and administrative difficulties were far greater on the spot than we could measure at home. Moreover, naval officers, even when granted the fullest authority, are chary of giving orders to the Army about purely military matters. This would be even more true if the positions were reversed. We had hoped that by relieving General Mackesy from direct major responsibility we should make him feel more free to adopt bold tactics. The result was contrary to this expectation. He continued to use every argument, and there was no lack of them, to prevent drastic action. Things had changed to our detriment in the week that had passed since the idea of an improvised assault upon Narvik Town had been rejected. The 2,000 German soldiers were no doubt working night and day at their defences, and these and the town all lay hidden under a pall of snow. The enemy had no doubt by now also organised two or three thousand sailors who had escaped from the sunken destroyers. Their arrangements for bringing air-power to bear improved every day, and both our ships and landed troops endured increasing bombardment. On the 21st Lord Cork wrote to me as follows:

I write to thank you for the trust you have reposed in me. I shall certainly do my best to justify it. The inertia is difficult to overcome, and of course the obstacles to the movement of troops are considerable, particularly the snow, which on northern slopes of hills is still many feet deep. I myself have tested that, and as it has been snowing on and off for two days the position has not improved. The initial error was that the original force started on the assumption they would meet with no resistance, a mistake we often make—*e.g.*, Tanga.* As it is, the soldiers have not yet got their reserves of small arms ammunition, or water, but tons of stuff and personnel they do not want. . . .

What is really our one pressing need is fighters; we are so overmatched in the air. There is a daily inspection of this place, and they come when there are transports or steamers to bomb. Sooner or later they must get a hit. I flew over Narvik yesterday, but it was very difficult to see much. The rocky cliff is covered with snow, except for rock outcrops, round which the drifts must be deep. It is snow down to the water's edge, which makes it impossible to see the nature of the foreshore.

While waiting for the conditions necessary for an attack we are isolating the town from the world by breaking down the railway culverts, etc., and the large ferry steamer has been shelled and burnt. . . . It is exasperating not being able to get on, and I quite understand your wondering why we do not, but I assure you that it is not from want of desire to do so.

Lord Cork decided upon a reconnaissance in force, under cover of a naval bombardment, but here General Mackesy interposed. He stated that before the proposed action against Narvik began he felt it his duty to represent that there was no officer or man in his command who would not feel ashamed for himself and his country if thousands of Norwegian men, women, and children in Narvik were subjected to the proposed bombardment. Lord Cork contented himself with forwarding this statement without comment. Neither the Prime Minister nor I could be present at

* The landing at Tanga, near Zanzibar, in 1914.

the Defence Committee meeting on April 22, as we had to attend the Supreme War Council in Paris on that day. Before leaving I had drafted a reply which was approved by our colleagues:

I presume that Lord Cork has read the Bombardment Instructions issued at the outbreak of war. If he finds it necessary to go beyond these instructions on account of the enemy using the shelter of buildings to maintain himself in Narvik, he may deem it wise to give six hours' warning by every means at his disposal, including, if possible, leaflets, and to inform the German commander that all civilians must leave the town, and that he would be held responsible if he obstructed their departure. He might also offer to leave the railway line unmolested for a period of six hours to enable civilians to make good their escape by that route.

The Defence Committee endorsed this policy, strongly expressing the view that "it would be impossible to allow the Germans to convert Norwegian towns into forts by keeping the civilians in the towns to prevent us from attacking."

* * * * *

We arrived in Paris with our minds oppressed by the anxieties and confusion of the campaign in Norway, for the conduct of which the British were responsible. But M. Reynaud, having welcomed us, opened with a statement on the general military position which by its gravity dwarfed our joint Scandinavian excursions. Geography, he said, gave Germany the permanent advantage of interior lines. She had 190 divisions, of which 150 could be used on the Western Front. Against these the Allies had 100, of which 10 were British. In the previous war, Germany, with a population of 65 millions, had raised 248 divisions, of which 207 fought on the Western Front. France on her part had raised 118 divisions, of which 110 had been on the Western Front, and Great Britain 89 divisions, of which 63 had been on the Western Front, giving a total of 173 Allied against 207 German divisions in the West. Equality had been attained only when the Americans arrived with their 34 divisions. How much worse was the position to-day! The German population was now 80 millions, from which she could conceivably raise 300 divisions. France could hardly expect that there would be 20 British divisions in the West by the end of the year. We must therefore face a large and increasing numerical superiority, which was already three to two and would presently rise to two to one. As for equipment, Germany had the advantage both in aviation and aircraft equipment and also in artillery and stocks of ammunition. Thus Reynaud.

To this point then had we come from the days of the Rhineland occupation in 1936, when a mere operation of police would have sufficed; or since Munich, when Germany, occupied with Czechoslovakia, could spare but thirteen divisions for the Western Front; or even since September 1939, when, while the Polish resistance lasted, there were but forty-two German divisions in the West. All this terrible superiority had grown up because at no moment had the once victorious Allies dared to take any effective step, even when they were all-powerful, to resist repeated aggressions by Hitler and breaches of the treaties.

* * * * *

After this sombre overture, of the gravity of which we were all conscious, we turned to the Scandinavian tangle. The Prime Minister explained the position with clarity. We had landed 13,000 men at Namsos and Andalsnes without loss. Our forces had pushed forward farther than had been expected. On finding that the direct attack on Trondheim would demand a disproportionate amount of naval force, it had been decided to make a pincers movement from the north and south instead. But in the last two days these new plans had been rudely interrupted by a heavy air attack on Namsos. As there had been no anti-aircraft fire to oppose them the Germans had bombed at will. Meanwhile all German warships at Narvik had been destroyed. But the German troops there were strongly fortified, so that it had not yet been possible to attack them by land. If our first attempt did not succeed it would be renewed.

About Central Norway Mr. Chamberlain said that the British command were anxious to reinforce the troops who had gone there, to protect them against the German advance from the south, and to co-operate subsequently in the capture of Trondheim. It was already certain that reinforcements would be required. 5,000 British, 7,000 French, 3,000 Poles, three British mechanised battalions, one British light tank battalion, three French light divisions, and one British Territorial division were to be available in the near future. The limitation would not be the number of troops provided, but the number that could be landed and maintained in the country. M. Reynaud said that four French light divisions would be sent.

I now spoke for the first time at any length in these conferences, pointing out to the French the difficulties of landing troops and stores in the face of enemy aircraft and U-boats. Every single ship had to be convoyed by destroyers, every landing port continuously guarded by cruisers or destroyers, not only during the landing, but till A.A. guns could be mounted ashore. So far the Allied ships had been extraordinarily lucky and had sustained very few hits. The tremendous difficulties of the operation would be understood. Although 13,000 men had now been safely landed, the Allies had as yet no established bases, and were operating inland with weak and slender lines of communication, practically unprovided with artillery or supporting aircraft. Such was the position in Central Norway. At Narvik the Germans were less strong, the port far less exposed to air attack, and once the harbour had been secured it would be possible to land at a very much faster rate. Any forces which could not be landed at ports farther south should go to Narvik. Among the troops assigned to the Narvik operation, or indeed in Great Britain, there were none able to move across country in heavy snow. The task at Narvik would be not only to free the harbour and the town, nor even to clear the whole district of Germans, but to advance up the railway to the Swedish frontier in strength commensurate with any further German designs. It was the considered view of the British command that this could be done without slowing down the rate of landing at other ports beyond the point to which it was already restricted by the difficulties described.

We were all in full agreement on the unpleasantness of our plight and the little we could do at the moment to better it. The Supreme War Council agreed that the immediate military objectives should be

(a) the capture of Trondheim, and
(b) the capture of Narvik, and the concentration of an adequate Allied force on the Swedish frontier.

The next day we talked about the dangers to the Dutch and Belgians and their refusal to take any common measures with us. We were very conscious that Italy might declare war upon us at any time, and various naval measures were to be concerted in the Mediterranean between Admiral Pound and Admiral Darlan. To our meeting General Sikorski, the head of the Polish Government, also was invited. He declared his ability to constitute a force of a hundred thousand men within a few months. Active steps were also being taken to recruit a Polish division in the United States.

At this meeting it was agreed also that if Germany invaded Holland the Allied armies should at once advance into Belgium without further approaches to the Belgian Government; and that the R.A.F. could bomb the German marshalling yards and the oil refineries in the Ruhr.

★   ★   ★   ★   ★

When we got back from the conference I was so much concerned at the complete failure not only of our efforts against the enemy, but of our method of conducting the war, that I wrote as follows to the Prime Minister:

Being anxious to sustain you to the best of my ability, I must warn you that you are approaching a head-on smash in Norway.

I am very grateful to you for having at my request taken over the day-to-day management of the Military Co-ordination [Committee], etc. I think I ought however to let you know that I shall not be willing to receive that task back from you without the necessary powers. At present no one has the power. There are six Chiefs [and Deputy Chiefs] of the Staff, three Ministers, and General Ismay, who all have a voice in Norwegian operations (apart from Narvik). But no one is responsible for the creation and direction of military policy except yourself. If you feel able to bear this burden, you may count upon my unswerving loyalty as First Lord of the Admiralty. If you do not feel you can bear it, with all your other duties, you will have to delegate your powers to a deputy who can concert and direct the general movement of our war action, and who will enjoy your support and that of the War Cabinet unless very good reason is shown to the contrary.

Before I could send it off I received a message from the Prime Minister saying that he had been considering the position in Scandinavia and felt it to be unsatisfactory. He asked me to call on him that evening at Downing Street after dinner to discuss the whole situation in private.

I have no record of what passed at our conversation, which was of a most friendly character. I am sure I put the points in my unsent letter, and that the Prime Minister agreed with their force and justice. He had every wish to give me the powers of direction for which I asked, and there was no kind of personal difficulty between us. He had however to consult and persuade a number of important personages, and it was not till May 1 that he was able to issue the following Note to the Cabinet and those concerned.

May 1, 1940

I have been examining, in consultation with the Ministers in charge of the Service departments, the existing arrangements for the consideration and decision of Defence questions, and I circulate for the information of my colleagues a Memorandum describing certain modifications which it has been decided to make in these arrangements forthwith. The modifications have been agreed to by the three Service Ministers. With the approval of the First Lord of the Admiralty, Major-General H. L. Ismay, C.B., D.S.O., has been appointed to the post of Senior Staff Officer in charge of the Central Staff which, as indicated in the Memorandum, is to be placed at the disposal of the First Lord. Major-General Ismay has been nominated, while serving in this capacity, an additional member of the Chiefs of Staff Committee. **N. C.**

DEFENCE ORGANISATION

In order to obtain a greater concentration of the direction of the war, the following modifications of present arrangements will take effect:

The First Lord of the Admiralty will continue to take the chair at all meetings of the Military Co-ordination Committee at which the Prime Minister does not preside himself, and in the absence of the Prime Minister will act as his deputy at such meetings on all matters delegated to the Committee by the War Cabinet.

He will be responsible on behalf of the Committee for giving guidance and direction to the Chiefs of Staff Committee, and for this purpose it will be open to him to summon that Committee for personal consultation at any time when he considers it necessary.

The Chiefs of Staff will retain their responsibility for giving their collective views to the Government, and, with their respective staffs, will prepare plans to achieve any objectives indicated to them by the First Lord on behalf of the Military Co-ordination Committee, and will accompany their plans by such comments as they consider appropriate.

The Chiefs of Staff, who will in their individual capacity remain responsible to their respective Ministers, will at all times keep their Ministers informed of their conclusions.

Where time permits, the plans of the Chiefs of Staff, with their comments and any comments by the First Lord, will be circulated for approval to the Military Co-ordination Committee, and, unless the Military Co-ordination Committee is authorised by the War Cabinet to take final decision, or in the case of disagreement on the Military Co-ordination Committee, circulated to the War Cabinet.

In urgent cases it may be necessary to omit the submission of plans to a formal meeting of the Committee, but in such cases the First Lord will no doubt find means of consulting the Service Ministers informally, and in the case of dissent the decision will be referred to the Prime Minister.

In order to facilitate the general plan outlined above and to afford a convenient means of maintaining a close liaison between the First Lord and the Chiefs of Staff, the First Lord will be assisted by a suitable Central Staff (distinct from the Admiralty Staff), under a Senior Staff Officer, who will be an additional member of the Chiefs of Staff Committee.

I accepted this arrangement, which seemed an improvement. I could now convene and preside over the meetings of the Chiefs of Staff Committee, without whom nothing could be done, and I was made responsible formally "for giving guidance and direction" to them. General Ismay, the Senior Staff Officer in charge of the Central Staff, was placed at my disposal *as my Staff Officer and representative*, and in this capacity was made a full member of the Chiefs of Staff Committee. I had known Ismay for many years, but now for the first time we became hand-in-glove, and much more. Thus the Chiefs of Staff were to large extent made responsible to me in their collective capacity, and as a deputy of the Prime Minister I could nominally influence with authority their decisions and policies. On the other hand, it was only natural that their primary loyalties should be to their own Service Ministers, who would have been less than human if they had not felt some resentment at the delegation of a part of their authority to one of their colleagues. Moreover, it was expressly laid down in the Memorandum that my responsibilities were to be discharged *on behalf of* the Military Co-ordination Committee. I was thus to have immense responsibilities, without effective power in my own hands to discharge them. Nevertheless I had a feeling that I might be able to make the new organisation work. It was destined to last only a week. But my personal and official connection with General Ismay and his relation to the Chiefs of Staff Committee was preserved unbroken and unweakened from May 1, 1940, to July 27, 1945, when I laid down my charge.

*   *   *   *   *

It is now necessary to recount the actual course of the fighting for Trondheim. Our northern force, from Namsos, was 80 miles from the town, and our southern force, from Andalsnes, was 150 miles away. The central attack through the fiord ("Hammer") had been abandoned, partly through fear of its cost and partly through hopes of the flanking movements. Both these movements now failed utterly. The Namsos force, commanded by Carton de Wiart, hastened forward in accordance with his instructions against the Norwegian snow and the German air. A brigade reached Verdal, fifty miles from Trondheim, at the head of the fiord, on the 19th. It was evident to me, and I warned the staffs, that the Germans could send in a single night a stronger force by water from Trondheim to chop them. This occurred two days later. Our troops were forced to withdraw some miles to where they could hold the enemy. The intolerable snow conditions, now sometimes in thaw, and the fact that the Germans who had come across the inner fiord were, like us, destitute of wheeled transport, prevented any serious fighting on the ground; and the small number of scattered troops plodding along the road offered little target to the unresisted air-power. Had Carton de Wiart known how limited were the forces he would have, or that the central attack on Trondheim had been abandoned—a vital point of which our staff machinery did not inform him—he would no doubt have made a more methodical advance. He acted in relation to the main objective as it had been imparted to him.

In the end nearly everybody got back exhausted, chilled, and resentful to Namsos, where the French Chasseur Brigade had remained; and Carton de Wiart, whose opinion on such issues commanded respect, declared that there was nothing for it but evacuation. Preparations for this were at once made by the Admiralty. On April 28 the evacuation of Namsos was ordered. The French contingent would re-embark before the British, leaving some of their ski troops to work with our rearguard.

The probable dates for leaving were the nights of the 1st and 2nd of May. Eventually the withdrawal was achieved in a single night. All the troops were re-embarked on the night of the 3rd, and were well out to sea when they were sighted by the German air reconnaissance at dawn. From eight o'clock in the morning to three in the afternoon wave after wave of enemy bombers attacked the warships and the transports. As no British air forces were available to protect the convoy we were lucky that no transport was hit. The French destroyer *Bison* and H.M.S. *Afridi*, which carried our rearguard, were "sunk fighting to the end".

★   ★   ★   ★   ★

A different series of misfortunes befell the troops landed at Andalsnes; but here at least we took our toll of the enemy. In response to urgent appeals from General Ruge, the Norwegian Commander-in-Chief, Brigadier Morgan's 148th Infantry Brigade had hastened forward as far as Lillehammer. Here it joined the tired-out, battered Norwegian forces whom the Germans, in the overwhelming strength of three fully-equipped divisions, were driving before them along the road and railway from Oslo towards Dombas and Trondheim. Severe fighting began. The ship carrying Brigadier Morgan's vehicles, including all artillery and mortars, had been sunk, but his young Territorials fought well with their rifles and machine-guns against the German vanguards. who were armed not only with 5.9 howitzers, but many heavy mortars and some tanks. On April 24 the leading battalion of the 15th Brigade, arriving from France, reached the crumbling front. General Paget, who commanded these Regular troops, learned from General Ruge that the Norwegian forces were exhausted and could fight no more until they had been thoroughly rested and re-equipped. He therefore assumed control, brought the rest of this brigade into action as fast as they arrived, and faced the Germans with determination in a series of spirited engagements. By adroit use of the railway, which fortunately remained unbroken, Paget extricated his own troops, Morgan's brigade, which had lost 700 men, and some Norwegian units. For one whole day the bulk of the British force hid in a long railway tunnel, fed by their precious supply train, and were thus completely lost to the enemy and his all-seeing air. After fighting five rearguard actions, in several of which the

Germans were heavily mauled, and having covered over a hundred miles, he reached the sea again at Andalsnes. This small place, like Namsos, had been flattened out by bombing; but by the night of May 1 the 15th Brigade, with what remained of Morgan's 148th Brigade, had been taken on board British cruisers and destroyers, and reached home without further trouble. General Paget's skill and resolution during these days opened his path to high command as the war developed.

A forlorn, gallant effort to give support from the air should be recorded. The only landing "ground" was the frozen lake of Lesjeskogen, forty miles from Andalsnes. There a squadron of Gladiators, flown from the *Glorious*, arrived on April 24. They were at once heavily attacked. The Fleet Air Arm did their best to help them; but the task of fighting for existence, of covering the operations of two expeditions 200 miles apart, and of protecting their bases was too much for a single squadron. By April 26 it could fly no more. Long-range efforts by British bombers, working from England, were also unavailing.

* * * * *

Our withdrawal enforced by local events had conformed to the decision already taken by the War Cabinet on the advice of the Military Co-ordination Committee, with the Prime Minister presiding. We had all come to the conclusion that it was beyond our power to seize and hold Trondheim. Both claws of the feeble pincers were broken. Mr. Chamberlain announced to the Cabinet that plans must be made for evacuating our forces both from Namsos and Andalsnes, though we should in the meanwhile continue to resist the German advance. The Cabinet was distressed at these proposals, which were however inevitable.

* * * * *

In order to delay to the utmost the northward advance of the enemy towards Narvik, we were now sending special companies raised in what was afterwards called "Commando" style, under an enterprising officer, Colonel Gubbins, to Mosjoen, 120 miles farther up the coast. I was most anxious that a small part of the Namsos force should make their way in whatever vehicles were available along the road to Grong. Even a couple of hundred would have sufficed to fight small rearguard actions. From Grong they would have to find their way on foot to Mosjoen.

I hoped by this means to gain the time for Gubbins to establish himself so that a stand could be made against the very small numbers which the enemy could as yet send there. I was repeatedly assured that the road was impassable. General Massy from London sent insistent requests. It was replied that even a small party of French Chasseurs, with their skis, could not traverse this route. "It seemed evident," wrote General Massy a few days later in his dispatch, "that if the French Chasseurs could not retire along this route the Germans could not advance along it. . . . This was an error, as the Germans have since made full use of it and have advanced so rapidly along it that our troops in Mosjoen have not had time to get properly established, and it is more than likely that we shall not be able to hold the place." This proved true. The destroyer *Janus* took a hundred Chasseurs Alpins and two light A.A. guns round by sea, but they left again before the Germans came.

*   *   *   *   *

We have now pursued the Norwegian campaign to the point where it was overwhelmed by gigantic events. The superiority of the Germans in design, management, and energy were plain. They put into ruthless execution a carefully-prepared plan of action. They comprehended perfectly the use of the air arm on a great scale in all its aspects. Moreover, their individual ascendancy was marked, especially in small parties. At Narvik a mixed and improvised German force barely six thousand strong held at bay for six weeks some twenty thousand Allied troops, and, though driven out of the town, lived to see them depart. The Narvik attack, so brilliantly opened by the Navy, was paralysed by the refusal of the military commander to run what was admittedly a desperate risk. The division of our resources between Narvik and Trondheim was injurious to both our plans. The abandonment of the central thrust on Trondheim wears an aspect of vacillation in the British High Command for which not only the experts but the political chiefs who yielded too easily to their advice must bear a burden. At Namsos there was a muddy waddle forward and back. Only in the Andalsnes expedition did we bite. The Germans traversed in seven days the road from Namsos to Mosjoen, which the British and French had declared impassable. At Bodo and Mo during the retreat of Gubbins' force to the north we were each time just too late, and the enemy, although they had to overcome hundreds of miles of rugged, snow-clogged country, drove us back in spite of gallant episodes. We, who had the command of the sea and could pounce anywhere on an undefended coast, were out-paced by the enemy moving by land across very large distances in the face of every obstacle. In this Norwegian encounter some of our finest troops, the Scots and Irish Guards, were baffled by the vigour, enterprise and training of Hitler's young men.

We tried hard at the call of duty to entangle and imbed ourselves in Norway. We thought fortune had been cruelly against us. We can now see that we were well out of it. Meanwhile we had to comfort ourselves as best we might by a series of successful evacuations. Failure at Trondheim! Stalemate at Narvik! Such in the first week of May were the only results we could show to the British nation, to our Allies, and to the neutral world, friendly or hostile. Considering the prominent part I played in these events and the impossibility of explaining the difficulties by which we had been overcome, or the defects of our staff and govermental organisation and our methods of conducting war, it was a marvel that I survived and maintained my position in public esteem and Parliamentary confidence. This was due to the fact that for six or seven years I had predicted with truth the course of events, and had given ceaseless warnings, then unheeded but now remembered.

* * * * *

Twilight War ended with Hitler's assault on Norway. It broke into the glare of the most fearful military explosion so far known to man. I have described the trance in which for eight months France and Britain had been held while all the world wondered. This phase proved most harmful to the Allies. From the moment when Stalin made terms with Hitler the Communists in France took their cue from Moscow and denounced the war as "an imperialist and capitalist crime against democracy". They did what they could to undermine morale in the Army and impede production in the workshops. The morale of France, both of her soldiers and her people, was now in May markedly lower than at the outbreak of war.

Nothing like this happened in Britain, where Soviet-directed Communism, though busy, was weak. Nevertheless we were still a party Government, under a Prime Minister from whom the

Opposition was bitterly estranged, and without the ardent and positive help of the trade union movement. The sedate, sincere, but routine character of the Administration did not evoke that intense effort, either in the governing circles or in the munitions factories, which was vital. The stroke of catastrophe and the spur of peril were needed to call forth the dormant might of the British nation. The tocsin was about to sound.

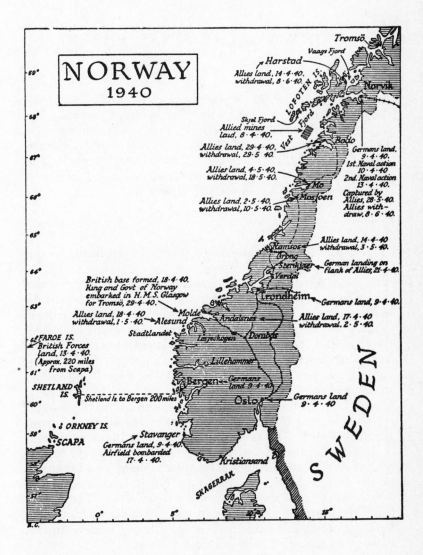

NORWAY
1940

Tromsö

Vaags Fjord

Harstad
Allies land, 14·4·40.
withdrawal, 8·6·40.

Norvik

LOFOTEN IS.

Skjel Fjord
Allied mines
laid, 8·4·40.

Vest Fjord

Bodo

Allies land, 29·4·40.
withdrawal, 29·5·40.

Germans land,
9·4·40.
1st. Naval action
10·4·40
2nd. Naval action
13·4·40.
Captured by
Allies, 28·5·40.
Allies with-
draw, 8·6·40.

Allies land, 4·5·40.
withdrawal, 18·5·40.

Mo
Mosjoen

Allies land, 2·5·40,
withdrawal, 10·5·40.

Allies land, 14·4·40
withdrawal, 3·5·40.

Namsos
Grong
Sternkjaer
Verdal

German landing on
flank of Allies, 21·4·40.

British base formed, 18·4·40.
King and Govt of Norway
embarked in H.M.S. Glasgow
for Tromsö, 29·4·40.

Trondheim

Germans land, 9·4·40.

Allies land, 18·4·40
withdrawal, 1·5·40.

Molde
Alesund

Andalsnes

Allies land, 17·4·40
withdrawal, 2·5·40.

Stadtlandet

Lesjeskogen

Dombas

FAROE IS.
British Forces
land, 13·4·40.
(Approx. 220 miles
from Scapa)

Lillehammer

SHETLAND
IS.

Shetland Is. to Bergen 200 miles

Bergen

Germans
land 9·4·40.

SWEDEN

Oslo

Germans land
9·4·40

ORKNEY IS.

SCAPA

Stavanger
Germans land, 9·4·40.
Airfield bombarded
17·4·40.

Kristiansand

SKAGERRAK

R.C.

0°   5°   10°   15°

585

CHAPTER XXXVII

NORWAY: THE FINAL PHASE

Immediate Assault on Narvik Abandoned – The Landings in May – General Auchinleck Appointed to the Chief Military Command – The Capture of the Town, May 28 – The Battle in France Dominates All – Evacuation – The Homeward Convoys – Apparition of the German Battle-Cruisers – The Loss of the "Glorious" and "Ardent" – The Story of the "Acasta" – Air Attack on German Ships at Trondheim – One Solid Result – The German Fleet Ruined.

IN DEFIANCE of chronology, it is well to set forth here the end of the Norwegian episode.

After April 16 Lord Cork was compelled to abandon the idea of an immediate assault on Narvik. A three hours' bombardment on April 24, carried out by the battleship *Warspite* and three cruisers, was not effective in dislodging the garrison. I had asked the First Sea Lord to arrange for the replacement of the *Warspite* by the less valuable *Resolution*, which was equally useful for bombarding purposes. Meanwhile the arrival of French and Polish troops, and still more the thaw, encouraged Lord Cork to press his attack on the town. The new plan was to land at the head of the fiord beyond Narvik, and thereafter to attack Narvik across Römbaks Fiord. The 24th Guards Brigade had been drawn off to stem the German advance from Trondheim; but by the beginning of May three battalions of Chasseurs Alpins, two battalions of the French Foreign Legion, four Polish battalions, and a Norwegian force of about 3,500 men were available. The enemy had for their part been reinforced by portions of the 3rd Mountain Division, which had either been brought by air from Southern Norway or smuggled in by rail from Sweden.

The first landing, under General Béthouart, the commander of the French contingent, took place on the night of May 12–13 at

Bjerkvik, with very little loss. General Auchinleck, whom I had sent to command all the troops in Northern Norway, was present and took charge the next day. His instructions were to cut off the iron ore supplies and to defend a foothold in Norway for the King and his Government. The new British commander naturally asked for very large additions to bring his force up to seventeen battalions, two hundred heavy and light anti-aircraft guns, and four squadrons of aeroplanes. It was only possible to promise about half these requirements.

But now tremendous events became dominant. On May 24, in the crisis of shattering defeat, it was decided, with almost universal agreement, that we must concentrate all we had in France and at home. The capture of Narvik had however to be achieved, both to ensure the destruction of the port and to cover our withdrawal. The main attack across Römbaks Fiord was begun on May 27 by two battalions of the Foreign Legion and one Norwegian battalion under the able leadership of General Béthouart. It was entirely successful. The landing was effected with practically no loss and the counter-attack beaten off. Narvik was taken on May 28. The Germans, who had so long resisted forces four times their strength, retreated into the mountains, leaving four hundred prisoners in our hands.

We now had to relinquish all that we had won after such painful exertions. The withdrawal was in itself a considerable operation, imposing a heavy burden on the Fleet, already fully extended by the fighting both in Norway and in the Narrow Seas. Dunkirk was upon us, and all available light forces were drawn to the south. The Battle Fleet must itself be held in readiness to resist invasion. Many of the cruisers and destroyers had already been sent south for anti-invasion duties. The Commander-in-Chief had at his disposal at Scapa the capital ships *Rodney*, *Valiant*, *Renown*, and *Repulse*. These had to cover all contingencies.

Good progress in evacuation was made at Narvik, and by June 8 all the troops, French, British, and Polish, amounting to 24,000 men, together with large quantities of stores and equipment, were embarked and sailed in four convoys without hindrance from the enemy, who indeed now amounted on shore to no more than a few thousand scattered, disorganised, but victorious individuals. During these last days valuable protection was afforded against the German Air Force not only by naval aircraft, but by a shore-based squadron of Hurricanes. This squadron had been ordered to keep in action till the end, destroying their aircraft if necessary. However, by their skill and daring these pilots performed the unprecedented feat—their last—of flying their Hurricanes on board the carrier *Glorious*, which sailed with the *Ark Royal* and the main body.

To cover all these operations Lord Cork had at his disposal, in addition to the carriers, the cruisers *Southampton* and *Coventry* and sixteen destroyers, besides smaller vessels. The cruiser *Devonshire* was meanwhile embarking the King of Norway and his staff from Tromsö, and was therefore moving independently. Lord Cork informed the Commander-in-Chief of his convoy arrangements, and asked for protection against possible attack by heavy ships. Admiral Forbes dispatched the *Valiant* on June 6 to meet the first convoy of troopships and escort it north of the Shetlands and then return to meet the second. Despite all other preoccupations, he had intended to use his battle-cruisers to protect the troopships, but on June 5 reports had reached him of two unknown ships apparently making for Iceland, and later of an enemy landing there. He therefore felt compelled to send his battle-cruisers to investigate these reports, which proved to be false. Thus on this unlucky day our available forces in the north were widely dispersed. The movement of the Narvik convoys and their protection followed closely the method pursued without mishap during the past six weeks. It had been customary to send transports and warships, including aircraft-carriers, over this route with no more than anti-submarine escort. No activity by German heavy ships had hitherto been detected. Now, having repaired the damage they had suffered in the earlier encounters, they suddenly appeared off the Norwegian coast.

The battle-cruisers *Scharnhorst* and *Gneisenau*, with the cruiser *Hipper* and four destroyers, left Kiel on June 4, with the object of attacking shipping and bases in the Narvik area and thus providing relief for what was left of their landed forces. No hint of our intended withdrawal reached them till June 7. On the news that British convoys were at sea the German admiral decided to attack them. Early the following morning, the 8th, he caught a tanker with a trawler escort, an empty troopship *Orama*, and the hospital ship *Atlantis*. He respected the immunity of the *Atlantis*. All the rest were sunk. That afternoon the *Hipper* and the destroyers returned to Trondheim, but the battle-cruisers, continuing their search for prey, were rewarded when at 4 p.m. they sighted the smoke of the aircraft-carrier *Glorious*, with her two escorting destroyers, the *Acasta* and *Ardent*. The *Glorious* had been detached early that morning to proceed home independently owing to shortage of fuel, and by now was nearly two hundred miles ahead of the main convoy. This explanation is not convincing. The *Glorious* presumably had enough fuel to steam at the speed of the convoy. All should have kept together.

The action began about 4.30 p.m. at over 27,000 yards. At this range the *Glorious*, with her 4-inch guns, was helpless. Efforts were made to get her torpedo-bombers into the air, but before this could be done she was hit in the forward hangar, and a fire began which destroyed the Hurricanes and prevented torpedoes being got up from below for the bombers. In the next half-hour she received staggering blows which deprived her of all chance of escape. By 5.20 she was listing heavily, and the order was given to abandon ship. She sank about twenty minutes later.

Meanwhile her two destroyers behaved nobly. Both made smoke in an endeavour to screen the *Glorious*, and both fired their torpedoes at the enemy before being overwhelmed. The *Ardent* was soon sunk. The story of the *Acasta*, commanded by Commander C. E. Glasfurd, R.N., now left alone at hopeless odds, has been told by the sole survivor, Leading-Seaman C. Carter.

On board our ship, what a deathly calm, hardly a word spoken, the ship was now steaming full speed away from the enemy, then came a host of orders, prepare all smoke floats, hose-pipes connected up, various other jobs were prepared, we were still stealing away from the enemy, and making smoke, and all our smoke floats had been set going. The Captain then had this message passed to all positions: "You may think we are running away from the enemy, we are not, our chummy ship [*Ardent*] has sunk, the *Glorious* is sinking, the least we can do is make a show, good luck to you all." We then altered course into our own smoke-screen. I had the order stand by to fire tubes 6 and 7, we then came out of the smoke-screen, altered course to starboard firing our torpedoes from port side. It was then I had my first glimpse of the enemy, to be honest it appeared to me to be a large one [ship] and a small one, and we were very close. I fired my two torpedoes from my tubes [aft], the foremost tubes fired theirs, we were all watching results. I'll never forget that cheer that went up;

on the port bow of one of the ships a yellow flash and a great column of smoke and water shot up from her. We knew we had hit, personally I could not see how we could have missed so close as we were. The enemy never fired a shot at us, I feel they must have been very surprised. After we had fired our torpedoes we went back into our own smoke-screen, altered course again to starboard. "Stand by to fire remaining torpedoes"; and this time as soon as we poked our nose out of the smoke-screen, the enemy let us have it. A shell hit the engine-room, killed my tubes' crew, I was blown to the after end of the tubes, I must have been knocked out for a while, because when I came to, my arm hurt me; the ship had stopped with a list to port. Here is something believe it or believe it not, I climbed back into the control seat, I see those two ships, I fired the remaining torpedoes, no one told me to, I guess I was raving mad. God alone knows why I fired them, but I did. The *Acasta's* guns were firing the whole time, even firing with a list on the ship. The enemy then hit us several times, but one big explosion took place right aft, I have often wondered whether the enemy hit us with a torpedo, in any case it seemed to lift the ship out of the water. At last the Captain gave orders to abandon ship. I will always remember the Surgeon Lt.,* his first ship, his first action. Before I jumped over the side, I saw him still attending to the wounded, a hopeless task, and when I was in the water I saw the Captain leaning over the bridge, take a cigarette from a case and light it. We shouted to him to come on our raft, he waved "Good-bye and good luck"—the end of a gallant man.

Thus perished 1,474 officers and men of the Royal Navy and forty-one of the Royal Air Force. Despite prolonged search, only thirty-nine were rescued and brought in later by a Norwegian ship. In addition six men were picked up by the enemy and taken to Germany. The *Scharnhorst*, heavily damaged by the *Acasta's* torpedo, made her way to Trondheim.

While this action was going on the cruiser *Devonshire*, with the King of Norway and his Ministers, was about a hundred miles to the westward. The *Valiant*, coming north to meet the convoy, was still a long way off. The only message received from the *Glorious* was corrupt and barely intelligible, which suggests that her main wireless equipment was destroyed from an early stage. The *Devonshire* alone received this message, but as its importance was not apparent she did not break wireless silence to pass it on, as to do so would have involved serious risk of revealing her position, which in the circumstances was highly undesirable. Not

* Temporary Surgeon-Lieutenant H. J. Stammers, R.N.V.R.

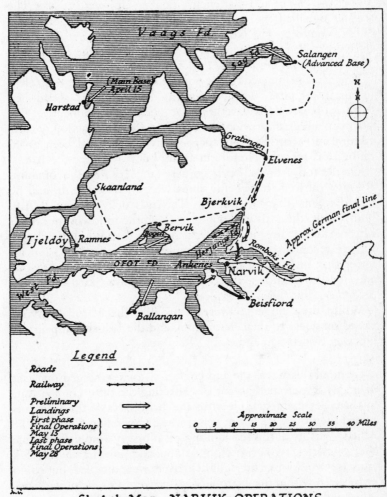

Sketch Map NARVIK OPERATIONS

until the following morning were suspicions aroused. Then the *Valiant* met the hospital ship *Atlantis*, who informed her of the loss of the *Orama* and that enemy capital ships were at sea. The *Valiant* signalled the information and pressed on to join Lord Cork's convoy. The Commander-in-Chief, Admiral Forbes, at once proceeded to sea with the only ships he had, the *Rodney*, the *Renown*, and six destroyers.

The damage inflicted on the *Scharnhorst* by the heroic *Acasta* had important results. The two enemy battle-cruisers abandoned further operations and returned at once to Trondheim. The German High Command were dissatisfied with the action of their admiral in departing from the objective which had been given him. They sent the *Hipper* out again; but it was then too late.

On the 10th Admiral Forbes ordered the *Ark Royal* to join him. Reports showed that enemy ships were in Trondheim, and he hoped to make an air attack. This was delivered by R.A.F. bombers on the 11th without effect. On the following morning fifteen Skuas from the *Ark Royal* made a dive-bombing attack. Enemy reconnaissance gave warning of their approach, and no fewer than eight were lost. To add one last misfortune to our tale, it is now known that one bomb from a Skua struck the *Scharnhorst* but failed to explode.

Whilst these tragedies were in progress the Narvik convoys passed on safely to their destination, and the British campaign in Norway came to an end.

*   *   *   *   *

From all this wreckage and confusion there emerged one fact of major importance potentially affecting the future of the war. In their desperate grapple with the British Navy the Germans ruined their own, such as it was, for the impending climax. The Allied losses in all this sea-fighting off Norway amounted to one aircraft-carrier, two cruisers, one sloop, and nine destroyers. Six cruisers, two sloops, and eight destroyers were disabled, but could be repaired within our margin of sea-power. On the other hand, at the end of June 1940, a momentous date, the effective German Fleet consisted of no more than *one 8-inch-gun cruiser, two light cruisers, and four destroyers.* Although many of their damaged ships, like ours, could be repaired, the German Navy was no factor in the supreme issue of the invasion of Britain.*

* See Appendix R.

CHAPTER XXXVIII

THE FALL OF THE GOVERNMENT

Debate of May 7 – A Vote of Censure Supervenes – Lloyd George's Last Parliamentary Stroke – I Do My Best with the House – My Advice to the Prime Minister – Conferences of May 9 – The German Onslaught – A Conversation with the Prime Minister, May 10 – The Dutch Agony – Mr. Chamberlain Resigns – The King Asks Me to Form a Government – Accession of the Labour and Liberal Parties – Facts and Dreams.

THE MANY disappointments and disasters of the brief campaign in Norway caused profound perturbation at home, and the currents of passion mounted even in the breasts of some of those who had been most slothful and purblind in the years before the war. The Opposition asked for a debate on the war situation, and this was arranged for May 7. The House was filled with Members in a high state of irritation and distress. Mr. Chamberlain's opening statement did not stem the hostile tide. He was mockingly interrupted, and reminded of his speech of April 4, when in quite another connection he had incautiously said, "Hitler missed the bus." He defined my new position and my relationship with the Chiefs of Staff, and in reply to Mr. Herbert Morrison made it clear that I had not held those powers during the Norwegian operations. One speaker after another from both sides of the House attacked the Government, and especially its chief, with unusual bitterness and vehemence, and found themselves sustained by growing applause from all quarters. Sir Roger Keyes, burning for distinction in the new war, sharply criticised the Naval Staff for their failure to attempt the capture of Trondheim. "When I saw," he said, "how badly things were going I never ceased importuning the Admiralty and War

Cabinet to let me take all responsibility and lead the attack." Wearing his uniform as Admiral of the Fleet, he supported the complaints of the Opposition with technical details and his own professional authority in a manner very agreeable to the mood of the House. From the benches behind the Government Mr. Amery quoted, amid ringing cheers, Cromwell's imperious words to the Long Parliament: "You have sat too long here for any good you have been doing. Depart, I say, and let us have done with you. In the name of God, go!" These were terrible words, coming from a friend and colleague of many years, a fellow Birmingham Member, and a Privy Counsellor of distinction and experience.

On the second day, May 8, the debate, although continuing upon an Adjournment Motion, assumed the character of a Vote of Censure, and Mr. Herbert Morrison, in the name of the Opposition, declared their intention to have a vote. The Prime Minister rose again, accepted the challenge, and in an unfortunate passage appealed to his friends to stand by him. He had a right to do this, as these friends had sustained his action, or inaction, and thus shared his responsibility in "the years which the locusts had eaten" before the war. But to-day they sat abashed and silenced, and some of them had joined the hostile demonstrations. This day saw the last decisive intervention of Mr. Lloyd George in the House of Commons. In a speech of not more than twenty minutes he struck a deeply-wounding blow at the head of the Government. He endeavoured to exculpate me: "I do not think that the First Lord was entirely responsible for all the things which happened in Norway." I immediately interposed, "I take complete responsibility for everything that has been done by the Admiralty, and I take my full share of the burden." After warning me not to allow myself to be converted into an air-raid shelter to keep the splinters from hitting my colleagues, Mr. Lloyd George turned upon Mr. Chamberlain. "It is not a question of who are the Prime Minister's friends. It is a far bigger issue. He has appealed for sacrifice. The nation is prepared for every sacrifice so long as it has leadership, so long as the Government show clearly what they are aiming at, and so long as the nation is confident that those who are leading it are doing their best." He ended, "I say solemnly that the Prime Minister should give an example of sacrifice, because there is nothing which can contribute more to victory in this war than that he should sacrifice the seals of office."

As Ministers we all stood together. The Secretaries of State for War and Air had already spoken. I had volunteered to wind up the debate, which was no more than my duty, not only in loyalty to the chief under whom I served, but also because of the exceptionally prominent part I had played in the use of our inadequate forces during our forlorn attempt to succour Norway. I did my very best to regain control of the House for the Government in the teeth of continuous interruption, coming chiefly from the Labour Opposition benches. I did this with good heart when I thought of their mistakes and dangerous pacifism in former years, and how only four months before the outbreak of the war they had voted solidly against conscription. I felt that I, and a few friends who had acted with me, had the right to inflict these censures, but they had not. When they broke in upon me I retorted upon them and defied them, and several times the clamour was such that I could not make myself heard. Yet all the time it was clear that their anger was not directed against me, but at the Prime Minister, whom I was defending to the utmost of my ability and without regard for any other considerations. When I sat down at eleven o'clock the House divided. The Government had a majority of 81, but over 30 Conservatives voted with the Labour and Liberal Oppositions, and a further 60 abstained. There was no doubt that in effect, though not in form, both the debate and the division were a violent manifestation of want of confidence in Mr. Chamberlain and his Administration.

After the debate was over he asked me to go to his room, and I saw at once that he took the most serious view of the sentiment of the House towards himself. He felt he could not go on. There ought to be a National Government. One party alone could not carry the burden. Someone must form a Government in which all parties would serve, or we could not get through. Aroused by the antagonisms of the debate, and being sure of my own past record on the issues at stake, I was strongly disposed to fight on. "This has been a damaging debate, but you have a good majority. Do not take the matter grievously to heart. We have a better case about Norway than it has been possible to convey to the House. Strengthen your Government from every quarter, and let us go on until our majority deserts us." To this effect I spoke.

But Chamberlain was neither convinced nor comforted, and I left him about midnight with the feeling that he would persist in his resolve to sacrifice himself if there was no other way, rather than attempt to carry the war further with a one-party Government.

I do not remember exactly how things happened during the morning of May 9, but the following occurred. Sir Kingsley Wood was very close to the Prime Minister as a colleague and a friend. They had long worked together in complete confidence. From him I learned that Mr. Chamberlain was resolved upon the formation of a National Government, and if he could not be the head he would give way to anyone commanding his confidence who could. Thus by the afternoon I became aware that I might well be called upon to take the lead. The prospect neither excited nor alarmed me. I thought it would be by far the best plan. I was content to let events unfold. In the afternoon the Prime Minister summoned me to Downing Street, where I found Lord Halifax, and after a talk about the situation in general we were told that Mr. Attlee and Mr. Greenwood would visit us in a few minutes for a consultation.

When they arrived we three Ministers sat on one side of the table and the Opposition Leaders on the other. Mr. Chamberlain declared the paramount need of a National Government, and sought to ascertain whether the Labour Party would serve under him. The Conference of their party was in session at Bournemouth. The conversation was most polite, but it was clear that the Labour leaders would not commit themselves without consulting their people, and they hinted, not obscurely, that they thought the response would be unfavourable. They then withdrew. It was a bright, sunny afternoon, and Lord Halifax and I sat for a while on a seat in the garden of Number 10 and talked about nothing in particular. I then returned to the Admiralty, and was occupied during the evening and a large part of the night in heavy business.

*   *   *   *   *

The morning of the 10th of May dawned, and with it came tremendous news. Boxes with telegrams poured in from the Admiralty, the War Office, and the Foreign Office. The Germans had struck their long-awaited blow. Holland and Belgium were both invaded. Their frontiers had been crossed at numerous points. The whole movement of the German Army upon the invasion of the Low Countries and of France had begun.

At about ten o'clock Sir Kingsley Wood came to see me, having just been with the Prime Minister. He told me that Mr. Chamberlain was inclined to feel that the great battle which had broken upon us made it necessary for him to remain at his post. Kingsley Wood had told him that, on the contrary, the new crisis made it all the more necessary to have a National Government, which alone could confront it, and he added that Mr. Chamberlain had accepted this view. At eleven o'clock I was again summoned to Downing Street by the Prime Minister. There once more I found Lord Halifax. We took our seats at the table opposite Mr. Chamberlain. He told us that he was satisfied that it was beyond his power to form a National Government. The response he had received from the Labour leaders left him in no doubt of this. The question therefore was whom he should advise the King to send for after his own resignation had been accepted. His demeanour was cool, unruffled, and seemingly quite detached from the personal aspect of the affair. He looked at us both across the table.

I have had many important interviews in my public life, and this was certainly the most important. Usually I talk a great deal, but on this occasion I was silent. Mr. Chamberlain evidently had in his mind the stormy scene in the House of Commons two nights before, when I had seemed to be in such heated controversy with the Labour Party. Although this had been in his support and defence, he nevertheless felt that it might be an obstacle to my obtaining their adherence at this juncture. I do not recall the actual words he used, but this was the implication. His biographer, Mr. Feiling, states definitely that he preferred Lord Halifax. As I remained silent a very long pause ensued. It certainly seemed longer than the two minutes which one observes in the commemorations of Armistice Day. Then at length Halifax spoke. He said that he felt that his position as a Peer, out of the House of Commons, would make it very difficult for him to discharge the duties of Prime Minister in a war like this. He would be held responsible for everything, but would not have the power to guide the assembly upon whose confidence the life of every Government depended. He spoke for some minutes in this sense, and by the time he had finished it was clear that the duty would fall upon me—had in fact fallen upon me. Then for the first time I spoke. I said I would have no communication with either of the Opposition parties until I had the King's Commission to form a Government. On this the momentous conversation came to an end, and we reverted to our ordinary easy and familiar manners of men who had worked for years together and whose lives in and out of office had been spent in all the friendliness of British politics. I then went back to the Admiralty, where, as may well be imagined, much awaited me.

The Dutch Ministers were in my room. Haggard and worn, with horror in their eyes, they had just flown over from Amsterdam. Their country had been attacked without the slightest pretext or warning. The avalanche of fire and steel had rolled across the frontiers, and when resistance broke out and the Dutch frontier guards fired an overwhelming onslaught was made from the air. The whole country was in a state of wild confusion. The long-prepared defence scheme had been put into operation; the dykes were opened, the waters spread far and wide. But the Germans had already crossed the outer lines, and were now streaming down the banks of the Rhine and through the inner Gravelines defences. They threatened the causeway which encloses the Zuyder Zee. Could we do anything to prevent this? Luckily, we had a flotilla not far away, and this was immediately ordered to sweep the causeway with fire and take the heaviest toll possible of the swarming invaders. The Queen was still in Holland, but it did not seem she could remain there long.

As a consequence of these discussions, a large number of orders were dispatched by the Admiralty to all our ships in the neighbourhood, and close relations were established with the Royal Dutch Navy. Even with the recent overrunning of Norway and Denmark in their minds, the Dutch Ministers seemed unable to understand how the great German nation, which up to the night before had professed nothing but friendship, should suddenly have made this frightful and brutal onslaught. Upon these proceedings and other affairs an hour or two passed. A spate of telegrams pressed in from all the frontiers affected by the forward heave of the German armies. It seemed that the old Schlieffen plan, brought up to date with its Dutch extension, was already in full operation. In 1914 the swinging right arm of the German invasion had swept through Belgium but had stopped short of

Holland. It was well known then that had that war been delayed for three or four years the extra army group would have been ready and the railway terminals and communications adapted for a movement through Holland. Now the famous movement had been launched with all these facilities and with every circumstance of surprise and treachery. But other developments lay ahead. The decisive stroke of the enemy was not to be a turning movement on the flank, but a break through the main front. This none of us or the French, who were in responsible command, foresaw. Earlier in the year I had, in a published interview, warned these neutral countries of the fate which was impending upon them, and which was evident from the troop dispositions and road and rail development, as well as from the captured German plans. My words had been resented.

In the splintering crash of this vast battle the quiet conversations we had had in Downing Street faded or fell back in one's mind. However, I remember being told that Mr. Chamberlain had gone, or was going, to see the King, and this was naturally to be expected. Presently a message arrived summoning me to the Palace at six o'clock. It only takes two minutes to drive there from the Admiralty along the Mall. Although I suppose the evening newspapers must have been full of the terrific news from the Continent, nothing had been mentioned about the Cabinet crisis. The public had not had time to take in what was happening either abroad or at home, and there was no crowd about the Palace gates.

I was taken immediately to the King. His Majesty received me most graciously and bade me sit down. He looked at me searchingly and quizzically for some moments, and then said, "I suppose you don't know why I have sent for you?" Adopting his mood, I replied, "Sir, I simply couldn't imagine why." He laughed and said, "I want to ask you to form a Government." I said I would certainly do so.

The King had made no stipulation about the Government being National in character, and I felt that my commission was in no formal way dependent upon this point. But in view of what had happened, and the conditions which had led to Mr. Chamberlain's resignation, a Government of National character was obviously inherent in the situation. If I found it impossible to come to terms with the Opposition parties, I should not have been constitutionally debarred from trying to form the strongest Government possible of all who would stand by the country in the hour of peril, provided that such a Government could command a majority in the House of Commons. I told the King that I would immediately send for the leaders of the Labour and Liberal Parties, that I proposed to form a War Cabinet of five or six Ministers, and that I hoped to let him have at least five names before midnight. On this I took my leave and returned to the Admiralty.

Between seven and eight, at my request, Mr. Attlee called upon me. He brought with him Mr. Greenwood. I told him of the authority I had to form a Government, and asked if the Labour Party would join. He said they would. I proposed that they should take rather more than a third of the places, having two seats in the War Cabinet of five, or it might be six, and I asked Mr. Attlee to let me have a list of men so that we could discuss particular offices. I mentioned Mr. Bevin, Mr. Alexander, Mr. Morrison, and Mr. Dalton as men whose services in high office were immediately required. I had, of course, known both Attlee and Greenwood for a long time in the House of Commons. During the ten years before the outbreak of war I had in my more or less independent position come far more often into collision with the Conservative and National Governments than with the Labour and Liberal Oppositions. We had a pleasant talk for a little while, and they went off to report by telephone to their friends and followers at Bournemouth, with whom of course they had been in the closest contact during the previous forty-eight hours.

I invited Mr. Chamberlain to lead the House of Commons as Lord President of the Council, and he replied by telephone that he accepted, and had arranged to broadcast at nine that night, stating that he had resigned, and urging everyone to support and aid his successor. This he did in magnanimous terms. I asked Lord Halifax to join the War Cabinet while remaining Foreign Secretary. At about ten I sent the King a list of the five names, as I had promised. The appointment of the three Service Ministers was vitally urgent. I had already made up my mind who they should be. Mr. Eden should go to the War Office, Mr. Alexander should come to the Admiralty, and Sir Archibald Sinclair, Leader of the Liberal Party, should take the Air Ministry. At the same time I assumed the office of Minister of Defence, without however atte.npting to define its scope and powers.

Thus, then, on the night of the 10th of May, at the outset of this mighty battle, I acquired the chief power in the State, which henceforth I wielded in ever-growing measure for five years and three months of world war, at the end of which time, all our enemies having surrendered unconditionally or being about to do so, I was immediately dismissed by the British electorate from all further conduct of their affairs.

During these last crowded days of the political crisis my pulse had not quickened at any moment. I took it all as it came. But I cannot conceal from the reader of this truthful account that as I went to bed at about 3 a.m. I was conscious of a profound sense of relief. At last I had the authority to give directions over the whole scene. I felt as if I were walking with destiny, and that all my past life had been but a preparation for this hour and for this trial. Ten years in the political wilderness had freed me from ordinary party antagonisms. My warnings over the last six years had been so numerous, so detailed, and were now so terribly vindicated, that no one could gainsay me. I could not be reproached either for making the war or with want of preparation for it. I thought I knew a good deal about it all, and I was sure I should not fail. Therefore, although impatient for the morning, I slept soundly and had no need for cheering dreams. Facts are better than dreams.

I. MISCELLANEOUS

A. A CONVERSATION WITH COUNT GRANDI.

B. MY NOTE ON THE FLEET AIR ARM.

C. A NOTE ON SUPPLY ORGANISATION.

D. STATEMENT ON THE OCCASION OF THE DEPUTATION OF CONSERVATIVE MEMBERS OF BOTH HOUSES TO THE PRIME MINISTER, JULY 28, 1936.

E. COMPARATIVE OUTPUT OF FIRST-LINE AIRCRAFT.

F. TABLES OF NAVAL STRENGTH, SEPTEMBER 3, 1939.

G. PLAN "CATHERINE", SEPTEMBER 12, 1939.

H. NEW CONSTRUCTION AND RECONSTRUCTION, OCTOBER 8 AND 21, 1939.

I. NEW CONSTRUCTION PROGRAMMES, 1939–40.

J. FLEET BASES, NOVEMBER 1, 1939.

K. NAVAL AID TO TURKEY, NOVEMBER 1, 1939.

L. THE BLACK-OUT, NOVEMBER 20, 1939.

M. THE MAGNETIC MINE, 1939–40.

N. EXTRACT FROM WAR DIARY OF U.47, NOVEMBER 28, 1939.

O. CULTIVATOR NO. 6, NOVEMBER 1939.

P. BRITISH MERCHANT VESSELS LOST BY ENEMY ACTION, SEPTEMBER 1939 TO APRIL 1940.

Q. OPERATION "ROYAL MARINE", MARCH 4, 1940.

R. NAVAL LOSSES IN THE NORWEGIAN CAMPAIGN, APRIL–JUNE, 1940.

APPENDICES

APPENDICES

APPENDIX A

A CONVERSATION WITH COUNT GRANDI*

Mr. Churchill to Sir Robert Vansittart　　　　　　　*September 28, 1935*

Though he pleaded the Italian cause with much address, he of course realises the whole position. . . .

I told him that since Parliament rose there had been a strong development of public opinion. England, and indeed the British Empire, could act unitedly on the basis of the League of Nations, and all parties thought that that instrument was the most powerful protection against future dangers wherever they might arise. He pointed out the injury to the League of Nations by the loss of Italy. The fall of the *régime* in Italy would inevitably produce a pro-German Italy. He seemed prepared for economic sanctions. They were quite ready to accept life upon a communal basis. However poor they were, they could endure. He spoke of the difficulty of following the movements of British public opinion. I said that no foreign ambassador could be blamed for that, but the fact of the change must be realised. Moreover, if fighting began in Abyssinia, cannons fired, blood was shed, villages were bombed, etc., an almost measureless rise in the temperature must be expected. He seemed to contemplate the imposition of economic sanctions which would at first be ineffective, but gradually increase until at some moment or other an event of war would occur.

I said the British Fleet was very strong, and, although it had to be rebuilt in the near future, it was good and efficient at the present moment, and it was now completely ready to defend itself; but I repeated that this was a purely defensive measure in view of our Mediterranean interests, and did not in any way differentiate our position from that of other members of the League of Nations. He accepted this with a sad smile.

I then talked of the importance of finding a way out: "He that ruleth his spirit is better than he that taketh a city." He replied that they would feel that everywhere except in Italy. They had to deal with two hundred thousand men with rifles in their hands. Mussolini's dictatorship was a popular dictatorship, and success was the essence of its strength. Finally, I said that I was in favour of a meeting between the political chiefs of the three countries. . . . The three men together could carry off something that one could never do by himself. After all, the claims of Italy to primacy in the Abyssinian sphere and the imperative need of internal reform [in Abyssinia] had been fully recognised by England and France. I told him I should support such an idea if it were agreeable. The British public would be willing to

* See p. 156.

607

try all roads to an honourable peace. I thought there should be a meeting of three. Any agreement they reached would of course be submitted to the League of Nations. It seemed to me the only chance of avoiding the destruction of Italy as a powerful and friendly factor in Europe. Even if it failed no harm would have been done, and at present we were heading for an absolute smash.

APPENDIX B

MY NOTE ON THE FLEET AIR ARM*

WRITTEN FOR SIR THOMAS INSKIP, MINISTER FOR THE CO-ORDINATION OF DEFENCE, IN 1936

1. It is impossible to resist an admiral's claim that he must have complete control of, and confidence in, the aircraft of the battle fleet, whether used for reconnaissance, gun-fire, or air attack on a hostile fleet. These are his very eyes. Therefore the Admiralty view must prevail in all that is required to secure this result.

2. The argument that similar conditions obtain in respect of Army co-operation aircraft cannot be countenanced. In one case the aircraft take flight from aerodromes and operate under precisely similar conditions to those of normal independent Air Force action. Flight from warships and action in connection with naval operations is a totally different matter. One is truly an affair of co-operation only, the other an integral part of modern naval operations.

3. A division must therefore be made between the Air Force controlled by the Admiralty and that controlled by the Air Ministry. This division does not depend upon the type of the undercarriage of the aircraft, nor necessarily the base from which it is flown. It depends upon the function. Is it predominantly a naval function or not?

4. Most of these defence functions can clearly be assigned. For instance, all functions which require aircraft of any description (whether with wheels, floats, or boats; whether reconnaissance, spotters or fighters, bombers or torpedo-seaplanes) to be carried regularly in warships or in aircraft-carriers naturally fall to the naval sphere.

5. The question thus reduces itself to the assignment of any type operating over the sea from shore bases. This again can only be decided in relation to the functions and responsibilities placed upon the Navy. Aircraft borne afloat could discharge a considerable function of trade protection. This would be especially true in the broad waters, where a

* See p. 143.

squadron of cruisers with their own scouting planes or a pair of small aircraft-carriers could search upon a front of a thousand miles. But the Navy could never be required—nor has it ever claimed—to maintain an air strength sufficient to cope with a concentrated attack upon merchant shipping in the Narrow Waters by a large hostile Air Force of great power. In fact, the maxim must be applied of Air Force *versus* Air Force and Navy *versus* Navy. When the main hostile Air Force or any definite detachment from it is to be encountered, it must be by the British Royal Air Force.

6. In this connection it should not be forgotten that a ship or ships may have to be selected and adapted for purely Air Force operations, like a raid on some deep-seated enemy base or vital centre. This is an Air Force operation, and necessitates the use of types of aircraft not normally associated with the Fleet. In this case the *rôles* of the Admiralty and the Air Ministry will be reversed, and the Navy would swim the ship in accordance with the tactical or strategic wishes of the Air Ministry. Far from becoming a baffle, this special case exemplifies the logic of the "division of command according to function".

7. What is conceded to the Navy should, within the limits assigned, be fully given. The Admiralty should have plenary control and provide the entire personnel of the Fleet Air Arm. Officers, cadets, petty officers, artificers, etc., for this force would be selected from the Royal Navy by the Admiralty. They would then acquire the art of flying and the management of aircraft in the R.A.F. training schools— to which perhaps naval officers should be attached—but after acquiring the necessary degree of proficiency as air chauffeurs and mechanics they would pass to shore establishments under the Admiralty for their training in Fleet Air Arm duties, just as the pilots of the Royal Air Force do to their squadrons at armament schools to learn air fighting. Thus, the personnel employed upon fleet air functions will be an integral part of the Navy, dependent for discipline and advancement as well as for their careers and pensions solely upon the Admiralty. This would apply to every rank and every trade involved, whether afloat or ashore.

8. Coincident with this arrangement whereby the Fleet Air Arm becomes wholly a naval Service, a further rearrangement of functions should be made, whereby the Air Ministry becomes responsible for active anti-aircraft defence. This implies, in so far as the Navy is concerned, that, at every naval port, shore anti-aircraft batteries, lights, aircraft, balloons, and other devices will be combined under one operational control, though the officer commanding would of course, with his command, be subordinate to the Fortress Commander.

9. In the same way, the control of the air defences of London and

of such other vulnerable areas as it may be necessary to equip with anti-air defences on a considerable scale should also be unified under one command and placed under the Air Ministry. The consequent control should cover not only the operations, but, as far as may conveniently be arranged, the training, the raising and administration of the entire personnel for active air defence.

10. The Air Ministry have as clear a title to control active anti-air defence as have the Navy to their own "eyes". For this purpose a new department should be brought into being in the Air Ministry, to be called "Anti-Air", to control all guns, searchlights, balloons, and personnel of every kind connected with this function, as well as such portion of the Royal Air Force as may from time to time be assigned to it for this duty. Under this department there will be Air Force officers, assisted by appropriate staffs, in command of all active air defences in specified localities and areas.

11. It is not suggested that the Air Ministry or Air Staff are at present capable of assuming unaided this heavy new responsibility. In the formation of the Anti-Air Command recourse must be had to both the older Services. Well-trained staff officers, both from the Army and the Navy, must be mingled with officers of the existing Air Staff.

N.B.—The question of the recruitment and of the interior administration of the units handed over to the Anti-Air Command for operations and training need not be a stumbling-block. They could be provided from the present sources unless and until a more convenient solution was apparent.

12. This memorandum has not hitherto dealt with *matériel*, but that is extremely simple. The Admiralty will decide upon the types of aircraft which their approved functions demand. The extent of the inroad which they require to make upon the finances and resources of the country must be decided by the Cabinet, operating through a Priorities Committee under the Minister for the Co-ordination of Defence. At the present stage this Minister would no doubt give his directions to the existing personnel, but in the event of war or the intensification of the preparations for war he would give them to a Ministry of Supply. There could of course be no question of Admiralty priorities being allowed to override other claims in the general sphere of air production. All must be decided from the supreme standpoint.

13. It is not intended that the Admiralty should develop technical departments for aircraft design separate from those existing in the Air Ministry or under a Ministry of Supply. They would however be free to form a nucleus technical staff to advise them on the possibilities

610

of scientific development and to prescribe their special naval requirements in suitable technical language to the supply department.

14. To sum up therefore we have:

First—The Admiralty should have plenary control of the Fleet Air Arm for all purposes which are defined as naval.

Secondly—A new department must be formed under the Air Ministry from the three Services for active anti-aircraft defence operations.

Thirdly—The question of *matériel* supply must be decided by a Priorities Committee under the Minister for the Co-ordination of Defence, and executed at present through existing channels, but eventually by a Ministry of Supply.

APPENDIX C

A NOTE ON SUPPLY ORGANISATION, JUNE 6, 1936*

1. The existing office of the Minister for the Co-ordination of Defence comprises unrelated and wrongly-grouped functions. The work of the Minister charged with strategic co-ordination is different, though not in the higher ranges disconnected, from the work of the Minister charged with (a) securing the execution of the existing programmes, and (b) planning British industry to spring quickly into war-time conditions and creating a high control effective for both this and the present purpose.

2. The first step therefore is to separate the functions of strategic thought from those of material supply in peace and war, and form the organisation to direct this latter process. An harmonious arrangement would be four separate departments—Navy, Army, Air Force, and Supply—with the Co-ordinating Minister at the summit of the four having the final voice upon priorities.

3. No multiplication of committees, however expert or elaborate, can achieve this purpose. Supply cannot be achieved without command. A definite chain of responsible authority must descend through the whole of British industry affected. (This must not be thought to imply State interference in the actual functions of industry.) At the present time the three Service authorities exercise separate command over their particular supply, and the fourth, or planning, authority is purely consultative, and that only upon the war need divorced from present supply. What is needed is to unify the supply command of

the three Service departments into an organism which also exercises command over the war expansion. (The Admiralty would retain control over the construction of warships and certain special naval stores.)

4. This unification should comprise not only the function of supply but that of design. The Service departments prescribe in general technical terms their need in type, quality, and quantity, and the supply organisation executes these in a manner best calculated to serve its customers. In other words, the Supply Department engages itself to deliver the approved types of war stores of all kinds to the Services when and where the latter require them.

5. None of this, nor the punctual execution of any of the approved programmes, can be achieved in the present atmosphere of ordinary peace-time preparation. It is neither necessary nor possible at this moment to take war-time powers and apply war-time methods. An intermediate state should be declared called (say) the period of emergency preparation.

6. Legislation should be drafted in two parts—first, that appropriate to the emergency preparation stage, and, second, that appropriate to a state of war. Part I should be carried out now. Part II should be envisaged, elaborated, the principles defined, the clauses drafted and left to be brought into operation by a fresh appeal to Parliament should war occur. The emergency stage should be capable of sliding into the war stage with the minimum of disturbance, the whole design having been foreseen.

7. To bring this new system into operation there should first be created a Minister of Supply. This Minister would form a Supply Council. Each member would be charged with the study of the four or five branches of production falling into his sphere. Thereafter, as soon as may be, the existing Service sub-departments of supply, design, contracts, etc., would be transferred by instalments to the new authority, who alone would deal with the Treasury upon finance. (By "finance" is meant payments within the scope of the authorised programmes.)

APPENDIX D

MY STATEMENT ON THE OCCASION OF THE
DEPUTATION OF CONSERVATIVE MEMBERS OF
BOTH HOUSES TO THE PRIME MINISTER,
JULY 28, 1936*

In time of peace the needs of our small Army, and to some extent of the Air Force and Admiralty, in particular weapons and ammunition are supplied by the War Office, which has for this purpose certain Government factories and habitual private contractors. This organisation is capable of meeting ordinary peace-time requirements, and providing the accumulating of reserves sufficient for a few weeks of war by our very limited regular forces. Outside this there was nothing until a few months ago. About three or four months ago authority was given to extend the scope of War Office orders in certain directions to ordinary civil industry.

On the other hand, in all the leading Continental countries the whole of industry has been for some time solidly and scientifically organised to turn over from peace to war. In Germany of course above all others this became the supreme study of the Government even before the Hitler *régime*. Indeed, under the impulse of revenge, Germany, forbidden by treaty to have fleets, armies, and an Air Force, concentrated with intense compression upon the perfecting of the transference of its whole industry to war purposes. We alone began seriously to examine the problem when everyone else had solved it. There was however still time in 1932 and 1933 to make a great advance. Three years ago, when Hitler came into power, we had perhaps a dozen officials studying the war organisation of industry, as compared with five or six hundred working continuously in Germany. The Hitler *régime* set all this vast machinery in motion. They did not venture to break the treaties about Army, Navy, and Air Force until they had a head of steam on in every industry which would, they hoped, speedily render them an armed nation unless they were immediately attacked by the Allies.

What is being done now? Nothing has been told to Parliament except some fragmentary items which by themselves are likely to mislead the ignorant. For instance, we were told last week that fifty-two firms had been inspected and offered contracts to make ammunition; that the old gun factory was to be reopened at Nottingham, and the Woolwich filling-station was to be moved to the West Coast. But no orders were given till three months ago, and none of this preparation

* See p. 204.

can reach a stage of mass deliveries for at least eighteen months from the date of the order. If by ammunition is meant projectiles (both bombs and shells) and cartridge-cases containing propellent, it will be necessary to equip all these factories with a certain amount of additional special-purpose machine-tools, and to modify their existing lay-out. In addition jigs and gauges for the actual manufacture must be made. ... The manufacture of these special machine-tools, jigs, and gauges will have to be done in most cases by firms quite different from those to whom the output of projectiles is entrusted. After the delivery of the special machine-tools a further delay is required while they are being set up in the producing factories, and while the process of production is being started. Then, and only then, at first in a trickle, then in a stream, and finally in a flood, deliveries will take place. Not till then can the accumulation of war resources begin. This inevitably lengthy process is still being applied on a relatively minute scale. The fifty-two firms have been offered contracts. Fourteen had last week accepted contracts. At the present moment it would be no exaggeration to state that the German ammunition plants may well amount to four or five hundred, already for very nearly two years in full swing.

Turning now to cannon: by cannon I mean guns firing explosive shells. The processes by which a cannon factory is started are necessarily lengthy; the special plants and machine-tools are more numerous, and the lay-out more elaborate. Our normal peace-time output of cannon in the last ten years has, apart from the Fleet, been negligible. We are therefore certainly separated by two years from any large deliveries of field guns or anti-aircraft guns. Last year it is probable that at least five thousand guns were made in Germany, and this process could be largely amplified in war. Surely we ought to call into being plant which would enable us, if need be, to create and arm a national army of a considerable size.

I have taken projectiles and cannon because these are the core of defence; but the same arguments and conditions, with certain modifications, apply over the whole field of equipment. The flexibility of British industry should make it possible to produce many forms of equipment—for instance, motor lorries and other kindred weapons, such as tanks and armoured cars—and many slighter forms of material necessary for an army, in a much shorter time if that industry is at once set going. Has it been set going? Why should we be told that the Territorial Army cannot be equipped until after the Regular Army is equipped? I do not know what is the position about rifles and rifle ammunition. I hope at least we have enough for a million men. But the delivery of rifles from new sources is a very lengthy process.

Even more pertinent is the production of machine-guns. I do not know at all what is the programme of Browning and Bren machine-guns. But if the orders for setting up the necessary plant were only given a few months ago one cannot expect any appreciable deliveries except by direct purchase from abroad before the beginning of 1938. The comparable German plants already in operation are capable of producing supplies limited only by the national manhood available to use them.

But this same argument can be followed out through all the processes of producing explosives, propellent, fuzes, poison gas, gas-masks, searchlights, trench-mortars, grenades, air-bombs, and all the special adaptations required for depth-charges, mines, etc., for the Navy. It must not be forgotten that the Navy is dependent upon the War Office and upon an expansion of national industry for a hundred and one minor articles, a shortage in any one of which will cause grave injury. Behind all this again lies of course the supply of raw materials, with its infinite complications.

What is the conclusion? It is that we are separated by about two years from any appreciable improvement in the material process of national defence, so far as concerns the whole volume of supplies for which the War Office has hitherto been responsible, with all the reactions that entails, both on the Navy and the War Office. But, upon the scale on which we are now acting, even at the end of two years the supply will be petty compared either with our needs in war or with what others have already acquired in peace.

Surely if these facts are even approximately true—and I believe they are mostly understatements—how can it be contended that there is no emergency; that we must not do anything to interfere with the ordinary trade of the country; that there is no need to approach the trade unions about dilution of trainees; that we can safely trust to what the Minister for Co-ordination of Defence described as "training the additional labour as required on the job"; and that nothing must be done which would cause alarm to the public, or lead them to feel that their ordinary habit of life was being deranged?

Complaint is made that the nation is unresponsive to the national need; that the trade unions are unhelpful; that recruiting for the Army and the Territorial Force is very slack, and even is obstructed by elements of public opinion. But as long as they are assured by the Government that there is no emergency these obstacles will continue.

I was given confidentially by the French Government an estimate of the German air strength in 1936. This tallies almost exactly with the figures I forecast to the Committee of Imperial Defence in December last. The Air Staff now think the French estimate too high. Personally I think it is too low. The number of machines which Germany could now put into action simultaneously may be nearer two thousand than fifteen hundred. Moreover, there is no reason to assume that they mean to stop at two thousand. The whole plant and lay-out of the German Air Force is on an enormous scale, and they may be already planning a development far greater than anything yet mentioned. Even if we accept the French figures of about fourteen hundred, the German strength at this moment is double that of our Metropolitan Air Force, judged by trained pilots and military machines that could go into action and be maintained in action. But the relative strength of two countries cannot be judged without reference to their power of replenishing their fighting force. The German industry is so organised that it can certainly produce at full blast a thousand a month and increase the number as the months pass. Can the British industry at the present time produce more than three hundred to three hundred and fifty a month? How long will it be before we can reach a war-potential output equal to the German? Certainly not within two years. When we allow for the extremely high rate of war wastage, a duel between the two countries would mean that before six months were out our force would be not a third of theirs. The preparation for war-time expansion at least three times the present size of the industry seems urgent in the highest degree. It is probable however that Germany is spending not less than one hundred and twenty millions on her Air Force this year. It is clear therefore that so far as this year is concerned we are not catching up. On the contrary, we are falling farther behind. How long will this continue into next year? No one can tell.

It has been announced that the programme of 120 squadrons and 1,500 first-line aircraft for Home Defence would be completed by April 1, 1937. Parliament has not been given any information how this programme is being carried out in machines, in personnel, in organisation, or in the ancillary supplies. We have been told nothing about it at all. I do not blame the Government for not giving full particulars. 't would be too dangerous now. Naturally however, in the absence of any information at all, there must be great anxiety and much private discussion. . . . I doubt very much whether by July next year we shall have thirty squadrons equipped with the new types. I understand that the deliveries of the new machines will not really begin to flow in large numbers for a year or fifteen months. Meanwhile we have very old-fashioned and obsolete tackle.

There is a second question about these new machines. When they begin to flow out of the factories in large numbers fifteen months hence, will they be equipped with all necessary appliances? Take, for instance, the machine-guns. If we are aiming at having a couple of thousand of the latest machines, i e., 1,500 and 500 reserve, in eighteen months from now, what arrangements have been made for their machine-guns? Some of these modern fighting machines have no fewer than eight machine-guns in their wings. Taking only an average of four, with proper reserves, that would require 10,000 machine-guns. Is it not a fact that the large-scale manufacture of the Browning and Bren machine-guns was only decided upon a few months ago?

Let us now try the aeroplane fleet we have built and are building by the test of bombing-power as measured by weight and range. Here I must again make comparison with Germany. Germany has the power at any time henceforward to send a fleet of aeroplanes capable of discharging in a single voyage at least 500 tons of bombs upon London. We know from our war statistics that one ton of explosive bombs killed ten people and wounded thirty, and did £50,000 worth of damage. Of course, it would be absurd to assume that the whole bombing fleet of Germany would make an endless succession of voyages to and from this country. All kinds of other considerations intervene. Still, as a practical measure of the relative power of the bombing fleets of the two countries, the weight of discharge per voyage is a very reasonable measure. Now, if we take the German potential discharge upon London at a minimum of 500 tons per voyage of their entire bombing fleet, what is our potential reply? *They* can do this from now on. What can we do? First of all, how could we retaliate upon Berlin? We have not at the present time a single squadron of machines which could carry an appreciable load of bombs to Berlin. What shall we have this time next year? I submit for your consideration that this time next year, when it may well be that the potential discharge of the German fleet is in the neighbourhood of a thousand tons, we shall not be able to discharge in retaliation more than sixty tons upon Berlin.

But leave Berlin out of the question. Nothing is more striking about our new fleet of bombers than their short range. The great bulk of our new heavy and medium bombers cannot do much more than reach the coasts of Germany from this Island. Only the nearest German cities would be within their reach. In fact, the retaliation of which we should be capable this time next year from this Island would be puerile judged by the weight of explosive dropped, and would be limited only to the fringes of Germany.

Of course, a better tale can be told if it is assumed that we can operate from French and Belgian jumping-off grounds. Then very large and vital industrial districts of Germany would be within reach of our machines. Our Air Force will be incomparably more effective if used in conjunction with those of France and Belgium than it would be in a duel with Germany alone.

I now pass to the next stage: our defence, passive and active, ground and air, at home. Evidently we might have to endure an ordeal in our great cities and vital feeding-ports such as no community has ever been subjected to before. What arrangements have been made in this field? Take London and its seven or eight million inhabitants. Nearly two years ago I explained in the House of Commons the danger of an attack by thermite bombs. These small bombs, little bigger than an orange, had even then been manufactured by millions in Germany. A single medium aeroplane can scatter five hundred. One must expect in a small raid literally tens of thousands of these bombs, which burn through from storey to storey. Supposing only a hundred fires were started and there were only ninety fire brigades, what happens? Obviously the attack would be on a far more formidable scale than that. One must expect that a proportion of heavy bombs would be dropped at the same time, and that water, light, gas, telephone systems, etc., would be seriously deranged. What happens then? Nothing like it has ever been seen in world history. There might be a vast exodus of the population, which would present to the Government problems of public order, of sanitation and food-supply, which would dominate their attention, and probably involve the use of all their disciplined forces.

What happens if the attack is directed upon the feeding-ports, particularly the Thames, Southampton, Bristol, and the Mersey, none of which are out of range? What arrangements have been made to bring in the food through a far greater number of subsidiary channels? What arrangements have been made to protect our defence centres? By defence centres I mean the centres upon which our power to continue resistance depends. The problem of the civil population and their miseries is one thing; the means by which we could carry on the war is another. Have we organised and created an alternative centre of government if London is thrown into confusion? No doubt there has been discussion of this on paper, but has anything been done to provide one or two alternative centres of command, with adequate deep-laid telephone connections and wireless, from which the necessary orders can be given by some coherent thinking-mechanism? . . .

APPENDIX E

COMPARATIVE OUTPUT OF FIRST-LINE AIRCRAFT*

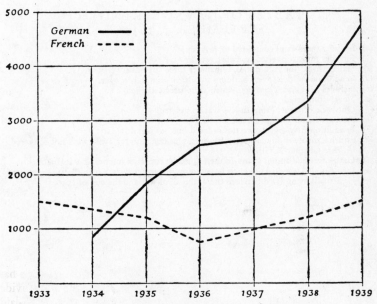

Note: German figures derived from captured documents;
French figures from a French source.

* See p. 213.

APPENDIX F

TABLES OF NAVAL STRENGTH
SEPTEMBER 3, 1939*

(*a*) Includes three ships converted to A.A. ships.
(*b*) Includes ships converted to escort vessels.
(*c*) Sixteen fitted for A.S. duties, remainder fitted for minesweeping.
(*d*) In addition six destroyers building for Brazil were taken over.
(*e*) Includes *Lion* and *Temeraire*, which were later cancelled.
(*f*) Never completed.
(*g*) Only one of these, *Prinz Eugen*, was completed.
(*h*) Includes training-cruiser *Emden*.
(*j*) In addition fifty-eight corvettes ordered but not laid down.
(*k*) British estimate at this date was fifty-nine, plus one built for Turkey but not delivered. (See Chapter XXIII.)
(*l*) Under war conditions many of these must be expected to complete in 1940.
(*m*) Includes all U-boats known to be building or projected on 3.9.39. Fifty-eight were actually completed between the outbreak of war and the end of 1940.

* See p. 367

BRITISH AND GERMAN FLEETS

	BRITISH INCLUDING DOMINIONS			GERMAN		

Type	Built	Building		Built	Building	
		Completing before 31.12.40	Completing after 31.12.40		Completing before 31.12.40	Completing after 31.12.40
Battleships ..	12	3	4(e)	—	2	2(f)
Battle-cruisers ..	3	—	—	2	—	—
"Pocket" Battleships	—	—	—	3	—	—
Aircraft-carriers ..	7	3	3	—	1(f)	1(f)
Seaplane-carriers ..	2	—	—	—	—	—
Cruisers:						
8-inch	15	—	—	2	2(g)	1(f)
6-inch or below	49(a)	13	6	6(h)	—	3(f)
Destroyers ..	184(b)	15(d)	17	22	3	13(l)
Sloops	38	4	—	—	—	—
Escort Destroyers	—	20	—	—	—	—
Corvettes (including patrol vessels) ..	8	3(j)	—	8	—	—
Torpedo-boats ..	—	—	—	30	4	6(l)
Minesweepers ..	42	—	—	32	10	—
Submarines ..	58	12	12	57(k)	40(m)	—
Monitors (15-inch)	2	—	—	—	—	—
Minelayers ..	7	2	2	—	—	—
River Gunboats ..	20	—	—	—	—	—
Trawlers	72(c)	20	—	—	—	—
Motor Torpedo-boats (including Motor Gunboats, etc.)	27	12	—	17	—	—

UNITED STATES

Strength of Fleet September 3, 1939
(excluding Coastguard Vessels)

Type	Completed	Under Construction and Projected	Estimated Date of Completion
Battleships	15	8	1 in 1941 1 in 1942 4 in 1943 2 later
Aircraft-carriers	5	2	1 in 1940 1 later
Aircraft Tenders	13	6	2 in 1941 4 later
Cruisers, 8-inch	18	—	—
,, 6-inch	18	7(a)	1 in 1939-40 6 in 1943
Destroyers	181(b)	42	11 in 1939 16 in 1940 15 in 1941
Destroyer Tenders	8	4	2 in 1940 2 later
Submarines	99(c)	15	4 in 1940 11 in 1941-42
Gunboats (including Patrol Vessels)	7	—	—
River Gunboats	6	—	—
Minelayers	10	1	1940
Minesweepers	26	3	1940
Submarine Tenders	6	2	1941
Submarine Chasers	14	16	4 in 1940 12 later
Motor Torpedo-boats	1	19	1939-40

NOTES

(a) Includes four ships mounting 5-inch guns.
(b) Includes 126 over age.
(c) Includes 65 over age.

FRANCE

September 3, 1939

Type	Completed	Building	Projected Date of Completion
Battleships	8 (incl. 1 Training Ship)	3	1 in 1940 1 in 1941 1 in 1943
Battle-cruisers	2	—	—
Aircraft-carriers	1	1	1 in 1942
Aviation Transport	1	—	—
Cruisers..	18	3	—
Light Cruisers .. .. .. (Contre-torpilleurs) ..	32	—	—
Destroyers .. .. .. (Torpilleurs)	28	24	6 in 1940
Motor Torpedo-boats ..	3	6	6 in 1940
Torpedo-boats	12	—	—
Cruiser Submarine	1	—	—
Submarines, 1st Class ..	38	3	—
,, 2nd Class ..	33	10	2 in 1940
Minelaying S.M.s ..	6	1	—
River Gunboats (incl. 2 ex-S.M. Chasers)	10	—	—
Net and Mine Layer	1	—	—
Minelayers	3	—	—
Minesweepers	26	7	—
Colonial Sloops	8	—	—
Submarine Chasers	13	8	5 in 1940

ITALY

September 3, 1939

Type	Completed	Building	Projected Date of Completion
Battleships	4	4	2 in 1940 2 in 1942
Cruisers, 8-inch	7	—	—
„ 6-inch	12	—	—
Old Cruisers	3	—	—
Cruisers, 5.3-inch	—	12	1942–43
Destroyers	59	8	1941–42
Torpedo-boats	69	4	1941–42
Submarines	105	14	10 in 1940 4 in 1941–42
Motor Torpedo-boats ..	69	—	—
Minelayers	16	—	—
Sloop	1	—	—
Seaplane Tender	1	—	—

JAPAN

September 3, 1939

Type	Completed	Building in 1939	Projected Date of Completion	Strength on entering War Dec. 7, 1941
Battleships ..	10	2	1 in 1941 1 in 1942	10
Aircraft-carriers ..	6	? 10	1 in 1940 4 in 1941 5 in 1942	11
Cruisers	18 8-inch 17 5.5-inch 3 old types	3 or 4	3 in 1940 1 in 1942	18 8-inch 20 5.5-inch 3 old types
Seaplane Tenders ..	2	2	2 in 1942	2
Minelayers	5	2	1 in 1939 1 in 1940	8
Destroyers ..	113	20	2 in 1939 10 in 1940 8 in 1941	129
Submarines ..	53	33	3 in 1940 11 in 1941 19 in 1942	67
Escort Vessels ..	4	—	—	4
Gunboats	10	3	2 in 1940 1 in 1941	13
Torpedo-boats ..	12	—	—	—

APPENDIX G

MINUTE OF SEPTEMBER 12, 1939

*PLAN "CATHERINE"**

PART I

(1) For a particular operation special tools must be constructed. D.N.C. thinks it would be possible to hoist an "R" [a battleship of the *Royal Sovereign* class] 9 ft., thus enabling a certain channel where the depth is only 26 ft. to be passed. There are at present no guns commanding this channel, and the States on either side are neutral. Therefore there would be no harm in hoisting the armour-belt temporarily up to the water level. The method proposed would be to fasten caissons [bulges] in two layers on the sides of the "R", giving the ship the enormous beam of 140 ft. No insuperable difficulty exists in fixing these, the inner set in dock and the outer in harbour. By filling or emptying these caissons the draught of the vessel can be altered at convenience, and, once past the shallow channel, the ship can be deepened again so as to bring the armour-belt comfortably below the waterline. The speed when fully hoisted might perhaps be 16 knots, and when allowed to fall back to normal draught 13 or 14. These speeds could be accepted for the operation. They are much better than I expected.

It is to be noted that the caissons afford admirable additional protection against torpedoes; they are in fact super-blisters.

It would also be necessary to strengthen the armour deck so as to give exceptional protection against air-bombing, which must be expected.

(2) The caissons will be spoken of as "galoshes", and the strengthening of the deck as the "umbrella".

(3) When the ice in the theatre concerned melts about March (?) the time for the operation would arrive. If orders are given for the necessary work by October 1, the designs being made meanwhile, we have six months, but seven would be accepted. It would be a great pity to waste the summer; therefore the highest priority would be required. Estimates of time and money should be provided on this basis.

(4) In principle two "Rs" should be so prepared, but of course three would be better. Their only possible antagonists during the summer of 1940 would be the *Scharnhorst* and *Gneisenau*. It may be taken for certain that neither of these ships, the sole resource of Germany, would expose themselves to the 15-inch batteries of the "Rs", who would shatter them.

* See p. 415.

(5) Besides the "Rs" thus prepared, a dozen mine-bumpers should be prepared. Kindly let me have designs. These vessels should be of sufficiently deep draught to cover the "Rs" when they follow, and be worked by a small engine-room party from the stern. They would have a heavy fore-end to take the shock of any exploding mine. One would directly precede each of the "Rs". Perhaps this requirement may be reduced, as the ships will go line ahead. I can form no picture of these mine-bumpers, but one must expect two or three rows of mines to be encountered, each of which might knock out one. It may be that ordinary merchant ships could be used for the purpose, being strengthened accordingly.

(6) Besides the above, it will be necessary to carry a three months, reasonable supply of oil for the whole expeditionary fleet. For this purpose turtle-back blistered tankers must be provided capable of going at least 12 knots. 12 knots may be considered provisionally as the speed of the passage, but better if possible.

Part II

(1) The objective is the command of the particular theatre [the Baltic], which will be secured by the placing [in it] of a battle squadron which the enemy heavy ships dare not engage. Around this battle squadron the light forces will act. It is suggested that three 10,000-ton 8-inch-gun cruisers and two 6-inch should form the cruiser squadron, together with two flotillas of the strongest combat destroyers, a detachment of submarines, and a considerable contingent of ancillary craft, including, if possible, depot ships and a fleet repair vessel.

(2) On the approved date the "Catherine" fleet would traverse the passage by night or day, as judged expedient, using, if desired, smoke-screens. The destroyers would sweep ahead of the fleet, the mine-bumpers would precede the "Rs", and the cruisers and lighter vessels would follow in their wake. All existing apparatus of paravanes and other precautions can be added. It ought therefore to be possible to overcome the mining danger, and there are no guns to bar the channel. A heavy attack from the air must be countered by the combined batteries of the fleet.

Note. An aircraft-carrier could be sent in at the same time, and kept supplied with reliefs of aircraft reaching it by flight.

Part III

It is not necessary to enlarge on the strategic advantages of securing the command of this theatre. It is the supreme naval offensive open to the Royal Navy. The isolation of Germany from Scandinavia would intercept the supplies of iron ore and food and all other trade.

The arrival of this fleet in the theatre and the establishment of command would probably determine the action of the Scandinavian States. They could be brought in on our side; in which case a convenient base could be found capable of being supplied overland. The difficulty is that until we get there they do not dare; but the three months' oil supply should give the necessary margin, and if the worst comes to the worst it is not seen why the fleet should not return as it came. The presence of this fleet in the theatre would hold all enemy forces on the spot. They would not dare to send them on the trade routes, except as a measure of despair. They would have to arm the whole northern shore against bombardment, or possibly even, if the alliance of the Scandinavian Powers was obtained, military descents. The influence of this movement upon Russia would be far-reaching, but we cannot count on this.

Secrecy is essential, as surprise must play its full part. For this purpose the term "Catherine" will always be used in speaking of the operation. The caissons will be explained as "additional blisters". The strengthening of the turtle-decks is normal A.A. precaution.

I commend these ideas to your study, hoping that the intention will be to solve the difficulties.

W. S. C.

APPENDIX H

NEW CONSTRUCTION AND RECONSTRUCTION*

First Lord to First Sea Lord and others   *October* 8, 1939

1. It is far more important to have some ships to fight with, and to have ships that Parliament has paid for delivered to date, than to squander effort upon remote construction which has no relation to our dangers!

2. A supreme effort must be made to finish *King George V* and *Prince of Wales* by their contract dates. The peace-time habit of contractors in booking orders and executing them when they please cannot be allowed to continue in time of war. Advise me of the penalties that may be enforced, in order that a case may be stated, if necessary, to the Law Officers of the Crown. Advise me also of the limiting factors. I suppose, as usual, the gun-mountings. It must be considered a marked failure by all concerned if these ships are not finished by their contract dates. I will myself inquire on Friday next into the condition of each of these ships, and will see the contractors personally at the Admiralty in your presence. Pray arrange these meetings from 5 p.m. onwards. It is no use the contractors saying it

* See p. 418.

cannot be done. I have seen it done when full pressure is applied and every resource and contrivance utilised. In short, we must have *K.G.V.* by July 1940, and *P. of W.* three months later. The ships we need to win the war with must be in commission in 1940.

Pray throw yourselves into this and give me your aid to smooth away the obstacles.

3. The above remarks apply also to the aircraft-carriers. *Illustrious* is to be five months late, and we know what that means. *Victorious* is even to be nine months late. *Formidable*, from the 1937 programme, is six months late, and *Indomitable* five months late. All these ships will be wanted to take part in the war, and not merely to sail the seas—perhaps under the German flag (!)—after it is over. Let me appeal to you to make this go. The later construction of aircraft-carriers will not save us if we are beaten in 1940.

4. Thirdly, there are the cruisers. Look, for example, at the *Dido*, which was contracted to be finished in June 1939 and is now offered to us in August 1940. What is the explanation of this fiasco?

5. We have at this moment to distinguish carefully between running an industry or a profession and winning the war. The skilled labour employed upon vessels which cannot complete during 1940 should, so far as is necessary or practicable, be shifted on to those that can complete in 1940. Special arrangements must be made as required to transfer the workmen from the later ships to those that are needed for the fighting. All ships finishing in 1941 fall into the shade, and those of 1942 into the darkness. We must keep the superiority in 1940.

6. The same principles apply even more strongly to destroyers and light craft; but these seem to be going on pretty well, and I have not yet had time to look in detail into their finishing dates. But we most urgently require two new battleships, four aircraft-carriers, and a dozen cruisers *commissioned and at work* before the end of 1940.

* * * * *

First Lord to First Sea Lord   *October* 21, 1939

I address this to you alone, because together we can do what is needful.

We must have a certain number of capital ships that are not afraid of a chance air-bomb. We have been able to protect them by bulges and Asdics against the U-boats. We must have them made secure against the air. It is quite true that it may well be a hundred to one against a hit with a heavy air-torpedo upon a ship, but the chance is always there, and the disproportion is grievous. Like a hero being stung by a malarious mosquito! We must work up to the old idea of a ship fit to lie the line against whatever may be coming.

To come to the point. I want four or five ships made into tortoises that we can put where we like and go to sleep content. There may be other types which will play their parts in the outer oceans; but we cannot go on without a squadron of heavy ships that can stand up to the battery from the air.

I wrote you this morning about the *Queen Elizabeth*. But we must make at least five other ships air-proof—*i.e.*, not afraid of a thousand-pound armour-piercing bomb, if by chance it should hit from ten thousand feet. This is not so large a structural rearrangement as might appear. You have got to pull a couple of turrets out of them, saving at least two thousand tons, and this two thousand tons has to be laid out in flat armour of six or seven inches, as high as possible, having regard to stability. The blank spaces of the turrets must be filled with A.A. guns. This means going down from eight guns to four. But surely four 15-inch can wipe out *Scharnhorst* or *Gneisenau*. Before the new German battleship arrives we must have *King George V* and *Prince of Wales*. Let us therefore concentrate on having five or six vessels which are not afraid of the air, and therefore can work in narrow waters, and keep the high-class stuff for the outer oceans. Pull the guns out and plaster the decks with steel. This is the war proposition of 1940.

How are you going to get these ships into dockyards' hands with all your other troubles?

Do not let us worry about the look of the ship. Pull the superimposed turrets out of them. Do one at Plymouth, one at Portsmouth, two on the Clyde, and one on the Tyne. These four-gun ships could be worked up to a very fine battery if the gunnery experts threw themselves into it. But, above all they must bristle with A.A., and they must swim or float wherever they choose. Here is the war *motif* of 1940, and we now have the time.

How all this reinforces our need for armoured ammunition ships, and armoured oilers, is easily seen. In all this we have not got to think so much of a sea action as of sea-power maintained in the teeth of air attack.

All this ought to be put in motion Monday, and enough information should be provided to enable us to take far-reaching decisions not later than Thursday. On that day let us have Controller, D.N.C., and D.N.O. and shift our fighting front from the side of the ship to the top. . . .

It looks to me as if the war would lag through the winter, with token fighting in all spheres, but that it will begin with mortal intensity in the spring.

Remember no one can gainsay what we together decide.

W. S. C.

APPENDIX I

NEW CONSTRUCTION PROGRAMMES, 1939–1940*

(EXCLUSIVE OF LIGHT COASTAL CRAFT)

I Type	II New Construction approved before Outbreak of War	III 1939 War Programmes	IV 1940 War Programmes	V Revised (War) Estimated Completion Dates (a) By end of 1940	V (b) By end of 1941 in addition to (a)	VI Actual Completions (a) By end of 1940	VI (b) By end of 1941 in addition to (a)
Battleships	9 (a)	—	1 (e)	2	2	1	2
Aircraft-carriers	6	—	—	3	2	2	2
Cruisers, 8-inch	—	—	—	—	—	—	—
" 6-inch and below	23 (b)	6	—	13	7	7	6
Fleet Destroyers	32	16	32	12	28	11 + 6 (g)	14
Escort Destroyers	20	36	30	26	34	25	25
Sloops	4	2	20	4	—	2	4
Corvettes (including Escort Vessels)	61 (c)	60	52 (f)	88	48	51	70
Submarines	12	19	49	22	23	19	19
Minelayers	4	—	—	2	—	—	4
Minesweepers	20 (d)	22	22	10	31	5	20
Trawlers (Anti-submarine)	20	32	100	42	50	30	53

NOTES: (a) Includes H.M. ships *Lion*, *Temeraire*, *Conqueror*, and *Thunderer*, which were subsequently cancelled.

(b) Includes four ships of 1939 programme not laid down on 3.9.39, two of which were subsequently cancelled.

(c) Fifty-eight corvettes ordered but not laid down on 3.9.39.

(d) Ordered but not laid down on 3.9.39.

(e) H.M.S. *Vanguard*.

(f) Twenty-seven of these were later named frigates.

(g) Six destroyers building for Brazil and taken over.

* See p. 419.

APPENDIX J

FLEET BASES*

First Lord to D.C.N.S. (to initiate action as last paragraph) and others   1.XI.39

It was arranged at a conference between the First Lord, the First Sea Lord, and, the C.-in-C. on *Nelson*, October 31, 1939, that the following arrangements should be made at Fleet bases:

1. Scapa cannot be available, except as a momentary refuelling base for the Fleet, before the spring. Work is however to proceed with all possible speed upon—

(*a*) Blockships in the exposed channels.

(*b*) Doubling the nets and placing them specially wherever required. They are to be at least as numerous and extensive as in the last war, plus the fact that the modern net is better. The routine of the gates is to be studied afresh with a view to briefer openings and greater security.

(*c*) The trawler and drifter fleet on the scale used in the Great War is to be earmarked for Scapa, and its disposition carefully considered by Plans Division. However, all these trawlers and drifters will be available for the Forth until it is time to use Scapa as a main base, *i.e.*, not before the end of February 1940.

(*d*) The work on the hutments is to proceed without intermission.

(*e*) Gun-platforms are to be made in concrete for the whole of the eighty guns contemplated for the defence of Scapa. The work on these is to proceed throughout the winter; but the guns will not be moved there or mounted until the spring, when everything must be ready for them.

(*f*) The aerodromes at Wick are to be increased to take four squadrons.

(*g*) The R.D.F. work is to be gone on with, but must take its turn with more urgent work.

Meanwhile Scapa can be used as a destroyer refuelling base, and the camouflaging of the oil tanks and the creation of dummy oil tanks should proceed as arranged. Staff at Scapa is not to be diminished, but there is no need to add to the oil storage there beyond the 120,000 tons already provided. The men now making the underground storage can be used for other work of a more urgent nature, even within the recent Board decision.

* See p. 443.

2. Loch Ewe. Port A is to be maintained in its present position, with its existing staff. A permanent boom and net is to be provided even before the Scapa nets are completed. The freshwater pipe is to be finished and any minor measures taken to render this base convenient as a concealed resting-place for the Fleet from time to time.

3. Rosyth is to be the main operational base of the Fleet, and everything is to be done to bring it to the highest possible efficiency. Any improvements in the nets should be made with first priority. The balloons must be supplied so as to give effective cover against low-flying attack to the anchorage below the bridge. The twenty-four 3.7 guns and the four Bofors which were lately moved to the Clyde are to travel back, battery by battery, in the next four days to the Forth, beginning after the Fleet has left the Clyde. It is not desired that this move should appear to be hurried, and the batteries may move as convenient and in a leisurely manner, provided that all are in their stations at the Forth within five days from the date of this minute. Strenuous effort with the highest priority for the R.D.F. installations which cover Rosyth must be forthcoming. Air Vice-Marshal Dowding is to-day conferring with the C.-in-C. Home Fleet upon the support which can be forthcoming from A.D.G.B. The arrangement previously reached with the Air Ministry must be regarded as the minimum, and it is hoped that at least six squadrons will be able to come into action on the first occasion the Fleet uses this base.

D.C.N.S. will kindly find out the upshot of the conference between the C.-in-C. and Vice-Marshal Dowding and report the results. We must certainly look forward to the Fleet being attacked as soon as it reaches the Forth, and all must be ready for that. Thereafter this base will continue to be worked up in every way until it is a place where the strong ships of the Fleet can rest in security. Special arrangements must be made to co-ordinate the fire of the ships with that of the shore batteries, observing that a 72-gun concentration should be possible over the anchorage.

4. The sixteen balloons now disposed at the Clyde should not be removed, as they will tend to mislead the enemy upon our intentions.

I should be glad if D.C.N.S. will vet this minute and make sure it is correct and solid in every detail, and, after obtaining the assent of the 1st S.L., make it operative in all departments.

W. S. C.

First Lord to First Sea Lord   3.1.40

SCAPA DEFENCES

1. When in September we undertook to man the Scapa batteries, etc., the numbers of Marines required were estimated at 3,000. This has now grown successively by War Office estimates to 6,000, to 7,000, to 10,000, or even 11,000. Of course, such figures are entirely beyond the capacity of the Royal Marines to supply.

2. Moreover, the training of the Royal Marines "hostilities only" men can only begin after March 1, when the necessary facilities can be given by the Army. Nothing had, in fact, been done since September except to gather together the nucleus of officers and N.C.O.s with about 800 men. These can readily be used by us either for the Marine striking force or the mobile defence force.

The War Office, on the other hand, have a surplus of trained men in their pools, and seem prepared to man the guns at Scapa as they are mounted at the rate of sixteen a month. As we want to use the base from March onwards, it is certain that this is the best way in which the need can be met.

3. If by any chance the War Office do not wish to resume the responsibility, then we must demand from them the training facilities from February 1 and their full assistance with all the technical ratings we cannot supply; and also make arrangements for the gradual handing over of the staff. It is clear, however, that the right thing is for them to do it, and we must press them hard.

4. I do not wish the Admiralty to make too great a demand upon the Army. It would seem that the numbers required could be substantially reduced if certain tolerances were allowed. The figure of thirty men per gun and fourteen per searchlight is intended to enable every gun and searchlight to be continuously manned at full strength, night and day, all the year round. But the Fleet will often be at sea, when a lower scale of readiness could be accepted. Moreover, one would not expect the guns to be continuously in action for very prolonged attack. If these attacks were made the Fleet would surely put to sea. It is a question whether the highest readiness might not be confined to a proportion of the guns, the others having a somewhat longer notice.

5. Is it really necessary to have 108 anti-aircraft lights? Is it likely that an enemy making an attack upon the Fleet at this great distance would do it by night? All their attacks up to the present have been by day, and it is only by day that precise targets can be hit.

6. When the Fleet is ready to use Scapa we must shift a large proportion, preferably half, of the guns and complements from Rosyth. We cannot claim to keep both going at the same time on the highest scale. Here is another economy.

7. It is suggested, therefore, by me that 5,000 men should be allotted to the Scapa defences, and that the Commander should be told to work up gradually the finest show of gun-power he can develop by

carefully studying local refinements which deal with each particular battery and post.

8. For a place like Scapa, with all this strong personnel on the spot, parachute landings or raids from U-boats may be considered most unlikely. There is therefore no need to have a battalion in addition to the artillery regiments. The Commander should make arrangements to have a sufficient emergency party ready to deal with any such small and improbable contingencies.

9. The case is different with the Shetlands, where we should be all the better for a battalion, though this need not be equipped on the Western Front scale.

W. S. C.

APPENDIX K

NAVAL AID TO TURKEY*

NOTE BY THE FIRST LORD OF THE ADMIRALTY, NOVEMBER 1, 1939

The First Sea Lord and I received General Orbay this afternoon, and informed him as follows:

In the event of Turkey being menaced by Russia His Majesty's Government would be disposed, upon Turkish invitation and in certain circumstances, to come to the aid of Turkey with naval forces superior to those of Russia in the Black Sea. For this purpose it was necessary that the anti-submarine and anti-aircraft defences of the Gulf of Smyrna and the Gulf of Ismid should be developed, British technical officers being lent if necessary. These precautions would be additional to the existing plans for placing anti-submarine nets in the Dardanelles and in the Bosphorus.

We were not now making a promise or entering into any military engagement; and it was probable that the contingency would not arise. We hoped that Russia would maintain a strict neutrality, or even possibly become friendly. However, if Turkey felt herself in danger, and asked for British naval assistance, we would then discuss the situation with her in the light of the Mediterranean situation and of the attitude of Italy, with the desire to enter into a formal engagement. It might be that the arrival of the British Fleet at Smyrna would in itself prevent Russia from proceeding to extremities, and that the advance of the British Fleet to the Gulf of Ismid would prevent a military descent by Russia on the mouth of the Bosphorus. At any

* See p. 402.

rate, it would be from this position that the operations necessary to establish the command of the Black Sea would be undertaken.

General Orbay expressed himself extremely gratified at this statement. He said that he understood perfectly there was no engagement. He would report to his Government on his return, and the necessary preparatory arrangements at the bases would be undertaken.

I did not attempt to enter into the juridical aspect, as that would no doubt be thrashed out should we ever reach the stage where a formal Convention had to be drawn up. It was assumed that Turkey would ask for British aid only in circumstances when she felt herself in grave danger, or had actually become a belligerent.

APPENDIX L

THE BLACK-OUT*

NOTE BY THE FIRST LORD OF THE ADMIRALTY, NOVEMBER 20, 1939

1. I venture to suggest to my colleagues that when the present moon begins to wane the black-out system should be modified to a sensible degree. We know that it is not the present policy of the German Government to indulge in indiscriminate bombing in England or France, and it is certainly not their interest to bomb any but a military objective. The bombing of military objectives can best be achieved, and probably only be achieved, by daylight or in moonlight. Should they change this policy, or should a raid be signalled, we could extinguish our lights again. It should have been possible by this time to have made arrangements to extinguish the street lighting on a Yellow Warning. However, so far as night bombing for the mere purpose of killing civilians is concerned, it is easy to find London by directional bearing and the map whether the city is lighted or not. There is no need to have the "rosy glow" as a guide, and it would not be a guide if it were extinguished before the raiders leave the sea. But there is not much in it anyway.

2. There is, of course, no need to turn on the full peace-time street-lighting. There are many modified forms. The system in force in the streets of Paris is practical and effective. You can see six hundred yards. The streets are light enough to drive about with safety, and yet much dimmer than in time of peace.

3. The penalty we pay for the present methods is very heavy. First, the loss of life. Secondly, as the Secretary of State for Air has protested,

* See p. 438.

the impediment to munitions output; and also work at the ports, even on the west coast. Thirdly, the irritating and depressing effect on the people, which is a drag upon their war-making capacity, and, because thought unreasonable, an injury to the prestige of His Majesty's Government. Fourthly, the anxieties of women and young girls in the darkened streets at night or in blacked-out trains. Fifthly, the effect on shopping and entertainments.

I would therefore propose that as from December 1:

(a) Street-lighting of a dimmed and modified character shall be resumed in the cities, towns, and villages.

(b) Motor-cars and railway trains shall be allowed substantially more light, even at some risk.

(c) The existing restrictions on blacking out houses, to which the public have adapted themselves, shall continue, but that vexatious prosecutions for minor infractions shall not be instituted. (I see in the newspapers that a man was prosecuted for smoking a cigarette too brightly at one place, and that a woman who turned on the light to tend her baby in a fit was fined in another.)

(d) The grant of these concessions should be accompanied by an effective propaganda, continuously delivered by the broadcast, and handed out to motorists at all refuelling stations, that on an air-raid warning all motorists should immediately stop their cars and extinguish their lights, and that all other lights should be extinguished. Severe examples should be made of persons who, after a warning has been sounded, show any light.

4. Under these conditions we might face the chances of the next three winter months, in which there is so much mist and fog. We can always revert to the existing practice if the war flares up, or if we do anything to provoke reprisals.

APPENDIX M

A NOTE ON THE MEASURES AGAINST THE MAGNETIC MINE*

Although the general characteristics of magnetic firing-devices for mines and torpedoes were well understood before the outbreak of war, the details of the particular mine developed by the Germans could not then be known. It was only after the recovery of a specimen at Shoeburyness on November 23, 1939, that we could apply the knowledge derived from past research to the immediate development of suitable counter-measures.

The first need was for new methods of minesweeping, the second was to provide passive means of defence for all ships against mines in unswept or imperfectly-swept channels. Both these problems were effectively solved, and the technical measures adopted in the earlier stages of the war are briefly described in the following paragraphs.

ACTIVE DEFENCE—MINESWEEPING METHODS

The Magnetic Mine

To sweep a magnetic mine it is necessary to create a magnetic field in its vicinity of sufficient intensity to actuate the firing mechanism and so detonate it at a safe distance from the minesweeper. A design for a mine-destructor ship had been prepared early in 1939, and such a ship was now brought into service experimentally, fitted with powerful electro-magnets capable of detonating a mine ahead of her as she advanced. She had some success early in 1940, but the method was not found suitable or sufficiently reliable for large-scale development.

At the same time various forms of electric sweep were developed for towing by shallow-draft vessels. Electro-magnetic coils carried in low-flying aircraft were also used, but this method presented many practical difficulties and involved considerable risk to the aircraft. Of all the methods tried that which came to be known as the L.L. sweep showed the most promise, and efforts were soon concentrated on perfecting this. The sweeping gear consisted of long lengths of heavy electric cable known as tails, towed by a small vessel, two or more of which operated together. By means of a powerful electric current passed through these tails at carefully adjusted time-intervals mines could be detonated at a safe distance astern of the sweepers. One of the difficulties which faced the designers of this equipment was that

* See p. 455.

of giving the cables buoyancy. The problem was solved by the cable industry, in the first instance by the use of a "sorbo" rubber sheath, but later the methods employed for sealing a tennis-ball were successfully adapted.

By the spring of 1940 the L.L. sweepers were coming into effective operation in increasing numbers. Thereafter the problem resolved itself into a battle of wits between the mine-designer and the minesweeping expert. Frequent changes were made by the Germans in the characteristics of the mine, each of which was in turn countered by readjustment of the mechanism of the sweep. Although the enemy had his successes and for a time might hold the initiative, the countermeasures invariably overcame his efforts in the end, and frequently it was possible to forecast his possible developments and prepare the counter in advance. Up to the end of the war the L.L. sweep continued to hold its own as the most effective answer to the purely magnetic mine.

The Acoustic Mine

In the autumn of 1940 the enemy began to use a new form of mine. This was the "acoustic" type, in which the firing mechanism was actuated by the sound of a ship's propellers travelling through the water. We had expected this development earlier, and were already well prepared for it. The solution lay in providing the minesweeper with means of emitting a sound of appropriate character and sufficient intensity to detonate the mine at a safe distance. Of the devices tried the most successful was the Kango vibrating hammer, fitted in a water-tight container under the keel of the ship. Effective results depended on finding the correct frequency of vibration, and, as before, this could only be achieved quickly by obtaining a specimen of the enemy mine. Once again we were fortunate; the first acoustic mine was detected in October 1940, and in November two were recovered intact from the mud flats in the Bristol Channel. Thereafter successful counter-measures followed swiftly.

Soon it transpired that both acoustic and magnetic firing-devices were being used by the enemy in the same mine, which would therefore respond to either impulse. In addition, many anti-sweeping devices appeared, designed to keep the firing mechanism inactive during the first or any predetermined number of impulses, or for a given period of time after the mine was laid. Thus a channel which had been thoroughly swept by our minesweepers, perhaps several times, might still contain mines which only "ripened" into dangerous activity later. Despite all these fruits of German ingenuity and a severe setback in January 1941, when the experimental station on the Solent was bombed and many valuable records destroyed, the ceaseless battle of wits continued to develop slowly in our favour. The eventual victory was a tribute to the tireless efforts of all concerned.

Passive Defence—Degaussing

It is common knowledge that all ships built of steel contain permanent and induced magnetism. The resulting magnetic field may be strong enough to actuate the firing mechanism of a specially designed mine laid on the sea-bed, but protection might be afforded by reducing the strength of this field. Although complete protection in shallow water could never be achieved it was evident that a considerable degree of immunity was attainable. Before the end of November 1939 preliminary trials at Portsmouth had shown that a ship's magnetism could be reduced by winding coils of cable horizontally round the hull and passing current through them from the ship's own electrical supply. The Admiralty at once accepted this principle; any ship with electric power could thus be given some measure of protection, and, whilst pressing on with further investigation to determine the more precise requirements, no time was lost in making large-scale preparations for equipping the Fleet with this form of defence. The aim was to secure immunity for any ship in depths of water over 10 fathoms, whilst minesweeping craft and other small vessels should be safe in much shallower depths. More extensive trials carried out in December showed that this "coiling" process would enable a ship to move with comparative safety in half the depth of water which would be needed without such protection. Moreover, no important interference with the ship's structure and no elaborate mechanism were involved, although many ships would require additional electric power plant. As an emergency measure temporary coils could be fitted externally on a ship's hull in a few days, but more permanent equipment, fitted internally, would have to be installed at the first favourable moment. Thus in the first instance there need be little delay in the normal turn-round of shipping. The process was given the name of "degaussing", and an organisation was set up, under Vice-Admiral Lane-Poole, to supervise the fitting of all ships with this equipment.

The supply and administrative problems involved were immense. Investigation showed that whereas the needs of degaussing would absorb 1,500 miles of suitable cable every week, the industrial capacity of the country could only supply about one-third of that amount in the first instance. Although our output could be stepped up, this could only be done at the expense of other important demands, and the full requirements could only be met by large imports of material from abroad. Furthermore, trained staffs must be provided at all our ports to control the work of fitting, determine the detailed requirements for each individual ship, and give technical advice to the many local authorities concerned with shipping movements. All this refers to the protection of the great mass of ships comprising the British and Allied merchant fleets.

By the first week of 1940 this organisation was gathering momentum. At this stage the chief preoccupation was to keep ships moving to and from our ports, particularly the East Coast ports, where the principal danger lay. All efforts were therefore concentrated on providing temporary coils, and the whole national output of suitable electric cable was requisitioned. Cable-makers worked night and day to meet the demand. Many a ship left port at this time with her hull encased in festoons of cable which could not be expected to survive the battering of the open sea, but at least she could traverse the dangerous coastal waters in safety, and could be refitted before again entering the mined area.

Wiping

Besides the method described above, another and simpler method of degaussing was developed which came to be known as "wiping". This process could be completed in a few hours by placing a large cable alongside the ship's hull and passing through it a powerful electric current from a shore supply. No permanent cables need be fitted to the ship, but the process had to be repeated at intervals of a few months. This method was not effective for large ships, but its application to the great multitude of small coasters which constantly worked in the danger zone gave much-needed relief to the organisation dealing with "coiling" and yielded immense savings in time, material, and labour. It was of particular value during the evacuation of Dunkirk, when so many small craft of many kinds not normally employed in the open sea were working in the shallow waters round the Channel coasts.

DEGAUSSING OF MERCHANT SHIPS.

MEMORANDUM BY THE FIRST LORD OF THE ADMIRALTY, MARCH 15, 1940

ADMIRALTY, *March* 15, 1940

My colleagues will be aware that one of our most helpful devices for countering the magnetic mine is the demagnetisation or degaussing of ships. This affords immunity in waters of over 10 fathoms.

The number of British ships trading to ports in the United Kingdom which require to be degaussed is about 4,300.

The work of degaussing began in the middle of January, and by the 9th of March 321 warships and 312 merchant vessels were completed. 219 warships and approximately 290 merchant vessels were in hand on the same date.

The supply of cable, which has up to the present governed the rate of equipment, is rapidly improving, and it is now the supply of labour in the shipyards which is likely to control the future rate of progress.

It would be a substantial advantage if part of the work of degaussing of British ships could be placed in foreign yards. The number of neutral ships engaged in trade with this country is about 700. Neutral crews, and in particular the crews of Norwegian ships, are beginning to be uneasy about the dangers from enemy mines on the trade ways to our ports. The importance to us of the safety of these neutral ships and of the confidence of their crews is a strong argument for disclosing to neutral countries the technical information which they require to demagnetise their ships which trade with this country.

Against the substantial advantages of arranging for some British ships to be demagnetised in foreign yards and of extending demagnetisation to neutral ships must be set any disadvantages of a loss of secrecy. If the enemy is informed of the measures which we are taking, he may (a) increase the sensitivity of his mines, or he may (b) mix mines of opposite polarity in the same field. If secrecy could be preserved its advantage would be to delay these reactions of the enemy. But technical details of our degaussing equipment have had to be given to all ship-repairing firms in this country. Information which has been so widely distributed almost certainly becomes quickly known to the enemy.

Moreover, (a) and (b) have the disadvantages to the enemy that—

(a) would make the mines easier to sweep and reduce the damage to non-degaussed ships by placing the explosion further forward or even ahead of the ships; and

(b) reversal of polarity would only be effective against certain ships which are difficult to demagnetise thoroughly, and would also require a sensitive setting of the mine.

The above position has altered since the arrival of the *Queen Elizabeth* at New York and the subsequent publicity given to the subject in the Press. The enemy now knows the nature of the protective measures we are taking, and, knowing the mechanism of his own mine, it will not be difficult for him to deduce the manner in which degaussing operates. He can therefore now adopt any counter-measures within his power. The Press notices have had the further effect of increasing demands for information from neutrals, and to continue to refuse such information conflicts with our general policy of encouraging neutral ships to trade in this country.

It is considered therefore by my advisers that we shall not be losing an advantage of any great importance by ceasing to treat the information as secret.

The Admiralty recommend therefore—

(i) that shipyards in neutral countries be used, if necessary, to supplement resources in this country for the degaussing of British merchant ships;

(ii) that technical information of our methods of demagnetisation be supplied as and when necessary to neutral countries for the degaussing of neutral ships trading with this country.

W. S. C.

APPENDIX N

EXTRACT FROM THE WAR DIARY OF U.47, NOVEMBER 28, 1939*

28.11.39
German Time.

1245. Pos^n. 60° 25′ N. 01° E.	Masts in sight bearing 120° (true).
1249. Wind NNW. 10–9. Sea 8. Cloudy.	I recognise a cruiser of the "London" class.
1334. 60° 24′ N. 01° 17′ E.	Range 8 hm. [approx. 880 yds.]. Estimated speed of cruiser 8 knots. 1 torpedo fired from No. 3 tube. After 1 min. 26 secs. an explosion heard. I can see the damage caused by the hit, aft of the funnel. The upper deck is buckled and torn. The starboard torpedo-tube mounting is twisted backwards over the ship side. The aircraft is resting on the tail unit. The cruiser appears to have a 5° list to starboard, as she disappears on a reciprocal course into a rain squall.
1403.	Surfaced. Set off in pursuit.
1420.	Cruiser again in sight bearing 090°. I dive to close her, but she disappears in another rain squall.
1451.	Surfaced and searched the area, but she could not be found.

* See p. 447.

On 29.XI.39 the following entry was made in the war diary of Admiral Doenitz: "Following the report that U.47 had torpedoed a cruiser, Propaganda claimed a sinking. From the service-man's point of view such inaccuracies and exaggerations are undesirable."

APPENDIX O

CULTIVATOR No. 6★

During these months of suspense and analysis I gave much thought and compelled much effort to the development of an idea which I thought might be helpful to the great battle when it began. For secrecy's sake this was called "White Rabbit No. 6", later changed to "Cultivator No. 6". It was a method of imparting to our armies a means of advance up to and through the hostile lines without undue or prohibitive casualties. I believed that a machine could be made which would cut a groove in the earth sufficiently deep and broad through which assaulting infantry and presently assaulting tanks could advance in comparative safety across No-man's-land and wire entanglements and come to grips with the enemy in his defences on equal terms and in superior strength. It was necessary that the machine cutting this trench should advance at sufficient speed to cross the distance between the two front lines during the hours of darkness. I hoped for a speed of three or four m.p.h., but even half a mile would be enough. If this method could be applied upon a front of perhaps twenty or twenty-five miles, for which two or three hundred trench-cutters might suffice, dawn would find an overwhelming force of determined infantry established on and in the German defences, with hundreds of lines-of-communication trenches stretching back behind them, along which reinforcements and supplies could flow. Thus we should establish ourselves in the enemy's front line by surprise and with little loss. This process could be repeated indefinitely.

When I had had the first tank made twenty-five years before, I turned to Tennyson d'Eyncourt, Director of Naval Construction, to solve the problem. Accordingly I broached the subject in November to Sir Stanley Goodall, who now held this most important office, and one of his ablest assistants, Mr. Hopkins, was put in charge, with a grant of £100,000 for experiments. The design and manufacture of a working model was completed in six weeks by Messrs. Ruston-Bucyrus, of Lincoln. This suggestive little machine, about three feet long, performed excellently in the Admiralty basement on a floor of sand.

★ See p. 497.

Having obtained the active support of the Chief of the Imperial General Staff, General Ironside, and other British military experts, I invited the Prime Minister and several of his colleagues to a demonstration. Later I took it over to France and exhibited it both to General Gamelin and later on to General Georges, who expressed approving interest. On December 6 I was assured that immediate orders and absolute priority would produce two hundred of these machines by March 1941. At the same time it was suggested that a bigger machine might dig a trench wide enough for tanks.

On February 7, 1940, Cabinet and Treasury approval were given for the construction of two hundred narrow "infantry" and forty wide "officer" machines. The design was so novel that trial units of the main components had first to be built. In April a hitch occurred. We had hitherto relied on a single Merlin-Marine type of engine, but now the Air Ministry wanted all these, and another heavier and larger engine had to be accepted instead. The machine in its final form weighed over a hundred tons, was seventy-seven feet long and eight feet high. This mammoth mole could cut in loam a trench five feet deep and seven and a half feet wide at half a mile an hour, involving the movement of eight thousand tons of soil. In March 1940 the whole process of manufacture was transferred to a special department of the Ministry of Supply. The utmost secrecy was maintained by the three hundred and fifty firms involved in making the separate parts, or in assembling them at selected centres. Geological analysis was made of the soil of Northern France and Belgium, and several suitable areas were found where the machine could be used as part of a great offensive battle plan.

But all this labour, requiring at every stage so many people to be convinced or persuaded, led to nothing. A very different form of warfare was soon to descend upon us like an avalanche, sweeping all before it. As will presently be seen, I lost no time in casting aside these elaborate plans and releasing the resources they involved. A few specimens alone were finished and preserved for some special tactical problem or for cutting emergency anti-tank obstacles. By May 1943 we had only the pilot model, four narrow and five wide machines made or making. After seeing the full-sized pilot model perform with astonishing efficiency, I minuted, "Cancel and wind up the four of the five 'officer' type, but keep the four 'infantry' type in good order. Their turn may come." These survivors were kept in store until the summer of 1945, when, the Siegfried line being pierced by other methods, all except one were dismantled. Such was the tale of "Cultivator No. 6". I am responsible but impenitent.

W. S. C.

APPENDIX P

BRITISH MERCHANT VESSELS LOST BY ENEMY ACTION DURING THE FIRST EIGHT MONTHS OF THE WAR*

(Numbers of ships shown in parentheses)

Enemy Agency	1939				1940				TOTAL Gross tons
	September	October	November	December	January	February	March	April	
U-boats	135,552 (26)	74,130 (14)	18,151 (5)	33,091 (6)	6,549 (2)	67,840 (9)	15,531 (3)	14,605 (3)	365,449 (68)
Mines	11,437 (2)	3,170 (2)	35,640 (13)	47,079 (12)	61,943 (11)	35,971 (9)	16,747 (8)	13,106 (6)	225,093 (63)
Surface Raider	5,051 (1)	27,412 (5)	706 (1)	21,964 (3)	—	—	—	5,207 (1)	60,340 (11)
Aircraft	—	—	—	487 (1)	23,296 (9)	—	5,439 (1)	—	29,222 (11)
Other and unknown causes	—	—	2,676 (3)	875 (1)	10,081 (2)	6,561 (3)	1,585 (1)	41,920 (9)†	63,698 (19)
TOTAL (gross tons) ..	152,040 (29)	104,712 (21)	57,173 (22)	103,496 (23)	101,869 (24)	110,372 (21)	39,302 (13)	74,838 (19)	743,802 (172)

* See p. 511.
† All these ships were sunk or seized by Germany in Norwegian ports.

APPENDIX Q

OPERATION "ROYAL MARINE"*

NOTE BY THE FIRST LORD OF THE ADMIRALTY

March 4, 1940

1. It will be possible to begin the naval operation at any time at 24 hours' notice after March 12. At that time there will, as planned, be available 2,000 fluvial mines of the naval type, comprising three variants. Thereafter a regular minimum supply of 1,000 per week has been arranged. The detachment of British sailors is on the spot, and the material is ready. All local arrangements have been made with the French through General Gamelin and Admiral Darlan. These mines will, it is believed, affect the river for the first hundred miles below Karlsruhe. There is always risk in keeping men and peculiar material teed-up so close (4-6 miles) to the enemy's front, although within the Maginot Line. The river is reported to be in perfect order this month. It will probably be deepened by the melting of the snows in April, involving some lengthening of the mine-tails; also the flow from the tributaries may be temporarily stopped, or even reversed.

2. The Air Force will not be ready till the moon is again good in mid-April. Therefore, unless our hand is forced by events, it would seem better to wait till then, so as to infest the whole river simultaneously, and thus also confuse the points of naval departure. By mid-April the Air Force should have a good supply of mines, which could be laid every night during the moon in the reaches between Bingen and Coblentz. All mines of both classes will become harmless before reaching the Dutch frontier. Before the end of April it is hoped that a supply of the special mines for the still-water canals may be ready, and by the May moon the mines for the mouths of the rivers flowing into the Heligoland Bight should be at hand.

3. Thus this whole considerable mining campaign could be brought into being on the following time-table:

Day 1.—Issue of proclamation reciting the character of the German attacks on the British coasts, shipping, and river-mouths, and declaring that henceforth (while this continues) the Rhine is a mined and forbidden area, and giving neutrals and civilians twenty-four hours' notice to desist from using it or crossing it.

Day 2.—After nightfall deposit as many mines as possible by both methods, and keep this up night after night. The supply by that time should be such as to keep all methods of discharge fully employed.

* See p. 517.

Day 28.—Begin the laying of the mines in the still-water canals and river-mouths, thereafter keeping the whole process working, as opportunity serves, until the kind of attacks to which we are being subjected are brought to an end by the enemy, or other results obtained.

4. The decisions in principle required are:

(a) Is this method of warfare justified and expedient in present circumstances?

(b) Must warning be given beforehand, observing that the first shock of surprise will be lost? However, this is not considered decisive, as the object is to prevent the use of the river and inland waterways rather than mere destruction.

(c) Should we wait till the Air Force are ready, or begin the naval action as soon as possible after March 12?

(d) What reprisals, if any, may be expected, observing that there is no natural or economic feature in France or Great Britain in any way comparable with the Rhine, except our coastal approaches, which are already beset.

5. It is desirable that the Fifth Sea Lord, who has the operation in charge, should go to Paris on Thursday, concert the details finally, and ascertain the reactions of the French Government. From the attitude of M. Daladier, General Gamelin, and Admiral Darlan it is thought these will be highly favourable.

APPENDIX R

NAVAL LOSSES IN
THE NORWEGIAN CAMPAIGN*

GERMAN NAVAL LOSSES, APRIL–JUNE 1940

SHIPS SUNK

Name	Type	Cause
Bluecher	8-inch Cruiser	Torpedo and gunfire by Norwegian coast defences, Oslo, April 9
Karlsruhe	Light Cruiser	Torpedoed by submarine *Truant* in Kattegat, April 9
Koenigsberg	Light Cruiser	Bombed by Fleet Air Arm, Bergen, April 10
Brunmer	Gunnery Training Ship	Torpedoed in Kattegat by submarine, April 15
Wilhelm Heidkamp	Destroyer	Torpedoed. First attack on Narvik, April 10
Anton Schmit	"	
Hans Ludemann	"	
Georg Thiele	"	
Bernd von Arnim	"	Destroyed by torpedo or gunfire. Second attack on Narvik, April 13 (five of these were damaged in the first attack on April 10)
Wolf Zenker	"	
Erich Geise	"	
Erich Koellner	"	
Hermann Kunne	"	
Dieter von Roeder	"	
Numbers 44, 64, 49, 1, 50, 54, 22, 13	U-boats	Various. 3 off Norway, 5 in North Sea
Albatross	Torpedo-boat	Wrecked, Oslo, April 9

In addition, three minesweepers, two patrol craft, eleven transports, and four fleet auxiliaries were sunk.

* See p. 592

SHIPS DAMAGED

Name	Type	Cause
Gneisenau	Battle-cruiser ..	Action with *Renown*, April 9. Torpedoed by submarine *Clyde*, June 20
Scharnhorst ..	Battle-cruiser	Torpedoed by *Acasta*, June 8
Hipper	8-inch Cruiser	Action with *Glowworm*, April 8
Luetzow	Pocket-battleship ..	Action with coastal batteries, Oslo, April 9. Torpedoed by submarine *Spearfish*, Kattegat, April 11
Emden	Light Cruiser	Action with coastal batteries, Oslo, April 9
Bremse	Gunnery Training Ship	Action with coastal batteries, Bergen, April 9

In addition, two transports were damaged and one captured.

SHIPS OUT OF ACTION DURING THE WHOLE PERIOD

Name	Type	Cause
Admiral Scheer ..	Pocket-battleship ..	Engine repairs
Leipzig	Light Cruiser	Torpedo damage repairs

GERMAN FLEET ON JUNE 30, 1940

Type	Effective	Remarks
Battle-cruisers ..	Nil	*Scharnhorst* and *Gneisenau* damaged
Pocket-battleships ..	Nil	*Admiral Scheer* under repair. *Luetzow* damaged
8-inch Cruiser ..	*Hipper*	
Light Cruisers ..	*Koeln*, *Nuernberg* ..	*Leipzig* and *Emden* damaged
Destroyers ..	*Schoemann*, *Lody*, *Ihn*, *Galster*	Six others under repair
Torpedo-boats ..	Nineteen	Six others under repair. Eight new craft working up

In addition, the two old battleships *Schlesien* and *Schleswig-Holstein* were available for coast defence.

ALLIED NAVAL LOSSES IN THE NORWEGIAN CAMPAIGN

SHIPS SUNK

Name	Type	Cause
Glorious	Aircraft-carrier	Gunfire, June 9
Effingham	Cruiser	Wrecked, May 17
Curlew	A.A. Cruiser	Bombed, May 26
Bittern	Sloop	Bombed, April 30
Glowworm	Destroyer	Gunfire, April 8
Gurkha	,,	Bombed, April 9
Hardy	,,	Gunfire, April 10
Hunter	,,	Gunfire, April 10
Afridi	,,	Bombed, May 3
Acasta	,,	Gunfire, June 9
Ardent	,,	Gunfire, June 9
Bison (French)	,,	Bombed, May 3
Grom (Polish)	,,	Bombed, May 4
Thistle	Submarine	U-boat, April 14
Tarpon	,,	Unknown, April 22
Sterlet	,,	Unknown, April 27
Seal	,,	Mined, May 5
Doris (French)	,,	U-boat, May 14
Orzel (Polish)	,,	Unknown, June 6

In addition, eleven trawlers, one loaded and two empty troop transports, and two supply ships were sunk.

SHIPS DAMAGED

(EXCLUDING MINOR DAMAGE)

Name	Type	Cause
Penelope	Cruiser	Grounding, April 11
Suffolk	,,	Bombed, April 17
Aurora	,,	Bombed, May 7
Curaçoa	A.A. Cruiser	Bombed, April 24
Cairo	,,	Bombed, May 28
Émile Bertin (French)	Cruiser	Bombed, April 19
Pelican	Sloop	Bombed, April 22
Black Swan	,,	Bombed, April 28
Hotspur	Destroyer	Gunfire, April 10
Eclipse	,,	Bombed, April 11
Punjabi	,,	Gunfire, April 13
Cossack	,,	Gunfire, April 13
Eskimo	,,	Torpedo, April 13
Highlander	,,	Grounding, April 13
Maori	,,	Bombed, May 2
Somali	,,	Bombed, May 15

II. FIRST LORD'S MINUTES

Short titles are frequently used in these Memoranda and Minutes when addressing members of the Board of Admiralty or heads of departments. For the convenience of the reader the corresponding full titles are tabulated below.

Short Title	Full Title
Controller	Controller and Third Sea Lord
Controller M.S.R.	Controller of Merchant Ship-building Repairs
D.C.N.S.	Deputy (later Vice) Chief of Naval Staff
A.C.N.S.	Assistant Chief of Naval Staff
D.N.I.	Director of Naval Intelligence
D.N.O.	Director of Naval Ordnance
D.T.D.	Director of Trade Division
D.N.C.	Director of Naval Construction
D.T.M.	Director of Torpedoes and Mining
D.S.R.	Director of Scientific Research

First Lord to Secretary and to all Departments   4.IX.39

To avoid confusion, German submarines are always to be described officially as U-boats in all official papers and communiqués.

First Lord to D.N.I. and Secretary   6.IX.39

1. This is an excellent paper, and the principles are approved. However, in the first place (say, September), when losses may be high, it is important that you show that we are killing U-boats. The policy of silence will come down later. The daily bulletin prepared by Captain Macnamara should, when possible, for the first week be shown to the First Lord, but should not be delayed if he is not available. It is of the highest importance that the Admiralty bulletin should maintain its reputation for truthfulness, and the tone should not be forced. The bulletin of to-day is exactly the right tone.

2. When Parliament is sitting, if there is anything worth telling, bad or good, the First Lord or Parliamentary Secretary will be disposed to make a statement to the House in answer to friendly private-notice questions.

These statements should be concerted with the Parliamentary Secretary, who advises the First Lord on Parliamentary business. Sensational or important episodes will require special attention of the First Lord or First Sea Lord.

3. Lord Stanhope, as Leader of the House of Lords, should always be made acquainted with the substance of any statement to be made in the House of Commons upon the course of the naval war.

Moreover, the First Lord wishes that his Private Secretary should keep Lord Stanhope informed during these early weeks upon matters in which his Lordship may have been interested. He should not be cut off from the course of events at the Admiralty, with which he has been so intimately concerned.

First Lord to D.N.I.   6.IX.39

(Secret.)

What is the position on the West Coast of Ireland? Are there any signs of succouring U-boats in Irish creeks or inlets? It would seem that money should be spent to secure a trustworthy body of Irish agents to keep most vigilant watch. Has this been done? Please report.

First Lord to D.C.N.S.   6.IX.39

Kindly give me report on progress of Dover barrage, and repeat weekly.

First Lord to Controller   6.IX.39

1. What are we doing about bringing out old merchant ships to replace tonnage losses? How many are there, and where? Kindly supply lists, with tonnage. Arrangements would have to be made to dock and clean all bottoms, otherwise speed will be grievously cut down.

2. I should be glad to receive proposals for acquiring neutral tonnage to the utmost extent.

First Lord to First Sea Lord, Controller, and others   6.IX.39

1. It is much too soon to approve additional construction of new cruisers, which cannot be finished for at least two years, even under war conditions. The matter can be considered during the next three months. Now that we are free from all Treaty restrictions, if any cruisers are built they should be of a new type, and capable of dominating the five German 8-inch cruisers now under construction.

2. Ask the D.N.C. at his convenience to give me a legend of a 14,000- or 15,000-ton cruiser carrying 9.2 guns with good armour against 8-inch projectiles, wide radius of action, and superior speed to any existing *Deutschland* or German 8-inch-gun cruisers. It would be necessary before building such vessels to carry the United States with us.

3. The rest of the programme is approved, as it all bears on U-boat hunting and ought to be ready within the year.

Pray let me have approximate estimates of delivery.

4. I shall be very glad to discuss the general questions of policy involved with the Board.

First Lord to Prime Minister   7.IX.39

It seems most necessary to drill the civil population in completely putting out their private lights, and the course hitherto followed has conduced to this. But surely the great installations of lights controlled from two or three centres are in a different category.

While enforcing the household black-outs, why not let the controllable lighting burn until an air-warning is received? Then when the hooters sound the whole of these widespread systems of lighting would go out at once together. This would reinforce the air-raid warning, and when the all-clear was sounded they would all go up together, telling everyone. Immense inconvenience would be removed, and the depressing effect of needless darkness; and as there are at least ten minutes to spare, there would be plenty of time to make the black-out complete.

Unless you have any objection, I should like to circulate this to our colleagues.

DATES OF COMPLETION FOR NAVAL CONSTRUCTION:
TABULAR STATEMENT PREPARED BY CONTROLLER

First Lord to Controller   9.IX.39

In peace-time vessels are built to keep up the strength of the Navy from year to year amid political difficulties. In war-time a definite tactical object must inspire all construction. If we take the Navies, actual and potential, of Germany and Italy, we can see clearly the exact vessels we have to cope with. Let me therefore have the comparable flotilla of each of these Powers, actual and prospective, up to 1941, so far as they are known. *Having regard to the U-boat menace, which must be expected to renew itself on a much larger scale towards the end of* 1940, the type of destroyer to be constructed must aim at numbers and celerity of construction rather than size and power. It ought to be possible to design destroyers which can be completed in under a year, in which case 50 at least should be begun forthwith. I am well aware of the need of a proportion of flotilla leaders and large destroyers capable of ocean service, but the arrival in our Fleet of 50 destroyers of the medium emergency type I am contemplating would liberate all larger vessels for ocean work and for combat.

Let me have the entire picture of our existing destroyer fleet, apart from the additions shown on this paper. Until I have acquainted myself with the destroyer power I will not try to understand the escort vessels, etc.

First Lord to Controller, D.N.C., and others   11.IX.39

The following ideas might be considered before our meeting at 9.30 Tuesday, September 12:

1. Suspend for a year all work on battleships that cannot come into action before the end of 1941. This decision to be reviewed every six months. Concentrate upon *King George V, Prince of Wales,* and *Duke of York,* and also upon *Jellicoe* if it can be pulled forward into 1941: otherwise suspend.

2. All aircraft-carriers should proceed according to accelerated programme.

3. Concentrate on the *Didos* which can be delivered before the end of 1941. By strong administrative action it should be possible to bring all the present programme within the sacred limit, to wit, ten ships. No new *Didos* till this problem has been solved.

4. *Fijis.* Please No! This policy of scattering over the seas weak cruisers which can neither fight nor flee the German 8-inch 10,000-ton cruisers—of which they will quite soon have five—should be abandoned. The idea of two *Fijis* fighting an 8-inch-gun cruiser will never

come off.* All experience shows that a cluster of weak ships will not fight one strong one. (*Vide* the escape of the *Goeben* across the mouth of the Adriatic, August 1914.)

5. I was distressed to see that till the end of 1940, *i.e.*, sixteen months, we only receive ten destroyers, and only seven this year, and that there is a gulf of nine months before the subsequent six are delivered. However, we have taken over the six Brazilians, which arrive during 1940 and mitigate this position. Let us go forward with all these to the utmost. These ships called "destroyers" have strayed far in design from their original *rôle* of "torpedo-boat destroyers", in answer to the French mosquito flotillas of the nineties. They are really small unarmoured cruisers with a far heavier stake in men and money than their capacity to stand the fire of their equals justifies. Nevertheless, for combat and for breasting ocean billows they have an indispensable part to play.

6. Fast escort vessels: I now learn these are really medium destroyers of 1,000 tons. The whole of this class should be pressed forward to the utmost.

7. We have also the whale-catcher type—but this is 940 tons, which is a great deal where numbers are required. I doubt whether our dollars will enable us to place forty of these in the United States. It would be much better to supplement them by a British-built programme of another type.

8. I would ask that a committee of, say, three sea-officers accustomed to flotilla work, plus two technicians, should sit at once to solve the following problems:

An anti-submarine and anti-air vessel which can be built within twelve months in many of the small yards of the country. One hundred should be built if the design is approved. The greatest simplicity of armament and equipment must be arrived at, and a constant eye kept upon mass production requirements. The *rôle* of these vessels is to liberate the destroyers and fast escort vessels for a wider range of action, and to take over the charge of the Narrow Seas, the Channel, the inshore Western Approaches, the Mediterranean, and the Red Sea, against submarine attack.

I hazard specifications only to have them vetted and corrected by the committee, viz.:

500 to 600 tons.
16 to 18 knots.

Two cannons around 4-inches, according as artillery may

* The *Fiji* class mounted 6-inch guns. None the less, the 6-inch cruisers *Ajax* and *Achilles* later fought a successful and glorious action with the *Graf Spee*, mounting 11-inch guns.

come to hand from any quarter, preferably of course firing high angle.

Depth-charges.

No torpedoes, and only moderate range of action.

These will be deemed the "Cheap and Nasties" (cheap to us, nasty to the U-boats). These ships, being built for a particular but urgent job, will no doubt be of little value to the Navy when that job is done—but let us get the job done.

9. The submarine programme is approved, as they still have a part to play.

I shall be very grateful if you will give me your views on these ideas, point by point, to-morrow night.

First Lord to First Sea Lord, Controller, and others   18.IX.39

As it is generally impossible to use the catapult aircraft in the open ocean, but nevertheless they would be a great convenience around the South American continental promontory, the question arises whether landing-grounds or smooth-water inlets cannot be marked down on uninhabited tracts or in the lee of islands, upon which aircraft catapulted from vessels in the neighbourhood could alight, claiming, if discovered, right of asylum. They could then be picked up by the cruiser at convenience. Perhaps this has already been done.

First Lord to First Sea Lord and others   20.IX.39

While I greatly desire the strengthening of this place against A.A. attack, and regard it as a matter of extreme urgency, I consider the scale of 80 3.7-inch guns goes beyond what is justified, having regard to other heavy needs. It is altogether out of proportion to lock up three regiments of A.A. artillery, etc. (comprising 6,200 men), for the whole war in Scapa. Scapa is no longer the base of the Grand Fleet, but only of three or four principal vessels. Alternative harbours can be used by these. The distance from Germany, 430 miles, is considerable. We must be very careful not to dissipate our strength unduly in passive defence.

I approve therefore of the additional 16 3.7-inch as a matter of the highest urgency. But I think they should be erected by the Admiralty to avoid the long delays and heavy charges of the War Office Ordnance Board.

The second 20 equipments should be considered in relation to the needs of Malta, as well as to the aircraft factories in England. This applies still more to the full scale of 3.7-inch guns, numbering 44. Their destination can only be considered in relation to the future war need.

The light A.A. guns seem to be excessive, having regard to the heavy pom-pom fire of the Fleet. The searchlights and balloons are most necessary, as are also the two Fighter Squadrons. Do we not require a more powerful R.D.F. station? And should there not be an additional R.D.F. station on the mainland?

In this case the urgency of getting something into position counts far more than making large-scale plans for 1940.

Let me have reduced proposals, with estimates of time and money, but without delaying action on the first instalments.

Also a report of the A.A. defences of Malta, and also of Chatham.

First Lord to First Sea Lord and others   21.IX.39

It was very pleasant to see the aircraft-carrier *Argus* in the basin at Portsmouth to-day. The boats of this vessel have been sent to the C.-in-C. Home Fleet, but no doubt they could easily be replaced, and various guns could be mounted. We are told that modern aircraft require a larger deck to fly on and off. In that case, would it not be well to build some aircraft suitable for the ship, as these can be made much quicker than a new aircraft-carrier? We ought to commission *Argus* as soon as possible, observing that the survivors of *Courageous* are available. Pray consider the steps that should be taken to this end. I am told she is a very strong ship underwater, but if not the bulkheads could be shored up or otherwise strengthened.★

First Lord to First Sea Lord and others   21.IX.39

D.C.N.S. and I were much impressed with the so-called Actæon net against torpedoes on which the *Vernon* are keen. This net was introduced at the end of the late war. It is a skirt or petticoat which is only effective when the vessel is in motion. The *Vernon* declare that a vessel can steam 18 knots with it on. The *Laconia* is to be tried out with one. The net is of thin wire and large mesh. It should be easy to make in large quantities very quickly. I suggest that this is a matter of the highest urgency and significance. It should be fitted on merchant ships, liners, and also—indeed, above all—upon ships of war having solitary missions without destroyer protection. Could not a committee be formed before the week is out which would grip this idea, already so far advanced by the naval authorities, and see whether it cannot be brought into the forefront of our immediate war preparations? If it is right it would require a very large-scale application.†

★ The *Argus* was commissioned and performed valuable service training pilots for the Fleet Air Arm in the Mediterranean.

† Many practical difficulties were encountered in the development of these nets. The early trials were unsuccessful, and it was not until 1942 that the equipment was perfected. Thereafter it was fitted in over 750 ships, with varying success. Ten ships are known to have been saved by this device.

First Lord to First Sea Lord and others   21.IX.39

The importance of using all available guns capable of firing at aircraft whether on ships in harbour or in the dockyard to resist an air attack should be impressed upon Commanders-in-Chief of home ports as well as upon officers at lesser stations. The concerting of the fire of these guns with the regular defences should be arranged. If necessary, the high-angle guns of ships in dry dock should be furnished with crews from the depots, and special arrangements made to supply the electrical power, even though the ship is under heavy repair. There must be many contrivances by which a greater volume of fire could be brought to bear upon attacking aircraft. We must consider the moonlight period ahead of us as one requiring exceptional vigilance. Please consider whether some general exhortations cannot be given.

First Lord to Admiral Somerville and Controller   23.IX.39

Let me have at your early convenience the programme of installation of R.D.F. in H.M. ships, showing what has been done up to date, and a forecast of future installations, with dates. Thereafter let me have a monthly return, showing progress. The first monthly return can be November 1.

First Lord to First Sea Lord and others   24.IX.39

A lot of our destroyers and small craft are bumping into one another under the present hard conditions of service. We must be very careful not to damp the ardour of officers in the flotillas by making heavy weather of occasional accidents. They should be encouraged to use their ships with war-time freedom, and should feel they will not be considered guilty of unprofessional conduct if they have done their best and something or other happens. I am sure this is already the spirit and your view, but am anxious it should be further inculcated by the Admiralty. There should be no general rule obliging a court-martial in every case of damage. The Board should use their power to dispense with this, so long as no negligence or crass stupidity is shown. Errors towards the enemy—*i.e.*, to fight—should be most leniently viewed, even if the consequences are not pleasant.

First Lord to First Sea Lord, D.C.N.S., and D.N.I.   24.IX.39
(*For general guidance.*)
(Most secret.)

1. Mr. Dulanty is thoroughly friendly to England. He was an officer under me in the Ministry of Munitions in 1917/18, but he has no control or authority in Southern Ireland (so-called Eire). He acts as a general smoother, representing everything Irish in the most

favourable light. Three-quarters of the people of Southern Ireland are with us, but the implacable, malignant minority can make so much trouble that de Valera dare not do anything to offend them. All this talk about partition and the bitterness that would be healed by a union of Northern and Southern Ireland will amount to nothing. They will not unite at the present time, and we cannot in any circumstances sell the loyalists of Northern Ireland. Will you kindly consider these observations as the basis upon which Admiralty dealings with Southern Ireland should proceed?

2. There seems to be a good deal of evidence, or at any rate suspicion, that the U-boats are being succoured from West of Ireland ports by the malignant section with whom de Valera dare not interfere. And we are debarred from using Berehaven, etc. If the U-boat campaign became more dangerous we should coerce Southern Ireland both about coast-watching and the use of Berehaven, etc. However, if it slackens off under our counter-attacks and protective measures the Cabinet will not be inclined to face the serious issues which forcible measures would entail. It looks therefore as if the present bad situation will continue for the present. But the Admiralty should never cease to formulate through every channel its complaints about it, and I will from time to time bring our grievances before the Cabinet. On no account must we appear to acquiesce in, still less be contented with, the odious treatment we are receiving.

First Lord to First Sea Lord and D.C.N.S.　　　　　　29.IX.39

While anxious not to fetter in any way the discretion of C.-in-C. Home Fleet, I think it might be as well for you to point out that the sending of heavy ships far out into the North Sea will certainly entail bombing attacks from aircraft, and will not draw German warships from their harbours. Although there were no hits on the last occasion, there might easily have been losses disproportionate to the tactical objects in view. This opinion was expressed to me by several Cabinet colleagues.

The first brush between the Fleet and the air has passed off very well, and useful data has been obtained, but we do not want to run unnecessary risks with our important vessels until their A.A. has been worked up to the required standard against aircraft flying 250 miles an hour.*

First Lord to Secretary　　　　　　30.IX.39

Surely the account you give of all these various disconnected

---

* This refers to an incident on September 26, when the Home Fleet was attacked by aircraft in the North Sea, without suffering damage. It was on this occasion that the *Ark Royal* was singled out for special attention. The Germans claimed she had been sunk, and the pilot who made the claim was decorated. For weeks afterwards the German wireless reiterated daily the question, "Where is the *Ark Royal?*"

Statistical Branches constitutes the case for a central body which should grip together all Admiralty statistics, and present them to me in a form increasingly simplified and graphic.

I want to know at the end of each week everything we have got, all the people we are employing, the progress of all vessels, works of construction, the progress of all munitions affecting us, the state of our merchant tonnage, together with losses, and numbers of every branch of the R.N. and R.M. The whole should be presented in a small book such as was kept for me by Sir Walter Layton when he was my statistical officer at the Ministry of Munitions in 1917 and 1918. Every week I had this book, which showed the past and the weekly progress, and also drew attention to what was lagging. In an hour or two I was able to cover the whole ground, as I knew exactly what to look for and when.

How do you propose this want of mine should be met?

OCTOBER 1939

First Lord to Secretary 9.X.39

The First Lord's Statistical Branch should consist of Professor Lindemann, who would do this besides his scientific activities. He would require a secretary who knows the Admiralty, a statistician, and a confidential typist who is also preferably an accountant. The duties of this branch will be:

1. To present to the First Lord a weekly picture of the progress of all new construction, showing delays from contract dates, though without inquiring into the cause, upon which First Lord will make his own inquiries.

2. To present returns of all British or British-controlled merchant ships, together with losses under various heads and new construction or acquisition—

   (a) during the week,
   (b) since the war began;

   also forecasts of new deliveries.

3. To record the consumption weekly and since war began of all ammunition, torpedoes, oil, etc., together with new deliveries, i.e., weekly and since the war began, monthly or weekly outputs and forecasts.

4. To keep a complete continuous statistical survey of Fleet Air Arm, going not only into aircraft, but pilots, guns, and equipment of all kinds, and point out all apparent lag.

5. To present a monthly survey of the losses of personnel of all kinds.
6. To keep records of inquiries, and any special papers relating to numbers and strength provided by First Lord.
7. To make special inquiries analysing for First Lord Cabinet papers and papers from other departments which have a statistical character, as requested by First Lord.

As soon as the personnel of the department is settled after discussion with Professor Lindemann, who should also advise on any additions to the above list of duties, a minute must be given to all departments to make the necessary returns to Statistical Branch (to be called "S") at the times required, and to afford any necessary assistance.

AIR SUPPLY

October 16, 1939

This most interesting paper is encouraging, but it does not touch the question on which the War Cabinet sought information—namely, the disparity between the monthly output of new aircraft and the number of squadrons comprising the first-line air strength of the R.A.F. We were told in 1937 that there would be 1,750 first-line aircraft modernly equipped by April 1, 1938 (see Sir Thomas Inskip's speeches). However, the House of Commons was content with the statement that this position had in fact been realised by April 1, 1939. We were throughout assured that reserves far above the German scale were the feature of the British system. We now have apparently only about 1,500 first-line aircraft with good reserves ready for action. On mobilisation the 125 squadrons of April 1, 1939, shrank to 96. It is necessary to know how many new squadrons will be fully formed during the months of November, December, January, and February. It is difficult to understand why, with a production of fighting machines which has averaged over 700 a month since May, and is now running even higher, only a handful of squadrons has been added to our first-line strength, and why that strength is below what we were assured was so reached in April of this year. One would have thought, with outputs so large and pilots so numerous, we should have been able to add ten or fifteen squadrons a month to our first-line air strength; and no explanation is furnished why this cannot happen. Ten squadrons of sixteen each, with 100 per cent. reserves, would only amount to 320 a month, or much less than half the output from the factories. The Cabinet ought to be told what are the limiting factors. They should be told this in full detail. Is it pilots or mechanics or higher ground staff or guns or instruments of any kind? We ought not, surely, to continue in ignorance of the reasons which prevent the heavy outputs of the factories

from being translated into a fighting front of first-line aircraft organised in squadrons. It may be impossible to remedy this, but at any rate
we ought to examine it without delay. It is not production that is
lagging behind, but the formation of fighting units with their full
reserve upon the approved scale.

D.S.R., Controller, and Secretary   16.X.39

1. I am very much obliged to the Director Scientific Research for
his interesting memorandum [on the Admiralty Research Depart-
ment], and I entirely agree with the principle that the first stage is the
formulation of a felt want by the fighting service. Once this is clearly
defined in terms of simple reality it is nearly always possible for the
scientific experts to find a solution. The Services should always be
encouraged to explain what it is that hurts or hinders them in any
particular branch of their work. For instance, a soldier advancing
across No-man's-land is hit by a bullet which prevents his locomotion
functioning further. It is no use telling him or his successor to be brave,
because that condition has already been satisfied. It is clear however
that if a steel plate or other obstacle had stood between the bullet and
the soldier the latter's powers of locomotion would not have been
deranged. The problem therefore becomes how to place a shield in
front of the soldier. It then emerges that the shield is too heavy for
him to carry; thus locomotion must be imparted to the shield; and
how? Hence the tanks. This is of course a simple example.

2. In your list of branches and departments very little seems to be
allowed for physical investigation, the bulk being concentrated upon
application and development. I am therefore very glad to know that
the Clarendon Laboratory will be utilised for this purpose, and I shall
be dealing with the paper on that subject later in the day.

First Lord to Controller and others   18.X.39

REQUISITIONING OF TRAWLERS

I have asked the Minister of Agriculture to bring Mr. Ernest Bevin
and his deputation to the Admiralty at 4.15 o'clock to-morrow after
they have explored the ground among themselves. Let all be notified,
and an official letter written to the Ministry of Agriculture inviting
them here. I will preside myself.

Meanwhile A.C.N.S., D.T.D., and Controller or Deputy-Controller
should, together with Financial Secretary, meet together this evening
to work out a plan, the object of which is the *Utmost Fish*, subject to
naval necessity. The immediate loss arising from our requisition should
be shared between ports, and the fact that a port has built the best kind
of trawlers must not lead to its being the worst sufferer. Side by side

with this equalisation process a type of trawler which can be built as quickly as possible and will serve its purpose should be given facilities in the shipyards. As soon as these trawlers flow in, they can either be added to the various ports or else be given to the ports from whom the chief requisition has been made, the equalising trawlers being restored after temporary use—this is for local opinion to decide. It is vital to keep the fish trade going, and we must fight for this part of our food supply as hard as we do against the U-boats.*

First Lord to First Sea Lord and D.C.N.S.   19.x.39
(Most secret.)

The Turkish situation has sharpened up. Suppose Turkey wanted us to put a fleet in the Black Sea sufficiently strong to prevent Russian military pressure upon the Bosphorus or other parts of the Turkish northern coast, and the Cabinet were satisfied that this might either keep Russia from going to war or, if she were at war, prevent her attacking Turkey, can the force be found?

What is the strength of the Russian Black Sea marine, and what would be sufficient to master them? Might this not be an area where British submarines with a few destroyers and a couple of protecting cruisers, all based on Turkish ports, would be able to give an immense measure of protection? Anyhow, the possibility should be studied in all its military bearings by the Naval Staff, and ways and means of finding and maintaining the force worked out.

Clearly, if Russia declares war upon us we must hold the Black Sea.

First Lord to First Sea Lord and Controller   23.x.39

Before going further into your paper on the Northern Barrage, I should like to know what amounts of explosives are involved, and how these could be provided without hampering the main fire of the armies. Perhaps the Controller could to-day discuss this point with Mr. Burgin or the head of his Chemical Department. I do not know what are the limiting factors in this field. I hear predictions that toluene may run short. I presume the output required for the barrage would be far outside the limits of the Admiralty cordite or explosive factories. I suggest that Controller has all this information collected informally, both from the Admiralty and the Ministry of Supply, and that we talk it over on our return.†

First Lord to First Sea Lord   23.x.39

I should be glad if you would arrange to discuss with the other Chiefs

---

* Throughout the war a special section of the Trade Division dealt with the needs of fishing vessels working round our coasts.
† See Chapter XXVIII.

of Staff this morning the question of raid or invasion, having regard to the position of the Fleet and the long dark nights. I frequently combated these ideas in the late war, but now the circumstances do not seem to be altogether the same. I have of course no knowledge of the military arrangements, but it seems to me there ought to be a certain number of mobile columns or organised forces that could be thrown rapidly against any descent. Of course, it may be that the air service will be able to assume full responsibility.

First Lord to First Sea Lord and D.C.N.S. 27.X.39

Pray consider this note which I wrote with the idea of circulating it to the Cabinet.

It is surely not our interest to oppose Russian claims for naval bases in the Baltic. These bases are only needed against Germany, and in the process of taking them a sharp antagonism of Russian and German interests becomes apparent. We should point out to the Finns that the preservation of their country from Russian invasion and conquest is the vital matter, and this will not be affected by Russian bases in the Gulf of Finland or the Gulf of Bothnia. Apart from Germany, Russian naval power in the Baltic could never be formidable to us. It is Germany alone that is the danger and the enemy there. There is indeed a common interest between Great Britain and Russia in forbidding as large a part of the Baltic as possible to Germany. It is quite natural that Russia should need to have bases which prevent German aggression in the Baltic Provinces or against Petrograd. If the above reasoning is right, we ought to let the Russians know what our outlook is, while trying to persuade the Finns to make concessions, and Russia to be content with strategic points.

First Lord to D.C.N.S. and Secretary 29.X.39

Arrange for a stand of arms to be placed in some convenient position in the basement, and let officers and able-bodied personnel employed in the Admiralty building have a rifle, a bayonet, and ammunition assigned to each. Fifty would be enough. Let this be done in forty-eight hours.

First Lord to General Smuts 29.X.39

(Personal and Private.)

Monitor Erebus is ready to sail for Capetown. As you know, we have never considered 15-inch guns necessary for defence of Capetown, but to please Pirow agreed to loan *Erebus* until those defences were modernised in view of his fear of attack by Japan. We realise the defences of Capetown remain weak, but the Germans have no battleships, and the only two battle-cruisers they possess, the *Scharnhorst* and *Gneisenau*, would be very unlikely to try to reach South African waters, or if they did so to risk damage far from a friendly dockyard from even weak defences. Should they break out a major naval operation would ensue, and we shall pursue them wherever they go with our most powerful vessels until they are hunted down. Therefore it seems to me you are unlikely to have the need of this ship. On the other hand, she would be most useful for various purposes in the shallows of the Belgian coast, especially if Holland were attacked. She was indeed built by Fisher and me for this very purpose in 1914. The question is therefore mainly political. Rather than do anything to embarrass you we would do without the ship. But if you can let us have her either by re-loan or re-transfer Admiralty will be most grateful, and would, of course, reimburse Union.*

All good wishes.

ADMIRALTY MINUTES

NOVEMBER 1939

First Lord to Secretary   4.XI.39

The French have a very complete installation in the country for all the business of their Admiralty, and have already moved there. Our policy is to stay in London until it becomes really impossible, but it follows from this that every effort must be made to bring our alternative installation up to a high level of efficiency.

Pray let me know how it stands, and whether we could in fact shift at a moment's notice without any break in control. Have the telephones, etc., been laid effectively? Are there underground wires as well as others? Do they connect with exchanges other than London, or are they dependent upon the main London exchange? If so, it is a great danger.

First Lord to First Sea Lord and others   9.XI.39

I am deeply concerned at the immense slowing down of trade, both in imports and exports, which has resulted from our struggle during the first ten weeks of the war. Unless it can be grappled with and the restriction diminished to, say, 20 per cent. of normal, very grave shortage will emerge. The complaints coming in from all the Civil Departments are serious. We shall have failed in our task if we merely substitute delays for sinkings. I frankly admit I had not appreciated this aspect, but in this war we must learn from day to day. We must

* General Smuts replied that of course he would do as we wished.

secretly loosen up the convoy system (while boasting about it publicly), especially on the outer routes. An intricate study must be made of the restrictions now imposed, and consequent lengthening of voyages, and a higher degree of risk must be accepted. This is possible now that so many of our ships are armed. They can go in smaller parties. Even across the Atlantic we may have to apply this principle to a certain degree. If we could only combine with it a large effective destroyer force, sweeping the Western Approaches as a matter of course instead of providing focal points on which convoys could be directed, we should have more freedom. This is no reversal or stultification of previous policy, which was absolutely necessary at the outset. It is a refinement and development of that policy so that its end shall not be defeated.

First Lord to D.C.N.S. 9.XI.39

It appears to me that St. Helena and Ascension must be made effectively secure against seizure by landing parties from, say, a *Deutschland*. We should look very foolish if we found them in possession of the two 6-inch guns with a supply ship in the harbour. I don't feel the garrisons there are strong enough.

First Lord to First Sea Lord 15.XI.39

Pray let me have details of the proposed first Canadian convoy. How many ships, which ships, how many men in each ship, what speed will convoy take, escort both A.S. and anti-raider? Place of assembly and date of departure should be mentioned verbally.

First Lord to Secretary and A.C.N.S. 16.XI.39

Have you made sure that the intake of air to Admiralty basement is secure? Are there alternative intakes in case of the present one being damaged by a bomb? What would happen in the case of fire in the courtyard?

There seem to be heaps of rubbish, timber, and other inflammable material lying about, not only in the courtyard, but in some of the rooms underneath them. All unnecessary inflammable material should be removed forthwith.

First Lord to First Sea Lord 20.XI.39

Nothing can be more important in the anti-submarine war than to try to obtain an independent flotilla which could work like a cavalry division on the approaches, without worrying about the traffic or U-boat sinkings, but could systematically search large areas over a wide front. In this way these areas would become untenable to U-boats, and many other advantages would flow from the manœuvre.*

* This policy did not become possible until a later phase in the war.

First Lord to First Sea Lord and others 22.XI.39

1. When a sudden emergency, like this magnetic mine stunt, arises it is natural that everyone who has any knowledge or authority in the matter should come together, and that a move should be got on in every direction. But do you not think we now want to bring into being a special section for the job, with the best man we can find at the head of it working directly under the Staff and the Board? Such a branch requires several subdivisions; for instance, one lot should be simply collecting and sifting all the evidence we have about these mines from their earliest effort on the West Coast, and interviewing survivors, etc., so that everything is collected and focused.

2. The second lot would deal with the experimental side, and the *Vernon* would be a part of this. I am told Admiral Lyster is doing something here; he has a plan of his own which he is working, but it is desirable that a general view should prevail.

3. The third section is concerned with action in the shape of production, and getting the stuff delivered for the different schemes; while the fourth, which is clearly operational, is already in existence.

It is not suggested that this organisation should be permanent, or that all those who take part in it should be working whole-time. It should be a feature in their daily duties, and all should be directed and concerted from the summit.

Pray consider this, and make out a paper scheme into which all would fit.

First Lord to First Sea Lord and others 23.XI.39

1. I approve the appointment of Admiral Wake-Walker to concert the magnetic mine business. But it is necessary that he should have precise functions and instructions. (1) He will assemble all the information available. (2) He will concert and press forward all the experiments, assigning their priority. (3) He will make proposals for the necessary production. (4) He will offer advice to the Naval Staff upon the operational aspect, which nevertheless will proceed independently from hour to hour under the Naval Staff and the C.-in-C. of the Nore. In all the foregoing he will of course act under the Board.

2. Let me see a chart of duties divided between these various branches, and make it clear that the officers of the various technical departments in the Admiralty shall be at Admiral Wake-Walker's service from time to time as may be needed. You will no doubt consult him in making this plan.

3. It is essential that Admiral Drax should be in on all this from the beginning, and also in touch with C.-in-C. Nore, so that he comes into full understanding and operation from December 1.*

* See Chapter XXVIII and Appendix M, dealing with the magnetic mine problem.

First Lord to First Sea Lord and others   27.XI.39

1. We must arrive at clear ideas about the Swedish iron ore for Germany. Doubt has been thrown on whether it is important to stop this or not. I am informed by M. of E.W. that on the contrary nothing would be more deadly, not only to German war-making capacity but to the life of the country, than to stop for three or even six months this import.

2. The suggestion has been made verbally to me by the Naval Staff that when Luleå freezes we should violate Norwegian neutrality by landing a force, or perhaps stationing a ship in territorial waters at Narvik. I am opposed to both these alternatives.

3. Pray examine and advise upon a proposal to establish a minefield, blocking Norwegian territorial waters at some lonely spot on the coast as far north as convenient. If the Norwegians will do this themselves, well and good. Otherwise a plan must be made for us to do it. Doubt has been thrown upon our ability to maintain the necessary watch on this minefield, or to intercept vessels laden with ore which go outside it. But this is surely ill-founded. The mere fact that we had laid the minefield and were known to be watching and blockading would deter the ore-ships, and the process would not be too onerous for the C.-in-C. Home Fleet. However, let me have your final view.

4. It must be remembered that, in addition to the ore-ships, much merchandise valuable to Germany is coming down the Norwegian Leads. A statement was shown me by the D.N.I. that five ore-ships had already, in November, gone from Narvik to Germany, and that empty ships are going up now to receive the ore. What do the M. of E.W. say to this? We must know what the facts are, and have agreement between the departments.

5. Meanwhile the Russians have notified us that their gigantic Arctic ice-breaker is almost immediately to come down the Norwegian territorial waters on her way nominally to Kronstadt. But at the same time we hear that the Russians are hiring this ice-breaker to Germany to break the ice up to Luleå. If this were done, and no other counter-measures taken, the whole flow of ore into Germany would proceed at its present rate of nearly a million tons a month, thus completely frustrating all our policies. How are we to deal with this? I will make you a suggestion verbally; but meanwhile the Foreign Office must be consulted on the whole position.

First Lord to Secretary   27.XI.39

I notice that in the Air Ministry every room is provided with candles and matches for use in emergency.

Pray take steps immediately to make similar provision in the Admiralty.

First Lord to D.C.N.S. and First Sea Lord   30.XI.39

I should be glad if you would consider whether it is not possible to add a third vessel to the Australasian escorts. Perhaps the Australians will offer another of their cruisers, but, if not, cannot we find another 6-inch-gun ship with a catapult? This would leave *Ramillies* freer to engage the enemy if an attack should be made by surface ships. It enables also scouting to be done far ahead and to the flanks of the convoy, thus giving ample warning. If such a cruiser could be found in China or in Indian waters, fitted with an Asdic and depth-charges, one would at least have some apparent answer to a U-boat. The transportation of the Australian divisions is an historic episode in Imperial history. An accident would be a disaster. Perhaps one of our detached submarines in the Indian Ocean could also help.

DECEMBER 1939

First Lord to Controller and others   3.XII.39
  (Secret.)

I was much interested in D.C.N.S.'s remark about the possibility of making a new battleship with the four spare 15-inch-gun turrets. Such a vessel would be of the battleship-cruiser type, heavily armoured and absolutely proof against air attack. Pray let me have a legend, with estimates in money and time. This ship could come in after the *K.G.V.* batch are finished and before *Temeraire* and *Lion.**

First Lord to Secretary, D.C.N.S., and First Sea Lord   12.XII.39

1. In view of the danger of surprise attacks at a time when the enemy may expect to find us off our guard, there must be no break or holiday period at Christmas or the New Year. The utmost vigilance must be practised at the Admiralty and in all naval ports. On the other hand, it should be possible between now and February 15 to give a week's leave to almost every officer concerned in staff duties. I am very glad to hear this is being planned at the Admiralty, and it will, I presume, be imitated as far as possible at the naval ports.

2. Every effort should be made to ease the strain upon the destroyer crews. At Devonport I am told admirable arrangements are made to relieve the flotilla complements as they come in from patrols, and that two or three days' rest in port brings them round in a wonderful manner. Similar arrangements are in force at Rosyth and Scapa, but

---

 * Plans for this ship went forward. She became H.M.S. *Vanguard.*

I am told that the amenities of Scapa are so much below those of the naval ports that the men are deeply disappointed when their brief spell of rest takes place there. No doubt in some cases this is inevitable, but I trust the whole question will be reviewed with the intention of comforting these crews to the utmost extent that operations will permit.

First Lord to D.C.N.S., Admiral Wake-Walker   24.XII.39
(to initiate action), and D.S.R.

I suppose you are already looking ahead to a possible change by the enemy from magnetic mines to acoustic or supersonic. Pray let me have a note at your convenience.

First Lord to Secretary, D.C.N.S., and First Sea Lord   28.XII.39

It should be explained to the Foreign Office that the six-mile limit in Italian waters was instituted by the Admiralty as a voluntary and self-denying ordinance at the outset of the war. It was never communicated to the Italians, nor made public to the world. It therefore forms no part of any bargain or agreement. It was simply a convenient guide for British naval authorities at a particular juncture. It has now become onerous, and possibly deeply injurious to the blockade, and in these circumstances the Admiralty would propose as a departmental matter to notify the C.-in-C. Mediterranean that the three-mile limit only need be observed. They will at the same time renew their injunctions to treat Italian shipping with special leniency, and to avoid causes of friction or complaint with that favoured country.

Let me see draft.

JANUARY 1940

First Lord to Secretary   4.I.40

Can anything be done to utilise the canal system to ease the transport of coal, north and south? Pray let me have a note on this at my return.

First Lord to First Sea Lord, Controller, D.T.M.,   12.I.40
Rear-Admiral A. H. Walker, and Professor Lindemann

OPERATION "ROYAL MARINE"

1. This matter was fully discussed in France with high military authorities, and various arrangements have been made. Captain Fitzgerald and Major Jefferis have seen the necessary people and should now furnish me with reports of their work. The French military men point out that they control the head-waters of the Saar and the Moselle,

in addition to the Rhine, and that many possibilities are open there. All are convinced that we should not act until a really large supply of the needful is in hand. Not only must the first go-off be on the largest scale at all points, but the daily and weekly supply thereafter must be such as to keep the tension at the highest pitch indefinitely.

2. It is of course understood that while all action is to be prepared the final decision rests with the Governments.

3. In all circumstances I am prepared to postpone the date from the February moon to the March moon. Meanwhile every exertion is to be made to perfect the plan and accumulate the greatest store.

4. A meeting of all concerned will be held in my room on Monday night at 9.30 p.m. By this time everyone should be able to report progress and everything should be concerted. I am asking the Secretary of State for Air to be present to hear the reports. These may be individually presented, but those concerned are to consult together in the interval. Above all, any obstacle or cause of undue delay is to be reported, so that the operations can be brought to full readiness as soon as possible. We may be forced to act before the March moon.*

First Lord to First Sea Lord, Controller, D.C.N.S.,   12.1.40
Secretary, and A.C.N.S.

The First Lord wishes to congratulate all those concerned in dealing with magnetic mines on the success which has so far been achieved.

First Lord to Admiral Usborne   13.1.40

"U.P." WEAPON

Your report dated 12.1.40. Everything seems to be going all right except the bombs, which are the only part of this process not under our control. I note that Messrs. Venner have fallen behind in respect of one component of these. But are you satisfied that the Air Ministry have done their part with the bombs?

Pray let me have a special report on the subject, and also let me know whether I should not write to the Secretary of State for Air, asking to have this part of the business handed over to us like the rest has been. These U.P. experiments are of immense importance. The whole security of H.M. warships and merchant ships may be enhanced by this development. I am counting on you to make sure that all is concerted and brought forward together, and that we shall go into mass production on a large scale at the earliest moment.

I am sorry that the experiments to-day with the ejection trials were

* See Chapters XXVIII and XXXII.

not completed, though I understand from Professor Lindemann that they were in principle satisfactory.

Pray press on with these with the utmost speed.

I think the time is coming when a report of progress should be furnished to the Air Ministry and the War Office, who have entrusted their interest in this matter to me. Perhaps therefore you would prepare a compendious statement, showing position to date and future prospects.*

First Lord to Controller   13.1.40

I am very glad to receive your paper on concrete ships. I am not at all satisfied that the idea has been sufficiently explored. Great progress has been made since the last war in ferro-concrete. Quite a different class of workman and materials would be called into being, and the strain on our ordinary shipbuilding plans proportionately relieved. In these circumstances, I think an effort should be made to make one seagoing ship at once. †

First Lord to Naval Secretary   14.1.40

Perhaps you will see Mr. Cripps (brother of Sir Stafford Cripps), who had a very good record in the last war and is a brave and able man. There must be many openings in some of our minesweepers.

[Enclosure: *Letter from Mr. Frederick Cripps asking "could he be used for minesweeping?"*]

First Lord to First Sea Lord   16.1.40

A.A. DEFENCES OF SCAPA

Surely it would be better to have a conference as I suggested and talk matters over round a table than that I should have to prepare a paper and raise the matter as a Cabinet issue? The squandering of our strength proceeds in every direction, everyone thinking he is serving the country by playing for safety locally. Our Army is puny as far

---

* This minute refers to the Unrotated Projectile (rocket propulsion), which was then being developed for use against low-flying aircraft. The device consisted of a battery of rockets which, on reaching a predetermined height, released long trailing wires, each carrying a small bomb at the end, and supported by a parachute. An aircraft fouling one of these wires would draw the bomb into its wing, where it would explode.

This device was a stop-gap necessitated by our grievous shortage of short-range weapons. Later on it was superseded by more effective weapons.

† The development of concrete ships promised important relief to our vital war industries. It seemed that they could be built quickly and cheaply by types of labour not required in normal shipbuilding and would save large quantities of steel. These claims were found on examination to be based on false assumptions, and many unforeseen technical difficulties arose. An experimental ship of 2,000 tons was built, but was a failure, and although experimental work continued the use of concrete hulls was only successful in barges up to about 200 tons.

as the fighting front is concerned; our Air Force is hopelessly inferior to the Germans; we are not allowed to do anything to stop them receiving their vital supplies of ore; we maintain an attitude of complete passivity, dispersing our forces ever more widely; the Navy demands Scapa and Rosyth both to be kept at the highest point. Do you realise that perhaps we are heading for *defeat*? I feel I must do my duty, even in small things, in trying to secure effective concentration upon the enemy and in preventing needless dispersion.

First Lord to First Sea Lord 19.1.40

FLEET AIR ARM—ESTIMATED COST DURING THE FIRST TWELVE MONTHS OF THE WAR

1. I have been increasingly disquieted about the demand which the Fleet Air Arm involves upon British war-making resources. None the less this estimate is a surprise to me, as I had not conceived how enormous was the charge involved. I have always been a strong advocate of the Fleet Air Arm; in fact, I drafted for Sir Thomas Inskip the compromise decision to which he eventually came in 1938. I feel all the more responsible for making sure that the Fleet Air Arm makes a real contribution to the present war in killing and defeating Germans.

2. When some years ago the Fleet Air Arm was being discussed the speed of carrier-borne and shore-based aircraft was not unequal; but since then the shore-based development has been such as to make it impossible for carrier-borne aircraft to compete with shore-based. This left the Fleet Air Arm the most important duties of reconnaissance in the ocean spaces, of spotting during an action with surface ships and launching torpedo seaplane attacks upon them. However, there are very few surface ships of the enemy, and one can only consider the possible break-out of a German raider or fast battleship as potential targets. Provision must be made for this; but certainly it does not justify anything like this immense expenditure.

3. On the other hand, our Air Force has fallen far behind that of Germany, and under present conditions the air menace to this Island, its factories, its naval ports and shipping, as well as to the Fleet in harbour, must be considered as the only *potentially mortal* attack we have to fear and face. I am most anxious therefore to liberate the R.A.F. from all ordinary coastal duties in the Narrow Waters and the North Sea, and to assume this responsibility for the Fleet Air Arm, which then, and then alone, would have a task proportioned to its cost and worthy of its quality.

4. Some time ago the Air Ministry were making their way in the

world and were very jealous of their sphere, but now that a prime importance has come to them, equal in many ways to that of the Royal Navy, they are much more tolerant; moreover, they are deeply anxious to increase their own disposable strength. They have recently allowed us to form two shore-based squadrons for the Orkneys, etc., and I believe that, with tact, and in the present good atmosphere, this principle might be applied all along the East Coast. We have, I suppose, an unequalled class of pilots and observers for such purposes, and the advantage to both services would be unquestionable.

5. I propose therefore, in principle for your consideration, that a plan should be drawn up by the First Sea Lord to save 100 to 150 pilots from the Fleet Air Arm, together with mechanics and administrative staff, in order to form six, seven, or eight shore-based naval squadrons, and that the complements of the aircraft-carriers, especially the unarmoured aircraft-carriers, should be reduced as much as is necessary. For reconnaissance in the outer seas we should have to content ourselves with very small complements. When the armoured carriers are complete their complement must be considered in the light of the conditions prevailing then in the North Sea. The F.A.A. training schools and other establishments must be rigorously combed to provide these new fighting forces.

6. If the details of this plan are worked out, I would approach the Air Ministry and offer to relieve them of the whole coastal work in home waters *without adding to the cost to the public*. We should make a smaller demand on future deliveries for carrier-borne aircraft, and ask in return to be given a supply of fighters or medium bombers, perhaps not at first of the latest type, but good enough for short-range action. We should then take over the whole responsibility as a measure of war emergency, and leave the future spheres of the department to be settled after the war is over.

Pray let me have your thought upon this.*

First Lord to D.C.N.S., D.N.I., and Secretary   31.1.40

Thirty years ago I was shown Foreign Office confidential books printed on paper so inflammable that they could be almost immediately destroyed. Since then all this business has advanced. It would be possible to print books on cellulose nitrate, which would almost explode on being lighted. Existing books could be photographed on

---

* This plan was swept away by events. The Fleet Air Arm made its contribution to the R.A.F. during the Battle of Britain. Later the development of the U-boat war taxed to the utmost the resources of Coastal Command, which itself drew heavily on Bomber Command to meet its ever-growing commitments.

Later again, in 1941, the advent of the "escort carrier" type enabled the Fleet Air Arm to play a conspicuous part in the defeat of the U-boats operating beyond the range of normal shore-based aircraft.

to this with great facility. Alternatively, or conjointly, these books could be reduced to tiny proportions and read by a small projecting apparatus. Let a small committee be formed on this question. Pray propose me names. Professor Lindemann will represent me.

First Lord to First Sea Lord and D.C.N.S.   31.I.40

Pictures have been published in many newspapers of the Australian troops marching through Sydney, etc., before starting for the war. Thus the enemy must know that convoys will be approaching the entrance to the Red Sea and the neighbourhood of Socotra. Although there is no intelligence of any U-boat in the Indian Ocean, how can we be quite sure one has not made its way up from Madagascar, where there was a rumour, to the Red Sea, and been oiled from some Italian or Arabian port? I must say I should feel more comfortable if anti-submarine escort could be provided from the neighbourhood of Socotra. This could be done by sending the destroyer *Vendetta* from Haifa to rendezvous, say 200 miles east of Socotra, with the destroyer *Westcott*, which is already following up the convoy from Singapore. The presence of these two Asdic-fitted destroyers would give complete assurance, and only one of them has to go far out of her way.

Pray let me have a note on this.

FEBRUARY 1940

First Lord to First Sea Lord   9.II.40

LEGEND OF PARTICULARS OF THIRD WAR EMERGENCY FLOTILLA

Destroyers of 1,650 tons almost amount to small cruisers. These unarmoured vessels with nearly 200 men on board become, as *Grenville* and *Exmouth* have shown, a prize and a target for a U-boat in themselves. In this case the destroyers are within 10 tons of the flotilla leader. By steadily increasing the size and cost of destroyers we transfer them gradually from the class of the hunters to that of the hunted. It is unsound to place so large a human stake in an unarmoured, highly vulnerable vessel. The length of time in building vessels of this class makes it unlikely they will take part in the present war. What we require are larger numbers of smaller vessels more quickly delivered. It will be necessary to keep the number of these very large destroyers at a minimum. The simplified armament and extra endurance are good features.

First Lord to First Sea Lord (with papers), D.C.N.S.,   11.11.40
D.N.I., Controller, and Secretary

JAPANESE STRENGTH—N.I.D. 02242/39

1. It is of the greatest importance to form a true opinion about present and prospective Japanese building. Before I can put this case to the Cabinet I must be satisfied that there is solid evidence of the ability of Japan to construct a Navy superior to the present Navies of Britain and the United States, built and building. The financial condition of Japan has lamentably deteriorated. She has for two and a half years been engaged in a most ruinous war in China; between one and one and a half millions of Japanese soldiers have had to be maintained in the field. No decisive progress has been made. On the contrary, it is believed the Chinese are gaining strength. Certainly there is a marked reaction in Japan, and the internal tension is very great.

2. We must look at the kind of statements which are made about their new shipbuilding intentions in the light of these facts. They have to buy a large proportion of their materials for warship construction from over the seas, and this, with the drain of the China war, must greatly affect their foreign exchange. What would be the cost of the programme set out in the First Sea Lord's table in yen, in sterling, and in dollars? It seems to me that they are going into figures of naval expense never attempted before at a time when their finances are rapidly deteriorating.

3. What is their steel capacity of production? What is their consuming power of steel? If my recollection serves me, the Japanese consuming power of steel is in the neighbourhood of 3 million tons a year, compared to British 15 and American 54. Yet such a programme as Japan is said to be embarking on would be, and is, a heavy drain on British or American strength. No doubt the heavy building in America and Britain will impose an additional effort on Japan. Whether they can go the pace is quite another question. I do not feel that mere rumours of ships they are said to have laid down form a sufficient basis. Has Major Morton's branch or committee which studies the military capacities of enemy or potentially enemy countries been consulted?

In short, I am extremely sceptical of the Japanese power to build a fleet equal to the present built and building fleets of either Britain or the United States.

First Lord to First Sea Lord   20.11.40

In view of yesterday's Cabinet decision all preparations should be made to carry out the operation referred to as soon as possible.

Pray let me have your proposals.

I consider the matter is most urgent, as it must be linked with the *Altmark*. The operation, being minor and innocent, may be called "Wilfred".*

First Lord to First Sea Lord and others   24.II.40

Let me have an early report on condition of *Exeter* and time likely for her repairs. Every effort should be made to keep the crew together. If *Exeter* repairs take more than three or four months, what are the other cruisers coming along in the interval which *Exeter's* crew could be taken on with their present captain? In the Army it would be thought madness to break up a unit like this, and I do not see why the same moral consideration should not affect the Navy too. †

First Lord to Controller and others   25.II.40

RECLASSIFICATION OF SMALLER WAR VESSELS

Director of Plan's remark that the term "destroyer has by association come to imply a particular type of vessel whose principal weapon is the torpedo" ignores the whole story of the destroyer, whose chief function was to destroy the torpedo-boat with superior gun-fire. The idea of destruction is not confined to destruction by torpedo; it may equally be expressed by depth-charges or gun-fire.

I agree with First Sea Lord about the needlessness of repeating the word "vessel", and his wish to simplify all titles to one word.

I should like the word "destroyer" to cover ships formerly described as "fast escort vessels", which are in fact medium destroyers. I do not like the word "whaler", which is an entire misnomer, as they are not going to catch whales, and I should like to have some suggestions about this. What is in fact the distinction between an "escorter", a "patroller", and a "whaler" as now specified? It seems most important to arrive at simple conclusions quickly on this subject, and enforce them from March 1 on all commands and departments. Let me see a list of the vessels built and building which will fall in the various categories.‡

---

* This refers to the mining of the Norwegian Leads. Owing to many political complications, referred to in Chapter XXXII, the operation did not take place until April 8.

† In Chapter XXIX my minutes are recorded dealing with the difficulties which arose over bringing the *Exeter* home after the River Plate action. She now remained under repair for many months.

‡ The "fast escort vessels" became known as "Hunt" class destroyers, as their names were all selected from famous packs of hounds. Large numbers were built, and they served with distinction both in the anti-U-boat war and in our amphibious operations. Later ancient names were revived.

The "whalers" became known as "corvettes", and later types were called "frigates". Escort vessels became "sloops".

MARCH 1940

First Lord to First Sea Lord and Secretary   1.III.40
A plan should be prepared for a battleship concentration in the Mediterranean (with other craft), supposing trouble should arise in March. I do not expect trouble; but it would be well to have all the combinations surveyed in advance.*

First Lord to First Sea Lord, Controller, and others   5.III.40
After the air attack on the Fleet on September 26 we all thought it most necessary to train the A.A. gunners against faster targets than those hitherto provided. Ideas were suggested by Professor Lindemann, experiments were made, and other ideas for flares, etc., put forward by the *Vernon*. What has happened about all this? Of course the weather has been terribly against it, but I fear there have virtually been no practices in home waters at high-speed targets. Five months have passed, and it is very serious if we have not been able to develop an effective system of fast targets, and obtain the necessary machines so that the Fleet can work up.

We must have this now that the weather is improving and the Fleet back at Scapa. An improvement in the gunnery of H.M. ships is of the utmost importance to their safety.

First Lord to First Sea Lord and Controller   5.III.40
1. Repairing ships is better than new building. A strong effort should be made to turn this 8,000-ton ship *Domala* into an effective cargo-carrying bottom. Immediately she could be seized upon, and repaired in the plainest way for the roughest work.
2. Are we doing enough about salvage? Let me have a return of the vessels now beached on our coasts, and a report on the measures taken to fit them again for sea. The very minimum should be done to them, compatible to life and navigation. *There ought to be a tremendous move-on in the salvage and repair departments.* The tonnage working on any given day ranks above the rate of new merchant shipbuilding.

First Lord to First Sea Lord   6.III.40
I think it would be only prudent for you to concert with the French the necessary regroupings of the Allied Fleets which would be appropriate to a hostile or menacing Italian attitude. Perhaps you will let me know about this on my return.

---

* As a result of these deliberations the battleship *Warspite* was ordered to return to the Mediterranean, but with the opening of the Norwegian campaign she was recalled to home waters, and did not reach the Mediterranean until May. Before the Italian declaration of war in June the *Malaya*, *Ramillies*, and *Royal Sovereign* had also joined the Mediterranean Fleet from convoy duty in the Atlantic.

(In the Train)
First Lord to Parliamentary Secretary   11.III.40

I am very glad you have had a considerable measure of success in your parleys with the trade unions. Be careful about the "Ministry of Labour Training Centres". As hitherto organised these have been nothing but quasi-philanthropic institutions to tone up the unfortunate people in the derelict areas. They have never been organised to make skilled tradesmen out of semi-skilled. In their present condition they are a snare so far as we are concerned. We have got to get competent people to learn new trades. The Minister of Labour has always said that his training centres cannot touch any but the unemployed, meaning thereby the peace-time unemployed. What we have to cater for is a far livelier class who are changing their occupations in consequence of the war.

I think you must rely on training in the dockyards and in special training schools established by the Admiralty.

Speak to me about this, as it seems to me to be a serious flaw.

First Lord to First Sea Lord and others   14.III.40

Now that we are not allowed to interfere with the Norwegian Corridor, would it not be possible to have one or two merchant ships of sufficient speed, specially strengthened in the bows and if possible equipped with a ram? These vessels would carry merchandise and travel up the Leads looking for German ore ships or any other German merchant vessels, and then ram them by accident. This is only another development of the "Q" ship idea.

First Lord to D.C.N.S., D.N.I. (to initiate action)   22.III.40
(Secret.)

Mr. Shinwell declares that in Vigo there are still a number of German merchant ships, many of whose crews are non-German, and among the Germans many non-Nazis. He suggests that with a little money and some organisation it would be possible to get these crews to take the ships to sea, when they could be picked up by our ships, and those who had brought them out suitably rewarded. Is there anything in this?

First Lord to D.C.N.S. and First Sea Lord   30.III.40

CUTTING FROM D.T. 29.3.40. TWENTY NAZI SHIPS GET READY TO SAIL. ATTEMPTS TO RUN THE BLOCKADE (AMSTERDAM, FRIDAY). ELSTER REPORTED AT ROTTERDAM

The reason why I cut this from the *Daily Telegraph* and asked my question of the D.N.I. is because an exodus of German ships from

Dutch ports might well be a danger sign in respect of Holland herself. I have no doubt the same thought has occurred to you.

First Lord to Secretary   31.III.40

WAR CABINET—SUB-COMMITTEE ON RESERVED OCCUPATIONS NOTE BY TREASURY

While there are nearly 1,500,000 unemployed and no serious drain of casualties from the Army, I propose to resist the disturbance of Admiralty work by movement of men we need from the dockyard. The matter must be settled by Cabinet decision. You should let Sir Horace Wilson know how much I regret I cannot meet his views.

APRIL 1940

First Lord to Controller   1.IV.40

Where are the facts about the return of the 40 destroyers which are in hospital to their duty? And can anything be done to speed up new destroyers, especially those of the 40th Flotilla, by leaving out some of the final improvements and latest additions, which take so much time? The great aim must be to have the maximum numbers during these coming summer months. They can go back to have further treatment when we have a larger margin.

First Lord to First Sea Lord and others   4.IV.40

While I do not see any adverse change in the Italian situation, I presume that the appropriate departments of the Admiralty Staff are at work upon, or already have completed, a plan of naval operations in the Mediterranean against Italy, should she force us into war with her. We might be asked for this by the Cabinet, and I should be glad to see it as soon as possible, at any rate during the course of the next four or five days.

First Lord to Controller   12.IV.40

The most intense efforts should be concentrated upon *Hood*, as we may need all our strength to meet an Italian threat or attack.

Pray let me have a time-table showing when she will be ready for sea.

First Lord to D.C.N.S.   12.IV.40

Are there any other Danish islands besides the Faroes which require attention?

Will you also kindly ask the Staff to examine the position at Curaçao, in case Holland should be overrun. The Fourth Sea Lord spoke to me on the oil supplies dependent upon Curaçao refineries. I should like a short paper upon the subject.

First Lord to Controller (M.S.R.)   12.IV.40

WEEKLY STATEMENT OF SHIPYARD WORKERS, DATED 9.IV.40

This report is much more favourable, and for the first time shows
a lift on new merchant construction. Altogether we have added
15,000 men since February 1, when we took over. Are you satisfied
that all arrangements made by the late Parliamentary Secretary are
completed, and working satisfactorily? We shall want another 30,000
men, and the most strenuous efforts must be made to procure them.
Can anything else be done now?

Has not the time arrived when you will be ready with your report
for the Cabinet, which I rather hoped to have sent them last week?
I should like to be able to have it ready for them next week. Will
you kindly let me see it in outline first?

First Lord to D.C.N.S.   13.IV.40

One of the branches under your control should make a careful study
of Spanish islands, in case Spain should be drawn into a breach of
neutrality.

First Lord to Controller, First Sea Lord, and Secretary   13.IV.40

CONTROLLER'S MINUTE OF APRIL 13 ABOUT "HOOD"*

This is a very different story to what was told me when it was
proposed to repair this ship at Malta. I was assured that the whole
operation would take thirty-five days, and that the ship would never be at
more than thirty-five days' notice, and that only for a short time. When
I asked the other day how long it would take to bring *Hood* back into
service I was told fourteen days. I take it therefore she has been above
twenty days under repair at present, to which must now be added
seventeen days more in April and thirty-one in May—total [sixty-]
eight days—or [nearly] double what I was told before this vital ship
was laid up in this critical period. Pray give me an explanation of this
extraordinary change. Moreover, after these [sixty-]eight days there
are to be fourteen days repairing her reserve feed tanks—total therefore [eighty-] two days, or [nearly] three months at the most critical
period in the war.

The engineer in charge of the *Hood* assured me when I was last at
Scapa that they had found out the way to nurse her defective condenser
tubes so as to get twenty-seven knots, and that there was no reason
why she should not remain in commission and carry on for six months.

I much regret not to have been more accurately informed, in view
of the Italian attitude.

* See also First Lord's minute of April 12 above.

First Lord to First Sea Lord and others   14.IV.40

On the assumption that Narvik falls into our hands in the near future we must consider the uses to which we intend to put it. First we want to make it a convenient oiling base, where our flotillas acting on the Norwegian coast can refuel at the highest economy. Secondly, we require to ship the masses of ore there to this country in a very active manner.

For these purposes we must have a moderate garrison, say about a thousand Territorial troops. A few efficient A.A. batteries, both high-and low-ceiling; a well-netted, boomed, and perhaps partially mined barrier; and a good supply of oil in tankers. Is there plenty of fresh water?

We must expect sporadic attacks from the air. A few coast defence guns should be mounted to protect the approaches. The sunken German torpedo-boats might perhaps supply some of these. Their salvage and repair must be explored, and the port got working as soon as possible. Some of the working party of Marines now being raised might well be sent to Narvik. There are, I believe, good shops where repairs can be effected. A portion of the staff, I suppose Plans Division, should begin work on this question to-day and formulate requirements. Our object must be to make Narvik self-supporting and self-defended at the earliest moment after we have it in our power, as we shall want all our stuff lower down the coast. The necessary guns (A.A.) may be taken from A.D.G.B.*

First Lord to Civil Lord   16.IV.40

FAROES

With your experience and connections in the department, you should now assume the duty of concerting the action to make the Faroes satisfactory for our purposes. D.C.N.S. will supply you with requirements. Pray make a weekly report. We must have an aerodrome and an R.D.F. at the very earliest moment, together with a certain amount of A.A. defence, and a few coast guns. This will be a very tempting base for a raider.

First Lord to Prime Minister   18.IV.40

COMMENTARY ON GERMAN REPORT OBTAINED BY THE FRENCH ON AMMUNITION

It is an error to suppose that an offensive can be maintained merely by the unlimited use of artillery ammunition. The creation of a labyrinth or zone of crater fields becomes itself an obstacle, of great

* Air Defence Great Britain.

difficulty to the attacking army. The moment must come when the infantry advance into this zone and have to fight hand to hand with the defenders. Meanwhile, so far as expenditure of ammunition is concerned, the defence can reserve its power till the enemy's infantry advance, and thus economise to an enormous extent. There is no truth in the statement that "all great offensives always came to a stop solely because the attacking armies did not have sufficient of ammunition". The impulse of an offensive dies away as the fighting troops become more distant from their point of departure. They thus get ahead of their supplies, whether ammunition or food. The more they have pulverised the intervening ground with their artillery, the more difficult it is to bring supplies of ammunition, even if they have them in their original forward dumps, up to the fighting troops. It is at such moments that the opportunity to deliver the counter-strokes arises.

Altogether this paper, which is most interesting, gives me the impression of being written by someone high up in the munitions department of Germany, who naturally thinks in terms only of shell. Shell is very important, and we are not likely to have too much of it, but there is not the slightest reason for supposing that unlimited artillery ammunition can win victory on a great scale in modern war. The transportation of the ammunition to the guns in the various phases of the battle remains, as heretofore, the limiting factor upon the artillery.

First Lord to Admiral Somerville   21.IV.40
Pray give me a short note upon the present position of R.D.F. so far as it concerns the Navy and Coast Defence, showing weak points and anything you wish done to remedy them.

First Lord to First Sea Lord and V.C.N.S.   25.IV.40
The reason why I am worrying about these minefields on the approaches to Narvik is that now *Warspite* has quitted, and we have an uncocked-up ship in *Resolution* only, this ship might be at a disadvantage in range should *Scharnhorst* or *Gneisenau* turn up one fine morning. Perhaps however it is possible to shelter in a fiord so as to avoid long-range fire, and force action at reduced ranges, or perhaps *Resolution* could be careened. Anyhow, I think it indispensable that we should reach certainty so far as the defence of Narvik from a surface raid is concerned.*

(Action this Day.)
First Lord to First Sea Lord and others   28.IV.40
In view of the bad reports from the Faroes about aircraft or seaplane

---

* Our ships were using Skjel Fiord, in the Lofoten Islands, as an advanced base. This covered the approach to Narvik through West Fiord.

bases and the fact that we must reckon with the Germans all along the Norwegian coast, it seems indispensable that we have a base in Iceland for our flying-boats and for oiling the ships on the Northern Patrol. Let a case be prepared for submission to the Foreign Office. The sooner we let the Icelanders know that this is what we require the better.*

First Lord to Sir James Lithgow and Controller   30.IV.40

These figures of our shipping gains from the German aggression against Norway and Denmark amount roughly to 750 ships, aggregating 3,000,000 tons. The effect of this upon our shipping and ship-building position requires to be considered. Clearly, we have obtained an easement we never foresaw when we embarked upon our present programme. I should be glad to know your reaction, and in particular how the latest paper prepared by Sir James Lithgow is affected.

SOME QUESTIONS ABOUT PERSONNEL

First Lord to First Sea Lord, Second Sea Lord,   18.IX.39
and Secretary

I have just approved the message to the Northern Patrol.

About the Newfoundland fishermen: the boatwork of the Newfoundlanders was an important thing to render this effective in the stormy winter months. These men are the hardiest and most skilful boatmen in rough seas who exist. They long for employment. Please propose me measures at once to raise 1,000 R.N.V.R. in Newfoundland, drafting the necessary letter to the Dominions Office and outlining terms and conditions. They have nothing to learn about the sea, but almost immediately some method of training and discipline could be brought into play. In ten days at the outside this should be working in Newfoundland.

First Lord to Second Sea Lord   21.IX.39

In conversation with the Commander-in-Chief Home Fleet I have promised to look into the question of providing a theatre and cinema ship for the Home Fleet and Northern Patrol at Scapa.

I think it much more desirable to use a ship than shore facilities. I have in mind the arrangements made for the Grand Fleet during the last war, when s.s. *Gurko* was used.

The ship should contain a large N.A.A.F.I. shop, as well as cinema and theatrical facilities, and possibly could be combined with a refrigerator storage ship.

* Iceland was occupied by British forces on May 10.

Pray let me have your plans for implementing this most important adjunct of naval life at Scapa.

First Lord to Second Sea Lord and Secretary   29.IX.39

LEAKAGE OF INFORMATION
(Secret.)

This is a proposal to dismiss from the Royal Navy, without trial, without formulating a charge, or even questioning, a Petty Officer who is identified from half a dozen of the same name by the fact that he has very white teeth, and who is reported to have been at a dinner at some unspecified date at which presumably indiscreet talk occurred. There is no suggestion that he was paid money, or that there was any treasonable intention. I do not find in these papers the slightest evidence that could be adduced before any court against this man, nor does the Director of Public Prosecutions. Yet, without being given any chance of defending himself, he is to be cast from the Service at the outset of a great war, with the kind of suspicion hanging over him for the rest of his life of having been a spy or a traitor.

Such processes cannot be allowed. If it is thought worth while to pursue these not very serious though annoying leakages into the sphere of penal action, the man must plainly be charged with some definite offence known to the Naval Discipline Act and brought before a court-martial, which can alone pronounce upon his guilt or innocence.

With regard to the dockyard employees and others, against whom the evidence is also vague and flimsy, no such procedure is necessary. It might perhaps be permissible, as a matter of administration, to move them about a little.

First Lord to Secretary   4.X.39

Let me have a list at once of the branches to which promotion from the lower deck still does not apply. What proportion do these branches bear to the other branches?

First Lord to Second Sea Lord, Parliamentary Secretary,   7.X.39
and Secretary

Will you kindly explain to me the reasons which debar individuals in certain branches from rising by merit to commissioned rank? If a cook may rise, or a steward, why not an electrical artificer or an ordnance rating or a shipwright? If a telegraphist may rise, why not a painter? Apparently there is no difficulty about painters rising in Germany!

First Lord to Secretary 7.X.39

ADMIRALS OF THE FLEET

This matter does not require verbal treatment. Kindly draft minutes *f.m.s.* [for my signature] to First and Second Sea Lords in the sense of surmounting the difficulties. I am very clear that the Admirals of the Fleet should remain on the Active List like Field-Marshals, and should not be penalised for winning promotion unduly young. You might explain to the Treasury privately that no money is involved. What is the value of being made Admiral of the Fleet if it is only to hoist the Union flag for one day and retire to Cheltenham, writing occasional letters to the *Times*?

First Lord to Second Sea Lord and others concerned, 14.X.39
and Secretary

There must be no discrimination on grounds of race or colour [in the employment of Indians or Colonial natives in the Royal Navy]. In practice much inconvenience would arise if this theoretical equality had many examples. Each case must be judged on its merits, from the point of view of smooth administration. I cannot see any objection to Indians serving on H.M. ships where they are qualified and needed, or, if their virtues so deserve, rising to be Admirals of the Fleet. But not too many of them, please.

First Lord to First Sea Lord 24.X.39

I see no reason to suspend these enlistments or bar the Navy door to the Dominions in time of war. Most particularly am I concerned with Newfoundland, about which I have given special directions. The Newfoundlanders are certainly not to be "left to find their own way to this country" from Newfoundland. Care and pains are to be taken to recruit, train, and convey to the United Kingdom as many as possible. I hope we shall get 1,000. I understand this is in progress, and let me have a report saying exactly what is being done in Newfoundland.

With regard to the other Dominions, suitable enlistments should be accepted whether for hostilities only or for permanent service. These ratings can be trained at the naval ports in the Dominions: at Sydney, at Halifax and Esquimalt, and at Simonstown. Opportunity will then be given to transport the men in batches to this country or draft them on to His Majesty's ships visiting the Dominions.

Pray let a scheme on these lines be put forward with a view to surmounting the difficulties.

First Lord to Fourth Sea Lord   12.XII.39

I am told that the minesweeper men have no badge. If this is so it must be remedied at once. I have asked Mr. Bracken to call for designs from Sir Kenneth Clark within one week, after which production must begin with the greatest speed, and distribution as the deliveries come to hand.

First Lord to Naval Secretary and others concerned   19.XII.39

"SALMON'S" WAR PATROL NARRATIVE

I am in entire accord with the Second Sea Lord's minute of yesterday. I shall be most willing to concur in the promotion and honours proposed, both to the officers and to the men. I await the proposals of the Sea Lords in respect of the promotion. Naval Secretary should prepare submissions for the honour, to the King, and, if possible, these should be published, both as to officers and men, before the *Salmon* sails again. Perhaps His Majesty would like himself to see the officer [Lt.-Commander Bickford], and conclude the audience by pinning on the D.S.O. Naval Secretary might find out what they think about this at the Palace. It seems probable that similar, though not necessarily the same, awards will be required in the case of the Commander of the *Ursula*, and here again the crew must participate. Every effort must be made to announce the awards to the men at the same time as the officers. The whole of this should be put through in forty-eight hours at the latest.

First Lord to Secretary   8.II.40

SPECIAL ENTRY CADETSHIPS

It seems very difficult to understand why this candidate should have been so decisively rejected, in view of his high educational qualifications, his Service connections, and his record as set out by his father in his letter of January 4. One has to be particularly careful that class prejudice does not enter into these decisions, and unless some better reasons are given to me I shall have to ask my Naval Secretary to interview the boy on my behalf, before assuming responsibility for writing to his father as proposed.

First Lord to Secretary   25.II.40

CANDIDATE FOR THE NAVY ENTRANCE EXAMINATION, NOVEMBER 1939, WHO FAILED

I do not at all mind "going behind the opinion of a board duly constituted", or even changing the board or its chairman if I think injustice has been done. How long is it since this board was remodelled? I could not help being unfavourably struck with the aspect of the Dartmouth cadets whom I saw marching by the other day. On the other hand, I was enormously impressed with the candidates for commission from the ranks whom I saw drilling and being trained on the parade-ground at Portsmouth. They were of course much older, but a far finer-looking type.

Not only shall my Naval Secretary see the boy, but I shall hope to have time to see him myself. Who are the naval representatives on the board of selection? Naval officers should be well represented.

Action accordingly.

Let me have a list of the whole board—with the full records of each member and the date of his appointment.

First Lord to First Sea Lord and D.C.N.S.　　　　　　25.II.40

1. I should like *Salmon* to go to Devonport as you suggested as an extra practice submarine for a few months after the severe and distinguished service she has rendered. There would be advantages in having Commander Bickford in the Plans Division of the Admiralty for, say, six months in order to bring them in close and direct contact with the very latest conditions prevailing in Heligoland Bight. This officer seems to me very able, and he has many things to say about anti-U-boat warfare which I trust will be gathered at the earliest opportunity.

2. Is there any reason why *Ursula* should not go on escort to the Norwegian convoy?

3. There may be other vessels which R.A.S. [Rear-Admiral Submarines] would say have also had heavy strain. Perhaps this might be looked into later.

4. If the war were general and everybody engaged to the hilt there would be no need to consider these variations of duty. But considering that the peculiar brunt falls upon very few at the present time, and that nothing is comparable to submarine work amid the minefields and all its increasing dangers, I am strongly of the opinion that we should keep a rotation, shifting boats and crews which have had a particularly hard time, or have distinguished themselves, to easier duties, and letting others have a chance of winning renown. Is there any possibility of arranging a certain number of relief crews for submarines, suitable for the Bight, so as to divide the strain among a larger proportion of the personnel? I should like this to be studied.

5. Have the men of the *Salmon* and *Ursula* received their medals and honours? The officers have already been decorated. Let special measures be taken to ensure that the men have these rewards before they go to sea again.

First Lord to Second Sea Lord and Fourth Sea Lord 24.III.40

Backgammon would be a good game for wardroom, gunroom, and warrant officers' mess, and I have no doubt it would amuse the sailors. What happened to the £1,000 Lord Rothermere gave me for various kinds of amusements? Is it all expended, and how? I have no doubt I could get some more if necessary. Backgammon is a better game than cards for the circumstances of war-time afloat, because it whiles away twenty minutes or a quarter of an hour, whereas cards are a much longer business.

First Lord to First Sea Lord and Second Sea Lord 25.III.40

I see charges of looting preferred against our men in the German Press. I should not think it necessary to mention this but for the fact that it has come to my notice that the captain of the *Altmark's* watch, chronometer, and Iron Cross were stolen, and are now in the hands of some of the sailors as souvenirs. Anything of this kind must be stopped with the utmost strictness. No souvenir of any value can be preserved without being reported and permission obtained. Personal property of enemies may be confiscated by the State, but never by individuals.

First Lord to Second Sea Lord 7.IV.40

I have seen the three candidates. Considering that these three boys were 5th, 8th, and 17th in the educational competitive examination out of more than ninety successful, 320 qualified, and 400 who competed, I see no reason why they should have been described as unfit for the naval service. It is quite true that A. . . . has a slightly cockney accent, and that the other two are the sons of a Chief Petty Officer and an engineer in the merchant service. But the whole intention of competitive examination is to open the career to ability, irrespective of class or fortune. Generally speaking, in the case of candidates who do exceptionally well in the examination the presumption should be that they will be accepted. Similarly, those who do very badly in the educational examination may nevertheless in a few cases be fit to serve. But the idea of rejecting boys at the very top of the list, unless some very grave defect presents itself, is wholly contrary to the principles approved by Parliament.

I am sure if the Committee, when they had these boys before them, had known that they were among the cleverest in the whole list they would not have taken so severe a view and ruled them out altogether on the personal interview. It seems to me that in future the Committee ought to conduct the interview *after* the examination, and with the results of it before them. Furthermore, it is wrong that a boy should be allowed to sit for examination, with all the stress and anxiety attached to it, when it has already been settled that even if he is first on the list he has already been ruled out.

I also feel that there is no need for any mention of a disqualifying standard for interview and record. The Interview Board should also be instructed that they may award different marks to the same candidate for different branches of the Service. It is obvious that a boy may be much more suitable for the Paymaster than the Executive Branch, and the Committee should be able to differentiate accordingly.

There will of course be no need for the Interview Committee to see all the candidates. There must be a qualifying educational standard. This is 400 marks at present, out of a total of 1,350. I notice that all the successful boys in the last examination had well over 600 marks. Surely it would ease the work of the Interview Committee if the qualifying educational standard were raised?

Pray make me proposals for rearranging the present system so as to achieve the above conditions. Cadetships are to be given in the three cases I have mentioned.

INDEX

reoccupation of Rhineland, 174; conversation of, with Churchill, 219–21;
misled by Hitler, 225; on need of
arrangement with Russia, 333–4; asks
question on Anglo-Polish guarantee,
336; declares Chamberlain should go,
594–5; mentioned, 19, 20, 29, 33, 166,
406

Georges, General, 96; shows Churchill
Rhine front, 342; shows figures of
French and German Armies, 357; against
extension of defence plan, 434; interested
in Cultivator No. 6, 654

German Air Force, re-creation of, 43–4, 99
et seq., 170; prohibition of, 68; approaches parity with British, 105–7;
reaches parity, 108–9, 114–16; officially
constituted, 117; possible policy of, in
event of war, 136–7; enlargement of,
190; strongest in Europe, 207; estimates
of strength of, 211, 430, 615–16; compared with British, 303, 615–17; Poland
attacked by, 395–8; danger from, in
seeking mastery of Baltic, 414, 416; in
Norway, 557

German Army, Treaty limits to, 39–40;
plans to re-create, 40–2, 99; exceeds
Treaty limits, 42; equality of, with
French Army, 58, 66–70, 83; compulsory service for, 117, 128–9, 190; function of, in Nazi Germany, 128–9;
organisation of, 129; expansion of, 113–
114, 190; compared with French, 130,
302–3, 357; strongest in Europe, 207;
Churchill's estimates of strength of,
211–12; defects shown in, during invasion of Austria, 242; generals on deficiencies of, 281–2; disposition of, in
invasion of Poland, 395–6; armoured
troops of, 397; strength of (1939), 429–
430, 573–4; advance of, through Norway, 557–8, 580–3; strength of, in First
World War, 573; superiority of, in
Norway, 582–3

German colonies, question of return of,
200, 218, 225, 276–7, 299

German Navy, Treaty limits to, 14, 44, 68,
123; illicit building up of, 44, 99, 108–9;
exceeds Treaty limits, 123–4; as permitted by Anglo-German Naval Agreement, 124–6; battleship design in, 146;
composition of (1939), 368; Baltic control of, 368; commerce-destroyers of,
459–61, 474; escorts expedition to Norway, 532–6, 542; in action with British
ships, 534–5; weakening of, 592; inter-
cepts ships off Norway, 588–92; strength
of (1939), 621

German shipping, 379, 390, 510, 682–3
German Workers' Party, 48–9

Germany, after First World War, 4, 10–
12; France threatened by, 4–6; reparations from, 6, 8–9, 23; loans to, 8–9, 10,
29, 52; total disarmament of, 11, 4–16,
29, 43; economic chaos in, 11–12, 55–6,
58; British sympathy for, 12; Communism in, 13, 48–9; hideous crimes of,
15–16; Hindenburg's Presidency, 24–5;
suggests Rhineland Pact, 25; enters
League, 25–7; Arbitration Treaties of,
27; dissatisfied with Eastern frontiers,
28; withdrawal of Inter-Allied Control
Commission from, 43; rearming of, 41–
46, 64, 70, 76, 91, 101–2, 117, 128–31,
136–7, 170, 210–13, 613; anti-Semitism
in, 45, 48, 52, 69, 91; Hitler's conception of, 47–8, 50–1; rise of Nazi Party in,
49–50, 51–4; question of revival of
monarchy in, 56–7; "equality of armaments" between France and, 58, 66–70,
83; opposition to Hitler in, 63, 74, 86–7;
Hitler achieves power over, 63–4; demands right to rearm, 66, 68–9; quits
Disarmament Conference, 67, 70; quits
League, 70; Churchill visits, 75; deliveries of metals to, 77; absorbs Austria,
81, 93–4, 240–5, 273; approach to France
before First World War, 83; Nazi unity
preserved in, 90–1; concentration camp
system in, 91; Churchill on power of
dictatorship in, 92; opposes Eastern Pact,
95; Saar returns to, 97–8; relations of,
with Russia, 102, 171; approaches air
parity with Great Britain, 105–7; reaches
air parity, 108, 114; British naval agreement with, 124–8, 170; function of
Army in Nazi, 128–9; Italy on side of,
148, 163, 171; contempt for Britain felt
by, 167–8; prepares for war, 170, 190–1,
317–18, 344; suggests twenty-five-year
pact, 172; reoccupies Rhineland, 173–9;
Hitler's prestige enhanced in, 179, 184;
defies League, 182–3; fortifies Rhineland, 183–4, 196; draws up plans for reoccupation of Austria, 185, 232; British
duty to oppose, 188; subscribes to Non-
Intervention, 192; intervenes in Spanish
war, 193, 219; asks for free hand in
Eastern Europe, 200–1; "living space"
for, 200, 232, 251, 338, 497; expenditure
of, on armaments, 203–4, 301; strongest
military Power, 207; anti-aircraft guns
in, 209–10; absorption by, of all German
minorities, 251; prepares to attack
Czechoslovakia, 257; cession of Sudetenland to, 269–73, 275–7; Godesberg ultimatum of, 276–7, 283; poor morale in,
281–2; mobilisation of, denied, 282; re-
sents Britain's attitude of mistrust, 295,
301, 323; arbitrates on Czecho-Hun-